EXTENSIONS OF $f(R)$ GRAVITY

Recent cosmological observations have posed a challenge for traditional theories of gravity: what is the force driving the accelerated expansion of the universe? What if dark energy or dark matter do not exist and what we observe is a modification of the gravitational interaction that dominates the universe at large scales?

Various extensions to Einstein's General Theory of Relativity have been proposed, and this book presents a detailed theoretical and phenomenological analysis of several leading, modified theories of gravity. Theories with generalized curvature-matter couplings are first explored, followed by hybrid metric-Palatini gravity. This timely book first discusses key motivations behind the development of these modified gravitational theories, before presenting a detailed overview of their subsequent development, mathematical structure, and cosmological and astrophysical implications.

Covering recent developments and with an emphasis on astrophysical and cosmological applications, this is the perfect text for graduate students and researchers.

TIBERIU HARKO is Senior Honorary Research Associate in the Department of Mathematics at University College London, visiting Professor in the Department of Physics at the Sun-Yat Sen University in Guangzhou, China, and Associate Professor at the Babes-Bolyai University in Cluj, Romania. His research interests include general relativity and modified theories of gravity, cosmology, astrophysics, and applied mathematics.

FRANCISCO S. N. LOBO is Principal Investigator in the Science Faculty at the University of Lisbon, and leads the Cosmology Group at the university's Institute of Astrophysics and Space Sciences. His research is focused on gravitational physics, and several of his publications have received extensive news coverage from publications including the *New York Times*, *Scientific American*, and *National Geographic*.

CAMBRIDGE MONOGRAPHS ON MATHEMATICAL PHYSICS

General Editors: P. V. Landshoff, D. R. Nelson, S. Weinberg

S. J. Aarseth *Gravitational N-Body Simulations: Tools and Algorithms*[†]
J. Ambjørn, B. Durhuus and T. Jonsson *Quantum Geometry: A Statistical Field Theory Approach*[†]
A. M. Anile *Relativistic Fluids and Magneto-fluids: With Applications in Astrophysics and Plasma Physics*
J. A. de Azcárraga and J. M. Izquierdo *Lie Groups, Lie Algebras, Cohomology and Some Applications in Physics*[†]
O. Babelon, D. Bernard and M. Talon *Introduction to Classical Integrable Systems*[†]
F. Bastianelli and P. van Nieuwenhuizen *Path Integrals and Anomalies in Curved Space*[†]
D. Baumann and L. McAllister *Inflation and String Theory*
V. Belinski and M. Henneaux *The Cosmological Singularity*[†]
V. Belinski and E. Verdaguer *Gravitational Solitons*[†]
J. Bernstein *Kinetic Theory in the Expanding Universe*[†]
G. F. Bertsch and R. A. Broglia *Oscillations in Finite Quantum Systems*[†]
N. D. Birrell and P. C. W. Davies *Quantum Fields in Curved Space*[†]
K. Bolejko, A. Krasiski, C. Hellaby and M.-N. Célérier *Structures in the Universe by Exact Methods: Formation, Evolution, Interactions*
D. M. Brink *Semi-Classical Methods for Nucleus-Nucleus Scattering*[†]
M. Burgess *Classical Covariant Fields*[†]
E. A. Calzetta and B.-L. B. Hu *Nonequilibrium Quantum Field Theory*
S. Carlip *Quantum Gravity in 2+1 Dimensions*[†]
P. Cartier and C. DeWitt-Morette *Functional Integration: Action and Symmetries*[†]
J. C. Collins *Renormalization: An Introduction to Renormalization, the Renormalization Group and the Operator-Product Expansion*[†]
P. D. B. Collins *An Introduction to Regge Theory and High Energy Physics*[†]
M. Creutz *Quarks, Gluons and Lattices*[†]
P. D. D'Eath *Supersymmetric Quantum Cosmology*[†]
J. Dereziıski and C. Gérard *Mathematics of Quantization and Quantum Fields*
F. de Felice and D. Bini *Classical Measurements in Curved Space-Times*
F. de Felice and C. J. S Clarke *Relativity on Curved Manifolds*[†]
B. DeWitt *Supermanifolds, 2nd edition*[†]
P. G. O. Freund *Introduction to Supersymmetry*[†]
F. G. Friedlander *The Wave Equation on a Curved Space-Time*[†]
J. L. Friedman and N. Stergioulas *Rotating Relativistic Stars*
Y. Frishman and J. Sonnenschein *Non-Perturbative Field Theory: From Two Dimensional Conformal Field Theory to QCD in Four Dimensions*
J. A. Fuchs *Affine Lie Algebras and Quantum Groups: An Introduction, with Applications in Conformal Field Theory*[†]
J. Fuchs and C. Schweigert *Symmetries, Lie Algebras and Representations: A Graduate Course for Physicists*[†]
Y. Fujii and K. Maeda *The Scalar-Tensor Theory of Gravitation*[†]
J. A. H. Futterman, F. A. Handler, R. A. Matzner *Scattering from Black Holes*[†]
A. S. Galperin, E. A. Ivanov, V. I. Ogievetsky and E. S. Sokatchev *Harmonic Superspace*[†]
R. Gambini and J. Pullin *Loops, Knots, Gauge Theories and Quantum Gravity*[†]
T. Gannon *Moonshine beyond the Monster: The Bridge Connecting Algebra, Modular Forms and Physics*[†]
A. García-Díaz *Exact Solutions in Three-Dimensional Gravity*
M. Göckeler and T. Schücker *Differential Geometry, Gauge Theories, and Gravity*[†]
C. Gómez, M. Ruiz-Altaba and G. Sierra *Quantum Groups in Two-Dimensional Physics*[†]
M. B. Green, J. H. Schwarz and E. Witten *Superstring Theory Volume 1: Introduction*
M. B. Green, J. H. Schwarz and E. Witten *Superstring Theory Volume 2: Loop Amplitudes, Anomalies and Phenomenology*
V. N. Gribov *The Theory of Complex Angular Momenta: Gribov Lectures on Theoretical Physics*[†]
J. B. Griffiths and J. Podolský *Exact Space-Times in Einstein's General Relativity*[†]
T. Harko and F. Lobo *Extensions of* f(R) *Gravity: Curvature-Matter Couplings and Hybrid Metric-Palatini Gravity*
S. W. Hawking and G. F. R. Ellis *The Large Scale Structure of Space-Time*[†]
F. Iachello and A. Arima *The Interacting Boson Model*[†]
F. Iachello and P. van Isacker *The Interacting Boson-Fermion Model*[†]
C. Itzykson and J. M. Drouffe *Statistical Field Theory Volume 1: From Brownian Motion to Renormalization and Lattice Gauge Theory*[†]
C. Itzykson and J. M. Drouffe *Statistical Field Theory Volume 2: Strong Coupling, Monte Carlo Methods, Conformal Field Theory and Random Systems*[†]
G. Jaroszkiewicz *Principles of Discrete Time Mechanics*
G. Jaroszkiewicz *Quantized Detector Networks*

C. V. Johnson *D-Branes*†
P. S. Joshi *Gravitational Collapse and Spacetime Singularities*†
J. I. Kapusta and C. Gale *Finite-Temperature Field Theory: Principles and Applications, 2nd edition*†
V. E. Korepin, N. M. Bogoliubov and A. G. Izergin *Quantum Inverse Scattering Method and Correlation Functions*†
J. Kroon *Conformal Methods in General Relativity*
M. Le Bellac *Thermal Field Theory*†
Y. Makeenko *Methods of Contemporary Gauge Theory*†
S. Mallik and S. Sarkar *Hadrons at Finite Temperature*
N. Manton and P. Sutcliffe *Topological Solitons*†
N. H. March *Liquid Metals: Concepts and Theory*†
I. Montvay and G. Münster *Quantum Fields on a Lattice*†
P. Nath *Supersymmetry, Supergravity, and Unification*
L. O'Raifeartaigh *Group Structure of Gauge Theories*†
T. Ortín *Gravity and Strings, 2nd edition*
A. M. Ozorio de Almeida *Hamiltonian Systems: Chaos and Quantization*†
M. Paranjape *The Theory and Applications of Instanton Calculations*
L. Parker and D. Toms *Quantum Field Theory in Curved Spacetime: Quantized Fields and Gravity*
R. Penrose and W. Rindler *Spinors and Space-Time Volume 1: Two-Spinor Calculus and Relativistic Fields*†
R. Penrose and W. Rindler *Spinors and Space-Time Volume 2: Spinor and Twistor Methods in Space-Time Geometry*†
S. Pokorski *Gauge Field Theories, 2nd edition*†
J. Polchinski *String Theory Volume 1: An Introduction to the Bosonic String*†
J. Polchinski *String Theory Volume 2: Superstring Theory and Beyond*†
J. C. Polkinghorne *Models of High Energy Processes*†
V. N. Popov *Functional Integrals and Collective Excitations*†
L. V. Prokhorov and S. V. Shabanov *Hamiltonian Mechanics of Gauge Systems*
S. Raychaudhuri and K. Sridhar *Particle Physics of Brane Worlds and Extra Dimensions*
A. Recknagel and V. Schiomerus *Boundary Conformal Field Theory and the Worldsheet Approach to D-Branes*
M. Reuter and F. Saueressig *Quantum Gravity and the Functional Renormalization Group*
R. J. Rivers *Path Integral Methods in Quantum Field Theory*†
R. G. Roberts *The Structure of the Proton: Deep Inelastic Scattering*†
C. Rovelli *Quantum Gravity*†
W. C. Saslaw *Gravitational Physics of Stellar and Galactic Systems*†
R. N. Sen *Causality, Measurement Theory and the Differentiable Structure of Space-Time*
M. Shifman and A. Yung *Supersymmetric Solitons*
Y. M. Shnir *Topological and Non-Topological Solitons in Scalar Field Theories*
H. Stephani, D. Kramer, M. MacCallum, C. Hoenselaers and E. Herlt *Exact Solutions of Einstein's Field Equations, 2nd edition*†
J. Stewart *Advanced General Relativity*†
J. C. Taylor *Gauge Theories of Weak Interactions*†
T. Thiemann *Modern Canonical Quantum General Relativity*†
D. J. Toms *The Schwinger Action Principle and Effective Action*†
A. Vilenkin and E. P. S. Shellard *Cosmic Strings and Other Topological Defects*†
R. S. Ward and R. O. Wells, Jr *Twistor Geometry and Field Theory*†
E. J. Weinberg *Classical Solutions in Quantum Field Theory: Solitons and Instantons in High Energy Physics*
J. R. Wilson and G. J. Mathews *Relativistic Numerical Hydrodynamics*†

† Issued as a paperback

Extensions of $f(R)$ Gravity

Curvature-Matter Couplings and Hybrid Metric-Palatini Theory

TIBERIU HARKO
University College London

FRANCISCO S. N. LOBO
University of Lisbon

CAMBRIDGE
UNIVERSITY PRESS

CAMBRIDGE
UNIVERSITY PRESS

University Printing House, Cambridge CB2 8BS, United Kingdom

One Liberty Plaza, 20th Floor, New York, NY 10006, USA

477 Williamstown Road, Port Melbourne, VIC 3207, Australia

314–321, 3rd Floor, Plot 3, Splendor Forum, Jasola District Centre, New Delhi – 110025, India

79 Anson Road, #06–04/06, Singapore 079906

Cambridge University Press is part of the University of Cambridge.

It furthers the University's mission by disseminating knowledge in the pursuit of education, learning, and research at the highest international levels of excellence.

www.cambridge.org
Information on this title: www.cambridge.org/9781108428743
DOI: 10.1017/9781108645683

First published 2019

Printed and bound in Great Britain by Clays Ltd, Elcograf S.p.A

A catalogue record for this publication is available from the British Library.

Library of Congress Cataloging-in-Publication Data
Names: Harko, Tiberiu, 1953– author. | Lobo, Francisco S. N., author.
Title: Extensions of f(R) gravity : curvature-matter couplings and hybrid metric-Palatini theory / Tiberiu Harko (University College London), Francisco S.N. Lobo (Universidade de Lisboa).
Description: Cambridge : Cambridge University Press, [2018] | Series: Cambridge monographs on mathematical physics | Includes bibliographical references and index.
Identifiers: LCCN 2018022347 | ISBN 9781108428743 (hardback : alk. paper)
Subjects: LCSH: Expanding universe. | Curvature cosmology. | Large scale structure (Astronomy) | Gravitation. | Cosmology.
Classification: LCC QB991.E94 H37 2018 | DDC 523.1/8–dc23
LC record available at https://lccn.loc.gov/2018022347

ISBN 978-1-108-42874-3 Hardback

Contents

Preface

During the last few decades, cosmology has evolved from being mainly a theoretical area of physics to become a field supported by high-precision observational data. Recent experiments call upon state of the art technology in astronomy and astrophysics to provide detailed information on the contents and history of the universe, which has led to the measurement of parameters that describe our universe with increasing precision. The standard model of cosmology is remarkably successful in accounting for the observed features of the universe. However, a number of fundamental open questions remain at the foundation of the standard model. In particular, we lack a fundamental understanding of the recent acceleration of the universe [413, 436]. What is the so-called dark energy that is driving the cosmic acceleration? Is it vacuum energy or a dynamical field? Or is the acceleration due to infra-red modifications of Einstein's theory of General Relativity (GR)? How is structure formation affected in these alternative scenarios? What are the implications of this acceleration for the future of the universe?

The resolution of these fundamental questions is extremely important for theoretical cosmology. Dark energy models are usually assumed to be responsible for the acceleration of the cosmic expansion in most cosmological studies. However, it is clear that these questions involve not only gravity, but also particle physics. String theory provides a synthesis of these two branches of physics and is widely believed to be moving toward a viable quantum gravity theory. One of the key predictions of string theory is the existence of extra spatial dimensions. In the brane-world scenario, motivated by recent developments in string theory, the observed three-dimensional universe is embedded in a higher-dimensional spacetime [334]. The new degrees of freedom belong to the gravitational sector, and can be responsible for the late-time cosmic acceleration [162, 178]. On the other hand, generalizations of the Einstein–Hilbert Lagrangian, including quadratic Lagrangians, which involve second order curvature invariants, have also been extensively explored [103, 157, 324, 383, 471]. These modified theories of gravity not only provide an alternative explanation for the expansion history of the universe [99, 120, 377], but they also offer a paradigm fundamentally distinct from the simplest dark energy models of cosmic acceleration [140], even from those that perfectly mimic the same expansion history.

A large number of modified theories of gravity can be represented in a scalar-tensor formulation by means of appropriate metric rescalings and field redefinitions. It is, therefore, not surprising that we can think of scalar–tensor gravity theories as a first stepping stone to explore modifications of GR. They have the advantage of apparent simplicity and have been intensively analyzed in the literature. First proposed in its present form by Brans and Dicke for a single scalar field [88], they have been extensively generalized and have maintained the interest of researchers until the present day. For instance, with a conformal transformation, these theories can be recast as matter-interacting scalar fields in GR, where in this format, they play an important role in dark energy modeling. Recently, relative to scalar–tensor theory, much work has been invested in the Galileon models and their generalizations [161]. The latter models allow nonlinear derivative interactions of the scalar field in the Lagrangian and lead to second order field equations, thus removing any ghost-like instabilities. The Lagrangian was first written down by Horndeski in 1974 [254], which contains four arbitrary functions of the scalar field and its kinetic energy. The form of the Lagrangian is significantly simplified by requiring specific self-tuning properties, however, the screening is too effective, and will screen curvature from other matter sources as well as from the vacuum energy [130]. An alternative approach consists of searching for a de Sitter critical point for any kind of material content [342]. These models might alleviate the cosmological constant problem and can deliver a background dynamic that is compatible with the latest observational data.

Thus, a promising alternative to explain the late-time cosmic acceleration is to assume that at large scales Einstein's theory of GR breaks down, and a more general action describes the gravitational field. Thus, one may generalize the Einstein–Hilbert action by including second order curvature invariants such as R^2, $R^{\mu\nu}R_{\mu\nu}$, $R^{\mu\nu\alpha\beta}R_{\mu\nu\alpha\beta}$, $C^{\mu\nu\alpha\beta}C_{\mu\nu\alpha\beta}$, etc. Some of the physical motivations for these modifications of gravity were inspired on effective models raised in string theory, which indeed may lead to the possibility of a more realistic representation of the gravitational fields near curvature singularities [378]. Moreover, the quantization of fields in curved spacetimes tell us that the high-energy completion of the Einstein–Hilbert Lagrangian of GR involves higher-order terms on the curvature invariants above. This is in agreement with the results provided from the point of view of treating GR as an effective field theory [127]. Among these extensions of GR the so-called $f(R)$ gravity has drawn much attention over the last years, since it can reproduce the late-time cosmic acceleration, and, in spite of containing higher-order derivatives, it is free of the Ostrogradsky instability, as can be shown by its equivalence with scalar-tensor theories (for a review on $f(R)$ gravity see Refs. [103, 157, 324, 383, 471]).

Moreover, $f(R)$ gravity has also been proposed as a solution for the inflationary paradigm [37], where the so-called Starobinsky model is a successful proposal, since it satisfies the latest constraints released by Planck [10]. In addition, the equivalence of $f(R)$ gravity to some class of scalar–tensor theories has provided an extension of the so-called chameleon mechanism [279, 280] to $f(R)$ gravity, leading to some viable extensions of GR that pass the solar system constraints [260, 384]. Other alternative formulations for these extensions of GR have been considered in the literature, namely, the presence of nonminimal couplings between the scalar curvature and the matter Lagrangian density [44, 237]. Here, it was shown that an explicit coupling between an arbitrary function of the scalar curvature R and the Lagrangian density of matter generates a nonvanishing covariant derivative of the energy-momentum tensor, which implies non-geodesic motion and consequently leads to the appearance of an extra force [44]. These theories and generalizations will be extensively analyzed in Part II of the book. Another interesting approach to modified gravity involves the Palatini formalism, where the metric and affine connection are regarded as independent degrees of freedom, which yields an interesting phenomenology for cosmology [396]; and the metric-affine formalism, where the matter part of the action now depends and is varied with respect to the connection [472]. Recently, a novel approach to modified theories of gravity was proposed that consists of adding to the metric Einstein–Hilbert Lagrangian an $f(R)$ term constructed a la Palatini [224]. It was shown that the theory can pass the solar system observational constraints even if the scalar field is very light. This implies the existence of a long-range scalar field, which is able to modify the cosmological and galactic dynamics, but leaves the solar system unaffected. This hybrid metric-Palatini gravitational theory will be extensively analyzed in the Part III of the book.

Note that these modified theories of gravity are focused on extensions of the curvature-based Einstein–Hilbert action. Nevertheless, one could equally well modify gravity starting from its torsion-based formulation and, in particular, from the Teleparallel Equivalent of General Relativity (TEGR) [318]. The interesting point is that although GR is completely equivalent with TEGR at the level of the field equations, their modifications (for instance $f(R)$ and $f(T)$ gravities, where T is the torsion) are not equivalent, and they correspond to different classes of gravitational modifications. Hence, $f(T)$ gravity has novel and interesting cosmological implications, capable of describing inflation, the late-time acceleration, large scale structure, bouncing solutions, nonminimal couplings to matter, etc. [96, 245, 244].

An extremely important aspect of modern cosmology is the synergy between theory and observations. Dark energy models and modified gravity affect the geometry of the universe and cosmological structure formation, impacting the background expansion and leaving an imprint on the statistical properties of the large-scale structure. There are a number of well-established probes of cosmic evolution, such as type Ia supernovae, baryon acoustic oscillations (BAO), weak gravitational lensing, galaxy clustering, and galaxy clusters properties [504]. Different methods measure different observables, probing expansion and structure formation in different and often complementary ways and have different systematic effects. In particular, joint analyses with Cosmic Microwave Background (CMB) data are helpful in breaking degeneracies by constraining the standard cosmological parameters. Indeed, CMB has revolutionized the way we perceive the universe. The information encoded in its temperature and polarization maps provides one of the strongest evidences in favor of the hot Big Bang theory and has enabled ways to constrain cosmological models with unprecedented accuracy [11].

In this work, we perform a detailed theoretical and phenomenological analysis of specific modified theories of gravity and investigate their astrophysical and cosmological applications. The theories that we present are essentially two largely explored extensions of $f(R)$ gravity, namely: (i) modified gravity with curvature-matter coupling; and (ii) the hybrid metric-Palatini gravitational theory. The extensions can be read independently of one another. This book is outlined in the following manner:

1. As this work is also aimed at the advanced undergraduate level, in Part I, we present a review of the General Theory of Relativity and of $f(R)$ gravity. In Chapter 1 we present an Introduction to GR, and in Chapter 2, we review the mathematical formalism in order to describe the gravitational field in the framework of GR. The Einstein gravitational field equations, which are defined in a Riemannian geometry, establish a deep connection between the geometric properties of the spacetime, and its matter content, provide a full description of both the geometric characteristics of the spacetime and of the dynamics of particles. In Chapter 3, we derive the Einstein field equations, through the variational principle, and discuss their mathematical properties, as well as some of their physical implications, such as the weak field limit and the gravitational field equations in spherical symmetry. In Chapter 4, we review the fundamental observational tests of GR at the solar system level. Furthermore, we will discuss some of the basic properties of compact, stellar-type astrophysical objects, and we will provide a brief introduction to the description of the accretion processes. In Chapter 5, we review some of the basic results and models in modern cosmology. We start our investigation in the framework of Newtonian cosmology, whose results, interestingly enough, coincide, in some particular situations, with those of GR. We will then proceed to a brief presentation of the general relativistic cosmology, and of its more significant results, focusing on the problems of dark matter and of dark energy. Finally, in Chapter 6, we go beyond Einstein's General Relativity and briefly review the basic mathematical formalism of $f(R)$ modified gravity theory, and we will discuss some of its cosmological and

astrophysical applications. This chapter will serve as an introduction and as a bridge to the following parts of the book.

2. In Part II, we review a plethora of modified theories of gravity with generalized curvature-matter couplings. The explicit nonminimal couplings – for instance, between an arbitrary function of the scalar curvature R and the Lagrangian density of matter – induce a nonvanishing covariant derivative of the energy-momentum tensor, implying non-geodesic motion and consequently leading to the appearance of an extra force. Applied to the cosmological context, these curvature-matter coupling lead to interesting phenomenology, where one can obtain a unified description of the cosmological epochs. We also consider the possibility that the behavior of the galactic flat rotation curves can be explained in the framework of the curvature-matter coupling models, where the extra-terms in the gravitational field equations modify the equations of motion of test particles, and induce a supplementary gravitational interaction.

   Thus, we review these generalized curvature-matter coupling modified theories of gravity in Part II, which is outlined in the following manner: In Chapter 7, we extensively motivate the approach, and in Chapter 8, we introduce the linear curvature-matter coupling and present some of its interesting features. In Chapter 9, we generalize the latter linear curvature-matter coupling by considering the maximal extension of the Einstein–Hilbert action, which results in the $f(R, L_m)$ gravitational theory, and extend the theory with the inclusion of general scalar field and kinetic term dependencies. In Chapter 10, we consider another extension of GR, namely, $f(R,T)$ modified theories of gravity, where the gravitational Lagrangian is given by an arbitrary function of the Ricci scalar R and of the trace of the energy-momentum tensor T. Furthermore, we include an explicit invariant Ricci-energy-momentum tensor coupling, and explore some of its astrophysical and cosmological phenomenology. This latter theory is motivated as considering a traceless energy-momentum tensor, $T = 0$, whereby the gravitational field equations for the $f(R,T)$ theory reduce to that of $f(R)$ gravity and all nonminimal couplings of gravity to the matter field vanish, while the inclusion of the $R_{\mu\nu}T^{\mu\nu}$ term still allows a nonminimal coupling. The possibility of explaining dark matter as a consequence of the curvature-matter coupling is considered in Chapter 11. In Chapter 12, we consider the application of the thermodynamics of open systems for the physical interpretation of modified gravity with a curvature-matter coupling, namely, for $f(R, L_m)$ and $f(R,T)$ theories of gravity. We begin our analysis by presenting the basic ideas of the irreversible thermodynamics and of the description of matter creation in a full covariant formulation. In Chapter 13, we present a brief introduction to the study of the quantum cosmology of $f(R,T)$ gravity, and some of its physical and theoretical implications. In Chapter 14, we present the theoretical and cosmological consequences of the modification of gravity that can be obtained by considering the quantum fluctuations of the gravitational metric, where the basic idea of this approach is the assumption that the quantum metric can be generally represented as the sum of a classical and of a fluctuating part, the latter being of a quantum (or stochastic) nature. Finally, in Chapter 15, we conclude.

3. In Part III, we explore another extension of $f(R)$ gravity, which has been motivated by the possibility of accounting for the self-accelerated cosmic expansion without invoking dark energy sources. Indeed, it has been established that both metric and Palatini versions of $f(R)$ gravity have interesting features but also manifest severe and different downsides. A hybrid combination of theories, containing elements from both of these formalisms, turns out to be very successful in accounting for the observed phenomenology, and it is able to avoid some drawbacks of the original approaches. Thus, the second part of this work explores the formulation of this hybrid metric-Palatini approach and its main achievements in passing the local tests and in applications to astrophysical and cosmological scenarios, where it provides a unified approach to the problems of dark energy and dark matter. Thus, we review the formulation and the main applications of hybrid gravity models in late-time cosmological and astrophysical scenarios.

   In Chapter 16, we start the discussion, and in Chapter 17 we consider the action and the field equations of the hybrid metric-Palatini formalism. In particular, we discuss the scalar–tensor representation, the Cauchy problem, and more general hybrid theories. Chapter 18 is devoted to hybrid-gravity cosmology. We derive the Friedmann equations, construct the related dynamical system, and briefly consider some solutions. Furthermore, we analyze the cosmological perturbations in order to understand structure formation in these theories. We focus on the evolution of perturbations in the matter dominated era and vacuum fluctuations relevant to inflation. Chapter 19 is dedicated to the astrophysical applications of the hybrid metric-Palatini gravitational theory. More specifically, the weak field behavior that is crucial for the solar system precision tests of gravity is considered. We also discuss the galactic phenomenology and the astrophysical applications of hybrid gravity as an alternative to dark matter. In particular, we study the stellar dynamics and the theory of orbits, the generalization of the virial theorem, the flat rotation curves of spiral galaxies, and the galactic clusters starting from the relativistic Boltzmann equation for collisionless systems of particles. In Chapter 20, we investigate the properties of the relativistic high density compact astrophysical stellar objects, and in Chapter 21, we analyze wormhole geometries in the scalar–tensor formulation of hybrid metric-Palatini gravity, and conclude in Chapter 22.

The definitions used throughout this book are the following:
We consider the $(-,+,+,+)$ metric signature.

The *Riemann curvature tensor*, $R^{\sigma}_{\lambda\mu\nu}$, a fourth order tensor, is defined in terms of the Christoffel symbols as

$$R^{\sigma}{}_{\lambda\mu\nu} = \frac{\partial \Gamma^{\sigma}_{\lambda\nu}}{\partial x^{\mu}} - \frac{\partial \Gamma^{\sigma}_{\lambda\mu}}{\partial x^{\nu}} + \Gamma^{\tau}_{\nu\lambda}\Gamma^{\sigma}_{\mu\tau} - \Gamma^{\tau}_{\mu\lambda}\Gamma^{\sigma}_{\nu\tau}.$$

From the Riemann tensor, by contraction, we obtain a second order tensor $R_{\lambda\nu}$, called the *Ricci tensor*, given by

$$R_{\lambda\nu} = R^{\sigma}_{\lambda\sigma\nu} = \frac{\partial \Gamma^{\sigma}_{\lambda\nu}}{\partial x^{\sigma}} - \frac{\partial \Gamma^{\sigma}_{\lambda\sigma}}{\partial x^{\nu}} + \Gamma^{\tau}_{\nu\lambda}\Gamma^{\sigma}_{\sigma\tau} - \Gamma^{\tau}_{\sigma\lambda}\Gamma^{\sigma}_{\nu\tau}.$$

Contracting the Ricci tensor gives the *Ricci scalar*, defined as

$$R = R^{\lambda}_{\lambda} = g^{\lambda\mu}R_{\lambda\mu} = g^{\lambda\nu}g^{\mu\sigma}R_{\lambda\mu\nu\sigma}.$$

Acknowledgments

We would like to thank our collaborators who have accompanied us in the elaboration of the theories presented in this book, in particular, Orfeu Bertolami, Christian Boehmer, Salvatore Capozziello, Bogdan Danilă, Zahra Haghani, Tomi Koivisto, Zoltán Kovács, José Lemos, Shidong Liang, Xing Liu, Man Kwong Mak, José Pedro Mimoso, Shin'ichi Nojiri, Sergey D. Odintsov, Gonzalo J. Olmo, Jorge Páramos, Diego Pavón, João Luís Rosa, Shahab Shahidi, Sergey Sushkov, Stefano Vignolo, and Min-Xing Xu.

We are very grateful to Bogdan Danilă for his help in the preparation of the figures. TH would like to thank the Yat-Sen School, of the Sun Yat-Sen University in Guangzhou, P. R. China, for the kind hospitality offered during the last stages of the writing of the book. FSNL also acknowledges financial support of the Fundação para a Ciência e Tecnologia through an Investigador FCT Research contract, with reference IF/00859/2012, funded by FCT/MCTES (Portugal).

Part I

Review of General Relativity

1
Introduction

Einstein's GR is one of the greatest intellectual achievements ever conceived by the human mind. In General Relativity (GR) the basic assumption is that any gravitational field can be interpreted geometrically, and it is directly related to a significant variation of the space-time metric $g_{\mu\nu}$. Geometrically, the metric tensor provides the infinitesimal distance between two neighboring points of the space-time continuum. Therefore, in GR the gravitational field is fully determined by the quantities that describe the intrinsic geometrical properties and structure of the space-time. This important idea has the fundamental implication that the space-time geometry itself (such as its metric and its curvature) is locally induced by physical phenomena and processes (generally involving the mass distribution or the motion of gravitating objects), and that space and time are not a priori determined absolute concepts. For an arbitrary gravitational field, which generally varies in both space and time, the metric of the four-dimensional space-time is *non-Euclidean* (Riemannian). Therefore, its geometric properties cannot be described any longer by the simple and well known results of Euclidean geometry, which is constructed based on Euclid's fifth postulate of the parallels, which dictates that through an arbitrary point one can construct one and only one parallel to a given straight line.

The fifth postulate of the parallels, as formulated by Euclid, is a fundamental assertion in today's "elementary" geometry, and it cannot be reduced to a more basic axiom, or proven independently. It clearly differentiates the Euclidean space, and its underlying geometry, from other mathematical spaces that could be constructed from different geometrical considerations. The first examples of non-Euclidean spaces were discovered by János Bolyai (1802–1860) and Nikolai Lobachevski (1793–1856), and belong to what is called today hyperbolic geometry. A different class of geometries – the spherical geometries – was found by Georg Bernhard Riemann (1826–1866). All these early works were based on, and greatly influenced by, the profound geometric research of Karl Friederich Gauss (1777–1855). Presently all these different classes of geometric theories, based on a number of common axioms and principles, are called Riemannian geometry. One of the most remarkable evolutions in modern physics was the natural emergence of the ideas of the Riemannian geometry as introduced in the framework of gravitational fields, which took place in the years 1909–1916, and is due to Albert Einstein (1879–1955) and his coworker Marcel Grossmann (1878–1936).

Note that the earlier Newtonian model of the universe is based on the fundamental concepts of absolute space, obeying the axioms and postulates of Euclidean geometry, and of absolute time, dominated by a static gravitational force proportional to the inverse of the square of the distance. Thus, Newton's model is superseded, in Einstein's theory, by a four-dimensional complex space-time geometry where the rules of the Euclidean geometry do not hold any longer. The curved trajectories, which Newtonian gravity imposes upon free-falling bodies in the Euclidean–Newtonian representation, are interpreted in GR by the lines of minimal length in the curved space-time of the Riemannian–Einsteinian approach.

In order to describe physical phenomena we must first fix a reference frame, which can be chosen arbitrarily. But a fundamental requirement in physics is that the laws of nature must be written in such a way that they are valid in any four-dimensional system of coordinates, independently of the coordinates we are using. This fundamental statement can also be formulated through the requirement that all laws of nature must be written in a *covariant form*, making them independent of the reference frame. Of course, this general requirement does not allude to the physical equivalence of all possible systems of coordinates or reference frames (as in the case of the special theory of relativity, where all inertial frames of reference are indeed equivalent). Contrary to Special Relativity, in GR the specific occurrences of physical processes, encompassing the properties of the motion of bodies in different physical fields, vary in different systems of reference. Thus, the first step in building the theory of gravitation is to construct the mathematical framework that would allow the formulation of the laws of physics in a covariant manner. This is carried out in Chapter 2.

GR is based, from a physical point of view, on the equivalence principle, which states the equivalence of the non-inertial frames of reference with certain gravitational fields. However, similarly to Newtonian gravitational physics, in GR there is a fundamental difference between *real gravitational fields*, and *fields that are generated by the motion of non-inertial reference systems*. The most important property of the "true" gravitational fields is that no coordinate transformation can cancel them. Therefore, once a gravitational field is present, the geometry of the space-time has the fundamental property that the elements of the metric tensor $g_{\mu\nu}$ cannot be brought by any admissible transformation of coordinates to their constant Galilean values, over all space-time, with the metric tensor reduced to a diagonal form. Space-times in which the metric tensor cannot be reduced globally to a diagonal form with constant values of the components are called *curved*, in contrast to the case of the flat (Euclidian/Galilean) space-times, where such a transformation can always be performed.

However, by a specific choice of the coordinate system we can transform all the metric tensor components $g_{\mu\nu}$ to a diagonal Galilean form at any fixed point of the curved (Riemannian) space-time. From a mathematical point of view this amounts to the transformation to a diagonal form of a quadratic form characterized by constant coefficients (the components of the metric tensor $g_{\mu\nu}$ at the given point, having a pseudo-Euclidian signature). After performing, at the given point, the transformation of the local metric tensor to the diagonal form, in our sign convention the matrix of the metric tensor components will have one negative and three positive principal values. We will call this configuration of signs the *signature* of the metric tensor, and of the associated matrix. From this mathematical result it follows immediately that the *determinant*, given by $g = \det |g_{\mu\nu}|$, constructed from the elements of the matrix of the metric tensor $g_{\mu\nu}$, is always *negative* for a physically acceptable space-time, so that $g < 0$.

The basic idea of GR is that once a gravitational field is present in a given region of the universe, the intrinsic geometry of the space-time is non-Euclidian (Riemannian). This is indeed the case for both true gravitational fields, whose properties are fully determined by the curvature of the space-time, as well as for fields described by a non-inertial reference frame. In the geometric description of gravity proposed by GR, the assignment of the system of coordinates to a given frame, in which the natural processes are described, is not restricted in any way, be it mathematical or physical. The set of the three space coordinates (x^1, x^2, x^3) can be chosen arbitrarily as quantities describing the position of physical objects in the ordinary space. The time coordinate x^0 can be introduced with the help of a clock running in an arbitrary way.

In order to describe the gravitational field in the framework of GR, the Einstein gravitational field equations are defined in a Riemannian geometry, and they establish a deep connection between the geometric properties of the space-time and its matter content, providing a full description of both geometric characteristics of the space-time, and of the dynamics. In Chapter 3, we will derive the Einstein field equations, and discuss their mathematical properties, as well as some of their physical implications.

Thus, GR provides an excellent description of the gravitational effects and phenomena, in particular, for the weak field limit within the boundaries of the solar system. More than 150 years ago, astronomical observations pointed out an anomaly in the motion of Mercury, the innermost planet, which seemed to defy the Newtonian laws of gravitation. The planetary orbits as derived from Newton's inverse square law are stable ellipses, while the orbit of Mercury presents a precession of its perihelion, which could not be explained by using the then "standard" Newtonian laws of motion. However, the precession can be fully explained once we adopt the geometric description of GR, according to which the planets move on geodesic orbits in the curved Riemannian space-time generated by the presence of the Sun. Not only do planets feel the change in geometry, but light, and more generally electromagnetic waves, are also affected by the curvature of the space-time when traveling in the vicinity of the Sun. The presence of the curvature induces an observable deflection of the light rays, and a similarly detectable time delay in the propagation of radar signals.

GR also has a profound implication on the properties of compact objects, like stars, whose interior geometry is very different from that of their Newtonian counterparts. Once a star reaches three solar masses it will collapse to form a new and strange object called a black hole, whose intriguing properties can be well understood and explained in the framework of Einstein's theory of GR. Black holes can absorb matter in an astrophysical process called accretion, and the radiation emitted by the accreted matter represents an important astrophysical evidence for the possible existence of black holes in the universe. In Chapter 4, we will review the fundamental observational tests of GR that have been carried out at the solar system level and discuss some of the basic properties of compact, stellar type astrophysical objects, such as the description of accretion processes.

Turning now to large scales, cosmology is the scientific (physical/mathematical) study of the universe, of its components, and of its history. The major questions one may ask in cosmology are: How did the universe come into being? How did it evolve in time, and what is its fate? As a science, cosmology has evolved tremendously in the past 20 years, with unexpected (and revolutionary) new information about the origin, structure, and evolution of the universe coming in at a high rate. This information was mostly acquired through recent and significant technological improvements in telescope designs, and from space missions. Today cosmology has basically become a search for the decoding of not only what composes of the universe (the astrophysical objects within it, and their material composition), but also of its overall architecture, and past and present history. The beginning of modern cosmology can be traced back to the observational and theoretical advances made in the early twentieth century. In those times astronomers generally adopted the view that our galaxy (the Milky Way) had the shape of a disk, and was an isolated object in an infinite universe. However, there were many visible celestial objects, called spiral nebulae, such as M31 (the Andromeda galaxy), whose positioning with respect to our galaxy were not precisely known. Were these nebulae located inside the Milky Way, or far away from it?

In 1912, Vesto Slipher (1875–1969) investigated the electromagnetic spectra from the spiral nebulae, and found that many of them were Doppler-shifted. This means that the frequency of the light emitted by these nebulae was influenced by the speed of the source (in the same way as the frequency of sound changes for a passing train). Then astronomers quickly became aware that not only were the spiral nebulae (galaxies) moving rapidly away from our galaxy, in which the Earth is located, but they were shifting away from each other as well. Hence, once this information became available, astronomers started to interpret these galactic motions in terms of a global expansion of the universe. By 1925, Slipher had investigated around 40 galaxies, and found that electromagnetic spectra displaying redshifted lines were much more abundant than those exhibiting blueshifted spectra. Thus, Slipher realized that almost all the galaxies he investigated were quickly moving away from our galaxy, although Andromeda's galaxy blueshifted spectral lines indicated that it was approaching our Milky Way galaxy at a speed of approximately 300 km/s.

A key advance in astronomy occurred in 1918, when powerful methods to measure the distances to the spiral nebulae (today known to be independent galaxies) were developed. To determine astronomical distances Harlow Shapley (1885–1972) introduced the use of Cepheids, bright stars whose periodic pulsations indicated periods ranging between a minimum of a few days to a maximum of one month. The period of the variability of the luminosity of these stars is very precisely related to their absolute luminosity, which can be calibrated with a high precision by using the known distance of the neighboring Large Magellanic Cloud, a satellite galaxy of the Milky Way, located at a distance of 50 kiloparsecs from it. During the years between 1923 and 1929, Edwin Hubble (1889–1953) was able to detect, with the use of the 100-inch" telescope at Mount Wilson in California, the Cepheid stars in M31. Hubble advanced a new astronomical distance measurement technique, by employing the observational data of the brightest stars in the more remote galaxies. Hubble supplemented his galactic distance determinations with Slipher's redshift data of the spiral nebulae (galaxies), to make one of the most amazing discoveries in the history of science, namely, that *all galaxies are receding from us*. Moreover, he also found the law describing the universal expansion, which is given by a simple proportionality relation relating the velocity v of the galaxy and its distance d to the Earth or, more exactly, by Hubble's law of galactic expansion $v = H_0 d$. In Hubble's expansion law of the universe the constant of proportionality H_0 is denoted today *Hubble's constant*, and it has units of km/s/Mpc. Its precise determination is one of the central issues of present day observational cosmology.

In 1917, Einstein proposed the first general relativistic cosmological model, corresponding to a static, homogeneous, and isotropic universe, having a spherical geometry [183]. This model raised a number of interesting theoretical questions. The gravitational pull of matter led to an instability (acceleration) in this model, something Einstein did not expect, and did not want on observational grounds, since at that time the expansion of the universe was still to be discovered. Thus, Einstein modified his equations for GR by introducing a new term, proportional to the metric tensor, with the constant scalar proportionality coefficient called *the cosmological constant*, and denoted by Λ. This new term counteracts the gravitational attraction of matter, and hence it can be interpreted as describing *a kind of antigravity effect*. But even after the introduction of the cosmological constant, it turns out that the Einstein static universe is still not stable against small perturbations. The first to investigate the geometrical and physical properties of Einstein's static universe was the astronomer Willem de Sitter (1872–1934) [165, 166]. By using the Einstein equations with the cosmological constant, and by adopting a different mathematical model, de Sitter obtained a new solution of the field equations for a vacuum universe with vanishing energy density and pressure. However, the de Sitter space is not static, and by means of a transformation of the coordinates [502], the de Sitter metric can be reformulated into a dynamical representation, which still plays an essential role in modern cosmology.

In the early 1920s, the Russian mathematician and meteorologist Alexander Friedmann (1889–1925) realized that Einstein's gravitational field equations have non-static solutions that could set out an expanding universe [196], whose size is a function of time. Friedmann's solutions showed that our universe was born in one single event [306], about 13 thousand million years ago. Even today all the galaxies are still traveling apart from us due of this initial "explosion." Hence the Friedmann solutions imply that all matter, the universe itself, and space and time themselves appeared at once, in a single instant. The British astronomer Fred Hoyle (1915–2001), proponent and strong supporter of an alternative cosmological theory called the *steady state theory*, derogatorily labeled this model as a "big bang," and under this name it became the standard cosmological paradigm of our times, according to which the universe was born in a single point, in a state of very high density and temperature. After learning about Friedmann's work, and the discovery of the expansion of the universe, Einstein immediately discarded the cosmological constant, regarding it as the biggest blunder of his life.

The first major confirmation of the big bang theory came in 1964, when, using a horn antenna (7.35 cm) at Bell Labs, Arno Penzias (1933–) and Robert Wilson (1936–) [411] accidentally detected an isotropic cosmological microwave background, a distant echo that survived up to the present day from the primeval big bang "explosion." The Cosmic Microwave Background radiation is the main observational evidence for the hot big bang model. It has a perfect blackbody spectrum, and its temperature today has been determined to be $T = 2.73$ K (that is, a rather cold $-270°$C).

April 1992 represents another milestone in the history of modern cosmology. The COBE satellite team released the results on the discovery of anisotropies in the Cosmic Microwave Background radiation (CMB) at the level of 10^{-5} K, or one part in 100,000. The CMB temperature measurements provide a snapshot of the primeval matter density fluctuations that eventually led to the formation of galaxies, a process that started when the universe was around $t \approx 400{,}000$ years old. The map of the sky obtained by COBE, and by the next satellite experiments, is also the best evidence for the cosmological principle, claiming that the universe possesses a high degree of isotropy (or spherical symmetry). The research initiated by COBE were extended and significantly improved by another satellite experiment, the Wilkinson Microwave Anisotropy Probe (WMAP). The

WMAP team also provided a detailed and significantly improved full-sky map of the oldest detectable electromagnetic radiation in the universe. WMAP recorded and analyzed in detail microwave radiation from 379,000 years after the big bang, that is, around 13 billion years ago. Recently, the Planck satellite mission [11] significantly improved the observations made by WMAP, and provided high precision determinations of several crucial cosmological parameters, including the estimation of the average density of ordinary matter (baryonic density parameter), and of the density parameter of the dark matter in the universe.

If the attractive force of gravity were the only force determining the expansion dynamics of the universe, then we would expect the universe to be quickly decelerating and, in the limiting situation of a universe having essentially zero total energy, having an expansion rate decreasing as $1/t$. That's why the observational discovery of the late-time acceleration of the universe, initially announced in the years 1998–1999 by two distinct research teams (led by Riess and Schmidt [436] and by Perlmutter [413]), astonished most cosmologists and general relativists. Much work has been devoted to this unexpected cosmological effect, and many observational investigations as well as theoretical studies performed in the past 20 years have confirmed this amazing phenomenon. Moreover, a number of other observations, including the Planck satellite data [11], have led to the astonishing result that the observed universe consists only in a proportion of 4–5% of ordinary matter, that is, matter composed of baryons (protons, neutrons, etc.), electrons, and the other known elementary particles. Around 95% of the energy-matter balance of the universe consists of two *basically unknown components*, *dark matter* (~25%) and *dark energy* (~70%), respectively. These amazing results have led to the formulation of another paradigm in modern cosmology, the ΛCDM (Λ Cold Dark Matter) paradigm, which assumes (cold) dark matter as the major matter component in a universe whose late-time dynamics is determined by Einstein's cosmological constant, also giving the preponderent contribution to the total energy balance of the universe. In Chapter 5, we will review some of the basic results and models in modern cosmology.

Thus, in this context, modern astrophysical and cosmological models are plagued with two severe theoretical problems, namely, the dark energy and the dark matter enigmas. Relative to the latter, the dynamics of test particles around galaxies, as well as the corresponding mass discrepancy in galactic clusters, is explained by postulating the existence of a hypothetical form of dark matter particle. Relative to the dark energy problem, as mentioned in the previous paragraph, high precision observational data has confirmed with startling evidence that the universe is undergoing a phase of accelerated expansion. This phase is one of the most important and challenging current problems in cosmology, and represents a new imbalance in the governing gravitational equations. Several candidates, responsible for this expansion, have been proposed in the literature, in particular, dark energy models and modified theories of gravity, among others. The simplest scenario to explain the late-time cosmic speedup is to invoke the ΛCDM paradigm. However, if we assume that the cosmological constant constitutes the vacuum energy of the gravitational field, we are faced with an extremely embarrassing discrepancy of 120 orders of magnitude between the observed value and that predicted by quantum field theory. This is the celebrated cosmological constant problem.

The physical motivations for these modifications of GR also consist in the possibility of a more realistic representation of the gravitational fields near curvature singularities, and to create some first order approximation for the quantum theory of gravitational fields. It is clear that these questions involve not only gravity, but also particle physics. String theory provides a synthesis of these two branches of physics and is widely believed to be moving toward a viable quantum gravity theory. However, when adopting string theory as a full theory of quantum gravity, one does not recover GR in the low-energy limit, but rather a scalar–tensor theory of gravity. The initial motivations for scalar–tensor theories arose from the need to implement the Machs principle, which was not fully incorporated in GR. In fact, pioneering renormalization approaches to GR clearly showed the need for the introduction of counterterms, implying the presence of extra degrees of freedom, in addition to the spin two massless gravitons. It was shown that the corrections introduced by renormalization are at least quadratic in the curvature invariants, which lead to extensions of the Einstein–Hilbert Lagrangian.

In Chapter 6, we will consider an extension of the Einstein–Hilbert action, namely, $f(R)$ gravity, which contains several appealing features, as it combines mathematical simplicity and a fair amount of generality. Chapter 6 will also serves as a bridge to the final parts of the book, where we consider an intensive analysis of two extensions to $f(R)$ gravity considered in the literature, namely, modified theories of gravity with couplings between curvature and matter, and hybrid metric-Palatini gravity. While these modified theories of gravity offer an alternative explanation to the standard cosmological model for the expansion history of the universe, it offers a paradigm for nature fundamentally distinct from dark energy models of cosmic acceleration, even those that perfectly mimic the same expansion history.

Thus, a goal of this work is to perform a theoretical and phenomenological analysis of specific infrared modifications of GR, and to find the consistency of the generalized curvature-matter couplings in modified gravity and the hybrid metric-Palatini theory. Finally, one of the expected outcomes and impact of this work is to deepen the theoretical understanding of the dynamics of the universe and the perplexing nature of gravity itself.

2

The Mathematical Foundations

In the present chapter, we will briefly review the mathematical formalism, and the basic results in Einstein's General Relativity, one of the greatest intellectual achievements in theoretical physics. Hence, the first step in building the theory of gravitation is to construct the mathematical framework that would allow the formulation of the laws of physics in a covariant manner. This can be done by formulating the laws of physics by using vectors and tensors, mathematical quantities whose transformation properties will automatically implement the idea of the covariance of the physical laws. Hence we begin our presentation of General Relativity by introducing the fundamental mathematical concepts of vectors and tensors, and the differential operations that can be constructed for these quantities.

In the presentation of the mathematical formalism of General Relativity we will follow the physicist's approach to tensor calculus and differential geometry, as pioneered by Einstein [185], and then further developed and extended in the classic book by Landau and Lifshitz [302]. Mathematical books that treat tensors and differential geometry, written in the same spirit, are the books by Raigorodski, Stavrions and Balan [427], by Fomenko and Mishchenko [194], and by Ibragimov [263], as well as the three volumes by Dubrovin, Fomenko and Novikov [174, 175, 176], whose study are strongly recommended for those who intend to master the mathematics of General Relativity. The books by Logunov [386] and by Grøn and Hervik [210] also present some good mathematical introductions to Riemannian geometry.

2.1 Basic Mathematical Objects of Physics: Vectors and Tensors

When we begin to study gravitational fields, and in order to obtain a consistent description of nature, we need to consider physical processes in arbitrary reference frames. To achieve this goal we must develop a four-dimensional geometry in arbitrary coordinate systems, in which the standard Cartesian coordinates may be replaced by arbitrary curvilinear ones. Consider a coordinate system x^α, $\alpha = 0, 1, 2, \ldots, n$, which, in order to maintain the generality of our approach, we define in an arbitrary n-dimensional geometrical space. As a substitute to this coordinate system we can always introduce another, which is defined with the help of the coordinate transformations [302, 330]

$$x'^\alpha = f^\alpha\left(x^0, x^1, \ldots, x^n\right) = f^\alpha\left(x^\beta\right), \qquad \alpha, \beta = 0, 1, \ldots, n. \tag{2.1}$$

The functions f^α must satisfy some basic mathematical requirements, that is, we require them to be continuous functions on their domain of definition. Moreover, their partial derivatives of order n must also be continuous functions. If the Jacobian J of the transformation, defined as [302],

$$J = \det\left|\frac{\partial f^\alpha}{\partial x^\beta}\right| = \left|\begin{matrix} \frac{\partial x'^0}{\partial x^0}\frac{\partial x'^0}{\partial x^1}\frac{\partial x'^0}{\partial x^2}\frac{\partial x'^0}{\partial x^3}\cdots \\ \cdots \\ \frac{\partial x'^3}{\partial x^0}\frac{\partial x'^3}{\partial x^1}\frac{\partial x'^3}{\partial x^2}\frac{\partial x'^3}{\partial x^3}\cdots \\ \cdots \end{matrix}\right|, \tag{2.2}$$

is different from zero at each point of the given space, $J \neq 0$, $\forall x^\alpha$, $\alpha = 1, 2, \ldots, n$, then the variables x'^α will be independent variables. This is an important condition, since it allows for the initial variables x^α to be uniquely expressed as functions of the new ones x'^α as

$$x^\alpha = g\left(x'^\beta\right), \tag{2.3}$$

where $g = f^{-1}$ must also be a continuous and differentiable function.

2.1.1 Scalars and Vectors

A basic principle in physics requires the independence of the physical quantities and physical laws with respect to the choice of the coordinate system. Therefore, they must be formulated and described by geometrical objects having specific transformation rules from one coordinate system to another arbitrary one. The simplest such geometrical object is a *scalar* Φ, which in a coordinate change transforms as

$$\Phi'\left(x'^\alpha\right) = \Phi\left(x^\alpha\left(x'^\alpha\right)\right). \tag{2.4}$$

In other words a scalar is an invariant quantity with respect to arbitrary coordinate transformations.

We now introduce an important mathematical convention, called *the Einstein summation convention*, which tells us that an index that appears twice in an expression in different positions is summed over. This convention allows us to eliminate the symbol $\sum$ from a summation. It also provides the possibility of writing mathematical expressions containing complicated summations in a very simple and elegant manner.

In an arbitrary coordinate transformation the gradient of a scalar function $\Phi(x)$ transforms, according to the chain rule, in the following way,

$$\frac{\partial \Phi'(x')}{\partial x'^{\alpha}} = \frac{\partial \Phi(x)}{\partial x^{\beta}} \cdot \frac{\partial x^{\beta}}{\partial x'^{\alpha}}. \tag{2.5}$$

Based on this transformation rule we can introduce the concept of a covariant four-vector. We define as a *covariant four-vector* a set of four functions $A_\alpha = (A_0, A_1, A_2, A_3)$, which under a general coordinate transformation has the same transformation properties as the *derivative of a scalar function* [302, 427]. Therefore a covariant four-vector obeys the transformation rule

$$A'_\alpha(x') = A_\beta(x)\frac{\partial x^{\beta}}{\partial x'^{\alpha}}. \tag{2.6}$$

Covariant four-vectors are not the only geometric objects we can construct. Another group of geometric objects can be built up as follows. Consider the transformation rule of the differentials of the coordinate functions x', given by [302, 427],

$$dx'^{\mu} = \frac{\partial x'^{\mu}}{\partial x^{\alpha}} dx^{\alpha}. \tag{2.7}$$

This transformation rule allows us to introduce the concept of the contravariant vector. We define the *contravariant four-vector* A^μ as a collection of four distinct functions A^μ, where the index μ takes the values $\mu = 0, 1, 2, 3$, and which under a general change of the coordinate system transforms according to the transformation rule of the *differentials of the coordinates*,

$$A'^{\mu}(x') = \frac{\partial x'^{\mu}}{\partial x^{\alpha}} A^{\alpha}(x). \tag{2.8}$$

2.1.2 Tensors

With the help of the contravariant and covariant vectors, we can easily construct higher-order quantities. We define the *contravariant tensor of the second-order* $T^{\mu\nu}$ as a collection (set) of sixteen quantities having, under a general coordinate transformation, the same transformation properties of *the products of the components of two contravariant four-vectors*, that is, according to the rule

$$T'^{\mu\nu}(x') = \frac{\partial x'^{\mu}}{\partial x^{\alpha}} \frac{\partial x'^{\nu}}{\partial x^{\beta}} T^{\alpha\beta}(x). \tag{2.9}$$

Similarly, a *covariant tensor of order two* $T_{\mu\nu}$ transforms according to the mathematical rule

$$T_{\mu\nu}(x') = \frac{\partial x^{\alpha}}{\partial x'^{\mu}} \frac{\partial x^{\beta}}{\partial x'^{\nu}} T_{\alpha\beta}(x). \tag{2.10}$$

A *mixed tensor* T^{μ}_{ν} transforms as

$$T^{\mu}_{\nu}(x') = \frac{\partial x'^{\mu}}{\partial x^{\alpha}} \frac{\partial x^{\beta}}{\partial x'^{\nu}} T^{\alpha}_{\beta}(x). \tag{2.11}$$

A special tensor, having the same values in all coordinate systems, is the unit four-tensor δ^{μ}_{ν} (the Kronecker symbol). It is defined according to

$$\delta^{\mu}_{\nu} = \begin{cases} 1, & \nu = \mu \\ 0, & \nu \neq \mu \end{cases}. \tag{2.12}$$

Let A^{μ} be an arbitrary four-vector. Upon multiplication by δ^{ν}_{μ} we obtain

$$A^{\mu}\delta^{\nu}_{\mu} = A^{\nu}, \tag{2.13}$$

which is another four-vector. This result shows that δ^{ν}_{μ} is a tensor.

It is a simple mathematical exercise to prove that the product of two arbitrary four-vectors (also called inner product or scalar product) is a scalar, that is, a quantity that is invariant with respect to all possible arbitrary coordinate transformations. The proof is as follows

$$\begin{aligned} A'^{\mu}(x')\, A'_{\mu}(x') &= \frac{\partial x'^{\mu}}{\partial x^{\alpha}} A^{\alpha}(x) A_{\beta}(x) \frac{\partial x^{\beta}}{\partial x'^{\mu}} \\ &= \frac{\partial x^{\beta}}{\partial x^{\alpha}} A^{\alpha}(x) A_{\beta}(x) = \delta^{\beta}_{\alpha} A^{\alpha}(x) A_{\beta}(x) \\ &= A^{\alpha}(x) A_{\alpha}(x), \end{aligned} \tag{2.14}$$

where we have used in the final step the relation between the partial derivatives of the coordinates and the unit four-tensor δ^{β}_{α}, given by $\partial x^{\beta}/\partial x^{\alpha} = \delta^{\beta}_{\alpha}$.

Two arbitrary second-order tensors $A_{\mu\nu}$ and $B^{\delta\lambda}$ are called reciprocal to each other if they satisfy the condition

$$A_{\mu\nu} B^{\nu\lambda} = \delta^{\lambda}_{\mu}. \tag{2.15}$$

From the general rules of the transformations of tensors follows the important property that if all components of a given tensor are equal to zero in one fixed coordinate system, then in any other coordinate system they will also be identically equal to zero. Another important property of tensors we can easily check is that the transformation rules of the second-order covariant and contravariant tensors have the group property [330]. This can be seen easily by means of the following calculations,

$$A'^{\mu}(x') = \frac{\partial x'^{\mu}}{\partial x^{\alpha}} A^{\alpha}(x), \qquad A''^{\nu}(x'') = \frac{\partial x''^{\nu}}{\partial x'^{\mu}} A'^{\mu}(x'), \tag{2.16}$$

$$A''^{\mu}(x'') = \frac{\partial x''^{\nu}}{\partial x'^{\mu}} \frac{\partial x'^{\mu}}{\partial x^{\alpha}} A^{\alpha}(x) = \frac{\partial x''^{\nu}}{\partial x^{\alpha}} A^{\alpha}(x). \tag{2.17}$$

We introduce now two important classes of tensors, the symmetric and antisymmetric tensors, respectively. An arbitrary tensor $T^{\mu\nu}$ is called *symmetric* if

$$T^{\mu\nu} = T^{\nu\mu}, \tag{2.18}$$

and *antisymmetric* if it has the following property

$$T^{\mu\nu} = -T^{\nu\mu}. \tag{2.19}$$

The antisymmetric tensors have the property that all their diagonal components (that is, the components $A^{00}, A^{11}, \ldots$), are zero. This can be proven immediately, since, for example, for an antisymmetric tensor we must have $A^{00} = -A^{00} = 0$, etc.

2.1.3 Tensor Algebra

In the framework of tensor algebra we will consider five operations: addition, multiplication, convolution, contraction, and permutation of indices, respectively.

Tensor Addition and Subtraction

Consider two tensors having identical algebraic structures, that is, tensors having the same number of both contravariant and covariant indices, such as

$$A^{\alpha\beta}_{\mu\nu\sigma}, \qquad B^{\alpha\beta}_{\mu\nu\sigma}. \tag{2.20}$$

Then, by using these two tensors we can obtain a new tensor $C^{\alpha\beta}_{\mu\nu\sigma}$, called the sum of the two tensors, and defined as

$$C^{\alpha\beta}_{\mu\nu\sigma} = A^{\alpha\beta}_{\mu\nu\sigma} + B^{\alpha\beta}_{\mu\nu\sigma}. \tag{2.21}$$

Tensor Multiplication

Another algebraic operation on tensors is tensor multiplication, in which tensors can be multiplied algebraically. This operation is independent on the structure of indices of the given tensors. For example,

$$C^{\alpha\beta\lambda}_{\mu\nu\sigma\rho} = A^{\alpha\beta}_{\mu\nu\sigma} \cdot B^{\lambda}_{\rho}. \tag{2.22}$$

However, when performing tensor multiplication we have to take into account both the order of multipliers and the order of indices.

Tensor Convolution

The four-dimensional unit tensor (Kronecker symbol) δ^{μ}_{ν} can be used to introduce the convolution operation of indices. For example, in the case of a fourth-order tensor, the convolution operation gives

$$A^{\alpha\beta}_{\mu\nu} \cdot \delta^{\nu}_{\sigma} = A^{\alpha\beta}_{\mu\sigma}. \tag{2.23}$$

Hence, as one can see from the above equation, after convolution a summation must be performed over identical indices coming from the Kronecker symbol.

The Contraction of Tensors

From the components of the four-dimensional second-order tensor $A^{\mu\nu}$ we can form a scalar quantity by constructing the sum given by

$$A^{\mu}_{\mu} = A^0_0 + A^1_1 + A^2_2 + A^3_3. \tag{2.24}$$

We call the sum A^{μ}_{μ} the *trace of the tensor*. The operation by which we perform it is known as *the contraction of tensors*. The scalar (inner) product of two arbitrary vectors A_{μ} and B^{ν} is in fact an algebraic contraction operation: through it we construct the scalar $A_{\mu}B^{\mu}$ from the tensor $A_{\mu}B^{\nu}$. In general, contracting on any pair of indices decreases the order of the tensor by two. For example, $A^{\alpha}_{\beta\alpha\gamma} = A_{\beta\gamma}$ is a tensor of second-order, $A^{\mu}_{\nu}B^{\nu} = C^{\mu}$ is a four-vector, $C^{\alpha\beta}_{\alpha\beta}$ is a scalar, etc.

The Operation of Permutation of the Indices

Once a tensor is given, through the permutation of its indices we can construct another tensor. Of course this operation can be performed only if the initial tensor is not symmetric over these indices. For example,

$$B^{\mu\nu}_{\lambda\sigma} \to A^{\mu\nu}_{\sigma\lambda}. \tag{2.25}$$

By using the permutation operation in combination with the tensorial addition, we can construct tensors that are symmetric over several indices. Thus, for example,

$$T_{(\mu\nu)} = \frac{1}{2}\left(T_{\mu\nu} + T_{\nu\mu}\right). \tag{2.26}$$

We can also construct tensors that are antisymmetric over several indices. Thus, in the case of a second-order tensor $A_{\mu\nu}$, we have

$$A_{[\mu\nu]} = \frac{1}{2}\left(A_{\mu\nu} - A_{\nu\mu}\right). \tag{2.27}$$

2.2 Essentials of Riemannian Geometry

2.2.1 The Metric Tensor

We define a Riemannian space V_n as a real differentiable manifold, with the basic property that at each of its points a tensor

$$g_{\mu\nu}(x) = g_{\mu\nu}\left(x^1, x^2, \ldots, x^n\right), \tag{2.28}$$

is given [176, 194, 427]. The tensor $g_{\mu\nu}$ is twice covariant, symmetric, and non-degenerate, so that

$$g_{\mu\nu} = g_{\nu\mu}, \qquad g = \det |g_{\mu\nu}| \neq 0. \tag{2.29}$$

We will call $g_{\mu\nu}$ the *metric tensor* of the given Riemannian geometry. The functions $g_{\mu\nu}$ must have the basic properties of being continuous and of having continuous derivatives with respect to all coordinate variables $(x^1, x^2, \ldots, x^n)$, up to the n-th order.

In a Riemannian space it is always possible to construct with the help of the metric tensor an invariant second-order differential form, denoted ds^2, which is called an *interval* [302], and is defined according to

$$ds^2 = g_{\mu\nu} dx^\mu dx^\nu. \tag{2.30}$$

It is a well-known result in linear algebra that a square matrix can be reduced to a diagonal form, which can be done by finding the eigenvalues of the matrix. A similar result is valid for ds^2, which with the help of an appropriate coordinate transformation can always be transformed to a diagonal form. However, after such a transformation, the diagonal components of the matrix $g_{\mu\nu}$ will not all be positive (like in the case of the standard Euclidean geometry), but may have different signs. The difference between the numbers of positive and negative diagonal components is called in differential geometry the *signature of the Riemannian metric tensor* [302, 330]. Another important result from linear algebra tells us that for general quadratic forms the signature is a constant, and it is an invariant property of the metric. In an arbitrary Riemannian space V_n the interval will have alternating signs, which can be chosen according to different conventions. Hence, for example, we can have metric signatures of the form $(+,-,-,-)$, or $(-,+,+,+)$. Throughout this book, we adopt the latter signature.

Contrary to the case of Euclidean geometry, in Riemannian geometry the interval can have any sign. The intervals having the invariant property $ds^2 < 0$ are labelled as *timelike*, if $ds^2 > 0$ they are *spacelike*, and if $ds^2 = 0$ they are called *null*. These terms are the generalization of the terminology initially used in Special Relativity, in which space and time are united into a single four-dimensional geometric structure, and where the interval has the form

$$ds^2 = -\left(dx^0\right)^2 + \left(dx^1\right)^2 + \left(dx^2\right)^2 + \left(dx^3\right)^2, \tag{2.31}$$

where we have introduced Cartesian (Galilean) coordinates, defined as $x^0 = ct$, $x^1 = x$, $x^2 = y$, and $x^3 = z$, respectively, with c denoting the speed of light. In arbitrary coordinates the interval assumes the general form (2.30). The metric (2.31) is called *the Minkowski metric*, and it provides the basic example of *the unification of space and time* in the geometric structure of the Special Theory of Relativity. As is well-known, the special-relativistic interval (2.31) is invariant with respect to the group of Lorentz transformations, given by

$$\begin{aligned} x' &= \gamma\,(x - vt)\,, \\ y' &= y, \\ z' &= z, \\ t' &= \gamma\,(t - vx/c^2)\,, \end{aligned}$$

where $\gamma = 1/\sqrt{1 - v^2/c^2}$, and v is the relative speed of the reference frames with respect to each other.

According to our fundamental assumption on the metric tensor its determinant must be non-zero, $|g_{\mu\nu}| \neq 0$. Therefore we can associate to $g_{\mu\nu}$ a contravariant metric tensor, denoted $g^{\mu\nu}$, whose components can be determined with the help of the equations

$$g_{\mu\sigma} g^{\sigma\nu} = \delta^{\nu}_{\mu}. \tag{2.32}$$

A given physical quantity can be represented in either contra-variant or covariant components or, if it is a tensor, in mixed components. The only (geometric) quantity that can determine the relation between the different representations of a physical variable is the metric tensor. This equivalence relation is given, in the case of vectors, by the expressions

$$A^{\nu} = g^{\nu\mu} A_{\mu}, \qquad A_{\nu} = g_{\nu\mu} A^{\mu}. \tag{2.33}$$

In the case of tensors the relation between the different representations of a given physical quantity is realized by making use of the metric tensor according to the formulae

$$A^{\mu}_{\nu} = g^{\mu\alpha} A_{\alpha\nu}, \qquad A^{\mu\nu} = g^{\mu\alpha} g^{\nu\beta} A_{\alpha\beta}, \quad \text{etc.} \tag{2.34}$$

2.2.2 The Equations of Motion and Geodesic Lines

Representing one of the fundamental interactions in nature, gravitational fields have the basic property that the motion of all objects in them takes place in the same dynamical way. The gravitational dynamics are independent of the masses of the bodies, provided that the given initial conditions of the motion are the same. For example, the laws of motion of the objects falling in the gravitational field created by the attraction of the earth are the same for all bodies, no matter what their mass is (as shown for the first time by Galileo Galilei in his famous Pisa tower experiment). *Independently on the value of their mass, all objects in a gravitational field acquire one and the same acceleration.*

This basic property of all gravitational fields gives us the opportunity of introducing a fundamental relation between the motion of an object in a given gravitational field, and the dynamics of an object located far away from any possible external field, but whose motion is analyzed in the framework of a *noninertial (accelerated) frame of reference.* This analysis, performed for the first time by Einstein in his famous elevator thought experiment, shows that the dynamical properties of motion in arbitrary noninertial systems are identical to those corresponding to the motion in an inertial reference frame in the presence of the gravitational force. We can also formulate this result by stating that a noninertial reference frame is equivalent to a certain real gravitational field. This fundamental result is called *the equivalence principle*, and it is the basic physical principle of the General Theory of Relativity.

An important concept in Riemannian geometry, as well as in its physical applications, is the concept of a geodesic line (curve). In a Riemann space the geodesic lines play the same role as the straight lines play in an Euclidean space. They are also called extremal lines, because they can be defined by a variational procedure. Hence, for introducing the extremal (geodesic) lines we will use the mathematical tools of the variational calculus, and obtain their analytic representation by using a variational principle.

Modern physics is essentially based on the Principle of Least Action, which states that all (conservative) natural processes must evolve in such a way that some quantity associated to them takes a minimum value. According to the formal definition of the Principle, the motion of a conservative dynamical system between two arbitrary points a and b takes place such that a function of the coordinates $x(t)$ and velocities $\dot{x}(t)$, called the action, has a minimum value with respect to all other trajectories between the given points that correspond to the same energy. The action of the physical system, which is a scalar quantity, is usually denoted by S, and it is defined as the time integral of the Lagrangian function L,

$$S = \int_{t_1}^{t_2} L\left(x(t), \dot{x}(t)\right) dt, \tag{2.35}$$

where L is defined as the difference between the kinetic T and potential energy V of the system,

$$L = T - V. \tag{2.36}$$

Therefore the problem of the motion of an arbitrary conservative physical system can be reduced to the problem of finding the path $x(t)$, $t_1 \leq t \leq t_2$, for which the action S takes a minimal value. From a purely mathematical point of view, this class of problems belongs to the domain of mathematics called variational calculus. The mathematical quantity defined like the action is called the functional. The essential feature of the variational calculus is represented by the generalization of the concepts of the extremum of functions. The basic problem in variational calculus is not related to finding the minimum or maximum of a function, but to find the extremum of a mathematical functional that, generally, is a "function of functions." Hence to solve variational problems one must find the functions that make the functional extremal. A functional will reach its extremal value when its "variation" is equal to zero. This situation is very similar to the case of a function having an extremal value when its derivative is equal to zero.

Motion of Nonrelativistic Particles

In classical physics, the properties and the dynamical evolution of a system is described by its evolution equations. The simplest such equations are given for the case of a nonrelativistic point particle with mass m, whose evolution in an external potential $V(x)$ is described by Newton's equation of motion

$$m\frac{d^2x}{dt^2} = -\frac{\partial V(x)}{\partial x}. \tag{2.37}$$

These equations follow from extremizing the action (2.35), where the Lagrangian of the system is defined as

$$L = T - V = \frac{1}{2}m\left(\frac{dx}{dt}\right)^2 - V(x). \tag{2.38}$$

More specifically, for the variation of the action we obtain

$$\begin{aligned} \delta S &= \delta\int_{t_1}^{t_2}\left[\frac{1}{2}m\left(\frac{dx}{dt}\right)^2 - V(x)\right]dt = \int_{t_1}^{t_2}\left[m\frac{dx}{dt}\frac{d\delta x}{dt} - \frac{\partial V}{\partial x}\delta x\right]dt \\ &= -\int_{t_1}^{t_2}\left[m\frac{d^2x}{dt^2} + \frac{\partial V}{\partial x}\right]\delta x dt + m\frac{dx}{dt}\delta x|_{t_1}^{t_2} \\ &= -\int_{t_1}^{t_2}\left[m\frac{d^2x}{dt^2} + \frac{\partial V}{\partial x}\right]\delta x dt = 0, \end{aligned} \tag{2.39}$$

from which the Newtonian equations of motion follow immediately, under the assumption that the variations of x vanish at the initial and final points a and b of the particle trajectory.

The Lagrange Equations

We consider now the problem of an arbitrary mechanical system evolving between two points (states) a and b, respectively. There are many trajectories joining the two points, but a conservative mechanical system that is moving between them always chooses the trajectory for which the action functional is extremal. Let's assume that the real trajectory of the body is given by $x = x(t)$. Let us also introduce a second trajectory, which is very nearby to the first one, and which is given by $x(t) + \epsilon h(t)$, where $h(t)$ is an arbitrary time dependent function, and ϵ is a constant parameter satisfying the condition $\epsilon \ll 1$. At the ends of all varied paths, $h(t)$ must satisfy the simple conditions

$$h(t_1) = h(t_2) = 0. \tag{2.40}$$

The action integral, which is a scalar, will take different values according to the different trajectories taken by the particle to go from a to b, and consequently

$$S\left[x(t)\right] \neq S\left[x(t) + \epsilon h(t)\right]. \tag{2.41}$$

To find the trajectory or trajectories that make the action extremal, we shall vary infinitesimally the path, and evaluate the action for the extremal and for the varied path. By using a Taylor series expansion, for the Lagrangian along the varied trajectory we find,

$$L\left(x + \epsilon h, \dot{x} + \epsilon \dot{h}\right) = L\left(x, \dot{x}\right) + \epsilon\left(\frac{\partial L}{\partial x} h + \frac{\partial L}{\partial \dot{x}} \dot{h}\right). \tag{2.42}$$

Therefore the variation of the action along the two nearby trajectories can be obtained as

$$S\left[x(t) + \epsilon h(t)\right] - S\left[x(t)\right] = \epsilon \int_{t_1}^{t_2} \left(\frac{\partial L}{\partial x} h + \frac{\partial L}{\partial \dot{x}} \dot{h}\right) dt. \tag{2.43}$$

The second term in the integrand can be transformed by using partial integration,

$$\begin{aligned}\int_{t_1}^{t_2} \frac{\partial L}{\partial \dot{x}} \dot{h} dt &= \frac{\partial L}{\partial \dot{x}} h(t)\Big|_{t_1}^{t_2} - \int_{t_1}^{t_2} \left(\frac{d}{dt}\frac{\partial L}{\partial \dot{x}}\right) h(t) dt \\ &= -\int_{t_1}^{t_2} \left(\frac{d}{dt}\frac{\partial L}{\partial \dot{x}}\right) h(t) dt,\end{aligned} \tag{2.44}$$

where we have used the conditions $h\left(t_1\right) = h\left(t_2\right) = 0$. Then the variation of the action integral for a very small ϵ is

$$\begin{aligned}\frac{\delta S}{\delta \epsilon} &= \lim_{\epsilon \to 0} \frac{S\left[x(t) + \epsilon h(t)\right] - S\left[x(t)\right]}{\epsilon} \\ &= \int_{t_1}^{t_2} \left(\frac{\partial L}{\partial x} - \frac{d}{dt}\frac{\partial L}{\partial \dot{x}}\right) h(t) dt.\end{aligned} \tag{2.45}$$

By taking into account that $h(t) \neq 0$ for $t_1 < t < t_2$, the trajectory $x(t)$ followed by the particle between the two fixed points a and b is an extremal of the action S if the quantity between the brackets in the integral be zero. This is a necessary and sufficient condition. Thus we obtain the equation

$$\frac{\partial L}{\partial x} - \frac{d}{dt}\frac{\partial L}{\partial \dot{x}} = 0. \tag{2.46}$$

This equation is called the Lagrange equation (or sometimes the Euler–Lagrange equation). From it the equation of motion of a particle under the action of a conservative force can be obtained, once the Lagrange function of the system is known.

The Free Relativistic Particle in the Minkowski Space

A free relativistic particle is a particle that is not under the influence of any external force. Moreover, we will analyze the motion the particle in the Minkowski geometry, where the metric can be written as

$$ds^2 = \eta_{\mu\nu}dx^\mu dx^\nu = -\left(dx^0\right)^2 + \left(dx^1\right)^2 + \left(dx^2\right)^2 + \left(dx^3\right)^2, \tag{2.47}$$

where we have introduced the Minkowskian metric $\eta_{\mu\nu} = \mathrm{diag}\,(-1, +1, +1, +1)$, and the three-velocity of the particle, defined as $\vec{v} = (dx/dt, dy/dt, dz/dt)$. Therefore the action for the motion of the free relativistic particle can be rewritten as

$$S = -\alpha \int ds = -\alpha \int \sqrt{-\eta_{\mu\nu}dx^\mu dx^\nu} = -\alpha \int c\sqrt{1 - \frac{\vec{v}^2}{c^2}}dt, \tag{2.48}$$

and we have used the relation

$$\begin{aligned}\sqrt{-\eta_{\mu\nu}dx^\mu dx^\nu} &= \sqrt{c^2dt^2\left[1 - \left(\frac{dx^1}{cdt}\right)^2 - \left(\frac{dx^2}{cdt}\right)^2 - \left(\frac{dx^3}{cdt}\right)^2\right]} \\ &= cdt\sqrt{1 - \frac{\vec{v}^2}{c^2}}\,.\end{aligned} \tag{2.49}$$

Generally, the action integral can also be represented as an integral with respect to the time of the Lagrange function $L = L\left(t, x^\mu, \dot{x}^\mu\right)$ of the system, $S = \int_{t_1}^{t_2} L\, dt$.

From the comparison of Eqs. (2.48) and (2.35) the form of the Lagrange function for a free particle relativistic particle immediately follows, as given by

$$L = -\alpha c\sqrt{1 - \frac{\vec{v}^2}{c^2}}. \tag{2.50}$$

To determine the coefficient α we use the correspondence principle, according to which in the limit $c \to \infty$, the relativistic expression of L must recover the classical expression $L = m_0\vec{v}^2/2$, where m_0 is the rest mass of the particle. By expanding the Lagrangian (2.50) in powers series of v/c we obtain

$$L \approx -\alpha c + \frac{\alpha\vec{v}^2}{2c}. \tag{2.51}$$

After omitting the constant term in the Lagrangian, since it does not influence the equation of motion, the comparison of the above equation with the classical Lagrangian gives $\alpha = m_0c$. Hence for the Lagrangian of the free relativistic point particle we obtain the expression

$$L = -m_0c^2\sqrt{1 - \frac{\vec{v}^2}{c^2}}. \tag{2.52}$$

We proceed now to the derivation of the equations of motion of the particle from the variational principle. By taking the variation of the action of the free relativistic point particle we obtain

$$\begin{aligned}\delta S &= -m_0 c\delta \int_a^b ds = -m_0 c \int_a^b \sqrt{dx_\mu dx^\mu} \\ &= -m_0 c \int_a^b \frac{(\delta dx_\mu)\, dx^\mu + dx_\mu\, (\delta dx^\mu)}{2\sqrt{dx_\mu dx^\mu}} \\ &= -m_0 c \int_a^b \frac{dx_\mu d\delta x^\mu}{ds} = -m_0 c \int_a^b U_\mu d\delta x^\mu, \end{aligned} \tag{2.53}$$

where we have assumed that the operators d and δ commute, and where we have also introduced the four-velocity U^μ of a particle, defined as

$$U^\mu = \frac{dx^\mu}{ds}. \tag{2.54}$$

In Minkowski space the covariant components of the four-velocity can be obtained as $U_\mu = \eta_{\mu\nu}U^\nu$. By integrating by parts we find

$$\delta S = -m_0 c \int_a^b U_\mu d\delta x^\mu = -m_0 c U_\mu \delta x^\mu|_a^b + m_0 c \int \frac{dU_\mu}{ds}\delta x^\mu ds. \tag{2.55}$$

But the variations of δx^μ at the two limits corresponding to endpoints a and b of the particle trajectory are zero, $(\delta x^\mu)|_a = (\delta x^\mu)|_b = 0$, and therefore we find the equation of motion of the free relativistic particle as given by

$$\frac{dU_\mu}{ds} = 0. \tag{2.56}$$

The Geodesic Equation

In an arbitrary Riemannian space, the distance between two nearby points is fully determined and described by the metric tensor $g_{\mu\nu}$ via the interval ds. On the other hand, it is important to note that the scalar quantity ds is not a total differential in a strict mathematical sense.

The integral of ds between points a and b gives a quantity S, which is the finite value of the interval between the two given points,

$$S = -\int_a^b ds = -\alpha \int_a^b \sqrt{g_{\mu\nu}dx^\mu dx^\nu} = -\alpha \int_a^b \sqrt{g_{\mu\nu}\frac{dx^\mu(s)}{ds}\frac{dx^\nu(s)}{ds}}ds, \tag{2.57}$$

where $\alpha \neq 0$ is a constant. The extremum of S is determined by the relation

$$\delta S = -\alpha\delta \int_a^b ds = -\alpha \int_a^b \delta\, ds = -\alpha\delta \int_a^b \sqrt{g_{\mu\nu}\frac{dx^\mu}{ds}\frac{dx^\nu}{ds}}ds = 0. \tag{2.58}$$

In space-time a point particle follows a trajectory that is called the *world line*. The equations of a world line can be expressed in a parametric form, with the coordinates x^μ given as functions of a parameter τ, so that $x^\mu = x^\mu(\tau)$.

The action (2.57) remains invariant if we replace the parameter s by another parameter $s' = f(s)$, where f is an arbitrary monotonic function. As a result of such a transformation we have first $ds' = (df(s)/ds)ds$, and

$$\frac{dx^\mu(s)}{ds} = \frac{dx^\mu(s')}{ds'}\frac{ds'}{ds} = \frac{dx^\mu(s')}{ds'}\frac{df(s)}{ds}. \tag{2.59}$$

Hence for the new action we obtain

$$\begin{aligned} S' &= -\alpha \int_a^b \sqrt{g_{\mu\nu}\frac{dx^\mu(s')}{ds'}\frac{dx^\nu(s')}{ds'}}ds' \\ &= -\alpha \int_a^b \sqrt{g_{\mu\nu}\frac{dx^\mu(s)}{ds}\frac{dx^\nu(s)}{ds}\left[\frac{df(s)}{ds}\right]^{-2}}\frac{df(s)}{ds}ds = S, \end{aligned} \tag{2.60}$$

which shows that indeed the action is invariant with respect to an arbitrary transformation of the parameter s.

Next, we will consider the motion of a particle in an arbitrary geometry, and we will look for functions $g_{\mu\nu}(x)$ that make the integral (2.57) extremal [302, 330]. In order to calculate the total variation of the interval we first obtain

$$\begin{aligned} \delta(ds^2) &= 2ds\delta(ds) = \delta(g_{\mu\nu}dx^\mu dx^\nu) \\ &= \frac{\partial g_{\mu\nu}}{\partial x^\sigma}\delta x^\sigma dx^\mu dx^\nu + 2g_{\mu\nu}(x)dx^\mu\delta(dx^\nu). \end{aligned} \tag{2.61}$$

At this moment we note that

$$\delta(dx^\nu) = d(\delta x^\nu). \tag{2.62}$$

Hence we have

$$\delta(ds) = \frac{1}{2}\frac{\partial g_{\mu\nu}}{\partial x^\sigma}U^\mu dx^\nu \delta x^\sigma + g_{\mu\nu}(x)U^\mu\delta(dx^\nu). \tag{2.63}$$

Substituting Eq. (2.63) into Eq. (2.58), we obtain

$$\delta S = \int_a^b \left[\frac{1}{2}\frac{\partial g_{\mu\nu}}{\partial x^\sigma}U^\mu U^\nu \delta x^\sigma + g_{\mu\nu}(x)U^\mu\frac{\delta(dx^\nu)}{ds}\right]ds = 0. \tag{2.64}$$

Since

$$g_{\mu\nu}(x)U^\mu\frac{\delta(dx^\nu)}{ds} = \frac{d}{ds}\left(g_{\mu\nu}(x)U^\mu\delta x^\nu\right) - \delta x^\nu\frac{d}{ds}\left(g_{\mu\nu}(x)U^\mu\right), \tag{2.65}$$

and at the integration limits $\delta x^\nu|_a^b = 0$, from Eq. (2.64) we find

$$\delta S = \int_a^b \left[\frac{1}{2}\frac{\partial g_{\mu\nu}}{\partial x^\sigma}U^\mu U^\nu - g_{\mu\sigma}\frac{dU^\mu}{ds} - \frac{\partial g_{\mu\sigma}}{\partial x^\lambda}U^\mu U^\lambda\right]ds\delta x^\sigma = 0. \tag{2.66}$$

We represent the last term in the above equation as

$$\frac{\partial g_{\mu\sigma}}{\partial x^{\lambda}} U^{\mu} U^{\lambda} = \frac{1}{2}\left(\frac{\partial g_{\mu\sigma}}{\partial x^{\lambda}} + \frac{\partial g_{\lambda\sigma}}{\partial x^{\mu}}\right) U^{\mu} U^{\lambda}. \tag{2.67}$$

Substituting Eq. (2.67) into Eq. (2.66) we obtain

$$\delta S = \int_a^b \left[\frac{1}{2}\left(\frac{\partial g_{\mu\sigma}}{\partial x^{\lambda}} + \frac{\partial g_{\lambda\sigma}}{\partial x^{\mu}} - \frac{\partial g_{\mu\lambda}}{\partial x^{\sigma}}\right) U^{\mu} U^{\lambda} + g_{\mu\sigma}\frac{dU^{\mu}}{ds}\right] ds\delta x^{\sigma} = 0. \tag{2.68}$$

Now we note that since the variations δx^{σ} of the coordinates are arbitrary, the integral is identically zero only if the condition

$$g_{\mu\sigma}\frac{dU^{\mu}}{ds} + \frac{1}{2}\left(\frac{\partial g_{\mu\sigma}}{\partial x^{\lambda}} + \frac{\partial g_{\lambda\sigma}}{\partial x^{\mu}} - \frac{\partial g_{\mu\lambda}}{\partial x^{\sigma}}\right) U^{\mu} U^{\lambda} = 0, \tag{2.69}$$

is satisfied. Multiplying Eq. (2.69) by $g^{\sigma\alpha}$, we obtain

$$\frac{dU^{\alpha}}{ds} + \Gamma^{\alpha}_{\mu\lambda} U^{\mu} U^{\lambda} = 0, \tag{2.70}$$

where we have introduced the Christoffel symbols defined as

$$\Gamma^{\alpha}_{\mu\lambda} = \frac{1}{2} g^{\alpha\sigma}\left(\frac{\partial g_{\mu\sigma}}{\partial x^{\lambda}} + \frac{\partial g_{\lambda\sigma}}{\partial x^{\mu}} - \frac{\partial g_{\mu\lambda}}{\partial x^{\sigma}}\right). \tag{2.71}$$

The Christoffel symbols *are not tensor quantities.* Equation (2.70) are the equations for a geodesic line in the Riemannian geometry. Writing them in components it is easy to see that they form a system of four nonlinear differential equations. However, not all equations in the system are independent, since the components of the velocity must also satisfy the following independent condition

$$g_{\mu\nu}(x) U^{\mu} U^{\nu} = -1. \tag{2.72}$$

Properties of the Christoffel Symbols

Along with the Christoffel symbols $\Gamma^{\alpha}_{\mu\lambda}$, we will also use the quantities $\Gamma_{\sigma,\mu\lambda}$, which are defined with the use of the following rule,

$$\Gamma_{\sigma,\mu\lambda} = g_{\sigma\alpha}\Gamma^{\alpha}_{\mu\lambda} = \frac{1}{2}\left(\frac{\partial g_{\sigma\mu}}{\partial x^{\lambda}} + \frac{\partial g_{\sigma\lambda}}{\partial x^{\mu}} - \frac{\partial g_{\mu\lambda}}{\partial x^{\sigma}}\right). \tag{2.73}$$

Conversely,

$$\Gamma^{\alpha}_{\mu\lambda} = g^{\alpha\sigma}\Gamma_{\sigma,\mu\lambda}. \tag{2.74}$$

We have already mentioned that the Christoffel symbols are not tensor quantities. However, their transformation formula does play an important role in many geometrical and physical applications. The rule of transformation of the Christoffel symbols can be obtained from the analysis of the transformation properties of the covariant derivative. By introducing a frame x', the covariant derivative of a vector A_{μ} with respect to this frame is given by $\nabla_{\nu} A'_{\mu} = \partial A_{\mu}/\partial x'^{\nu} - \Gamma'^{\rho}_{\mu\nu} A'_{\rho}$.

Taking into account that both $\nabla_\nu A'_\mu$ and A'_μ are tensors, and by using their transformation properties, after a simple calculation we obtain

$$\Gamma^{\alpha}_{\mu\lambda} = \Gamma'^{\rho}_{\sigma\omega}\frac{\partial x^{\alpha}}{\partial x'^{\rho}}\frac{\partial x'^{\sigma}}{\partial x^{\mu}}\frac{\partial x'^{\omega}}{\partial x^{\lambda}} + \frac{\partial^2 x'^{\beta}}{\partial x^{\mu}\partial x^{\lambda}}\frac{\partial x^{\alpha}}{\partial x'^{\beta}}. \tag{2.75}$$

From the above equation we can see that the Christoffel symbols $\Gamma^{\alpha}_{\mu\lambda}$ transform according to the rules of the tensor transformation only for the particular case of the linear coordinate transformations (in this case the last term in Eq. (2.75) identically vanishes). However, it turns out that this term is symmetric in the two indices μ and λ, and therefore it does not influence the transformation properties of the quantity $S^{\alpha}_{\mu\lambda} = \Gamma^{\alpha}_{\mu\lambda} - \Gamma^{\alpha}_{\lambda\mu}$. Hence it follows that the quantity $S^{\alpha}_{\mu\lambda}$ has tensorial transformation properties, and it transforms like a tensor, so that

$$S^{\alpha}_{\mu\lambda} = S'^{\rho}_{\sigma\omega}\frac{\partial x^{\alpha}}{\partial x'^{\rho}}\frac{\partial x'^{\sigma}}{\partial x^{\mu}}\frac{\partial x'^{\omega}}{\partial x^{\lambda}}. \tag{2.76}$$

In the following we will call $S^{\alpha}_{\mu\lambda}$ *the torsion tensor* of the space-time manifold. In a gravitational theory strictly based on the equivalence principle, the torsion tensor must vanish in all points of the space-time manifold. We can prove this result as follows. The equivalence principle imposes the necessity of the existence of a locally Galilean coordinate frame, in which all the components of the Christoffel symbol $\Gamma^{\alpha}_{\mu\lambda}$ vanish at a given point. Consequently, all the components of $S^{\alpha}_{\mu\lambda}$ must also be equal to zero in the arbitrarily chosen point. Since, as we have previously seen, $S^{\alpha}_{\mu\lambda}$ is a tensor, it has the basic property that if all of its components vanish in a particular coordinate system, they must also be equal to zero in all coordinate frames. This important result on the vanishing of the torsion of the space-time requires that the lower indices of the Christoffel symbols must be symmetric, that is,

$$\Gamma^{\alpha}_{\mu\lambda} = \Gamma^{\alpha}_{\lambda\mu}. \tag{2.77}$$

In an arbitrary four-dimensional space-time the Christoffel symbols $\Gamma^{\alpha}_{\mu\lambda}$ have generally 40 components.

We will now prove the important result according to which we can always adopt a local coordinate frame in which at a previously defined point P all the components of the Christoffel symbols $\Gamma^{\alpha}_{\mu\lambda}$ vanish identically. Such a reference frame, which has an important role in many applications of General Relativity, is designated *a locally inertial, or locally geodesic system* [302], and the corresponding coordinates are called geodesic, or Riemannian coordinates.

Let us choose the given point, with coordinates x^{μ}, as the origin of the initial coordinate system, and let the values of the Christoffel symbols in this system be $\left(\Gamma^{\mu}_{\alpha\beta}\right)_0$. In the vicinity of our point P we perform now the following coordinate transformation

$$x'^{\mu} = x^{\mu} + \frac{1}{2}\left(\Gamma^{\mu}_{\alpha\beta}\right)_0 x^{\alpha}x^{\beta} + \ldots \tag{2.78}$$

Then

$$\left(\frac{\partial^2 x'^\nu}{\partial x^\alpha \partial x^\beta}\frac{\partial x^\mu}{\partial x'^\nu}\right)_0 = \left(\Gamma^\mu_{\alpha\beta}\right)_0, \tag{2.79}$$

and according to the law of the transformation of the Christoffel symbols (2.75), all the $\Gamma'^\rho_{\sigma\omega}$ become equal to zero [302]. In can also be proven that by a transformation of the coordinates x^μ it is always possible to reduce the values of the Christoffel symbols to zero along each point of an arbitrary curve [330].

2.2.3 The Covariant Derivative

Now we proceed to the study of the differential operations in the Riemannian geometry. We first consider an arbitrary covariant vector A_λ, and contract it with the velocity four-vector U^λ. As a result a scalar quantity T is formed [320], as given by

$$T = A_\lambda U^\lambda. \tag{2.80}$$

We differentiate now T with respect to s, to obtain a scalar function dT/ds, which can be easily computed as

$$\frac{dT}{ds} = \frac{d}{ds}\left(A_\lambda U^\lambda\right) = \frac{dA_\lambda}{ds}U^\lambda + A_\nu \frac{dU^\nu}{ds} = \frac{\partial A_\lambda}{\partial x^\sigma}U^\sigma U^\lambda + A_\nu \frac{dU^\nu}{ds}. \tag{2.81}$$

Substituting for dU^ν/ds its expression obtained from Eq. (2.70), we find

$$\frac{d}{ds}\left(A_\lambda U^\lambda\right) = \left[\frac{\partial A_\lambda}{\partial x^\sigma} - \Gamma^\nu_{\sigma\lambda}A_\nu\right]U^\sigma U^\lambda. \tag{2.82}$$

Now (2.82) is a scalar quantity, while U^σ is a vector. Therefore it immediately follows that in the right-hand side of Eq. (2.82) we have constructed a second-order tensor, given by

$$A_{\lambda;\sigma} = \nabla_\sigma A_\lambda = \frac{DA_\lambda}{dx^\sigma} = \frac{\partial A_\lambda}{\partial x^\sigma} - \Gamma^\nu_{\sigma\lambda}A_\nu, \tag{2.83}$$

which we will designate as the *covariant derivative* of the vector A_λ [302, 330]. In the following we will denote the covariant derivative by using the symbol ∇. Thus we have introduced the important mathematical concept of the covariant derivative of the covariant vector A_λ, as defined by Eq. (2.83). We will call the quantity

$$DA_\lambda = dA_\lambda - \Gamma^\nu_{\sigma\lambda}A_\nu dx^\sigma, \tag{2.84}$$

the *covariant differential* of the vector A_λ [302].

We will now define the *covariant derivative of the contravariant vector* A^λ, representing another important mathematical concept. To achieve this goal we write the scalar quantity dT/ds in the form

$$\frac{dT}{ds} = \frac{d}{ds}\left(A^\mu U^\nu g_{\mu\nu}\right) = \frac{\partial A^\mu}{\partial x^\sigma} U^\sigma U^\nu g_{\mu\nu} + A^\mu g_{\mu\lambda}\frac{dU^\lambda}{ds} + A^\mu U^\nu U^\sigma \frac{\partial g_{\mu\nu}}{\partial x^\sigma}. \tag{2.85}$$

Substituting into the right-hand side of this equation the expression of dU^λ/ds as obtained from Eq. (2.70), we find

$$\frac{d}{ds}\left(A^\mu U^\nu g_{\mu\nu}\right) = U^\nu U^\sigma \left[g_{\mu\nu}\frac{\partial A^\mu}{\partial x^\sigma} - A^\mu g_{\mu\lambda}\Gamma^\lambda_{\sigma\nu} + A^\mu \frac{\partial g_{\mu\nu}}{\partial x^\sigma}\right]. \tag{2.86}$$

By taking into account the definition of the Christoffel symbols as given by Eq. (2.71), we have

$$\frac{d}{ds}\left(A^\mu U^\nu g_{\mu\nu}\right) = \left[g_{\mu\nu}\frac{\partial A^\mu}{\partial x^\sigma} + \frac{1}{2}\left(\frac{\partial g_{\mu\nu}}{\partial x^\sigma} + \frac{\partial g_{\sigma\nu}}{\partial x^\mu} - \frac{\partial g_{\sigma\mu}}{\partial x^\nu}\right)\right] U^\nu U^\sigma. \tag{2.87}$$

Representing U^ν as

$$U^\nu = U_\lambda g^{\lambda\nu}, \tag{2.88}$$

and after substitution into Eq. (2.87), we find

$$\frac{d}{ds}\left(A^\mu U^\nu g_{\mu\nu}\right) = \left[\frac{\partial A^\lambda}{\partial x^\sigma} + \Gamma^\lambda_{\sigma\mu} A^\mu\right] U^\sigma U_\lambda. \tag{2.89}$$

Obviously the expressions in both sides of the above equation are scalars. Then it immediately follows that the covariant derivative of the contravariant vector A^λ, introduced as

$$\nabla_\sigma A^\lambda = \frac{DA^\lambda}{dx^\sigma} = \frac{\partial A^\lambda}{\partial x^\sigma} + \Gamma^\lambda_{\sigma\mu} A^\mu, \tag{2.90}$$

is a second-order tensorial quantity. Thus, we have also introduced the concept of the covariant derivative of the contravariant vector A^λ, and obtained the rule according to which it can be calculated. The covariant differential of the vector A^λ is given by

$$DA^\lambda = dA^\lambda + \Gamma^\lambda_{\sigma\mu} A^\mu dx^\sigma. \tag{2.91}$$

With the use of the definitions and rules (2.83) and (2.90), we can also easily find the covariant derivatives of a second-order tensor as

$$\nabla_\sigma A_{\mu\nu} = \frac{\partial A_{\mu\nu}}{\partial x^\sigma} - \Gamma^\lambda_{\sigma\mu} A_{\lambda\nu} - \Gamma^\lambda_{\sigma\nu} A_{\lambda\mu}, \tag{2.92}$$

$$\nabla_\sigma A^{\mu\nu} = \frac{\partial A^{\mu\nu}}{\partial x^\sigma} + \Gamma^\mu_{\sigma\lambda} A^{\nu\lambda} + \Gamma^\nu_{\sigma\lambda} A^{\mu\lambda}, \tag{2.93}$$

$$\nabla_\sigma A^\nu_\rho = \frac{\partial A^\nu_\rho}{\partial x^\sigma} - \Gamma^\lambda_{\rho\sigma} A^\nu_\lambda + \Gamma^\nu_{\sigma\lambda} A^\lambda_\rho. \tag{2.94}$$

The covariant derivative of a tensor of arbitrary order can be similarly obtained by generalizing the rules we have just introduced for vectors and second-order tensors. This can be done by using the following *general rule for obtaining the covariant derivative* [176, 302, 427]. Hence, to calculate the covariant derivative of a tensor $A^{\cdots}_{\cdots}$ of arbitrary order with respect to the coordinate x^μ, we must proceed as follows: For each covariant index μ of the tensor $(A^{\cdots}_{\cdot\mu\cdot})$, we supplement the ordinary derivative $\partial A^{\cdots}_{\cdots}/\partial x^\mu$ by a term $-\Gamma^\alpha_{\mu\nu}A^{\cdots}_{\cdot\alpha\cdot}$, while for each contravariant index ν $(A^{\cdot\nu\cdot}_{\cdots})$, we supplement the ordinary derivative by a term $+\Gamma^\nu_{\sigma\mu}A^{\cdot\sigma\cdot}_{\cdots}$.

For a scalar function ϕ, its covariant derivative reduces to an ordinary derivative, that is, it gives the covariant vector

$$\nabla_k\phi = \frac{\partial\phi}{\partial x^k}. \tag{2.95}$$

We can easily check that the covariant derivative of a product of two vectors can be computed by using the same Leibniz rule valid for the ordinary derivative of the products of functions. Therefore, the covariant derivative of a product of two arbitrary vectors is

$$\nabla_\sigma\left(A_\mu A_\nu\right) = A_\nu\nabla_\sigma A_\mu + A_\mu\nabla_\sigma A_\nu. \tag{2.96}$$

With the use of Eq. (2.92) we can easily show that

$$\nabla_\sigma g_{\mu\nu} = 0, \tag{2.97}$$

from which a basic result in Riemnannian geometry follows, namely, that *in a Riemannian space-time the covariant derivative of the metric tensor is identically equal to zero.*

The Determinant and the Differential of the Metric Tensor

The determinant g of the metric tensor $g_{\mu\nu}$ of a Riemannian space, as well as its differential dg, appear and play a fundamental role in many general relativistic calculations. The differential of the determinant g can be computed in the following manner. We take the ordinary differential, defined by the operator d, of each component of the metric tensor $g_{\mu\nu}$, and multiply it by the minor $A^{\mu\nu}$ corresponding to each $g_{\mu\nu}$ [302],

$$dg = A^{\mu\nu}dg_{\mu\nu}. \tag{2.98}$$

Hence we have obtained the result that the minor $A^{\mu\nu}$, corresponding to the element $g_{\mu\nu}$, can be computed with the help of the relation

$$A^{\mu\nu} = \frac{\partial g}{\partial g_{\mu\nu}}. \tag{2.99}$$

From standard linear algebra it is known that the components of the tensor $g^{\mu\nu}$, reciprocal to the metric tensor $g_{\mu\nu}$, are given by the following rule: we compute first the minors of the element $g_{\mu\nu}$, and then we divide them by the determinant g, so that

$$g^{\mu\nu} = \frac{A^{\mu\nu}}{g}. \tag{2.100}$$

Therefore, for the components of the tensor $g^{\mu\nu}$ that are reciprocal to the metric tensor $g_{\mu\nu}$ ($g_{\mu\nu}g^{\nu\sigma} = \delta^{\sigma}_{\mu}$), we find the important computational rule given by

$$g^{\mu\nu} = \frac{1}{g}\frac{\partial g}{\partial g_{\mu\nu}} = \frac{\partial}{\partial g_{\mu\nu}} \ln |g|. \tag{2.101}$$

The differential dg of the determinant of the metric tensor $g_{\mu\nu}$ can be obtained from the following relation [176, 427]

$$dg = g g^{\mu\nu} dg_{\mu\nu} = -g g_{\mu\nu} dg^{\mu\nu}. \tag{2.102}$$

To obtain the second relation in the above equation we have used the equality $g_{\mu\nu}g^{\mu\nu} = \delta^{\mu}_{\mu} = 4$, which implies that $g^{\mu\nu}dg_{\mu\nu} = -g_{\mu\nu}dg^{\mu\nu}$.

The Contracted Christoffel Symbols, and the Divergence Relations

The contraction $\Gamma^{\alpha}_{\mu\alpha}$ of the Christoffel symbol plays an important role in many computations. From the definition (2.71) we have

$$\Gamma^{\alpha}_{\mu\alpha} = \frac{1}{2} g^{\alpha\sigma} \left(\frac{\partial g_{\mu\sigma}}{\partial x^{\alpha}} + \frac{\partial g_{\alpha\sigma}}{\partial x^{\mu}} - \frac{\partial g_{\mu\alpha}}{\partial x^{\sigma}} \right). \tag{2.103}$$

Changing the position of the indices α and σ in the first and third term in the parentheses, we see that these terms cancel each other, and therefore we find

$$\Gamma_{\mu} = \Gamma^{\alpha}_{\mu\alpha} = \frac{1}{2} g^{\alpha\sigma} \frac{\partial g_{\alpha\sigma}}{\partial x^{\mu}} = \frac{1}{2g}\frac{\partial g}{\partial x^{\mu}} = \frac{\partial \ln \sqrt{|g|}}{\partial x^{\mu}}. \tag{2.104}$$

It is also useful to find the expression of the quantity $g^{\mu\lambda}\Gamma^{\alpha}_{\mu\lambda}$, which can be obtained as follows

$$\begin{aligned} g^{\mu\lambda}\Gamma^{\alpha}_{\mu\lambda} &= \frac{1}{2} g^{\mu\lambda} g^{\alpha\sigma} \left(\frac{\partial g_{\mu\sigma}}{\partial x^{\lambda}} + \frac{\partial g_{\lambda\sigma}}{\partial x^{\mu}} - \frac{\partial g_{\mu\lambda}}{\partial x^{\sigma}} \right) \\ &= g^{\mu\lambda} g^{\alpha\sigma} \left(\frac{\partial g_{\mu\sigma}}{\partial x^{\lambda}} - \frac{1}{2}\frac{\partial g_{\mu\lambda}}{\partial x^{\sigma}} \right). \end{aligned} \tag{2.105}$$

With the use of Eq. (2.102), the above relation can be transformed into

$$g^{\mu\lambda}\Gamma^{\alpha}_{\mu\lambda} = -\frac{1}{\sqrt{|g|}} \frac{\partial \left(\sqrt{|g|} g^{\alpha\beta} \right)}{\partial x^{\beta}}. \tag{2.106}$$

With the help of the formulae that we have just obtained, we can easily derive the expression for $\nabla_{\mu}A^{\mu}$, which is the generalized divergence of a vector in arbitrary coordinates, and which has a very simple form. By using the definition of the covariant derivative, we have

$$\begin{aligned}\nabla_\mu A^\mu &= \frac{\partial A^\mu}{\partial x^\mu} + \Gamma^\alpha_{\alpha\mu} A^\mu \\ &= \frac{\partial A^\mu}{\partial x^\mu} + A^\mu \frac{\partial}{\partial x^\mu} \ln \sqrt{|g|},\end{aligned} \tag{2.107}$$

or finally

$$\nabla_\mu A^\mu = \frac{1}{\sqrt{|g|}} \frac{\partial \left(\sqrt{|g|} A^\mu\right)}{\partial x^\mu}. \tag{2.108}$$

Now we derive a similar expression for the divergence of an antisymmetric second-rank tensor $A^{\mu\nu}$. From the definition of the covariant derivative, we have

$$\nabla_\nu A^{\mu\nu} = \frac{\partial A^{\mu\nu}}{\partial x^\nu} + \Gamma^\mu_{\alpha\nu} A^{\alpha\nu} + \Gamma^\nu_{\alpha\nu} A^{\mu\alpha}. \tag{2.109}$$

But since $A^{\mu\nu} = -A^{\nu\mu}$, we have

$$\Gamma^\mu_{\alpha\nu} A^{\alpha\nu} = -\Gamma^\mu_{\nu\alpha} A^{\nu\alpha} = 0. \tag{2.110}$$

Substituting the expression of $\Gamma^\nu_{\alpha\nu}$, as given by Eq. (2.104), we finally obtain

$$\nabla_\nu A^{\mu\nu} = \frac{1}{\sqrt{|g|}} \frac{\partial \left(\sqrt{|g|} A^{\mu\nu}\right)}{\partial x^\nu}. \tag{2.111}$$

Next, we consider the divergence of the mixed components A^ν_μ of a symmetric tensor $A_{\mu\nu} = A_{\nu\mu}$. We obtain first

$$\begin{aligned}\nabla_\nu A^\nu_\mu &= \frac{\partial A^\nu_\mu}{\partial x^\nu} + \Gamma^\nu_{\alpha\nu} A^\alpha_\mu - \Gamma^\alpha_{\mu\nu} A^\nu_\alpha \\ &= \frac{1}{\sqrt{|g|}} \frac{\partial \left(\sqrt{|g|} A^\nu_\mu\right)}{\partial x^\nu} - \Gamma^\alpha_{\nu\mu} A^\nu_\alpha.\end{aligned} \tag{2.112}$$

For the last term we obtain

$$\begin{aligned}-\Gamma^\alpha_{\nu\mu} A^\nu_\alpha &= -\frac{1}{2}\left(\frac{\partial g_{\mu\alpha}}{\partial x^\nu} + \frac{\partial g_{\nu\alpha}}{\partial x^\mu} - \frac{\partial g_{\mu\nu}}{\partial x^\alpha}\right) A^{\nu\alpha} \\ &= -\frac{1}{2}\frac{\partial g_{\nu\alpha}}{\partial x^\mu} A^{\nu\alpha},\end{aligned} \tag{2.113}$$

where the first and the third term in the parentheses cancel due to the symmetry of $A^{\nu\alpha}$. Hence we obtain

$$\nabla_\nu A^\nu_\mu = \frac{1}{\sqrt{|g|}} \frac{\partial \left(\sqrt{|g|} A^\nu_\mu\right)}{\partial x^\nu} - \frac{1}{2}\frac{\partial g_{\nu\alpha}}{\partial x^\mu} A^{\nu\alpha}. \tag{2.114}$$

Finally, we consider the transformation to an arbitrary coordinate system of the sum $\partial^2\phi/\partial x_\mu \partial x^\mu = \Box$ of the second-order partial derivatives of a scalar function ϕ. In a system of arbitrary coordinates, this results in

$$\Box\phi = \nabla_\mu \nabla^\mu \phi = \nabla_\mu \left(g^{\mu\nu} \frac{\partial \phi}{\partial x^\nu} \right)$$
$$= \frac{1}{\sqrt{|g|}} \frac{\partial}{\partial x^\mu} \left(\sqrt{|g|} g^{\mu\nu} \frac{\partial \phi}{\partial x^\nu} \right). \tag{2.115}$$

2.3 The Riemann Curvature Tensor and Its Properties

One of the important features of Riemannian geometry is that the operation of covariant differentiation, performed with the help of the operator ∇_μ, is *noncommutative*. In the following we will consider this problem in detail. Let's construct the second covariant derivative of an arbitrary vector A_λ in two steps: first, we consider its covariant derivative with respect to the coordinate x^μ, and, secondly, with respect to the coordinate x^ν. Hence, applying twice the covariant differentiation operator leads to the following expression,

$$\nabla_\nu \nabla_\mu A_\lambda = \frac{\partial \left(\nabla_\mu A_\lambda \right)}{\partial x^\nu} - \Gamma^\tau_{\nu\lambda} \nabla_\mu A_\tau - \Gamma^\tau_{\mu\nu} \nabla_\tau A_\lambda. \tag{2.116}$$

On the other hand we have

$$\nabla_\mu A_\lambda = \frac{\partial A_\lambda}{\partial x^\mu} - \Gamma^\tau_{\lambda\mu} A_\tau, \quad \nabla_\mu A_\tau = \frac{\partial A_\tau}{\partial x^\mu} - \Gamma^\sigma_{\mu\tau} A_\sigma, \quad \nabla_\tau A_\lambda = \frac{\partial A_\lambda}{\partial x^\tau} - \Gamma^\sigma_{\lambda\tau} A_\sigma,$$

and after substituting in Eq. (2.116) we obtain

$$\nabla_\nu \nabla_\mu A_\lambda = \frac{\partial^2 A_\lambda}{\partial x^\mu \partial x^\nu} - \Gamma^\tau_{\lambda\mu} \frac{\partial A_\tau}{\partial x^\nu} - \Gamma^\tau_{\nu\lambda} \frac{\partial A_\tau}{\partial x^\mu} - \Gamma^\tau_{\mu\nu} \frac{\partial A_\lambda}{\partial x^\tau}$$
$$- A_\tau \frac{\partial \Gamma^\tau_{\lambda\mu}}{\partial x^\nu} + \Gamma^\tau_{\nu\lambda} \Gamma^\sigma_{\mu\tau} A_\sigma + \Gamma^\tau_{\mu\nu} \Gamma^\sigma_{\lambda\tau} A_\sigma. \tag{2.117}$$

Similarly we can calculate the quantity

$$\nabla_\mu \nabla_\nu A_\lambda = \frac{\partial \left(\nabla_\nu A_\lambda \right)}{\partial x^\mu} - \Gamma^\tau_{\mu\lambda} \nabla_\nu A_\tau - \Gamma^\tau_{\mu\nu} \nabla_\tau A_\lambda, \tag{2.118}$$

to obtain

$$\nabla_\mu \nabla_\nu A_\lambda = \frac{\partial^2 A_\lambda}{\partial x^\mu \partial x^\nu} - \Gamma^\tau_{\lambda\nu} \frac{\partial A_\tau}{\partial x^\mu} - \Gamma^\tau_{\mu\lambda} \frac{\partial A_\tau}{\partial x^\nu} - \Gamma^\tau_{\mu\nu} \frac{\partial A_\lambda}{\partial x^\tau}$$
$$- A_\tau \frac{\partial \Gamma^\tau_{\lambda\nu}}{\partial x^\mu} + \Gamma^\tau_{\mu\lambda} \Gamma^\sigma_{\nu\tau} A_\sigma + \Gamma^\tau_{\mu\nu} \Gamma^\sigma_{\lambda\tau} A_\sigma. \tag{2.119}$$

By subtracting Eqs. (2.117) and (2.119) it turns out that the commutator of the covariant derivative can be finally represented as

$$
\begin{aligned}
[\nabla_\mu, \nabla_\nu] A_\lambda &= (\nabla_\mu \nabla_\nu - \nabla_\nu \nabla_\mu) A_\lambda \\
&= \left[\frac{\partial \Gamma^\sigma_{\lambda\nu}}{\partial x^\mu} - \frac{\partial \Gamma^\sigma_{\lambda\mu}}{\partial x^\nu} + \Gamma^\tau_{\nu\lambda}\Gamma^\sigma_{\mu\tau} - \Gamma^\tau_{\mu\lambda}\Gamma^\sigma_{\nu\tau} \right] A_\sigma \\
&= R^\sigma_{\lambda\mu\nu} A_\sigma . \qquad (2.120)
\end{aligned}
$$

The purely geometric quantity $R^\sigma_{\lambda\mu\nu}$, defined as

$$
R^\sigma_{\lambda\mu\nu} = \frac{\partial \Gamma^\sigma_{\lambda\nu}}{\partial x^\mu} - \frac{\partial \Gamma^\sigma_{\lambda\mu}}{\partial x^\nu} + \Gamma^\tau_{\nu\lambda}\Gamma^\sigma_{\mu\tau} - \Gamma^\tau_{\mu\lambda}\Gamma^\sigma_{\nu\tau}, \qquad (2.121)
$$

is called the *Riemann curvature tensor*, and it is a fourth-order tensor. From this tensor, by contraction, we obtain a second-order tensor $R_{\lambda\nu}$, given by

$$
R_{\lambda\nu} = R^\sigma_{\lambda\sigma\nu} = \frac{\partial \Gamma^\sigma_{\lambda\nu}}{\partial x^\sigma} - \frac{\partial \Gamma^\sigma_{\lambda\sigma}}{\partial x^\nu} + \Gamma^\tau_{\nu\lambda}\Gamma^\sigma_{\sigma\tau} - \Gamma^\tau_{\sigma\lambda}\Gamma^\sigma_{\nu\tau}. \qquad (2.122)
$$

$R_{\lambda\nu}$ is called the *Ricci tensor*. Contracting the Ricci tensor gives the *Ricci scalar*, defined as

$$
R = R^\lambda_\lambda = g^{\lambda\mu} R_{\lambda\mu} = g^{\lambda\nu} g^{\mu\sigma} R_{\lambda\mu\nu\sigma}. \qquad (2.123)
$$

From its definition (2.121) one can easily observe that the Riemann curvature tensor has the important property of antisymmetry with respect to its last two indices μ and ν, respectively,

$$
R^\sigma_{\lambda\mu\nu} = -R^\sigma_{\lambda\nu\mu}. \qquad (2.124)
$$

One can lower the upper index of the curvature tensor to construct a purely covariant one, given by

$$
R_{\rho\lambda\mu\nu} = g_{\rho\sigma} R^\sigma_{\lambda\mu\nu}. \qquad (2.125)
$$

This tensor also has some important symmetry properties, namely,

$$
R_{\rho\lambda\mu\nu} = -R_{\lambda\rho\mu\nu} = -R_{\rho\lambda\nu\mu}, \qquad R_{\rho\lambda\mu\nu} = R_{\mu\nu\rho\lambda}. \qquad (2.126)
$$

Hence we have obtained the results that the covariant Riemann curvature tensor $R_{\rho\lambda\mu\nu}$ obeys some interesting symmetry properties: it is antisymmetric with respect to the first pair of indices $\rho\lambda$, as well as with respect to the second one, $\mu\nu$. Moreover, it is symmetric if we perform the interchange of the two pair of indices $\rho\lambda$ and $\mu\nu$ with one another.

The Bianchi Identity

As we have already discussed, in Riemannian geometry we can always find a local coordinate system (called the locally geodesic coordinate system), which has the property that in a given point P the first derivatives of all the metric tensor $g_{\mu\nu}$ components cancel out, being identically equal to zero. Hence, in this coordinate frame, all the Christoffel symbols are also identically equal to zero. The use of the locally geodesic coordinate systems provide very powerful tools for finding and proving tensor identities, since if one can prove that in a geodesic coordinate system a tensor is zero, then, due to the mathematical form of the rules of the transformations of tensors, the given tensor must be zero in any arbitrary coordinate system. Hence, by using this property of the locally geodesic coordinate system, it turns out that in such a system the Riemann curvature tensor has the simple form

$$R^{\sigma}_{\lambda\mu\nu}\big|_P = \frac{\partial\Gamma^{\sigma}_{\lambda\nu}}{\partial x^{\mu}} - \frac{\partial\Gamma^{\sigma}_{\lambda\mu}}{\partial x^{\nu}}. \tag{2.127}$$

Now we cyclically transpose the indices μ, ν, ρ in the above definition, and then we sum up the obtained expressions. Hence, after a simple calculation, and after substituting the ordinary derivatives with the covariant ones, we obtain the important Bianchi identity, formulated in a general reference frame as

$$\nabla_\rho R^{\sigma}_{\lambda\mu\nu} + \nabla_\nu R^{\sigma}_{\lambda\rho\mu} + \nabla_\mu R^{\sigma}_{\lambda\nu\rho} = 0. \tag{2.128}$$

The Einstein Tensor

In the Bianchi identity (2.128) we perform the contraction of the indices σ and ν. Thus we obtain

$$-\nabla_\rho R_{\lambda\mu} + \nabla_\sigma R^{\sigma}_{\lambda\rho\mu} + \nabla_\mu R_{\lambda\rho} = 0. \tag{2.129}$$

By multiplying the above expression by $g^{\lambda\sigma}$ we find

$$-\nabla_\rho R^{\alpha}_{\mu} + \nabla_\sigma\left(g^{\lambda\alpha}R^{\sigma}_{\lambda\rho\mu}\right) + \nabla_\mu R^{\alpha}_{\rho} = 0. \tag{2.130}$$

In obtaining the above relation we have taken into account the previous result according to which the covariant divergence of the metric tensor $g_{\mu\nu}$ identically vanishes. We contract now the indices ρ and α to obtain

$$-\nabla_\rho R^{\rho}_{\mu} + \nabla_\sigma\left(g^{\lambda\rho}R^{\sigma}_{\lambda\rho\mu}\right) + \nabla_\mu R = 0. \tag{2.131}$$

The term under the second derivative can be transformed by using the already known symmetry properties of the Riemann curvature tensor, as well as the definition of the Ricci tensor. Thus, we obtain

$$g^{\lambda\rho}R^{\sigma}_{\lambda\rho\mu} = g^{\lambda\rho}g^{\nu\sigma}R_{\nu\lambda\rho\mu} = g^{\nu\sigma}g^{\lambda\rho}R_{\lambda\nu\mu\rho} = g^{\nu\sigma}R^{\rho}_{\nu\mu\rho} = -R^{\sigma}_{\mu}. \tag{2.132}$$

Hence, Eq. (2.131) becomes

$$\nabla_\rho\left(R^{\rho}_{\mu} - \frac{1}{2}R\delta^{\rho}_{\mu}\right) \equiv 0. \tag{2.133}$$

The tensor

$$G^{\rho}_{\mu} = R^{\rho}_{\mu} - \frac{1}{2}R\delta^{\rho}_{\mu}, \tag{2.134}$$

is called the Einstein tensor, and it is the fundamental tensor quantity used in the description of the physical and geometrical properties of the gravitational field.

The Variation of the Ricci Tensor: The Palatini Identity

The variation of the Ricci tensor plays an important role in obtaining the gravitational field equations from the Hilbert–Einstein variational principle. The variation of $R_{\mu\nu}$, $\delta R_{\mu\nu}$ obeys an important relation, called the *Palatini identity.* To obtain it we start from the variation of the Riemann tensor, which can be written as

$$\delta R^{\sigma}_{\lambda\mu\nu} = \frac{\partial\left(\delta\Gamma^{\sigma}_{\lambda\nu}\right)}{\partial x^{\mu}} - \frac{\partial\left(\delta\Gamma^{\sigma}_{\lambda\mu}\right)}{\partial x^{\nu}} + \left(\delta\Gamma^{\tau}_{\nu\lambda}\right)\Gamma^{\sigma}_{\mu\tau} + \Gamma^{\tau}_{\nu\lambda}\left(\delta\Gamma^{\sigma}_{\mu\tau}\right) - \left(\delta\Gamma^{\tau}_{\mu\lambda}\right)\Gamma^{\sigma}_{\nu\tau} - \Gamma^{\tau}_{\mu\lambda}\left(\delta\Gamma^{\sigma}_{\nu\tau}\right). \tag{2.135}$$

As one can see from Eq. (2.75), the variation of the Christoffel symbols $\delta\Gamma^{\sigma}_{\lambda\nu}$ is a tensor. Therefore we can write down its covariant derivative as

$$\nabla_{\mu}\left(\delta\Gamma^{\sigma}_{\lambda\nu}\right) = \frac{\partial\left(\delta\Gamma^{\sigma}_{\lambda\nu}\right)}{\partial x^{\mu}} + \Gamma^{\sigma}_{\tau\mu}\delta\Gamma^{\tau}_{\lambda\nu} - \Gamma^{\tau}_{\mu\lambda}\delta\Gamma^{\sigma}_{\tau\nu} - \Gamma^{\tau}_{\mu\nu}\delta\Gamma^{\sigma}_{\lambda\tau}. \tag{2.136}$$

By simple inspection it follows that the variation of the Riemann tensor is equal to the difference of the covariant derivatives of the variation of the Christoffel symbols,

$$\delta R^{\sigma}_{\lambda\mu\nu} = \nabla_{\mu}\left(\delta\Gamma^{\sigma}_{\lambda\nu}\right) - \nabla_{\nu}\left(\delta\Gamma^{\sigma}_{\lambda\mu}\right). \tag{2.137}$$

By contracting now the above identity we obtain for the variation of the Ricci tensor

$$\delta R_{\lambda\nu} = \delta R^{\sigma}_{\lambda\sigma\nu} = \nabla_{\sigma}\left(\delta\Gamma^{\sigma}_{\lambda\nu}\right) - \nabla_{\nu}\left(\delta\Gamma^{\sigma}_{\lambda\sigma}\right). \tag{2.138}$$

This relation is called the Palatini identity.

The Invariance of the Volume Element

Finally, we will study the behavior under arbitrary coordinate transformations of the four-dimensional volume in the Riemannian geometry,

$$v' = \int \sqrt{-g'}\, dx'^{0}dx'^{1}dx'^{2}dx'^{3}, \tag{2.139}$$

and we will show that this geometrical quantity is an invariant [330]. Under arbitrary coordinate transformations the metric tensor transforms as

$$g'_{\mu\nu}\left(x'\right) = g_{\lambda\sigma}\left(x\right)\frac{\partial x^{\lambda}}{\partial x'^{\mu}}\frac{\partial x^{\sigma}}{\partial x'^{\nu}} = \frac{\partial x^{\lambda}}{\partial x'^{\mu}}\cdot g_{\lambda\sigma}\left(x\right)\frac{\partial x^{\sigma}}{\partial x'^{\nu}}. \tag{2.140}$$

In the new coordinate system the determinant of the metric tensor $g' = \det g'_{\mu\nu}$ becomes

$$\begin{aligned} g' &= \det\left(g_{\lambda\sigma}\left(x\right)\frac{\partial x^{\lambda}}{\partial x'^{\mu}}\right)\det\left(\frac{\partial x^{\sigma}}{\partial x'^{\nu}}\right) \\ &= \det\left(g_{\lambda\sigma}\left(x\right)\right)\det\left(\frac{\partial x^{\lambda}}{\partial x'^{\mu}}\right)\det\left(\frac{\partial x^{\sigma}}{\partial x'^{\nu}}\right). \end{aligned} \tag{2.141}$$

Hence, we find the relation

$$g' = gJ^2, \tag{2.142}$$

where J is the coordinate transformation Jacobian,

$$J = \frac{\partial\left(x^0, x^1, x^2, x^3\right)}{\partial\left(x'^0, x'^1, x'^2, x'^3\right)}. \tag{2.143}$$

Substituting the expression of g' into Eq. (2.139) we have

$$v' = \int \sqrt{-g}\frac{\partial\left(x^0, x^1, x^2, x^3\right)}{\partial\left(x'^0, x'^1, x'^2, x'^3\right)}dx'^0dx'^1dx'^2dx'^3 = \int \sqrt{-g}dx^0dx^1dx^2dx^3. \tag{2.144}$$

But

$$v = \int \sqrt{-g}dx^0dx^1dx^2dx^3 = \int \sqrt{-g}d^4x, \tag{2.145}$$

represents the four-dimensional volume, and thus we have established the equality

$$v' = v. \tag{2.146}$$

Hence, we have shown that under arbitrary coordinate transformations the quantity $\sqrt{-g}d^4x$ is an invariant, and it has the same value in all frames of reference.

2.4 Summary

In the present chapter we have introduced some of the basic mathematical concepts and tools necessary for the understanding of the General Theory of Relativity, as well as of its theoretical extensions, generalizations, and modifications. The presentation is in no way exhaustive, and the reader interested in a rigorous and detailed presentation of the tensor calculus and of Riemannian geometry should consult the books presented in the References. Our main goal was to introduce the central mathematical concepts that are used in the mathematical formalism of the General Theory of Relativity in a simple and comprehensible way, and to present and summarize the basic definitions of the geometric quantities that will be used in the next chapters. We have also emphasized the calculational aspects of differential geometry, by presenting in detail the derivation of some of the basic results in tensor calculus and differential geometry. In particular, we have presented the aspects of having direct applications in obtaining the Einstein equations of the gravitational field by using the Hilbert variational principle, in obtaining the equations of particle motion in the framework of General Relativity, and for exploring their physical applications and consequences.

3
The Gravitational Field Equations

In order to describe the gravitational field in the framework of GR, a set of differential equations, called the *Einstein field equations*, are used. These equations are defined in a Riemannian geometry, and they establish a deep connection between the geometric properties of the space-time, and its matter content, providing a full description of both geometric characteristics of the space-time, and of the dynamics of the particles. In the present chapter, we will derive the Einstein field equations, discuss their mathematical properties, as well as some of their physical implications.

There are many good books on GR, and a lot of internet resources that could be used by the readers who are interested in the derivation of the Einstein equations, and their applications. The classic books by Weyl [507], Pauli [408], Eddington [182], Schrödinger [455], and Dirac [170] offer a wealth of knowledge and deep physical and geometrical interpretation from which all readers could maximally benefit. The book by Weinberg [501] contains a very detailed presentation of the physical and mathematical foundations of GR. The recent advances in theoretical gravity, differential geometry and their connections are covered in the books by Wald [493], Grøn and Hervik [210], and Carroll [119].

3.1 The Variational Principle and the Field Equations

The basic physical principles on which GR is based can be formulated as follows:

a. The special relativistic Minkowski metric, defined in a pseudo-Euclidian geometry, cannot explain the observed gravitational redshift of spectral lines of the radiation emitted by compact stars. This imposes the necessity of a transition from the pseudo-Euclidian geometry of Special Relativity to a Riemannian geometry, in which the metric tensor is a function of all the coordinates. Moreover, in Riemannian geometry the principle of equivalence can be implemented easily, since from a geometric point of view any infinitesimal region of a Riemannian space-time can be locally described with the help of a Minkowski-type metric. As a direct physical consequence of this result, it turns out that in this region, and at a given point, the special relativistic description of physical phenomena remains valid.

b. The second fundamental assumption in GR is that the basic field variables describing all properties of the gravitational fields are the metric tensor components $g_{\mu\nu}$. In the most important case of the four-dimensional space-time, and by taking into account that $g_{\mu\nu}$ is symmetric, the number of distinct components of the metric tensor is ten. The metric tensor components are independent, but they can be subject to an arbitrary transformation of coordinates. Therefore, we require that general coordinate transformations do not change the physics and the dynamical properties of natural processes. This important physical idea can be reformulated in mathematical terms as the *principle of covariance*, which requires *the frame independence of physical laws*. The principle of covariance is automatically satisfied if the equations describing natural phenomena can be formulated mathematically as *tensor equations*. Hence, in order to build a rigorous theory of gravity we must cast the basic field equations of gravity in a tensorial formulation, which is valid in all reference frames.

c. The Newtonian laws of gravitation have been very successful in the description of the planetary and galactic dynamics. Hence Newton's laws of gravitation and the law of motion must be recovered from any new theory of gravity at some level of approximation. In Newtonian gravity the gravitational potential ϕ is described by the Poisson equation, $\Delta\phi = 4\pi G\rho$, where ρ is the matter density. The Poisson equation is a partial differential equation of second order, so we require that the new field equations of gravity must also be of second order in the time and space derivatives.

d. One of the fundamental laws of nature is the conservation of energy. In the special theory of relativity, energy and momentum can be combined into a single four-vector (in the case of a single particle) or, for complex physical systems, in a second-order tensor, called the energy-momentum tensor T_{μ}^{ν}. The conservation laws of the momentum and energy are expressed by the equation of conservation of the energy-momentum tensor, which in arbitrary coordinates can be formulated as $\nabla_{\nu}T_{\mu}^{\nu} = 0$. The matter conservation equation should be derivable from the field equations.

A very elegant and powerful mathematical method for obtaining the gravitational field equations in the framework of the General Theory of Relativity was introduced, and fully developed, by David Hilbert, in 1915. In the variational procedure proposed by Hilbert, one starts, as usual, with an appropriate action S, assumed to describe all properties of the given physical system located in the four-dimensional space-time. In the case of gravity, one should take into account two major components in the action: the first corresponds to the gravitational field, as described by the geometric properties of the space-time, and is denoted by S_g, while the other component of the action S_m corresponds to the matter fields, which are distinct from gravitation, and act as a source of the gravitational field. Therefore, the total action for gravity and any other physical fields present in the system is written down generally as

$$S = S_g + S_m. \tag{3.1}$$

To construct an action for a gravitational system one must explicitly obtain the functional forms of S_g and S_m. There are a number of simple physical requirements that help us in performing this task, by means of taking into account two important properties: a) the action must have the physical dimension of mass $\times$ $\left(\text{length}^2\right)$ $\times$ $\left(\text{time}^{-1}\right)$; b) the action must be invariant under the transformation $x \to x'$, which expresses the principle of physical covariance.

The metric tensor $g_{\mu\nu}$ is a dimensionless quantity, since $ds^2 = g_{\mu\nu}dx^\mu dx^\nu$, and both ds and dx^μ must have the physical dimensions of length. As a consequence of this fact the Ricci tensor $R_{\mu\nu}$ has dimensions $(\text{length})^{-2}$. Then it follows that R (the Ricci scalar $R = R^\mu_\mu = g^{\mu\nu}R_{\mu\nu}$) also has dimensions $(\text{length})^{-2}$. In addition to $g_{\mu\nu}$ and its first derivatives, the Ricci scalar R also contains the second derivatives of $g_{\mu\nu}$. However, R is linear in these derivatives. Thus, *we postulate* that the gravitational action can be given by the following relation,

$$S_g = -\frac{c^3}{16\pi G}\int R\sqrt{-g}d^4x, \tag{3.2}$$

where c is, as usual, the speed of light in vacuum, the Newtonian gravitational constant G is given by $G = 6.67 \times 10^{-8}\ \text{cm}^3/\text{g}\,\text{s}^2$, and we have denoted $d^4x = dx^0dx^1dx^2dx^3$. We shall write the action of the matter in the following general form [302]

$$S_m = \frac{1}{c}\int L_m\left(g^{\mu\nu}, \frac{\partial g^{\mu\nu}}{\partial x^\lambda}\right)\sqrt{-g}d^4x. \tag{3.3}$$

Therefore for a physical system consisting of the gravitational field and matter, the total action can be represented as

$$S = -\frac{c^3}{16\pi G}\int R\sqrt{-g}d^4x + \frac{1}{c}\int L_m\left(g^{\mu\nu}, \frac{\partial g^{\mu\nu}}{\partial x^\lambda}\right)\sqrt{-g}d^4x. \tag{3.4}$$

We already know that the quantities that determine the physical and geometrical properties of the gravitational field are the components $g_{\mu\nu}$ of the metric tensor. Hence, in the application of the principle of least action to the gravitational field we have to resort to the variation of the metric tensor components $g_{\mu\nu}$ describing the geometric properties of the space-time. Therefore, the final field equations for gravity will be obtained by taking the variation of S so that

$$\delta S = \delta\left(S_g + S_m\right) = \delta S_g + \delta S_m = 0. \tag{3.5}$$

3.1.1 The Variation of the Gravitational Action

In order to deduce the geometric part of the gravitational field equations we vary first the gravitational field itself, that is, the components $g_{\mu\nu}$ of the metric tensor. To calculate the variation of S_g we proceed as follows

$$\begin{aligned}\delta\int R\sqrt{-g}d^4x &= \delta\int g^{\mu\nu}R_{\mu\nu}\sqrt{-g}d^4x \\ &= \int\left[R_{\mu\nu}\sqrt{-g}\delta g^{\mu\nu} + g^{\mu\nu}R_{\mu\nu}\delta\left(\sqrt{-g}\right) + g^{\mu\nu}\left(\delta R_{\mu\nu}\right)\sqrt{-g}\right]d^4x. \qquad (3.6)\end{aligned}$$

We can easily find the variation of the determinant g of the metric tensor by the following simple computation

$$\delta\sqrt{-g} = -\frac{1}{2\sqrt{-g}}\delta g = -\frac{1}{2}\sqrt{-g}g_{\mu\nu}\delta g^{\mu\nu}. \qquad (3.7)$$

Substituting the above relation into the variation (3.6) we have

$$\delta\int R\sqrt{-g}d^4x = \int\left(R_{\mu\nu} - \frac{1}{2}g_{\mu\nu}R\right)\delta g^{\mu\nu}\sqrt{-g}d^4x + \int g^{\mu\nu}\left(\delta R_{\mu\nu}\right)\sqrt{-g}d^4x. \qquad (3.8)$$

Variation of the Gravitational Action in a Locally Geodesic Frame

The calculation of the second integral can be done easily if we choose a locally geodesic (inertial) system of coordinates, at a fixed point P. Then at that point all the Christoffel symbols cancel, so that $\Gamma^{\lambda}_{\mu\nu} = 0$. By taking into account that the Christoffel symbols vanish (but their derivatives do not), we find that at P the Ricci tensor can be represented as

$$R_{\mu\nu}\big|_P = \frac{\partial\Gamma^{\sigma}_{\mu\nu}}{\partial x^{\sigma}} - \frac{\partial\Gamma^{\sigma}_{\nu\sigma}}{\partial x^{\mu}}. \qquad (3.9)$$

Then we obtain

$$\begin{aligned}\left[g^{\mu\nu}\left(\delta R_{\mu\nu}\right)\right]\big|_P &= \left[g^{\mu\nu}\left(\frac{\partial\delta\Gamma^{\sigma}_{\mu\nu}}{\partial x^{\sigma}} - \frac{\partial\delta\Gamma^{\lambda}_{\nu\lambda}}{\partial x^{\mu}}\right)\right]\Bigg|_P \\ &= \left(g^{\mu\nu}\frac{\partial\delta\Gamma^{\sigma}_{\mu\nu}}{\partial x^{\sigma}} - g^{\sigma\nu}\frac{\partial\delta\Gamma^{\lambda}_{\nu\lambda}}{\partial x^{\sigma}}\right)\Bigg|_P = \frac{\partial w^{\sigma}}{\partial x^{\sigma}}\Bigg|_P, \qquad (3.10)\end{aligned}$$

where

$$w^{\sigma} = \left(g^{\mu\nu}\delta\Gamma^{\sigma}_{\mu\nu} - g^{\sigma\nu}\Gamma^{\lambda}_{\nu\lambda}\right)\big|_P. \qquad (3.11)$$

Since w^{σ} is a vector, we can write Eq. (3.10) in an arbitrary reference system as

$$g^{\mu\nu}\left(\delta R_{\mu\nu}\right) = \frac{1}{\sqrt{-g}}\frac{\partial}{\partial x^{\sigma}}\left(\sqrt{-g}w^{\sigma}\right). \qquad (3.12)$$

Consequently, the second integral in Eq. (3.8) takes the form

$$\int g^{\mu\nu}\left(\delta R_{\mu\nu}\right)\sqrt{-g}d^4x = \int \frac{\partial}{\partial x^\sigma}\left(\sqrt{-g}w^\sigma\right)d^4x. \tag{3.13}$$

With the use of Gauss's theorem, the second integral in Eq. (3.13) can be transformed into a surface integral of the vector w^σ over the hypersurface surrounding the entire four-volume encompassing the considered physical system. But by hypothesis the variations of the gravitational field (and of the metric tensor components) are zero at the integration limits, hence it follows that these terms drop out. Thus, it turns out that the variation of the gravitational (geometric) part of the total action S is given by

$$\delta S_g = -\frac{c^3}{16\pi G}\int\left(R_{\mu\nu} - \frac{1}{2}g_{\mu\nu}R\right)\delta g^{\mu\nu}\sqrt{-g}d^4x. \tag{3.14}$$

Variation of the Gravitational Action in an Arbitrary Frame

The second term in Eq. (3.8), $\int g^{\mu\nu}\left(\delta R_{\mu\nu}\right)\sqrt{-g}d^4x$, can also be evaluated in an arbitrary frame. This can be done easily with the use of the Palatini identity (2.138)

$$\delta R_{\mu\nu} = \nabla_\rho\left(\delta\Gamma^\rho_{\mu\nu}\right) - \nabla_\nu\left(\delta\Gamma^\rho_{\rho\mu}\right). \tag{3.15}$$

Hence, we obtain

$$\begin{aligned} g^{\mu\nu}\delta R_{\mu\nu} &= g^{\mu\nu}\left[\nabla_\rho\left(\delta\Gamma^\rho_{\mu\nu}\right) - \nabla_\nu\left(\delta\Gamma^\rho_{\rho\mu}\right)\right] = \nabla_\rho\left(g^{\mu\nu}\delta\Gamma^\rho_{\mu\nu}\right) - \nabla_\nu\left(g^{\mu\nu}\delta\Gamma^\rho_{\rho\mu}\right) \\ &= \nabla_\rho\left[g^{\mu\nu}\delta\Gamma^\rho_{\mu\nu} - g^{\mu\rho}\delta\Gamma^\sigma_{\sigma\mu}\right] = \nabla_\rho A^\rho = \frac{1}{\sqrt{-g}}\frac{\partial}{\partial x^\rho}\left(\sqrt{-g}w^\rho\right). \end{aligned} \tag{3.16}$$

Therefore

$$\int g^{\mu\nu}\left(\delta R_{\mu\nu}\right)\sqrt{-g}d^4x = \int \frac{\partial}{\partial x^\rho}\left(\sqrt{-g}w^\rho\right)d^4x, \tag{3.17}$$

and after using Gauss's theorem, this term can be discarded from the variation of the gravitational action that, in an arbitrary reference frame, takes the form (3.14).

The Palatini Formalism

An alternative method of obtaining the gravitational field equations is represented by the Palatini approach, which we will present briefly in the following. In this mathematical formalism one adopts as independent variables both the metric $g_{\mu\nu}$ and the connection (Christoffel symbols) $\Gamma^\rho_{\mu\nu}$. Moreover, we will not assume the a priori existence of any explicit relationship between the metric and the Christoffel symbols, but we will assume that the Riemannian manifold is torsionless, that is, the Christoffel symbols are symmetric in their lower indices, $\Gamma^\rho_{\mu\nu} = \Gamma^\rho_{\nu\mu}$. For simplicity, in the following we will restrict our analysis to the vacuum case. The basic idea in the Palatini formalism is to consider the gravitational action as a function of the metric, of the connection, and of the first derivatives of the connection, so that

$$\begin{aligned} S_g &= -\frac{c^3}{16\pi G}\int L_{EH}\left(g,\Gamma,\partial\Gamma\right)\sqrt{-g}d^4x \\ &= -\frac{c^3}{16\pi G}\int L_{EH}\left(g_{\mu\nu},\Gamma^{\sigma}_{\mu\nu},\frac{\partial}{\partial x^{\rho}}\Gamma^{\sigma}_{\mu\nu}\right)\sqrt{-g}d^4x, \end{aligned} \tag{3.18}$$

where the Einstein–Hilbert Lagrangian density is given by

$$L_{EH} = R = g^{\mu\nu}R_{\mu\nu}\left(\Gamma,\partial\Gamma\right). \tag{3.19}$$

We vary now the action with respect to the metric only. Thus, we obtain first

$$\begin{aligned} \delta S_g &= -\frac{c^3}{16\pi G}\delta\int R\left(\Gamma,\partial\Gamma\right)\sqrt{-g}d^4x \\ &= -\frac{c^3}{16\pi G}\delta\int g^{\mu\nu}R_{\mu\nu}\left(\Gamma,\partial\Gamma\right)\sqrt{-g}d^4x \\ &= -\frac{c^3}{16\pi G}\int\left[R_{\mu\nu}\left(\Gamma,\partial\Gamma\right)-\frac{1}{2}R\left(\Gamma,\partial\Gamma\right)g_{\mu\nu}\right]\delta g^{\mu\nu}\sqrt{-g}d^4x \\ &= 0, \end{aligned} \tag{3.20}$$

which immediately gives the Einstein field equation in vacuum,

$$R_{\mu\nu}\left(\Gamma,\partial\Gamma\right)-\frac{1}{2}R\left(\Gamma,\partial\Gamma\right)g_{\mu\nu} = 0. \tag{3.21}$$

Consider now the variation of the action with respect to the connection, which yields

$$\begin{aligned} \delta S_g &= -\frac{c^3}{16\pi G}\int \delta R_{\mu\nu}\left(\Gamma,\partial\Gamma\right)g^{\mu\nu}\sqrt{-g}d^4x \\ &= -\frac{c^3}{16\pi G}\int\left[\nabla_{\nu}\left(\delta\Gamma^{\sigma}_{\mu\sigma}\right)-\nabla_{\sigma}\left(\delta\Gamma^{\sigma}_{\mu\nu}\right)\right]g^{\mu\nu}\sqrt{-g}d^4x, \end{aligned} \tag{3.22}$$

where we have used the Palatini identity given by Eq. (3.15). Equation (3.22) can be reformulated in an equivalent form as

$$\begin{aligned} \delta S_g = &-\frac{c^3}{16\pi G}\int\nabla_{\nu}\left(g^{\mu\nu}\delta\Gamma^{\sigma}_{\mu\sigma}-g^{\mu\sigma}\delta\Gamma^{\nu}_{\mu\sigma}\right)\sqrt{-g}d^4x \\ &+\frac{c^3}{16\pi G}\int\left[\left(\nabla_{\rho}g^{\mu\nu}\right)\delta\Gamma^{\rho}_{\mu\nu}-\left(\nabla_{\nu}g^{\mu\nu}\right)\delta\Gamma^{\rho}_{\mu\rho}\right]\sqrt{-g}d^4x. \end{aligned} \tag{3.23}$$

With the use of Gauss's theorem, the first integral in the above expression can be transformed into a surface integral over the boundary, which vanishes under the assumption that the variation of the connection vanishes at the boundary. Therefore for the variation of the gravitational action we obtain

$$\delta S_g = \frac{c^3}{16\pi G}\int\left(\delta^{\nu}_{\rho}\nabla_{\sigma}g^{\mu\sigma}-\nabla_{\rho}g^{\mu\nu}\right)\delta\Gamma^{\rho}_{\mu\nu}\sqrt{-g}d^4x = 0, \tag{3.24}$$

which gives, by taking into account that the variation of the connection is arbitrary,

$$\delta_{\rho}^{\nu}\nabla_{\sigma}g^{\mu\sigma}-\nabla_{\rho}g^{\mu\nu}=0. \tag{3.25}$$

Due to our assumption that the manifold is torsionless, the variation in the connection, although arbitrary, must be also symmetric in its lower two indices. Consequently, the condition $\delta S_g = 0$ requires that only the symmetric part of the term in the variation of the action must vanish. On the other hand, the contraction with $\delta\Gamma^{\rho}_{\mu\nu}$ of the antisymmetric term will automatically give zero. By contracting Eq. (3.25) by taking $\nu=\rho$ we obtain

$$4\nabla_{\sigma}g^{\mu\sigma}-\nabla_{\sigma}g^{\mu\sigma}=0, \tag{3.26}$$

which gives $\nabla_{\sigma}g^{\mu\sigma}=0$. Hence, from Eq. (3.25) we obtain

$$\nabla_{\rho}g^{\mu\nu}=0. \tag{3.27}$$

Therefore, by imposing the condition of the stationarity of the Einstein–Hilbert action with respect to variations in the connection (assumed to be symmetric), we have found the result that the covariant derivative of the metric tensor must vanish. This also immediately leads to representation of the Christoffel symbols in terms of the derivatives of the metric tensor components.

3.1.2 The Variation of the Matter Action

As a next step in the derivation of the full set of gravitational field equations we consider now the variation of the action for the matter

$$S_m=\frac{1}{c}\int L_m\left(g^{\mu\nu},\frac{\partial g^{\mu\nu}}{\partial x^{\lambda}}\right)\sqrt{-g}d^4x. \tag{3.28}$$

To perform the variation, consider in the action a transformation from the coordinates x^{μ} to the coordinates $x'^{\mu}=x^{\mu}+\xi^{\mu}$, where the variations ξ^{μ} are small quantities $\xi^{\mu}\ll x^{\mu}$. Under this infinitesimal coordinate transformation, the contravariant components $g^{\mu\nu}$ of the metric tensor transform as follows

$$\begin{aligned}g'^{\mu\nu}\left(x'^{\lambda}\right)&=g^{\alpha\beta}\left(x^{\lambda}\right)\frac{\partial x'^{\mu}}{\partial x^{\alpha}}\frac{\partial x'^{\nu}}{\partial x^{\beta}}=g^{\alpha\beta}\left(x^{\lambda}\right)\left(\delta^{\mu}_{\alpha}+\frac{\partial\xi^{\mu}}{\partial x^{\alpha}}\right)\left(\delta^{\nu}_{\beta}+\frac{\partial\xi^{\nu}}{\partial x^{\beta}}\right)\\&\approx g^{\mu\nu}\left(x^{\lambda}\right)+g^{\mu\alpha}\frac{\partial\xi^{\nu}}{\partial x^{\alpha}}+g^{\nu\beta}\frac{\partial\xi^{\mu}}{\partial x^{\beta}}.\end{aligned}$$

As one can easily verify in the above equation the components $g'^{\mu\nu}\left(x'^{\lambda}\right)$ of the metric tensor are functions of the transformed (new) coordinates x'^{λ}, while the tensor $g^{\mu\nu}\left(x^{\lambda}\right)$ is a function of the initial coordinates x^{λ}. But in order to obtain a meaningful relation between the different representations it is necessary to write down all terms in the equation as a function of one and the same coordinate variable. In order to achieve this goal we expand the metric tensor $g'^{\mu\nu}\left(x^{\lambda}+\xi^{\lambda}\right)$

in powers of ξ^λ. Moreover, in the following we will neglect terms of higher order in ξ^λ in the series expansion, and therefore with a very good approximation we can replace $g'^{\mu\nu}$ by $g^{\mu\nu}$ in all terms containing ξ^λ. Hence we obtain

$$g'^{\mu\nu}\left(x^\lambda\right) \approx g^{\mu\nu}\left(x^\lambda\right) - \xi^\lambda \frac{\partial g^{\mu\nu}}{\partial x^\lambda} + g^{\mu\lambda}\frac{\partial \xi^\nu}{\partial x^\lambda} + g^{\nu\lambda}\frac{\partial \xi^\mu}{\partial x^\lambda}. \tag{3.29}$$

It is easy to see that the last three terms in Eq. (3.29) can be combined into $\nabla^\mu \xi^\nu + \nabla^\nu \xi^\mu$. Hence, Eq. (3.29) can be reformulated as

$$g'^{\mu\nu} = g^{\mu\nu} + \delta g^{\mu\nu}, \qquad \delta g^{\mu\nu} = \nabla^\mu \xi^\nu + \nabla^\nu \xi^\mu, \tag{3.30}$$

and

$$g'_{\mu\nu} = g_{\mu\nu} + \delta g_{\mu\nu}, \qquad \delta g_{\mu\nu} = -\nabla_\mu \xi_\nu - \nabla_\nu \xi_\mu, \tag{3.31}$$

respectively.

We take now the variation with respect to the metric tensor $g^{\mu\nu}$ of Eq. (3.28), and obtain the following

$$\begin{aligned}
\delta S_m &= \frac{1}{c}\delta \int L_m\left(g^{\mu\nu}, \frac{\partial g^{\mu\nu}}{\partial x^\lambda}\right)\sqrt{-g}d^4x \\
&= \frac{1}{c}\int \left[\frac{\partial\left(\sqrt{-g}L_m\right)}{\partial g^{\mu\nu}}\delta g^{\mu\nu} + \frac{\partial\left(\sqrt{-g}L_m\right)}{\partial \frac{\partial g^{\mu\nu}}{\partial x^\lambda}}\delta\frac{\partial g^{\mu\nu}}{\partial x^\lambda}\right]d^4x \\
&= \frac{1}{c}\int \left[\frac{\partial\left(\sqrt{-g}L_m\right)}{\partial g^{\mu\nu}}\delta g^{\mu\nu} + \frac{\partial}{\partial x^\lambda}\left(\frac{\partial\left(\sqrt{-g}L_m\right)}{\partial \frac{\partial g^{\mu\nu}}{\partial x^\lambda}}\delta g^{\mu\nu}\right)\right. \\
&\quad \left. -\frac{\partial}{\partial x^\lambda}\left(\frac{\partial\left(\sqrt{-g}L_m\right)}{\partial \frac{\partial g^{\mu\nu}}{\partial x^\lambda}}\right)\delta g^{\mu\nu}\right]d^4x \\
&= \frac{1}{c}\int \left[\frac{\partial\left(\sqrt{-g}L_m\right)}{\partial g^{\mu\nu}} - \frac{\partial}{\partial x^\lambda}\left(\frac{\partial\left(\sqrt{-g}L_m\right)}{\partial \frac{\partial g^{\mu\nu}}{\partial x^\lambda}}\right)\right]\delta g^{\mu\nu}d^4x,
\end{aligned} \tag{3.32}$$

where we have used again Gauss's theorem, and set $\delta g^{\mu\nu} = 0$ at the integration limit. At this moment we introduce the notation

$$\frac{1}{2}\sqrt{-g}T_{\mu\nu} = \frac{1}{2}\sqrt{-g}T_{\nu\mu} = \frac{\partial\left(\sqrt{-g}L_m\right)}{\partial g^{\mu\nu}} - \frac{\partial}{\partial x^\lambda}\left(\frac{\partial\left(\sqrt{-g}L_m\right)}{\partial \frac{\partial g^{\mu\nu}}{\partial x^\lambda}}\right). \tag{3.33}$$

The quantity $T_{\mu\nu}$, a second-order symmetric tensor, is called *the energy-momentum tensor of matter*, and it plays a fundamental role in the description of gravitational phenomena. Equation (3.33) provides a very powerful and efficient way for calculating the energy-momentum tensor of any physical system by simply taking the derivative of the matter action L_m with respect to the components $g^{\mu\nu}$ of the metric tensor and its derivatives. The second-order tensor obtained via this procedure is *symmetric*.

Therefore the variation of the matter action takes the form

$$\delta S_m = \frac{1}{2c}\int T_{\mu\nu}\delta g^{\mu\nu}\sqrt{-g}d^4x = -\frac{1}{2c}\int T^{\mu\nu}\delta g_{\mu\nu}d^4x. \tag{3.34}$$

To obtain the second representation of the variation of the matter action we have used the identities $g^{\mu\nu}\delta g_{\mu\nu} = -g_{\mu\nu}\delta g^{\mu\nu}$ and $T^{\mu\nu}\delta g_{\mu\nu} = -T_{\mu\nu}\delta g^{\mu\nu}$, respectively.

The Conservation of the Energy-Momentum Tensor

We substitute now into Eq. (3.34) the expression (3.30), giving the variation of $\delta g^{\mu\nu}$ of the metric tensor in an arbitrary coordinate transformation. By taking into account the symmetry of the matter energy-momentum tensor, we obtain

$$\delta S_m = \frac{1}{2c}\int T_{\mu\nu}\left(\nabla^\mu\xi^\nu + \nabla^\nu\xi^\mu\right)\sqrt{-g}d^4x = \frac{1}{c}\int T_{\mu\nu}\nabla^\mu\xi^\nu\sqrt{-g}d^4x. \tag{3.35}$$

We can easily transform the above expression in the following way,

$$\delta S_m = \frac{1}{c}\int \nabla_\nu\left(T_\mu^\nu\xi^\mu\right)\sqrt{-g}d^4x - \frac{1}{c}\int\left(\nabla_\nu T_\mu^\nu\right)\xi^\mu\sqrt{-g}d^4x. \tag{3.36}$$

With the use of the expression of the covariant divergence of an arbitrary four-vector, the first integral can be reformulated as

$$\frac{1}{c}\int \nabla_\nu\left(T_\mu^\nu\xi^\mu\right)\sqrt{-g}d^4x = \frac{1}{c}\int\frac{\partial}{\partial x^\nu}\left(\sqrt{-g}T_\mu^\nu\xi^\mu\right)d^4x. \tag{3.37}$$

We transform now the volume integral, with the use of Gauss's theorem, into an integral over a hypersurface encompassing the given volume. Since the variations of the ξ^μ's must vanish at the limits of integration, the integral also identically vanishes. Thus, by taking the variation of the matter action as zero, we find

$$\delta S_m = -\frac{1}{c}\int\left(\nabla_\nu T_\mu^\nu\right)\xi^\mu\sqrt{-g}d^4x = 0. \tag{3.38}$$

Since the variations ξ^μ are arbitrary, it then immediately follows that the energy-momentum tensor of any physical system obeys the important property

$$\nabla_\nu T_\mu^\nu = 0, \tag{3.39}$$

meaning that the energy-momentum tensor T_μ^ν is a *conserved quantity.*

Finally, we would like to mention that for a macroscopic object described by two thermodynamic parameters, the energy density ϵ and the thermodynamic pressure p, respectively, the energy-momentum tensor can be obtained as

$$T_{\mu\nu} = (\epsilon + p)\,U_\mu U_\nu + p\,g_{\mu\nu}, \tag{3.40}$$

where U_μ denotes the four-velocity of the system.

3.1.3 The Einstein Gravitational Field Equations

From the principle of least action (3.5), by taking into account the previously obtained results on the variation of the gravitational and matter action, respectively, we immediately find the total variation of the action as

$$\delta S = -\frac{c^3}{16\pi G}\int \left(R_{\mu\nu} - \frac{1}{2}Rg_{\mu\nu} - \frac{8\pi G}{c^4}T_{\mu\nu}\right)\delta g^{\mu\nu}\sqrt{-g}d^4x = 0. \tag{3.41}$$

Since the variations $\delta g^{\mu\nu}$ of the metric tensor are arbitrary, we obtain

$$G_{\mu\nu} = R_{\mu\nu} - \frac{1}{2}Rg_{\mu\nu} = \frac{8\pi G}{c^4}T_{\mu\nu}, \tag{3.42}$$

which are called the Einstein field equations, or simply the Einstein equations. They provide a beautiful description of the gravitational field properties in geometric terms, and point out the existence of a deep connection between geometry, gravity, and space-time properties. In mixed components the Einstein equations take the form

$$G_\mu^\nu = R_\mu^\nu - \frac{1}{2}R\delta_\mu^\nu = \frac{8\pi G}{c^4}T_\mu^\nu. \tag{3.43}$$

By taking the covariant divergence of the previous two equations, since the covariant divergence of the Einstein tensor G_μ^ν identically vanishes, we easily find

$$\nabla_\nu G_\mu^\nu \equiv 0 = \nabla_\nu T_\mu^\nu. \tag{3.44}$$

Therefore the gravitational field equations automatically incorporate the important law of the conservation of the matter energy-momentum tensor $T_{\mu\nu}$.

By contracting the Einstein field equations (3.43) over the indices μ and ν, by taking into account that $\delta_\mu^\mu = 4$, and denoting by $T = T_\nu^\nu$ the trace of the energy-momentum tensor, we obtain

$$R = -\frac{8\pi G}{c^4}T. \tag{3.45}$$

Hence the Einstein field equations can be reformulated in the following equivalent form

$$R_{\mu\nu} = \frac{8\pi G}{c^4}\left(T_{\mu\nu} - \frac{1}{2}Tg_{\mu\nu}\right). \tag{3.46}$$

In a vacuum all the components of the matter energy-momentum tensor identically vanish, $T_{\mu\nu} \equiv 0$. Hence in this case the gravitational field equations take the simple form

$$R_{\mu\nu} = 0. \tag{3.47}$$

At this moment we need to point out that Eqs. (3.42) *do not represent the most general form of the Einstein field equations.* The conservation law of the energy-momentum tensor is also satisfied if we add to the left hand side (geometric part) of the field equations a term of the form $\Lambda g_{\mu\nu}$, where Λ is *an arbitrary (real) constant*, called *the cosmological constant*. This follows from the fundamental property $\nabla_\lambda g_{\mu\nu} \equiv 0$ of the metric tensor in a Riemannian geometry. Therefore, the simplest extension of the Einstein equations (3.42) can be formulated as

$$R_{\mu\nu} - \frac{1}{2} R g_{\mu\nu} + \Lambda g_{\mu\nu} = \frac{8\pi G}{c^4} T_{\mu\nu}. \tag{3.48}$$

Note that in the presence of a cosmological constant Λ the Einstein field equations can be derived from the variational principle

$$S = -\frac{c^3}{16\pi G} \int (R - 2\Lambda) \sqrt{-g} d^4x + \frac{1}{c} \int L_m \left(g^{\mu\nu}, \frac{\partial g^{\mu\nu}}{\partial x^\lambda} \right) \sqrt{-g} d^4x. \tag{3.49}$$

3.1.4 Particle Motion and the Newtonian Limit

The Geodesic Equation

According to the basic physical principle of GR, a particle in a gravitational field moves in the four-dimensional Riemannian space-time in such a way that its trajectory represents an *extremal*, or a *geodesic line* in the given geometry. One can derive the geodesic equations of motion of a point particle in a gravitational field by generalizing to the curved case the special relativistic differential equation describing the free motion of a particle in the absence of the gravitational field. In a Galilean (geometrically flat) four-dimensional coordinate system the equations of motion of a relativistic particle are

$$\frac{dU^\mu}{ds} = 0, \tag{3.50}$$

where $U^\mu = dx^\mu / ds$ denotes the relativistic four-velocity of the point particle. Equivalently, we can write the equation of motion as $dU^\mu = 0$. In a Riemannian space-time, the equation of motion of a free particle is generalized by substituting the differential operator d with the covariant differential operator D, so that

$$DU^\mu = dU^\mu + \Gamma^\mu_{\alpha\beta} U^\alpha dx^\beta = 0. \tag{3.51}$$

Dividing this equation by ds we obtain the equation of motion (the geodesic equation) as

$$\frac{dU^\mu}{ds} + \Gamma^\mu_{\alpha\beta} U^\alpha U^\beta = 0, \tag{3.52}$$

or, equivalently,

$$\frac{d^2x^\mu}{ds^2} + \Gamma^\mu_{\alpha\beta} \frac{dx^\alpha}{ds} \frac{dx^\beta}{ds} = 0. \tag{3.53}$$

From the geodesic equation we see that the motion of the particle in a gravitational field, taking place in a Riemannian geometric setting, is fully determined by the Christoffel symbols $\Gamma^{\mu}_{\alpha\beta}$, which thus also acquire a physical interpretation. The quantity d^2x^μ/ds^2 is the four-acceleration of the particle, defined in the standard way. Therefore the function $-m\Gamma^{\mu}_{\alpha\beta}U^\alpha U^\beta$ represents the gravitational four-force acting on the point particle, which is a velocity-dependent quantity. Moreover, we can see that the metric tensor $g_{\mu\nu}$ plays the important role of the *potential of the gravitational field*, since its derivatives with respect to the coordinates determine the strength of the gravitational field acting on the particle via the Christoffel symbols $\Gamma^{\mu}_{\alpha\beta}$. As we have already seen, by an appropriate assignment of the coordinate system to a given arbitrary point P of the space-time we can always determine all the components of the Christoffel symbol to vanish at P. The possibility of the choice of such a locally–inertial (geodesic) frame means that, according to the principle of equivalence, we can always eliminate the gravitational field in a small neighborhood of the space-time near the given point P. Note that the four-momentum of a particle in a gravitational field is defined as $p^\mu = mcU^\mu$, and its square is given by $p_\mu p^\mu = m^2c^2$.

The geodesic equation is not usable in its standard form for the study of the propagation of a light signal in a gravitational field, since the propagation of a light ray takes place so that the associated interval ds is zero. Hence, since $ds = 0$, all terms in Eq. (3.52) tend to infinity, and therefore the equation is inapplicable to electromagnetic radiation. To obtain the equation of motion for light we exploit the fact that the direction of propagation of the electromagnetic radiation is fully determined by the wave vector k^μ, defined as $k^\mu = dx^\mu/d\lambda$, which is tangent to the ray. In the definition of the wave vector we have introduced the parameter λ, which is an *arbitrary parameter* changing along the direction of propagation of the electromagnetic radiation. In the special relativistic approach the wave vector does not change along the trajectory of the light ray propagating in vacuum, that is, the wave vector has the property $dk^\mu = 0$. For the propagation of light in a gravitational field this equation must be generalized to the form

$$\frac{dk^\mu}{d\lambda} + \Gamma^{\mu}_{\alpha\beta}k^\alpha k^\beta = 0. \tag{3.54}$$

These equations will also determine the functional form of the parameter λ.

As we have already shown in the previous chapter, the geodesic equation of motion of a particle in a gravitational field can also be derived with the use of the variational principle $\delta S = \delta \int \sqrt{g_{mu\nu}dx^\mu dx^\nu} = 0$, leading to the same result as given by Eq. (3.52).

The Newtonian Limit

One of the basic requirements of the general relativistic theory of gravitation is that it satisfies the *principle of correspondence*, that is, Newtonian gravity must be recovered from it in the particular limit of *small particle velocities* and *weak gravitational fields*. To investigate this transition we will only need the 00 component of the metric tensor, g_{00}, which we will assume to be given by

$$g_{00} = -\left(1 + \frac{2\phi}{c^2}\right), \tag{3.55}$$

where ϕ is the *Newtonian gravitational potential*. By neglecting any pressure effects in our system, we can approximate the matter energy-momentum tensor by the expression

$$T_\mu^\nu = \rho c^2 U_\mu U^\nu, \tag{3.56}$$

where ρ represents the energy density of the gravitating object. Under our assumption of slow macroscopic motion, we can neglect all space components of the four-velocity, and keep only the time component. Thus, for the four-velocity components we have

$$U^0 = 1, \qquad U_0 = -1, \qquad U^i \approx 0, \qquad i = 1, 2, 3. \tag{3.57}$$

Then we obtain the result that the only remaining component of the energy-momentum tensor is

$$T_0^0 = -\rho c^2, \tag{3.58}$$

giving $T = T_0^0 = -\rho c^2$. With the use of the approximations (3.57) and (3.58), the Einstein gravitational field equations, written in the form (3.46), reduce to a single equation, given by

$$R_0^0 = \frac{4\pi G}{c^2}\rho. \tag{3.59}$$

To calculate R_0^0 we first see that the terms in the definition of the Ricci tensor that contain products of the Christoffel symbols are always quantities of the second order. Moreover, the terms containing derivatives with respect to the time coordinate $x^0 = ct$ involve powers of $1/c$, and consequently they are negligibly small as compared with the terms containing derivatives with respect to the spatial coordinates $x^i, i = 1, 2, 3$. Hence, in the Newtonian approximation the only surviving component of the Ricci tensor is obtained as

$$R_{00} = R_0^0 \approx \frac{\partial \Gamma_{00}^i}{\partial x^i}, \qquad i \in \{1, 2, 3\}. \tag{3.60}$$

A simple calculation using the definition of the Christoffel symbols gives

$$\Gamma_{00}^i \approx -\frac{1}{2} g^{ij} \frac{\partial g_{00}}{\partial x^j} = \frac{1}{c^2}\frac{\partial \phi}{\partial x^i}. \tag{3.61}$$

Therefore, we find

$$R_0^0 = \frac{1}{c^2}\sum_{i=1}^{3} \frac{\partial}{\partial x^i}\frac{\partial \phi}{\partial x^i} = \frac{1}{c^2}\Delta\phi, \tag{3.62}$$

where Δ is the three-dimensional Laplace operator. Thus the Einstein equations reduce to

$$\Delta\phi = 4\pi G\rho. \tag{3.63}$$

This is the Poisson equation, determining the distribution of the gravitational potential in Newtonian gravity. Its solution can be expressed in the form

$$\phi = -G\int \frac{\rho dV}{R}, \tag{3.64}$$

where dV is the volume element.

3.2 The Spherically Symmetric Gravitational Field

The determination of the distribution of the gravitational field around a spherically symmetric massive self-gravitating object at rest is a fundamental theoretical and practical problem in both astronomy and astrophysics. For a spherically symmetric mass distribution the gravitational field itself must similarly have spherical symmetry. In the following we will limit our investigations to the case of *static gravitational fields*, that is, fields that are both *time independent* and *time symmetric*, that is, gravitational fields *unchanged by time reversal*, i.e., $t \rightarrow -t$. If the gravitational field is only time independent, it is called *stationary*. An example of stationary fields are the gravitational fields of rotating stars.

The spherical symmetry of the gravitational field has the important implication that the space-time metric $g_{\mu\nu}$, describing the basic properties of the gravitational field, as well as the interval ds, must have the same numerical values for all points in the four-dimensional space-time, located *at the same distance from the center* of the gravitating object. In a Galilean (pseudo-Euclidian) space this distance is nothing but the radius vector r. However, in a Riemannian geometry, induced by the presence of a gravitational field, one cannot find a quantity having the same properties as the Galilean (pseudo-Euclidian) radius vector.

3.2.1 The Metric for a Static and Spherically Symmetric Field

In the following, we consider spherical coordinates (r, θ, φ) in order to analyze gravitational fields with spherical symmetry. The definition of the angular spherical coordinates are unambiguous, since they can be measured by dividing in equal parts a circumference located at the same distance from the origin (concentric with the origin). However, the radial coordinate is ambiguous, since its relation to the measurement of distance is unknown for the time being. Therefore we shall treat it simply as a parameter. The time coordinate is also affected by ambiguities of a similar nature. The spherical symmetry as well as the static character of the gravitational field determine some strong restrictions on the possible form of the space-time interval. In order to implement spherical symmetry, we require that the interval is a function of only those combinations of the coordinates that are invariant with respect to the spatial rotations of the coordinates. One can show that the only such combinations are

$$\sqrt{x^2+y^2+z^2}=r, \qquad xdx+ydy+zdz=rdr$$
$$dx^2+dy^2+dz^2=dr^2+r^2d\theta^2+r^2+r^2\sin^2\theta d\varphi^2. \tag{3.65}$$

After an appropriate choice of the coordinate system, and after subjecting the coordinates to some transformation that keep the spherical symmetry of the field, one obtains for the interval ds^2 of *a static and spherically symmetric gravitational field* the representation

$$ds^2=-e^{\nu(r)}c^2dt^2+e^{\lambda(r)}dr^2+r^2\left(d\theta^2+\sin^2\theta d\varphi^2\right), \tag{3.66}$$

where the two independent metric tensor components ν and λ are functions of the radial coordinate r only.

3.2.2 The Gravitational Field Equations in Spherical Symmetry

For the interval (3.66), the covariant components of the metric tensor $g_{\mu\nu}$ are given by

$$g_{00}=-e^{\nu(r)}, \qquad g_{11}=e^{\lambda(r)}, \qquad g_{22}=r^2, \qquad g_{33}=r^2\sin^2\theta. \tag{3.67}$$

The contravariant components can be obtained as $g^{\mu\nu}=1/g_{\mu\nu}$. The Christoffel symbols can be easily calculated, and are given by

$$\Gamma^1_{11}=\frac{\lambda'}{2}, \quad \Gamma^0_{10}=\frac{\nu'}{2}, \quad \Gamma^0_{11}=\frac{\nu'}{2}e^{\nu-\lambda}, \quad \Gamma^1_{22}=-re^{-\lambda}, \quad \Gamma^2_{12}=\Gamma^3_{13}=\frac{1}{r}, \tag{3.68}$$

$$\Gamma^2_{33}=-\sin\theta\cos\theta, \quad \Gamma^3_{23}=\cot\theta, \quad \Gamma^1_{33}=-r\sin^2\theta e^{-\lambda}. \tag{3.69}$$

All the other components vanish identically. To write down the Einstein field equations of the gravitational field we must now calculate all the nonzero components of the Ricci tensor R^ν_μ. A simple but lengthy calculation leads to the following field equations, describing the gravitational properties of the static and spherically symmetric gravitating objects,

$$-\frac{1}{r^2}\frac{d}{dr}\left(re^{-\lambda}\right)+\frac{1}{r^2}=\frac{8\pi G}{c^4}T^0_0, \tag{3.70}$$

$$-e^{-\lambda}\left(\frac{\nu'}{r}+\frac{1}{r^2}\right)+\frac{1}{r^2}=\frac{8\pi G}{c^4}T^1_1, \tag{3.71}$$

$$-\frac{1}{2}e^{-\lambda}\left(\nu''+\frac{\nu'^2}{2}+\frac{\nu'-\lambda'}{r}-\frac{\nu'\lambda'}{2}\right)=\frac{8\pi G}{c^4}T^2_2=\frac{8\pi G}{c^4}T^3_3, \tag{3.72}$$

where a prime denotes the derivative with respect to the radial coordinate r.

3.2.3 The Schwarzschild Solution

In the extremely important case of a spherically symmetric gravitational field, the field equations (3.70) and (3.71) can be exactly integrated for the case of vacuum, that is, outside the masses producing the field. In this case all the components of the energy-momentum tensor $T_{\mu\nu}$ vanish identically, and the gravitational field equations take the simple form

$$-\frac{1}{r^2}\frac{d}{dr}\left(re^{-\lambda}\right)+\frac{1}{r^2}=0, \tag{3.73}$$

$$-e^{-\lambda}\left(\frac{\nu'}{r}+\frac{1}{r^2}\right)+\frac{1}{r^2}=0. \tag{3.74}$$

In the following, we do not need Eq. (3.72), since it can be shown that it is a consequence of Eqs. (3.73) and (3.74). By adding Eqs. (3.73) and (3.74) we obtain

$$\nu'+\lambda'=0, \tag{3.75}$$

or, equivalently,

$$\nu+\lambda=C=\text{constant}. \tag{3.76}$$

In the interval as given by Eq. (3.66), we still have the possibility of an arbitrary rescaling of the time coordinate t of the form $t=f(t')$. Through such a rescaling we can add to ν an arbitrary function of the time t, or, equivalently, an arbitrary constant, whose presence does not change the physics of the problem. Therefore we can choose, without any loss of generality, the constant C as zero, $C=0$, and hence

$$\nu+\lambda=0, \qquad \nu=-\lambda. \tag{3.77}$$

Then Eq. (3.73) can be written as

$$\frac{1}{r^2}\frac{d}{dr}\left(re^{-\lambda}\right)-\frac{1}{r^2}=0, \tag{3.78}$$

and can be solved immediately to yield

$$e^{-\lambda}=1+\frac{C_1}{r}, \tag{3.79}$$

where by C_1 we have denoted an arbitrary constant of integration. To determine the value of C_1, we compare the value of $e^{-\lambda}$, as given in Eq. (3.79), with its weak field limit, Eq. (3.55), where the potential has its Newtonian expression, $\phi=-GM/r$, and M is the object creating the gravitational field. Then, we immediately obtain

$$C_1=-\frac{2GM}{c^2}=-r_g, \tag{3.80}$$

where the quantity r_g, called the gravitational radius of the body, has the physical dimension of length. Thus, we have obtained the general relativistic interval in the vacuum outside a spherically symmetric mass distribution as given by the following relation

$$ds^2 = -\left(1 - \frac{2GM}{c^2 r}\right) c^2 dt^2 + \frac{dr^2}{1 - \frac{2GM}{c^2 r}} + r^2 \left(d\theta^2 + \sin^2 \theta d\varphi^2\right). \qquad (3.81)$$

This metric was first obtained by Karl Schwarzschild in 1916, and is called the Schwarzschild metric. In the limit $r \to \infty$, that is, far away from the gravitating objects creating the field, the metric automatically takes the Galilean (Minkowskian) form. The spatial metric is determined by the expression

$$dl^2 = \frac{dr^2}{1 - \frac{2GM}{c^2 r}} + r^2 \left(d\theta^2 + \sin^2 \theta d\varphi^2\right). \qquad (3.82)$$

We can easily see that in the Schwarzschild metric (3.81), the length (circumference) of a circle is given by the standard expression $2\pi r$. However, the radial distance l between two points r_1 and r_2, located along the same radius, is given by

$$l = \int_{r_1}^{r_2} \frac{dr}{\sqrt{1 - r_g/r}} > r_2 - r_1. \qquad (3.83)$$

On the other hand, since $g_{00} \leq 1$, it follows that in the Schwarzschild geometry, for observers located at finite distances from the gravitating objects creating a gravitational field, there is a slowing down of running clocks, and of the time measured by them, as compared with the time observed by the same clocks at infinity.

3.3 Summary and Conclusions

In the geometric description of the gravitational field the Einstein equations represent the basic theoretical tool for the investigation of the properties of the gravitational field, and of the geometric features associated to them. These equations establish a deep connection between the geometrical attributes of the space-time, and the matter distribution generating the field. From a purely mathematical point of view the Einstein field equations are a set of second order, strongly nonlinear differential equations. Except for a few simple cases (which are fortunately of major physical/astrophysical/cosmological interest), the Einstein field equations cannot be solved exactly, and in many situations one must resort to numerical methods for the investigation of the properties of their solutions.

In deriving the Einstein equations a basic requirement is the conservation of the matter energy-momentum tensor. This requirement fixes uniquely the geometric part of the equations, as a sum of two terms: the Einstein tensor, and a second term, proportional to the metric tensor multiplied by an arbitrary constant Λ. The divergence of these terms is zero, both individually and collectively, when summed up. The presence of the cosmological constant in the Einstein field equations has caused much debate over the past century. Its numerical value cannot be predicted by the theory, and cosmological observations show that it has a very small and fine-tuned numerical value. Hence in any fundamental general relativistic investigation of gravity we must ask, and try to answer, the question about the role of the cosmological constant in the description of gravity. Is it just an artifact of the mathematical formalism, or does it have a deep physical (or perhaps geometrical) meaning? Can its numerical value be derived from a fundamental physical theory that goes beyond GR? These are just a few questions we will address in the next chapters of this book.

4

The Solar System Tests and Astrophysical Applications

In the present chapter, we will first briefly review the fundamental observational tests of General Relativity that have been carried out at the solar system level. Next we will discuss some of the basic properties of compact, stellar-type astrophysical objects, and we will provide a brief introduction to the description of the accretion processes. In the study of the solar system tests and of the accretion we will present a general formalism that is adapted to the study of all similar phenomena/processes in arbitrary spherically symmetric geometries.

4.1 Classical Tests of General Relativity

To test the theoretical predictions of GR at the solar system level, three fundamental observational astronomical tests can be performed. These tests can also provide significant evidence for modified theories of gravity going beyond GR. These important solar system tests are the *perihelion precession of Mercury*, the *deflection of light by the Sun*, and the *radar echo delay observations*, respectively. They have been employed to successfully verify the predictions of the Schwarzschild solution of GR, as well as to test at the solar system level some of its generalizations, obtained within modified theories of gravity. In the following, we will study these three fundamental general relativistic effects in static spherically symmetric space-times [71, 76, 163, 230], and develop a general formalism that can be used for any static geometry. In each case we will also present the application of the general results to the specific case of the Schwarzschild metric.

4.1.1 The Perihelion Precession of Planetary Orbits

In an arbitrary static and spherically symmetric gravitational field, described in spherical coordinates (ct, r, θ, ϕ) by the metric

$$ds^2 = -e^{\nu(r)}c^2dt^2 + e^{\lambda(r)}dr^2 + r^2\left(d\theta^2 + \sin^2\theta d\phi^2\right), \tag{4.1}$$

the equations of motion of a test particle can be obtained from the Principle of Least Action

$$\delta\int\sqrt{-g_{\mu\nu}\frac{dx^\mu}{ds}\frac{dx^\nu}{ds}}=\delta\int\sqrt{e^{\nu}c^2\dot{t}^2-e^{\lambda}\dot{r}^2-r^2\left(\dot{\theta}^2+\sin^2\theta\dot{\phi}^2\right)}ds=0, \qquad (4.2)$$

where the derivative with respect to the interval ds is denoted by an overdot, so that $\dot{}=d/ds$. In the adopted coordinate system r is the radial coordinate, and the domains of the spherical coordinates (r,θ,ϕ) are $0\leq r\leq\infty$, $0\leq\theta\leq\pi$ and $0\leq\phi\leq 2\pi$, respectively,

It is easy to show that the trajectory of the motion is planar, and hence, without any loss of generality, we consider an equatorial plane, i.e., $\theta=\pi/2$, for the polar angle θ. Hence, in the following, we consider that ϕ is the only angular coordinate in the problem. As we can see immediately, neither ct nor ϕ appear explicitly in Eq. (4.2). This means that from their conjugate momenta we can obtain two constants of the motion, denoted E and L, respectively, and given by

$$e^{\nu}c^2\dot{t}=E=\text{constant},\qquad r^2\dot{\phi}=L=\text{constant}. \qquad (4.3)$$

The arbitrary constant E represents the conserved energy of the motion, while the constant L expresses the conservation of the angular momentum.

After division by ds^2, the line element Eq. (4.1) yields the following first-order nonlinear differential equation of motion for the radial coordinate r,

$$\dot{r}^2+e^{-\lambda}r^2\dot{\phi}^2=e^{-\lambda}\left(e^{\nu}c^2\dot{t}^2-1\right). \qquad (4.4)$$

Substituting now $\dot{t}$ and $\dot{\phi}$ from Eqs. (4.3) we obtain the relation

$$\dot{r}^2+e^{-\lambda}\frac{L^2}{r^2}=e^{-\lambda}\left(\frac{E^2}{c^2}e^{-\nu}-1\right). \qquad (4.5)$$

By changing the variable r according to $r=1/u$, and using the transformation of variables $d/ds=Lu^2d/d\phi$, Eq. (4.5) is transformed into the equivalent form

$$\left(\frac{du}{d\phi}\right)^2+e^{-\lambda}u^2=\frac{1}{L^2}e^{-\lambda}\left(\frac{E^2}{c^2}e^{-\nu}-1\right). \qquad (4.6)$$

Now, by representing the metric tensor component $e^{-\lambda}$ as $e^{-\lambda}=1-f(u)$, where $f(u)$ is an arbitrary function of $u=1/r$, we find

$$\left(\frac{du}{d\phi}\right)^2+u^2=f(u)u^2+\frac{E^2}{c^2L^2}e^{-\nu-\lambda}-\frac{1}{L^2}e^{-\lambda}\equiv G(u). \qquad (4.7)$$

The derivative of the above equation with respect to the angular coordinate ϕ gives

$$\frac{d^2u}{d\phi^2}+u=F(u), \qquad (4.8)$$

where we have introduced a new function $F(u)$ defined as

$$F(u)=\frac{1}{2}\frac{dG(u)}{du}. \tag{4.9}$$

The radius of a *circular* planar orbit $u=u_0=$ constant can be obtained by solving the algebraic equation $u_0=F(u_0)$. On the other hand, any departure

$$\delta=u-u_0 \tag{4.10}$$

from a circular orbit must obey the perturbation equation

$$\frac{d^2\delta}{d\phi^2}+\left[1-\left(\frac{dF}{du}\right)_{u=u_0}\right]\delta=O\left(\delta^2\right), \tag{4.11}$$

which can be easily derived by replacing u with $u=u_0+\delta$ in Eq. (4.8). Therefore, by keeping only first-order terms in δ, we obtain the perturbation δ of a given circular trajectory as

$$\delta=\delta_0\cos\left(\sqrt{1-\left(\frac{dF}{du}\right)_{u=u_0}}\phi+\beta\right), \tag{4.12}$$

where δ_0 and β denote two arbitrary constants of integration, which must be determined from the initial conditions of the motion. The angles for which r takes its minimum value are the angles of the perihelia of the particle motion orbit. Therefore it follows that for the minimum r, u, or δ must take their maximum values. Hence, the total change of the orbital angle ϕ from one planetary perihelion position to the next, can be obtained as

$$\phi=\frac{2\pi}{\sqrt{1-\left(\frac{dF}{du}\right)_{u=u_0}}}=\frac{2\pi}{1-\sigma}. \tag{4.13}$$

The quantity σ introduced above is called *the perihelion advance.* It gives us *the rate of progress of the perihelion* of the planet. As the planet moves forward in time through ϕ radians in its celestial orbit, its perihelion advances through $\sigma\phi$ radians. From Eq. (4.13), it follows that σ can be obtained as

$$\sigma=1-\sqrt{1-\left(\frac{dF}{du}\right)_{u=u_0}}, \tag{4.14}$$

or in the case of small values of the function $(dF/du)_{u=u_0}$, by the relation

$$\sigma=\frac{1}{2}\left(\frac{dF}{du}\right)_{u=u_0}. \tag{4.15}$$

For a complete rotation of the planet we obtain $\phi\approx 2\pi(1+\sigma)$, with the progress of the perihelion given by $\delta\phi=\phi-2\pi\approx 2\pi\sigma$.

To carry out explicit computations of the perihelion precession of a planet in an arbitrary space-time geometry, we must know the expression of the angular momentum L of the planet as a function of the parameters of the orbit. To obtain L we will consider a "Keplerian approximation" of the motion. For this, we study the movement of a planet on a Keplerian elliptic orbit, with semiaxis denoted by a and b, respectively, where b is given by $b = a\sqrt{1-e^2}$. By e we denote the eccentricity of the planetary path. The ellipse encompassed by the trajectory of the planet has a surface area that is given by πab. Moreover, the oriented elementary surface area of the ellipse can be obtained as $d\vec{\sigma} = \left(\vec{r} \times d\vec{r}\right)/2$, and therefore the areolar velocity of the celestial object is given by the equation

$$\left|\frac{d\vec{\sigma}}{dt}\right| = \frac{1}{2}\left|\vec{r} \times d\vec{r}\right| = \frac{1}{2}r^2\left(\frac{d\phi}{dt}\right) \approx \frac{\pi a^2}{T}\sqrt{1-e^2}, \tag{4.16}$$

where T is the period of the orbital motion. T can be obtained from Kepler's third law of planetary motion as

$$T^2 = \frac{4\pi^2 a^3}{GM}. \tag{4.17}$$

In the excellent approximation of the small planetary velocities we have $ds \approx cdt$, and the conservation of the relativistic angular momentum takes the form $r^2 d\phi/dt = cL$. Therefore, we find for the angular momentum the expression

$$L = 2\pi\frac{a^2\sqrt{1-e^2}}{cT}, \tag{4.18}$$

and the above relation immediately yields

$$\frac{1}{L^2} = \frac{c^2}{GMa\left(1-e^2\right)}. \tag{4.19}$$

Perihelion Precession in Schwarzschild Geometry

We compute now the perihelion precession of a solar system planet in the Schwarzschild geometry, with

$$e^{\nu} = e^{-\lambda} = 1 - \frac{2GM}{c^2 r} = 1 - \frac{2GM}{c^2}u. \tag{4.20}$$

Therefore in the Schwarzschild geometry the function $f(u)$ is given by $f(u) = \left(2GM/c^2\right)u$, with f being a linear function of u. Since for this geometry we also have the relation $\nu + \lambda = 0$, which follows from the definition of the metric coefficients, we obtain first

$$G(u) = \frac{2GM}{c^2}u^3 + \frac{1}{L^2}\left(\frac{E^2}{c^2} - 1\right) + \frac{2GM}{c^2 L^2}u, \tag{4.21}$$

and then

$$F(u) = 3\frac{GM}{c^2}u^2 + \frac{GM}{c^2 L^2}. \tag{4.22}$$

The radius u_0 of the circular orbit of the test body is found as the solution of the second-order algebraic equation,

$$u_0 = 3\frac{GM}{c^2}u_0^2 + \frac{GM}{c^2 L^2}, \tag{4.23}$$

having the only physical solution, corresponding to the minus sign in front of the square root, and given by

$$u_0 = \frac{1 \pm \sqrt{1 - 12G^2M^2/c^4L^2}}{6GM/c^2} \approx \frac{GM}{c^2L^2}. \tag{4.24}$$

Therefore for the perihelion advance of a planet in the Schwarzschild geometry we obtain

$$\delta\phi = \pi\left(\frac{dF}{du}\right)_{u=u_0} = \frac{6\pi GM}{c^2 a\left(1-e^2\right)}, \tag{4.25}$$

which is a standard result of General Relativity.

4.1.2 The Deflection of Light

As we have already shown, in a gravitational field a massless photon travels along a null geodesic, having the basic property $ds^2 = 0$. We have also seen that along the photon's trajectory for the affine parameter we can adopt any arbitrary parameter $\tilde{\lambda}$. In the following, the ordinary derivative with respect to the arbitrary affine parameter $\tilde{\lambda}$ will be denoted by an overdot. Due to the choice of the metric from the gravitational action we can obtain two conserved quantities for the motion, the energy E and the angular momentum L of the photon, given again by Eqs. (4.3).

Hence in a gravitational field the equation of motion of the photon can be written as

$$\dot{r}^2 + e^{-\lambda}r^2\dot{\phi}^2 = e^{\nu-\lambda}c^2\dot{t}^2, \tag{4.26}$$

With the use of the two constants of motion this equation can be rewritten as

$$\dot{r}^2 + e^{-\lambda}\frac{L^2}{r^2} = \frac{E^2}{c^2}e^{-\nu-\lambda}. \tag{4.27}$$

Now we change the radial variable r to $r = 1/u$, and to eliminate the derivative with respect to the affine parameter of r we employ the angular momentum conservation equation. Thus, we obtain

$$\left(\frac{du}{d\phi}\right)^2 + u^2 = f(u)u^2 + \frac{1}{c^2}\frac{E^2}{L^2}e^{-\nu-\lambda} \equiv P(u). \tag{4.28}$$

We take the derivative of the above equation with respect to the angular variable ϕ to find the equation of motion of the photon as given by

$$\frac{d^2u}{d\phi^2}+u=Q(u), \tag{4.29}$$

where we have denoted

$$Q(u)=\frac{1}{2}\frac{dP(u)}{du}. \tag{4.30}$$

First, we study the equation of motion in the zeroth order of approximation, in which the nonlinear term $Q(u)$ on the right hand side of Eq. (4.29) can be neglected. In this approximation, the solution of the equation of motion is a straight line, given by the equation

$$u=\frac{\cos\phi}{R}, \tag{4.31}$$

where R denotes the length (distance) of the *closest approach* to the gravitating massive object. We go now to the next order of approximation, in which we substitute Eq. (4.31) in the right-hand side of the equation of motion (4.29). Hence, we obtain a linear second-order inhomogeneous differential equation represented as

$$\frac{d^2u}{d\phi^2}+u=Q\left(\frac{\cos\phi}{R}\right). \tag{4.32}$$

Equation (4.32) has a solution, which in its more general form is given by $u=u\left(\phi\right)$. In the concrete astronomical situation in which the observations are performed we have a photon coming in from spatial infinity at the asymptotic angle $\phi=-\left(\pi/2+\varepsilon\right)$, and traveling to infinity at a different asymptotic angle $\phi=\pi/2+\varepsilon$. The deflection angle ε is obtained as a solution of the algebraic equation

$$u\left(\pi/2+\varepsilon\right)=0, \tag{4.33}$$

and thus the total deviation angle of the light ray can be computed as $\delta=2\varepsilon$.

The Deflection Angle of Light Rays in the Schwarzschild Geometry

For the Schwarzschild metric the metric tensor components satisfy the relation $\nu+\lambda=0$, and the function $f(u)$ is given by $f(u)=\left(2GM/c^2\right)u$. Therefore we immediately obtain

$$P(u)=\left(\frac{2GM}{c^2}\right)u^3, \tag{4.34}$$

and

$$Q(u)=\left(\frac{3GM}{c^2}\right)u^2, \tag{4.35}$$

respectively. In the lowest order of approximation by following the procedure outlined in the previous section, from Eqs. (4.31) and (4.32) we obtain for u the second-order linear inhomogeneous differential equation

$$\frac{d^2u}{d\phi^2}+u=\frac{3GM}{c^2R^2}\cos^2\phi=\frac{3GM}{2c^2R^2}\left(1+\cos 2\phi\right), \tag{4.36}$$

having the general solution given by

$$u(\phi)=\frac{\cos\phi}{R}+\frac{3GM}{2c^2R^2}\left(1-\frac{1}{3}\cos 2\phi\right). \tag{4.37}$$

By performing the substitutions $\phi=\pi/2+\varepsilon$, $u=0$ into Eq. (4.37) one finds

$$\varepsilon=\frac{2GM}{c^2R}, \tag{4.38}$$

where we have made use of the trigonometric relations $\cos\left(\pi/2+\varepsilon\right)=-\sin\varepsilon$, $\cos\left(\pi+2\varepsilon\right)=-\cos 2\varepsilon$, and the following approximations $\sin\varepsilon\approx\varepsilon$ and $\cos 2\varepsilon\approx 1$, respectively. Hence, the total deviation angle of a photon moving in the Schwarzschild geometry in the first order of approximation is obtained as

$$\delta=2\varepsilon=\frac{4GM}{c^2R}. \tag{4.39}$$

4.1.3 The Shapiro Radar Echo Delay Test

The third important solar system test of GR is the Shapiro radar echo delay [460, 461]. The basic physical concept of this test relies on the determination of the time necessary for electromagnetic (radar) waves to travel to an interior planet or satellite of the solar system in two different astronomical situations: first, when the radar signal goes very near the Sun and, secondly, when the electromagnetic wave does not travel in the vicinity of the Sun. The time necessary for light to travel between two solar system planets, situated at positions that are very far away from the Sun, is given by

$$T_0=\int_{-l_1}^{l_2}\frac{dy}{c}, \tag{4.40}$$

where by l_1 and l_2 we have denoted the distances of the planets to the Sun. On the other hand, if the electromagnetic signal passes nearby to the Sun, the time travel is modified by the curvature of the space, and it can be obtained as

$$T=\int_{-l_1}^{l_2}\frac{dy}{v}=\frac{1}{c}\int_{-l_1}^{l_2}e^{[\lambda(r)-\nu(r)]/2}dy, \tag{4.41}$$

where $v=ce^{(\nu-\lambda)/2}$ is the speed of light in the presence of the geometrical effects generated by the gravitational field of the Sun. The time difference between these two travels is

$$\delta T = T - T_0 = \frac{1}{c}\int_{-l_1}^{l_2} \left\{e^{[\lambda(r)-\nu(r)]/2} - 1\right\} dy. \tag{4.42}$$

Since $r = \sqrt{y^2 + R^2}$, we can compute δT as

$$\delta T = \frac{1}{c}\int_{-l_1}^{l_2} \left\{e^{\left[\lambda\left(\sqrt{y^2+R^2}\right)-\nu\left(\sqrt{y^2+R^2}\right)\right]/2} - 1\right\} dy. \tag{4.43}$$

Radar Echo Delay in the Schwarzschild Geometry

For the Schwarzschild metric $\lambda = -\nu$, we have

$$\exp(-\nu) = \left(1 - 2GM/c^2 r\right)^{-1} \approx 1 + 2GM/c^2 r, \tag{4.44}$$

and therefore

$$\delta T = \frac{2GM}{c^3}\int_{-l_1}^{l_2} \frac{dy}{\sqrt{y^2+R^2}} = \frac{2GM}{c^3}\ln\frac{\sqrt{R^2+l_2^2}+l_2}{\sqrt{R^2+l_1^2}-l_1}. \tag{4.45}$$

Since in the case of the solar system the conditions $R^2/l_1^2 \ll 1$ and $R^2/l_2^2 \ll 1$, respectively, represent a very good approximation, we obtain

$$\ln\frac{\sqrt{R^2+l_2^2}+l_2}{\sqrt{R^2+l_1^2}-l_1} \approx \ln\frac{4l_1 l_2}{R^2}. \tag{4.46}$$

Hence the time delay of radar signals traveling between two planets in the solar system is given by

$$\delta T = \frac{2GM}{c^3}\ln\frac{4l_1 l_2}{R^2}. \tag{4.47}$$

4.2 The Tolman–Oppenheimer–Volkoff Equation

In order to investigate astrophysical phenomena one can start in a first approach with the Newtonian equations of gravity and mechanics. This approximation is justified if the crucial factor GM/c^2R, where M is the mass of the gravitating object, and R is its radius, is much smaller as compared to unity, $GM/c^2R \ll 1$. However, if this factor would approach unity, the correct investigation of the gravitational phenomena requires the use of the theory of GR. The hydrostatic equilibrium equation, derived in the framework of GR, and describing the equilibrium properties of massive stellar objects, is known in the physical literature as the Tolman–Oppenheimer–Volkoff equation.

The derivation of the this equation is based on a number of assumptions, which are given as:

a. The matter content of compact gravitating objects can be described as a perfect fluid, with energy-momentum tensor

$$T_\mu^\nu = \left(\rho c^2 + p\right) U_\mu U^\nu + p\delta_\mu^\nu, \tag{4.48}$$

where ρ and p are the energy density and the pressure of the fluid, respectively, and U^{ν} is the four-velocity vector of the fluid.

b. The star has spherical symmetry, and the possible deviations from it are small, and therefore can be neglected. By using this assumption we can write the geometric metric inside the star in the standard form given by Eq. (4.1).

c. The self-gravitating system is in both mechanical and thermodynamical equilibrium. From this important assumption we can obtain two physical conditions. First, from the equilibrium requirement it follows that the metric coefficients inside the star are independent of the time t, so that the two metric tensor components ν and λ are functions of the radial coordinate r only. Secondly, in the adopted metric the four-velocity has no space-like components, so that its components are $U^0 = e^{-\nu/2}$, $U^1 = U^2 = U^3 = 0$. By using these values of U^{μ} it follows that the energy-momentum tensor of the matter inside the star has only diagonal nonvanishing components, $T_{\mu}^{\nu} = \mathrm{diag}\left(-\rho c^2, p, p, p\right)$.

Hence after adopting these assumptions, we can write down the gravitational field equations inside a massive gravitating object, which are given by the following system of strongly nonlinear differential equations,

$$-\frac{1}{r^2}\frac{d}{dr}\left(re^{-\lambda}\right)+\frac{1}{r^2}=\frac{8\pi G}{c^2}\rho, \tag{4.49}$$

$$e^{-\lambda}\left(\frac{\nu'}{r}+\frac{1}{r^2}\right)-\frac{1}{r^2}=\frac{8\pi G}{c^4}p, \tag{4.50}$$

$$p'=-\left(\rho c^2+p\right)\frac{\nu'}{2}, \tag{4.51}$$

where Eq. (4.51) follows from the conservation of the energy-momentum tensor, $\nabla_{\nu}T_{\mu}^{\nu}=0$. Equation (4.49) can be easily integrated with respect to the radial coordinate r to give

$$e^{-\lambda}=1-\frac{2Gm(r)}{c^2 r}, \tag{4.52}$$

where

$$m(r)=4\pi\int_0^r \rho r^2 dr, \tag{4.53}$$

is the stellar mass located within the radius r. After substituting the mass and ν' in Eq. (4.50), and solving for dp/dr, we obtain the equation describing the hydrostatic equilibrium properties of an isotropic general relativistic static fluid sphere as

$$\frac{dp(r)}{dr}=-\frac{\left(G/c^2\right)\left[\rho\left(r\right)c^2+p(r)\right]\left[4\pi r^3 p(r)/c^2+m(r)\right]}{r^2\left[1-2Gm(r)/c^2 r\right]}. \tag{4.54}$$

This equation is called the Tolman–Oppenheimer–Volkoff (TOV) equation. Together with the mass continuity equation

$$\frac{dm}{dr} = 4\pi\rho r^2, \tag{4.55}$$

the TOV equation gives the full description of the gravitational, mechanical, and hydrodynamical properties of general relativistic stars. The system of the two differential equations (4.54) and (4.55) must be integrated with the boundary conditions

$$m(0) = 0, \qquad \rho(R) = 0, \qquad p(R) = 0, \tag{4.56}$$

where R is the radius of the gaseous sphere. In order to close the system of equations an equation of state of the form $p = p(\rho)$ for the matter inside the star must also be provided. For most of the physically relevant equations of state of dense nuclear matter, the integration of the TOV and mass continuity equations must be done by numerical approaches.

The astrophysical applications of the TOV equation and of the mass continuity equation are considerably simplified if one introduces a dimensionless representation of the relevant physical quantities. To find a dimensionless form of the TOV equation (4.54) and of the mass continuity equation (4.55), we will introduce a new set of parameters consisting of the dimensionless independent variable η, and the dimensionless functions $\epsilon(\eta)$ (energy density), $P(\eta)$ (pressure), and $M(\eta)$ (mass), respectively, defined by means of the following rescalings of the physical quantities,

$$r = a\eta, \qquad \rho = \rho_c \epsilon(\eta), \qquad p = \rho_c c^2 P(\eta), \qquad m = M^* M(\eta). \tag{4.57}$$

Here a represents a scale factor (a characteristic length), ρ_c is the central density of the massive object, and M^* is a characteristic mass, respectively. With the help of the transformations (4.57), Eqs. (4.54) and (4.55) can be transformed to a dimensionless form given by

$$\frac{dM}{d\eta} = \eta^2 \epsilon, \qquad \frac{dP}{d\eta} = -\frac{(\epsilon + P)(P\eta^3 + M)}{\eta^2 (1 - 2M/\eta)}, \tag{4.58}$$

where we have taken

$$a^2 = \frac{c^2}{4\pi G\rho_c}, \qquad M^* = 4\pi\rho_c a^3 = \frac{c^3}{\sqrt{4\pi G^3 \rho_c}}. \tag{4.59}$$

In the new dimensionless variables the equation of state takes the form $P = P(\epsilon)$, while the boundary conditions that must be used when integrating the system are given by

$$M(0) = 0, \qquad \epsilon(\eta_S) = 0, \qquad P(\eta_S) = 0. \tag{4.60}$$

4.2.1 Constant Density General Relativistic Stars

We will investigate now the physical structure and the interior geometry of a homogeneous gaseous compact object. More exactly, we assume that the energy density of the dense matter is constant throughout the star, $\rho = \rho_c =$ constant. In this case for the dimensionless variable ϵ we have $\epsilon = 1, \forall \eta \in [0, \eta_S]$. The boundary conditions required to integrate the system of equations (4.58) are

$$M(0) = 0, \qquad P(0) = P_c, \qquad P(\eta_S) = 0, \tag{4.61}$$

where P_c is the central pressure of the star. By taking into account that the density is a constant throughout the star, the first of equations (4.58) can be immediately integrated to give the mass distribution as

$$M(\eta) = \frac{\eta^3}{3}. \tag{4.62}$$

Thus the hydrostatic equilibrium equation becomes

$$\frac{dP}{d\eta} = -\frac{\eta(1+P)(P+1/3)}{(1-2\eta^2/3)}, \tag{4.63}$$

and has the exact solution

$$P(\eta) = \frac{\frac{1}{3}(1+P_c) - \left(\frac{1}{3}+P_c\right)\sqrt{1-\frac{2}{3}\eta^2}}{\left(\frac{1}{3}+P_c\right)\sqrt{1-\frac{2}{3}\eta^2} - (1+P_c)}. \tag{4.64}$$

The radius R of the star can be obtained from the condition of the vanishing pressure at the vacuum boundary of the star, $P(\eta_S) = 0$, and is given by

$$R_S = a\frac{\sqrt{6P_c(2P_c+1)}}{3P_c+1}. \tag{4.65}$$

The total mass of the constant density star can be obtained as

$$M_S = \frac{4\pi}{3}\rho_c R_S^3 = \frac{1}{3}\frac{c^3}{\sqrt{4\pi G^3\rho_c}}\frac{[6P_c(2P_c+1)]^{3/2}}{(3P_c+1)^3}. \tag{4.66}$$

The radius R_S of the homogeneous stellar configuration can also be related to the gravitational (Schwarzschild) radius of the star $R_g = 2GM_S/c^2$, by the linear proportionality relation

$$R_S = \frac{(3P_c+1)^2}{2P_c[2(2P_c+1)]}R_g. \tag{4.67}$$

The values of the central pressure P_c depend on the equation of state of the dense matter, and on the physically allowed upper limits for the matter pressure. If we impose the classical restriction of GR on the pressure-density relation,

$$p \leq \frac{\rho c^2}{3}, \tag{4.68}$$

since $p \geq 0$, it follows that in the case of the constant density star $P_c \in [0, 1/3]$. However, we may also consider the more general restriction

$$p \leq \rho c^2, \tag{4.69}$$

with the upper limit giving the stiff equation of state of high density nuclear matter. Matter may be described by the stiff equation of state at densities higher than ten times the nuclear density, that is at densities higher than 10^{17} g/cm^3, and at temperatures $T = (\rho/\sigma)^{1/4} > 10^{13}$ K, where σ denotes the radiation constant. For this central pressure range it follows that $P_c \in [0, 1]$.

If we adopt the classical restriction of the central pressure as given by Eq. (4.68), we obtain the following stability criterion for a constant density star,

$$R \geq \frac{9}{5} R_g. \tag{4.70}$$

For constant density stars for which the more general restriction (4.69) is satisfied, the stability criterion can be formulated as

$$R \geq \frac{4}{3} R_g. \tag{4.71}$$

The metric tensor component $e^{-\lambda}$ inside the constant density star is given by the expression

$$e^{-\lambda} = 1 - \frac{1}{3}\frac{8\pi G\rho}{c^2} r^2, \tag{4.72}$$

while the metric tensor component e^{ν} can be obtained as a function of the dimensionless variable η as

$$e^{\nu(\eta)} = \frac{C}{\left[1 + P(\eta)\right]^2} = C\left[\left(\frac{1}{3} + P_c\right)\sqrt{1 - \frac{2}{3}\eta^2} - (1 + P_c)\right]^2, \tag{4.73}$$

where by C we have denoted an arbitrary constant of integration. For $r \geq R_S$ the geometry of the space-time is described by the Schwarzschild metric, given by

$$e^{\nu} = e^{-\lambda} = 1 - \frac{2GM_S}{c^2 r}. \tag{4.74}$$

In order to match the interior metric in the constant density star smoothly on its boundary surface with the exterior Schwarzschild metric, we have to impose the condition of the continuity of the gravitational potentials across the surface of the star. By matching the interior metric tensor component, given by Eq. (4.73), with the corresponding exterior metric tensor component given by Eq. (4.74) at the boundary $r = R_S$ we obtain the value of the integration constant as given by

$$C = \frac{1 - 2GM_S/c^2 R_S}{\left\{\left(\frac{1}{3} + P_c\right)\sqrt{1 - \frac{2}{3}\left(\frac{R_S}{a}\right)^2} - (1 + P_c)\right\}}. \tag{4.75}$$

Hence, we have completely solved the problem of the structure of the constant density general relativistic fluid spheres.

4.2.2 Exact Power Series Solutions of the Relativistic Structure Equations for a Linear Barotropic Fluid

Another possibility of bringing the general relativistic structure equations to a dimensionless form is represented by the transformations [247]

$$r = \eta R, \qquad \rho = \rho_c \varepsilon(\eta), \qquad p = \rho_c c^2 P(\eta), \qquad M = M^* m(\eta), \tag{4.76}$$

where R is the radius of the star, and ρ_c denotes again the central density. Then, in these variables, the TOV and the mass continuity equations take the form

$$\frac{dP}{d\eta} = -\frac{a\left[\varepsilon(\eta) + P(\eta)\right]\left[P(\eta)\eta^3 + m(\eta)\right]}{\eta^2\left[1 - 2am(\eta)/\eta\right]}, \tag{4.77}$$

$$\frac{dm}{d\eta} = \eta^2 \varepsilon(\eta), \tag{4.78}$$

respectively, where we have fixed the constants a and M^* by the relations

$$a = \frac{4\pi G \rho_c}{c^2} R^2, \qquad M^* = 4\pi \rho_c R^3. \tag{4.79}$$

As a function of the parameter a the radius R and the total mass M_S of the star are given by the relations

$$R = \sqrt{a}\frac{c}{\sqrt{4\pi G \rho_c}} = 10.3622\,\sqrt{a}\left(\frac{\rho_c}{10^{15}\ \mathrm{g/cm^3}}\right)^{-1/2}\ \mathrm{km}, \tag{4.80}$$

$$M_S = a^{3/2}\frac{c^3}{\sqrt{4\pi \rho_c G^3}} m(1) = 6.9910\, a^{3/2}\left(\frac{\rho_c}{10^{15}\ \mathrm{g/cm^3}}\right)^{-1/2} m(1)\, M_\odot, \tag{4.81}$$

while for the mass-radius ratio of the star we obtain

$$\frac{GM_S}{c^2 R} = am(1). \tag{4.82}$$

By eliminating the energy density $\varepsilon(\eta)$ between the Eqs. (4.78) and (4.77), we obtain the following second-order differential equations, which we denote the *the relativistic mass equation*

$$\eta\frac{d^2m(\eta)}{d\eta^2} - 2\frac{dm(\eta)}{d\eta} + \frac{a\left[\eta^3 P\left(\frac{m'(\eta)}{\eta^2}\right) + m(\eta)\right]\left[m'(\eta) + \eta^2 P\left(\frac{m'(\eta)}{\eta^2}\right)\right]}{\eta\left[1 - 2am(\eta)/\eta\right] P'\left(\frac{m'(\eta)}{\eta^2}\right)} = 0. \tag{4.83}$$

Alternatively, the relativistic mass equation takes the form

$$\eta\left[1-\frac{2am(\eta)}{\eta}\right]\left[\eta\frac{d^2m(\eta)}{d\eta^2}-2\frac{dm(\eta)}{d\eta}\right]P'\left[\frac{m'(\eta)}{\eta^2}\right]$$
$$+a\left\{m'(\eta)+\eta^2P\left[\frac{m'(\eta)}{\eta^2}\right]\right\}\left\{m(\eta)+\eta^3P\left[\frac{m'(\eta)}{\eta^2}\right]\right\}=0, \tag{4.84}$$

or equivalently

$$\left[1-\frac{2am(\eta)}{\eta}\right]\left[\eta\frac{d^2m(\eta)}{d\eta^2}-2\frac{dm(\eta)}{d\eta}\right]P'\left[\frac{m'(\eta)}{\eta^2}\right]$$
$$+a\eta^2\left\{\frac{m'(\eta)}{\eta^2}+P\left[\frac{m'(\eta)}{\eta^2}\right]\right\}\left\{\frac{m(\eta)}{\eta}+\eta^2P\left[\frac{m'(\eta)}{\eta^2}\right]\right\}=0. \tag{4.85}$$

Equation (4.83) must be integrated with the initial conditions $m(0)=0$, and $m'(0)=0$, respectively, and together with the equation of state of matter, $P=P(\varepsilon)=P\left[m'(\eta)/\eta^2\right]$.

The Linear Barotropic Fluid Star

We consider now the possibility of obtaining exact power series solutions of Eq. (4.83) for the linear barotropic equation of state. In the present section, we follow the approach introduced in [247]. In order to investigate the mathematical properties of the relativistic mass equation we introduce first the Cauchy convolution of the power series, defined as follows.

Definition. *Let* $f_1=\sum_{i_1=0}^{\infty}a_{1,i_i}x^{i_1}$, $f_2=\sum_{i_2=0}^{\infty}a_{2,i_2}x^{i_2}$, $f_3=\sum_{i_3=0}^{\infty}a_{3,i_3}x^{i_3}$, ..., $f_s=\sum_{i_s=0}^{\infty}a_{s,i_s}x^{i_s}$ *be* s *convergent power series,* $s\geq 2$. *Then we define the Cauchy product (convolution) of the* s *power series,* $s\geq 2$, *as*

$$f_1\circ f_2=\left(\sum_{i_1=0}^{\infty}a_{1,i_1}x^{i_1}\right)\left(\sum_{i_2=0}^{\infty}a_{2,i_2}x^{i_2}\right)=\sum_{i_1,i_2=0}^{\infty}a_{1,i_1}a_{2,i_2}x^{i_1+i_2}$$
$$=\sum_{j_2=0}^{\infty}\left(\sum_{i_1=0}^{j_2}a_{1,i_1}a_{2,,j_2-i_1}\right)x^{j_2}=\sum_{j_1=0}^{\infty}A_{2,j_2}x^{j_2}, \tag{4.86}$$

$$A_{2,j_2}=\sum_{i_1=0}^{j_2}a_{1,i_1}a_{2,j_2-i_1}, \tag{4.87}$$

$$f_1\circ f_2\circ f_3=\left(\sum_{i_1=0}^{\infty}a_{1,i_1}x^{i_1}\right)\left(\sum_{i_2=0}^{\infty}a_{2,i_2}x^{i_2}\right)\left(\sum_{i_3=0}^{\infty}a_{3,i_3}x^{i_3}\right)$$
$$=\sum_{i_1=0}^{\infty}a_{1,i_1}x^{i_1}\left(\sum_{i_2,i_3=0}^{\infty}a_{2,i_2}a_{3,i_3}x^{i_2+i_3}\right)$$

$$= \sum_{i_1=0}^{\infty} a_{1,i_1} x^{i_1} \left[\sum_{j_2=0}^{\infty} \left(\sum_{i_2=0}^{j_2} a_{2,i_2} a_{3,j_2-i_2} \right) x^{j_2} \right]$$

$$= \sum_{j_3=0}^{\infty} \left[\sum_{i_1=0}^{j_3} \sum_{i_2=0}^{j_3-i_1} a_{1,i_1} a_{2,i_2} a_{3,j_3-i_1-i_2} \right] x^{j_3}$$

$$= \sum_{j_3=0} A_{3,j_3} x^{j_3}, \tag{4.88}$$

$$A_{3,j_3} = \sum_{i_1=0}^{j_3} \sum_{i_2=0}^{j_3-i_1} a_{1,i_1} a_{2,i_2} a_{3,j_3-i_1-i_2}, \tag{4.89}$$

$$\ldots\ldots,$$

$$f_1 \circ f_2 \circ \ldots \circ f_s = \sum_{j_s=0}^{\infty} A_{s,j_s} x^{j_s}, \tag{4.90}$$

$$A_{s,j_s} = \sum_{i_1=0}^{j_s} \sum_{i_2=0}^{j_s-i_1} \ldots \sum_{i_{s-1}=0}^{j_s-i_1-\ldots-i_{s-1}} a_{1,i_1} a_{2,i_2} \ldots a_{s,j_s-i_1-\ldots-i_{s-1}}. \tag{4.91}$$

As a first example of an exact power series solution of the relativistic mass equation Eq. (4.83) we consider the case of the *linear barotropic equation of state* $p = \gamma\rho$, which we rewrite in the dimensionless form

$$P(\eta) = \gamma\varepsilon(\eta), \qquad \gamma = \text{constant}, \gamma \in [0,1]. \tag{4.92}$$

Then the TOV equation Eq. (4.77) becomes

$$\frac{d\varepsilon(\eta)}{d\eta} = -a\frac{\gamma+1}{\gamma}\frac{\varepsilon(\eta)\left[\gamma\varepsilon(\eta)\eta^3 + m(\eta)\right]}{\eta^2\left[1 - 2am(\eta)/\eta\right]}. \tag{4.93}$$

The relativistic mass equation Eq. (4.83) takes the form

$$\eta\left(1 - \frac{2am}{\eta}\right)\frac{d^2m}{d\eta^2} + \left[a\left(\frac{1}{\gamma}+5\right)\frac{m}{\eta} - 2\right]\frac{dm}{d\eta} + a(1+\gamma)\left(\frac{dm}{d\eta}\right)^2 = 0, \tag{4.94}$$

or equivalently

$$\eta\frac{d^2m}{d\eta^2} - 2\frac{dm}{d\eta} - 2am\frac{d^2m}{d\eta^2} + \alpha\frac{m}{\eta}\frac{dm}{d\eta} + \beta\left(\frac{dm}{d\eta}\right)^2 = 0, \tag{4.95}$$

where, for simplicity, we have introduced the coefficients α and β defined as $\alpha = a(1/\gamma + 5)$ and $\beta = a(1+\gamma)$, respectively.

Equations (4.94) or (4.95) must be solved with the boundary conditions $m(0) = 0$, and $(dm/d\eta)\,|_{\eta=0} = 0$. Note that Eqs. (4.94) and (4.95) are not in the autonomous form, that is, the coefficients of the derivative $dm/d\eta$ depend both on the mass function $m(\eta)$ and the dimensionless radius η. In order to solve Eq. (4.95), we will use the method of power series expansions.

Therefore we can state the following:

Theorem. *The relativistic mass equation (4.95) describing the interior of a star with matter content described by a linear barotropic equation of state* $P(\eta) = \gamma\varepsilon(\eta)$, $\gamma =$ constant, *has an exact non-singular convergent power series solution of the form*

$$m(\eta) = \sum_{n=1}^{\infty} c_{2n+1}\eta^{2n+1}, \eta \leq 1, \tag{4.96}$$

with the coefficients c_{2n+1} *obtained from the recursive relationship*

$$c_{2n+1} = -\frac{a}{2(n-1)(2n+1)\gamma}\sum_{i=1}^{n-1}(2n-2i+1)\times$$
$$\times\left[2\gamma(\gamma+3)i - 4\gamma n + \gamma^2 + 6\gamma + 1\right]c_{2i+1}c_{2n-2i+1}, \qquad n \geq 2. \tag{4.97}$$

Proof. In the following, we will look for a convergent power series solution of Eq. (4.94), by choosing $m(\eta)$ in the form given by Eq. (4.96). Then it is easy to prove the relations

$$\frac{dm}{d\eta} = \sum_{n=1}^{\infty}(2n+1)c_{2n+1}\eta^{2n}, \tag{4.98}$$

and

$$\frac{d^2m}{d\eta^2} = 2\sum_{n=1}^{\infty} n(2n+1)c_{2n+1}\eta^{2n-1}, \tag{4.99}$$

respectively. For the product of two power series we use the Cauchy convolution, so that

$$\left(\sum_{i=0}^{\infty} a_i\eta^i\right)\left(\sum_{j=0}^{\infty} b_j\eta^j\right) = \sum_{i,j=0}^{\infty} a_i b_j \eta^{i+j} = \sum_{n=0}^{\infty}\left(\sum_{i=0}^{n} a_i b_{n-i}\right)\eta^n. \tag{4.100}$$

Thus,

$$\left(\frac{dm}{d\eta}\right)^2 = \sum_{i,j=1}^{\infty}(2i+1)(2j+1)c_{2i+1}c_{2j+1}\eta^{2i+2j}$$
$$= \sum_{n=1}^{\infty}\left[\sum_{i=1}^{n}(2i+1)(2n-2i+1)c_{2i+1}c_{2n-2i+1}\right]\eta^{2n}, \tag{4.101}$$

$$\frac{m}{\eta}\frac{dm}{d\eta} = \sum_{i,j=1}^{\infty} c_{2i+1}(2j+1)c_{2j+1}\eta^{2i+2j}$$
$$= \sum_{n=1}^{\infty}\left[\sum_{i=1}^{n}(2n-2i+1)c_{2i+1}c_{2n-2i+1}\right]\eta^{2n}, \tag{4.102}$$

$$m\frac{d^2m}{d\eta^2} = \sum_{i,j=1}^{\infty} c_{2i+1}2j\left(2j+1\right)c_{2j+1}\eta^{2i+2j}$$
$$= \sum_{n=1}^{\infty}\left[\sum_{i=1}^{n} 2\left(n-i\right)\left(2n-2i+1\right)c_{2i+1}c_{2n-2i+1}\right]\eta^{2n}. \quad (4.103)$$

Hence, substitution of these results into Eq. (4.95) gives immediately

$$\sum_{n=1}^{\infty}\left\{2\left(n-1\right)\left(2n+1\right)c_{2n+1}+\sum_{i=1}^{n}\left[-4a\left(n-i\right)\left(2n-2i+1\right)\times\right.\right.$$
$$c_{2i+1}c_{2n-2i+1}+\alpha\left(2n-2i+1\right)c_{2i+1}c_{2n-2i+1}$$
$$\left.\left.+\beta\left(2i+1\right)\left(2n-2i+1\right)c_{2i+1}c_{2n-2i+1}\right]\right\}\eta^{2n}=0, \quad (4.104)$$

where we have transformed all the products of the power series by using the Cauchy convolution. Solving Eq. (4.104) for the coefficients c_{2n+1} gives the recursive relationship Eq. (4.97) for the coefficients of the series representation of the mass function. This ends the proof of the Theorem.

For the values of the coefficients c_{2n+1} we obtain the following explicit expressions

$$c_5 = -\frac{3a(\gamma+1)(3\gamma+1)}{10\gamma}c_3^2, \quad (4.105)$$

$$c_7 = \frac{3a^2(\gamma+1)(3\gamma+1)\left(15\gamma^2+9\gamma+4\right)}{140\gamma^2}c_3^3, \quad (4.106)$$

$$c_9 = -\frac{a^3(\gamma+1)(3\gamma+1)\left(945\gamma^4+864\gamma^3+618\gamma^2+200\gamma+61\right)}{2520\gamma^3}c_3^4, \quad (4.107)$$

$$c_{11} = \frac{a^4(\gamma+1)(3\gamma+1)}{184800\gamma^4}\left(85050\gamma^6+91665\gamma^5+80892\gamma^4\right.$$
$$\left.+38832\gamma^3+17936\gamma^2+4239\gamma+1258\right)c_3^5, \quad (4.108)$$

$$c_{13} = -\frac{a^5(\gamma+1)(3\gamma+1)}{12012000\gamma^5}\left(7016625\gamma^8+8057475\gamma^7+7978905\gamma^6+4456683\gamma^5\right.$$
$$\left.+2486451\gamma^4+839697\gamma^3+346075\gamma^2+61953\gamma+22952\right)c_3^6, \quad (4.109)$$

$$c_{15} = \frac{a^6(\gamma+1)(3\gamma+1)}{5045040000\gamma^6}\left(3831077250\gamma^{10}+4428596025\gamma^9+4702427055\gamma^8\right.$$
$$+2757559491\gamma^7+1705375683\gamma^6+636216069\gamma^5+311382965\gamma^4$$
$$\left.+72456873\gamma^3+36302375\gamma^2+3752022\gamma+2703152\right)c_3^7, \quad (4.110)$$

....

Using Eqs. (4.78) and (4.96), the energy density of the matter inside a general relativistic star described by a linear barotropic equation of state is obtained as

$$\varepsilon(\eta) = \sum_{n=1}^{\infty} (2n+1) c_{2n+1} \eta^{2n-2}. \tag{4.111}$$

By estimating the energy density at the center of the star $\eta = 0$ gives $\varepsilon(0) = 1 = 3c_3$, which fixes the value of the constant c_3 as $c_3 = 1/3$

The expression of the metric tensor component $e^{-\lambda}$ can be found as

$$e^{-\lambda(\eta)} = 1 - 2a\frac{m(\eta)}{\eta} = 1 - 2a\sum_{n=1}^{\infty} c_{2n+1}\eta^{2n}, \tag{4.112}$$

while from the conservation of the energy-momentum tensor we obtain

$$e^{\nu(\eta)} = A_0^2\, \varepsilon^{-\frac{2\gamma}{1+\gamma}}(\eta), \tag{4.113}$$

where A_0 is an arbitrary constant of integration. With the help of Eq. (4.111) we rewrite Eq. (4.113) in the form

$$e^{\nu(\eta)} = \left[\sum_{n=1}^{\infty} (2n+1) c_{2n+1}\eta^{2n-2}\right]^{-\frac{2\gamma}{1+\gamma}}, \qquad \gamma \neq -1, \tag{4.114}$$

where for simplicity we have removed the constant A_0 by the scaling process $dt \to A_0 dt$ in the line element (4.1). Thus the interior line element for a fluid sphere satisfying a linear barotropic equation of state takes the form

$$ds^2 = c^2\left[\sum_{n=1}^{\infty}(2n+1)c_{2n+1}\eta^{2n-2}\right]^{-\frac{2\gamma}{1+\gamma}} dt^2 - \frac{1}{1-2a\sum_{n=1}^{\infty} c_{2n+1}\eta^{2n}}dr^2 - r^2 d\Omega^2, \tag{4.115}$$

for $\gamma \neq -1$.

At the surface of the barotropic matter distribution $\varepsilon(1) = \varepsilon_S = \rho_S/\rho_c =$ constant, where $\rho_S = \rho(R)$ is the density of the barotropic fluid distribution on the boundary separating the two phases. Thus we obtain $e^{\nu(R)} = (\rho_S/\rho_c)^{-2\gamma/(1+\gamma)}$. Hence it follows that on the boundary $\eta = 1$ of the barotropic component the coefficients c_{2n+1} must satisfy the condition $\sum_{n=1}^{\infty}(2n+1)c_{2n+1} = \varepsilon_S$.

For the $e^{-\lambda}$ metric tensor component we obtain $e^{-\lambda} = 1 - 2GM_S/c^2R$.

The accuracy of these power series solutions was discussed in detail in [247], where from the comparisons between the series solution of the TOV and continuity equation, obtained via the solution of the relativistic mass equation, and the exact numerical solution, computed by numerically integrating the coupled system of Eqs. (4.78) and (4.93), it was found that the power series solution gives a very good approximation of the exact numerical solution. The approach based on the Cauchy convolution can be easily extended to other equations of state, like, for example, the polytropic equation of state [247], with equally good results.

4.2.3 The Buchdahl Limit in the Presence of a Cosmological Constant

An important result in relativistic astrophysics is the Buchdahl limit [94] that gives an absolute constraint of the maximally allowable mass M–radius R ratio for isotropic fluid spheres. This constraint was obtained with the use of the static and spherically symmetric gravitational field equations, and it is given by the relation

$$\frac{2M}{R} < \frac{8}{9}. \tag{4.116}$$

In the following we will use the natural system of units with $c = G = 1$. However, in order to obtain a more general description of stellar properties, we will investigate the maximum possible mass–radius ratio in the case of stable compact general relativistic objects by allowing for the presence of an effective cosmological constant Λ. In the spherically symmetric case the properties of compact objects can be described completely by the gravitational structure equations (the mass continuity and the TOV equation), which, in the presence of the cosmological constant, are given by [337]

$$\frac{dm}{dr} = 4\pi\rho r^2, \tag{4.117}$$

$$\frac{dp}{dr} = -\frac{(\rho+p)\left[m + 4\pi\left(p - \frac{2\Lambda}{3}\right)r^3\right]}{r^2\left(1 - \frac{2m}{r} - \frac{8\pi}{3}\Lambda r^2\right)}, \tag{4.118}$$

$$\frac{d\nu}{dr} = \frac{2\left[m + 4\pi\left(p - \frac{2\Lambda}{3}\right)r^3\right]}{r^2\left(1 - \frac{2m}{r} - \frac{8\pi}{3}\Lambda r^2\right)}, \tag{4.119}$$

where, as usual, $m(r)$ is the mass located within radius r. Equations (4.117)–(4.119) must be always considered together with an equation of state of the dense matter inside the star, $p = p(\rho)$, and together with the usual boundary conditions at the center and at the surface of the star, $p(R) = 0$, $p(0) = p_c$, and $\rho(0) = \rho_c$, respectively, where ρ_c and p_c are the central density and pressure.

With the use of Eqs. (4.117)–(4.119) one can easily check that the metric function $\zeta = e^{\nu/2} > 0$, $\forall r \in [0, R]$ satisfies the differential equation

$$\sqrt{1 - \frac{2m(r)}{r} - \frac{8\pi}{3}\Lambda r^2}\,\frac{1}{r}\frac{d}{dr}\left[\sqrt{1 - \frac{2m(r)}{r} - \frac{8\pi}{3}\Lambda r^2}\,\frac{1}{r}\frac{d\zeta}{dr}\right] = \frac{\zeta}{r}\frac{d}{dr}\frac{m(r)}{r^3}. \tag{4.120}$$

Now we introduce the fundamental assumption that the density ρ of the matter inside the star does not increase with increasing r. This implies that the mean density of the matter inside radius r, defined as $<\rho>= 3m(r)/4\pi r^3$, does not increase either. Therefore, we assume that inside a compact gravitating general relativistic high density stellar-type object, the mathematical condition

$$\frac{d}{dr}\frac{m(r)}{r^3} < 0, \tag{4.121}$$

holds for all r, and it is independent of the adopted equation of state of the dense matter.

By introducing a new independent variable ξ, defined as [94]

$$\xi = \int_0^r r' \left(1 - \frac{2m(r')}{r'} - \frac{8\pi}{3}\Lambda r'^2\right)^{-\frac{1}{2}} dr', \tag{4.122}$$

from Eq. (4.120) it follows that, for all stellar-type general relativistic matter distributions with negative density gradient, the following condition must be satisfied for all points inside the star,

$$\frac{d^2 e^{\frac{\nu(\xi)}{2}}}{d\xi^2} < 0, \qquad \forall r \in [0, R]\,. \tag{4.123}$$

With the use of the Mean Value Theorem we can conclude that the relation

$$\frac{de^{\frac{\nu(\xi)}{2}}}{d\xi} \leq \frac{e^{\frac{\nu(\xi)}{2}} - e^{\frac{\nu(0)}{2}}}{\xi}, \tag{4.124}$$

must hold for all ν and ξ. Taking into account that $e^{\frac{\nu(0)}{2}} > 0$, we find that

$$\frac{de^{\frac{\nu(\xi)}{2}}}{d\xi} \leq \frac{e^{\frac{\nu(\xi)}{2}}}{\xi}. \tag{4.125}$$

Reformulating the above relation in the initial variables we have

$$\frac{m(r) + 4\pi\left(p - \frac{2\Lambda}{3}\right)r^3}{r^3\sqrt{1 - \frac{2m}{r} - \frac{8\pi}{3}\Lambda r^2}} \leq \left[\int_0^r r' \left(1 - \frac{2m(r')}{r'} - \frac{8\pi}{3}\Lambda r'^2\right)^{-\frac{1}{2}} dr'\right]^{-1}. \tag{4.126}$$

For stable stellar-type compact objects the mean density m/r^3 must not increase outwards, and therefore

$$\frac{m(r')}{r'} \geq \frac{m(r)}{r}\left(\frac{r'}{r}\right)^2, \qquad \forall r' \leq r. \tag{4.127}$$

In the following, we will denote

$$\alpha(r) = 1 + \frac{4\pi}{3}\Lambda\frac{r^3}{m(r)}. \tag{4.128}$$

Moreover, we assume that in the presence of the cosmological constant λ the condition

$$\frac{\alpha(r')\,m(r')}{r'} \geq \frac{\alpha(r)\,m(r)}{r}\left(\frac{r'}{r}\right)^2, \tag{4.129}$$

or equivalently,

$$\left[1 + \frac{4\pi}{3}\Lambda\frac{r'^3}{m(r')}\right]\frac{m(r')}{r'} \geq \left[1 + \frac{4\pi}{3}\Lambda\frac{r^3}{m(r)}\right]\frac{m(r)}{r}\left(\frac{r'}{r}\right)^2, \tag{4.130}$$

is satisfied inside the compact self-gravitating high density star. In fact, the validity of Eq. (4.130) is independent of the cosmological constant Λ, and it holds for all compact matter distributions where the density is a decreasing function of the radial coordinate.

Therefore we can now estimate the right hand side of Eq. (4.126) as follows

$$\int_0^r \frac{r'}{\left[1-\frac{2m(r')}{r'}-\frac{8\pi}{3}\Lambda r'^2\right]^{\frac{1}{2}}}dr' = \int_0^r \frac{r'}{\left[1-\frac{2\alpha(r')m(r')}{r'}\right]^{\frac{1}{2}}}dr' \geq$$

$$\geq \int_0^r \frac{r'}{\left[1-\frac{2\alpha(r)m(r)}{r}\left(\frac{r'}{r}\right)^2\right]^{\frac{1}{2}}}dr' = \frac{r^3}{2\alpha(r)m(r)}\left[1-\left(1-\frac{2\alpha(r)m(r)}{r}\right)^{\frac{1}{2}}\right]. \tag{4.131}$$

By using Eq. (4.131), the inequality (4.126) takes the form

$$\frac{m(r)+4\pi\left(p-\frac{2\Lambda}{3}\right)r^3}{\sqrt{1-\frac{2m(r)}{r}-\frac{8\pi}{3}\Lambda r^2}} \leq \frac{2m(r)\left(1+\frac{4\pi}{3}\Lambda\frac{r^3}{m(r)}\right)}{1-\sqrt{1-\frac{2m(r)}{r}-\frac{8\pi}{3}\Lambda r^2}}. \tag{4.132}$$

The inequality (4.132) is valid for all points with the radial coordinate r inside the dense object. Its validity is independent on the sign of the cosmological constant Λ.

As a first application of (4.132) we consider the case $\Lambda = 0$. By evaluating (4.132) at the vacuum boundary $r = R$ of the star we obtain

$$\frac{1}{\sqrt{1-\frac{2M}{R}}} \leq 2\left[1-\left(1-\frac{2M}{R}\right)^{\frac{1}{2}}\right]^{-1}, \tag{4.133}$$

leading to the well-known Buchdahl bound [94]

$$\frac{2M}{R} \leq \frac{8}{9}. \tag{4.134}$$

For $\Lambda \neq 0$, the inequality (4.132) gives the following upper limit for the mass-radius ratio of high density astrophysical objects,

$$\frac{2M}{R} \leq \left(1-\frac{8\pi}{3}\Lambda R^2\right)\left[1-\frac{1}{9}\frac{\left(1-\frac{2\Lambda}{\bar{\rho}}\right)^2}{1-\frac{8\pi}{3}\Lambda R^2}\right], \tag{4.135}$$

where $\bar{\rho} = 3M/4\pi R^3$ is the mean density of the star.

The importance of the Buchdahl limit consists in the fact that it gives us a stability condition for high density general relativistic compact objects. No similar result exists in Newtonian gravity. In GR the maximum allowable mass is obtained for stars with constant density, and for which $dp/d\rho \to \infty$ at the center. If we introduce the supplementary restriction $dp/d\rho \leq 1$ the value of the limiting mass decreases to a value of around $3.2M_\odot$. This numerical value is usually adopted as the theoretical upper bound of neutron star masses.

The constraint $dp/d\rho \leq 1$ is referred to as the causality condition. It means that the violation of this condition could lead to the unphysical situation in which the velocity of compressional (sound) waves inside the star could exceed the speed of light, thus leading to the breakdown of the causality in the dense astrophysical matter.

4.3 Spherical Gravitational Collapse of Strange Quark Null Fluids

In the present section we present an exact non-static solution of the Einstein field equations corresponding to a collapsing charged null strange fluid in a Vaidya-type space-time, and investigate some of its singularity properties. The Vaidya geometry, permitting the incorporations of the effects of radiation, offers a more realistic background from the description of some astrophysical processes, as compared to static geometries, where all back reaction is ignored. In our presentation we follow the approach introduced and developed in [214, 221, 232].

4.3.1 Metric and Gravitational Field Equations

As a first step in our study we introduce ingoing Bondi coordinates (u, r, θ, φ) and the advanced Eddington time coordinate $u = t + r$, where $r \geq 0$ is the radial coordinate decreasing toward the future. Then, in these coordinates, the line element describing the radial collapse of a coherent stream of matter can be generally written as [232]

$$ds^2 = -\left[1 - \frac{2m(u,r)}{r}\right] du^2 + 2dudr + r^2\left(d\theta^2 + \sin^2\theta d\varphi^2\right), \tag{4.136}$$

where $m(u,r)$ denotes the mass function, which gives the gravitational mass within a given radius r.

In the following we will consider the matter content of our collapsing system as consisting of a mixture of a null-charged strange quark fluid and radiation. Quark matter, having equal numbers of u, d, and s quarks is usually neutral. However, rather extreme astrophysicals may lead to a charged strange quark matter configuration and, for the sake of generality, we shall also assume this possibility.

Hence for the matter energy-momentum tensor we adopt the expression [221]

$$T_{\mu\nu} = T_{\mu\nu}^{(n)} + T_{\mu\nu}^{(m)} + E_{\mu\nu}, \tag{4.137}$$

where

$$T_{\mu\nu}^{(n)} = \mu(u,r)\, l_\mu l_\nu, \tag{4.138}$$

is the component of the matter field that moves along the null hypersurfaces $u =$ constant. The term $T_{\mu\nu}^{(m)}$ is given by

$$T_{\mu\nu}^{(m)} = (\rho + p)(l_\mu n_\nu + l_\nu n_\mu) + p g_{\mu\nu}, \tag{4.139}$$

represents the energy-momentum tensor of the strange quark matter and

$$E_{\mu\nu} = \frac{1}{4\pi}\left(F_{\mu\alpha}F_\nu^\alpha - \frac{1}{4}g_{\mu\nu}F_{\alpha\beta}F^{\alpha\beta}\right), \tag{4.140}$$

is the electromagnetic contribution. The two null-vectors, l_μ and n_μ, are are given by

$$l_\mu = \delta_{(\mu)}^{(\mu)}, \tag{4.141}$$

$$n_\mu = \frac{1}{2}\left[1 - \frac{2m(u,r)}{r}\right]\delta_{(\mu)}^{(\mu)} - \delta_{(\mu)}^{(1)}, \tag{4.142}$$

respectively, so that $l_\alpha l^\alpha = n_\alpha n^\alpha = 0$ and $l_\alpha n^\alpha = -1$ (with $\delta_{(b)}^{(a)}$ the Kronecker symbol). The energy density and pressure in Eq. (4.139) have been obtained by diagonalizing the energy-momentum tensor. As for the electromagnetic tensor $F_{\mu\nu}$, it obeys the Maxwell equations [214]

$$\frac{\partial F_{\mu\nu}}{\partial x^\lambda} + \frac{\partial F_{\lambda\mu}}{\partial x^\nu} + \frac{\partial F_{\nu\lambda}}{\partial x^\mu} = 0, \tag{4.143}$$

$$\frac{1}{\sqrt{-g}}\frac{\partial}{\partial x^\mu}\left(\sqrt{-g}F^{\mu\nu}\right) = -4\pi j^\mu. \tag{4.144}$$

Without any loss of generality we can take the vector potential of the electromagnetic field as

$$A_\mu = \frac{q(u)}{r}\delta_{(\mu)}^{(\mu)}, \tag{4.145}$$

with $q(u)$ an arbitrary function of u. From the Maxwell Eqs. (4.143) and (4.144) it follows that the only nonvanishing components of $F_{\mu\nu}$ are

$$F_{ru} = -F_{ur} = \frac{q(u)}{r^2}. \tag{4.146}$$

Thus for the energy-momentum tensor of the electromagnetic field we find

$$E_\mu^\nu = \frac{q^2(u)}{r^4}\operatorname{diag}(-1, 1, -1, 1). \tag{4.147}$$

Then, with the use of the energy-momentum tensor (4.137) the gravitational field equations for the metric (4.136) take the form

$$\frac{1}{r^2}\frac{\partial m(u,r)}{\partial u} = 4\pi\mu(u,r), \tag{4.148}$$

$$\frac{2}{r^2}\frac{\partial m(u,r)}{\partial r} = 8\pi\rho(u,r) + \frac{q^2(u)}{r^4}, \tag{4.149}$$

$$-\frac{1}{r}\frac{\partial^2 m(u,r)}{\partial r^2} = 8\pi p(u,r) + \frac{q^2(u)}{r^4}. \tag{4.150}$$

As for the equation of state of the quark matter, we assume that it is given by the so-called MIT bag model equation of state [510]

$$p = \frac{1}{3}(\rho - 4B), \tag{4.151}$$

where B is the difference between the energy density of the perturbative and non-perturbative quantum chromodynamics (QCD) vacuum (the bag constant) and ρ, p are the energy density and thermodynamic pressure of the quark matter, respectively. Equation (4.151) is essentially the equation of state of a gas of massless particles, with corrections due to the QCD trace anomaly, and perturbative interactions [221]. The vacuum pressure B, which holds quark matter together, is a simple model for the long-range, confining interactions in QCD.

4.3.2 The General Solution of the Field Equations

With the use of the bag model equation of state of strange matter given by Eq. (4.151), the equation describing the dynamics of a mixture of null fluid, strange matter, and electric field can be obtained as

$$\frac{\partial^2 m(u,r)}{\partial r^2} = -\frac{2}{3r}\frac{\partial m(u,r)}{\partial r} - \frac{2}{3}\frac{q^2(u)}{r^3} + \frac{32\pi B}{3}r. \tag{4.152}$$

By means of the substitution

$$m(u,r) = m_0(u,r) + \frac{4\pi B}{3}r^3 - \frac{q^2(u)}{2r}, \tag{4.153}$$

we obtain for the new unknown function $m_0(u,r)$ the differential equation

$$\frac{\partial^2 m_0(u,r)}{\partial r^2} = -\frac{2}{3r}\frac{\partial m_0(u,r)}{\partial r}, \tag{4.154}$$

having the general solution

$$m_0(u,r) = C(u)\, r^{1/3} + F(u), \tag{4.155}$$

where $C(u)$ and $F(u)$ are two arbitrary integration functions. Hence the general solution of the gravitational field equations for collapsing strange quark matter in the Vaidya geometry (4.136) is given by [221]

$$m(u,r) = F(u) + C(u)\, r^{1/3} + \frac{4\pi B}{3}r^3 - \frac{q^2(u)}{2r}, \tag{4.156}$$

$$\mu(u,r) = \frac{1}{4\pi r^2}\left[\frac{dF(u)}{du} + \frac{dC(u)}{du}r^{1/3} - q(u)\frac{dq(u)}{du}\frac{1}{r}\right], \tag{4.157}$$

$$\rho(u,r) = \frac{1}{4\pi r^2}\left[\frac{1}{3}C(u)r^{-2/3} + \frac{q^2(u)}{2r^2} + 4\pi B r^2\right], \tag{4.158}$$

$$p(u,r) = \frac{1}{12\pi r^2}\left[\frac{1}{3}C(u)r^{-2/3} + \frac{q^2(u)}{2r^2} - 12\pi B r^2\right]. \tag{4.159}$$

The electromagnetic current can be obtained from the Maxwell Eq. (4.144), and is given by

$$j^{\mu}=\frac{1}{4\pi r^{2}}\frac{dq(u)}{du}l^{\mu}. \tag{4.160}$$

4.3.3 The Energy Conditions and the Singularity Properties

We discuss now the energy conditions for the obtained solution describing the collapse of a quark null fluid. The energy conditions are the weak, strong, and dominant energy conditions

$$\mu\geq 0,\qquad \rho\geq 0,\qquad p\geq 0,\qquad \rho\geq p\geq 0, \tag{4.161}$$

and they can be satisfied by appropriately choosing the arbitrary functions $F(u)$ and $C(u)$ that describe, from a physical point of view, the injection and initial distribution of mass, and of the function $q(u)$ that characterizes the distribution of the electric charge. The requirement $\mu\geq 0$ is equivalent to $dm/du\geq 0$ and leads to the condition

$$\frac{dF(u)}{du}+\frac{dC(u)}{du}r^{1/3}\geq q(u)\frac{dq(u)}{du}\frac{1}{r}, \tag{4.162}$$

which imposes a simultaneous constraint on all three functions $F(u), C(u)$, and $q(u)$.

For charged quark matter, and for small values of r, the right-hand side of Eq. (4.162) is very large, and therefore this energy condition may not hold. One possibility to satisfy Eq. (4.162) for all r is to assume that the function $q(u)$ behaves so that for $r\to 0$, we have $dq^{2}(u)/du\to 0$. This condition implies that the charge in the singular point $r=0$ is constant for all times. Alternatively, we may assume that in order to satisfy the energy conditions, at extremely small radii matter is continuously converted into strange quark matter. For neutral quark matter Eq. (4.162) is easily satisfied by choosing $dF(u)/du>0$ and $dC(u)/du>0$. To satisfy the condition $p\geq 0$ for large r, we must impose the constraint $C(u)\geq 36\pi Br^{7/2}\geq 0,\forall u$ on the function $C(u)$. With this choice the condition of the nonnegative energy density is also automatically satisfied. The MIT bag equation of state (4.151) leads to $\rho\geq p\geq 0$, and thus the dominant energy condition also holds.

The radii of the apparent horizon of the metric (4.136) are given by the solution of the equation $2m=r$. If $\lim_{u\to\infty}F(u)=F_0=$ constant, $\lim_{u\to\infty}C(u)=C_0=$ constant, and $\lim_{u\to\infty}q(u)=q_0=$ constant, then for the determination of the radii of the apparent horizons we obtain the algebraic equation

$$2F_0+2C_0r^{1/3}+\frac{8\pi B}{3}r^{3}-\frac{q_0^{2}}{r}=r, \tag{4.163}$$

which in general may have multiple solutions.

The nature of the singularities of the quark matter filled Vaidya-type space-time can be recognized from the behavior of the energy density and curvature scalars. Thus, for $R_{\alpha\beta}R^{\alpha\beta}$, we find

$$R_{\alpha\beta}R^{\alpha\beta} = \frac{8}{r^4}\left[\frac{1}{3}C(u)r^{-8/3} + \frac{q^2(u)}{r^4} + 4\pi B\right]^2, \tag{4.164}$$

which diverges for $r \to 0$.

4.3.4 The Formation of a Naked Singularity

The central shell-focusing singularity, that is, the one occurring at $r = 0$, is called *naked* if the radial null-geodesic equation admits one or more positive real root X_0. In the case of the simple Vaidya space-time, filled with radiation only, it has been shown that for a linear mass function $2m(u) = \lambda u$ the singularity at $r = 0$, $u = 0$ is naked for $\lambda \leq \frac{1}{8}$ (see [221] and references therein).

In order to simplify the mathematical formalism we restrict our investigations to some particular forms of the functions $F(u)$, $C(u)$, and $q(u)$. Thus for these functions we adopt the simple functional forms $F(u) = \alpha u/2$, $C(u) = \beta u^{2/3}$, and $q(u) = \gamma u$, respectively, with $\alpha > 0, \beta > 0$, and $\gamma \geq 0$ constants [221, 214, 232]. With these choices the equation of *the radially outgoing, future-directed null geodesic originating at the singularity* is

$$\frac{du}{dr} = \frac{2}{1 - \frac{\alpha u}{r} - \beta\left(\frac{u}{r}\right)^{2/3} - \gamma\left(\frac{u}{r}\right)^2 - \frac{8\pi B}{3}r^2}. \tag{4.165}$$

For the geodesic tangent to be uniquely defined and to exist at the singular point $r = 0$, $u = 0$ of Eq. (4.165), the following condition must hold [273],

$$\lim_{u,r\to 0}\frac{u}{r} = \lim_{u,r\to 0}\frac{du}{dr} = X_0. \tag{4.166}$$

When the limit exists, and X_0 is real and positive, there is *a future-directed, non-space-like geodesic originating from* $r = 0, u = 0$. In this case the singularity will be, at least, *locally naked.* For the null geodesic given by Eq. (4.165), the condition (4.166) gives the following algebraic equation,

$$\gamma X_0^3 + \alpha X_0^2 + \beta X_0^{5/3} - X_0 + 2 = 0. \tag{4.167}$$

With the help of the substitution $X_0 = y^3$, Eq. (4.167) becomes a ninth-order polynomial equation of the form $f(y) = \gamma y^9 + \alpha y^6 + \beta y^5 - y^3 + 2 = 0$. According to a theorem given by Poincaré [137], the number of positive roots of a polynomial equation equals the number of changes in sign in the sequence of the nonnegative coefficients of the polynomial $g(y) = (1+y)^k f(y)$. The Poincaré criterion indicates that Eq. (4.167) has two positive roots.

In order to obtain an estimate of the positive roots we shall consider the expression [137, 221]

$$g(y) = (1+y)^k f(y) = \sum_{\nu=0}^{k+9} b_k(\nu) y^\nu, \tag{4.168}$$

with $b_k(0) = \gamma > 0$. Let $\nu_k(1)$ denote the smallest integer for which $b_k(\nu_k(1)) \geq 0$ and $b_k(\nu_k(1)+1) < 0$. Then we obtain the numbers $\nu_k(1)$ and $\nu_k(2)$. These numbers satisfy the relations [221]

$$\frac{\nu_k(s)}{k-\nu_k(s)+1} \leq \xi(k,\nu,s) \leq \frac{\nu_k(s)+1}{k-\nu_k(s)}, \qquad s=1,2, \tag{4.169}$$

$$\lim_{k\to\infty} \frac{\nu_k(s)}{k-\nu_k(s)+1} = \lim_{k\to\infty} \xi(k,\nu,s) = y_s > 0, \qquad s=1,2, \tag{4.170}$$

$$f(y_s) = 0, \qquad s=1,2. \tag{4.171}$$

Therefore it is always possible to construct a convergent sequence to obtain the positive roots of Eq. (4.167). In the case of the neutral quark fluid, $q(u) \equiv 0$, Eq. (4.167) is reduced to a sixth-order algebraic equation, also having two positive roots. Hence, at least for our particular choice of the parameters, a naked singularity can be formed as the result of the collapse of strange quark matter. Depending on the initial distribution of density and velocity, and on the constitutive nature of the collapsing matter either a black hole, or a naked singularity is formed. The values of the parameters in the solution (4.156)–(4.159) determine which of these possibilities occurs. The solution describing the collapse of the quark matter is not asymptotically flat and this condition does not play any role in the formation of the naked singularity. Due to the presence of the bag constant B (playing the role of a cosmological constant, from a formal mathematical point of view) the mass function (4.156) gives a cosmological-type metric. The present discussion also illustrates the much richer interplay that can occur between particle physics and GR, when more involved quantum field theoretical models are included in the analysis of gravitational phenomena.

4.4 Steady Spherical Accretion in Arbitrary Spherically-Symmetric Space-times

In present day astrophysics it is generally accepted that most of the astrophysical objects, like stars, black holes, and even galaxies, grow substantially in mass via accretion processes. Around most of the active galactic nuclei (AGNs), black hole candidates, or even ordinary neutron stars, one can find large gas clouds surrounding the central astrophysical object, and forming an associated accretion disk, which can be investigated by astronomical observations. The accretion disks extend on a variety of length scales, ranging from a tenth of a parsec to a maximum of a few hundred parsecs. The gas clouds have a complex geometrical and physical structure, and they are usually supposed to form a torus (or warped disk), which is geometrically (and also optically) thick. The thick torus absorbs most of the soft X-rays and ultraviolet radiation. However, essential astrophysical information can be gained from the observation of the relativistic motion and flow of the gas clouds and streams in the curved geometry of the gravitational field of compact astrophysical objects. The history of the theoretical study of the accretion of an ideal fluid onto a compact astrophysical object in the framework of Newtonian gravity began with Bondi's classic paper [82]. In this section we will follow the approach to accretion introduced in [216].

4.4.1 The Conservation Equations

Let us consider the spherically symmetric and stationary (time-independent) accretion of an ideal fluid in an arbitrary static and spherically symmetric space-time, with the metric given by Eq. (4.1). In the following, as usual, r denotes the radial coordinate, while θ and φ are the ordinary angular spherical coordinates, respectively. The metric tensor components are assumed to be functions of the radial distance only, so that $\nu = \nu(r)$ and $\lambda = \lambda(r)$. For the energy-momentum tensor of the accreting fluid we adopt the ideal fluid expression, so that its components are given by

$$T_{\mu\nu} = \left(\mu c^2 + p\right) U_\mu U_\nu + p g_{\mu\nu}, \tag{4.172}$$

where μ is the total energy density, p is the thermodynamic pressure, and $U^\mu = dx^\mu/ds$ is the four-velocity. Generally, the total energy density of the fluid can be represented as $\mu c^2 = \rho c^2 + \varepsilon$, where ρ is the energy density and ε is the thermal energy. The pressure p is assumed to be an arbitrary function of the density, satisfying the general barotropic equation of state $p = p(\rho)$. In the case of a radial flow, the components of the four-velocity are $U^0 = cdt/ds$, $U_0 = -e^\nu U^0$, and $U^1 = U = dr/ds$. The components of the four-velocity are normalized so that $U_\mu U^\mu = -1$ or $U_0 U^0 + U_1 U^1 = -e^{-\nu} U_0^2 + e^\lambda U^2 = -1$.

The basic equations of motion of the fluid are the conservation of the mass flux $J^\mu = \rho U^\mu$, given by $\nabla_\mu J^\mu = 0$, and the conservation of the energy flux $\nabla_\mu T_0^\mu = 0$ [351]. For a steady spherically symmetric flow, the conservation equations have the form

$$\frac{d}{dr}\left(\rho c^2 U e^{\frac{\nu+\lambda}{2}} r^2\right) = 0, \tag{4.173}$$

and

$$\frac{d}{dr}\left[\left(\mu c^2 + p\right) U_0 u e^{\frac{\nu+\lambda}{2}} r^2\right] = 0, \tag{4.174}$$

respectively, giving

$$\rho c^2 U e^{\frac{\nu+\lambda}{2}} r^2 = C_1, \tag{4.175}$$

and

$$(\mu c^2 + p)\, U_0 U e^{\frac{\nu+\lambda}{2}} r^2 = C_2, \tag{4.176}$$

where C_1 and C_2 are constants of integration. By dividing Eq. (4.176) by Eq. (4.175) and squaring yields

$$\left(\frac{\mu c^2 + p}{\rho c^2}\right)^2 \left(e^{\nu} + e^{\nu+\lambda} U^2\right) = C_3, \tag{4.177}$$

where $C_3 = (C_2/C_1)^2$. By differentiating Eqs. (4.175) and (4.176), we obtain

$$\left(r\frac{\nu' + \lambda'}{2} + 2\right)\left\{2V^2 - \frac{r\left[\nu' e^{\nu} + (\nu' + \lambda')\, e^{\nu+\lambda} u^2\right]}{\left(r\frac{\nu'+\lambda'}{2} + 2\right)\left(e^{\nu} + e^{\nu+\lambda} U^2\right)}\right\}\frac{dr}{r}$$
$$+2\left[V^2 - \frac{e^{\nu+\lambda} u^2}{e^{\nu} + e^{\nu+\lambda} U^2}\right]\frac{dU}{U} = 0, \tag{4.178}$$

where we have denoted

$$V^2 = \frac{d\ln(\mu c^2 + p)}{d\ln \rho c^2} - 1. \tag{4.179}$$

In the case of accretion of a fluid by a compact object in the Schwarzschild metric, we have $e^{\nu} = e^{-\lambda} = 1 - 2m/r$, where $m = GM/c^2$ is the total mass of the accreting object. In this case, Eq. (4.178) reduces to the basic equation for steady spherically symmetric accretion onto compact objects, first derived in [351].

If in Eq. (4.178) one or the other of the bracketed factors vanish, we have a so-called turnaround (critical) point. This means that the corresponding solutions are doublevalued in either r or U. Only the solutions that pass through a turnaround point could describe astrophysical matter that falls into (or flows out of) the compact object, with a monotonically increasing velocity along the trajectory of the particles. The critical point of the motion (also called *the sonic point*) is located geometrically and physically where all bracketed factors in Eq. (4.178) vanish [351]. Thus, in an arbitrary space-time, the conditions for the existence of a critical point can be formulated as

$$2V^2\left(r\frac{\nu' + \lambda'}{2} + 2\right) - \frac{r\left[\nu' e^{\nu} + (\nu' + \lambda')\, e^{\nu+\lambda} U^2\right]}{\left(e^{\nu} + e^{\nu+\lambda} U^2\right)} = 0, \tag{4.180}$$

and

$$V^2 - \frac{e^{\nu+\lambda} U^2}{e^{\nu} + e^{\nu+\lambda} U^2} = 0, \tag{4.181}$$

respectively.

For the important case of the Schwarzschild metric, we obtain $U_c^2 = m/2r_c$ and $V_c^2 = U_c^2/(1-3u_c^2)$, respectively [351]. Substituting Eq. (4.181) into Eq. (4.180) gives the expression of the velocity at the sonic point as

$$U_c^2 = \left.\frac{r\nu' e^{-\lambda}}{4}\right|_{r=r_c}. \tag{4.182}$$

Then, Eq. (4.181) provides

$$V_c^2 = \frac{e^{\lambda(r_c)}U_c^2}{1+e^{\lambda(r_c)}U_c^2}. \tag{4.183}$$

It is interesting to note that the tangential velocity of a massive test particle in a stable circular orbit in the space-time with the metric given by Eq. (4.1) is given by $v_{tg}^2/c^2 = r\nu'/2$ [339]. Therefore, we obtain the following relations for the velocities at the sonic point and the tangential velocity of the particles:

$$U_c^2 = \left.\frac{v_{tg}^2 e^{-\lambda}}{2c^2}\right|_{r=r_c}, \tag{4.184}$$

and

$$V_c^2 = \frac{v_{tg}^2(r_c)/2c^2}{1+v_{tg}^2(r_c)/2c^2}, \tag{4.185}$$

respectively.

4.4.2 Accreting Fluids Satisfying a Polytropic Equation of State

In order to study the accretion processes onto compact objects we need to first specify the equation of state $p = p(\rho)$ of the inflowing matter, which we assume to be in the form of a gas, and the metric of the space-time. As for the equation of state of the incoming gas, we adopt the polytropic equation of state, so that

$$p = K\rho^{\Gamma}, \tag{4.186}$$

where K and Γ are constants [351]. The temperature T of the gas can be obtained from the ideal gas equation of state $p = \rho k_B T/\tilde{\mu} m_p$, where k_B is Boltzmann's constant, m_p is the mass of the proton, and $\tilde{\mu}$ is the mean molecular weight, respectively. As a function of temperature, the thermodynamic pressure and the density of the gas are given by

$$\rho = \left(\frac{c^2}{K}\right)^n T_p^n, \qquad p = K\left(\frac{c^2}{K}\right)^{n+1} T_p^{n+1}, \tag{4.187}$$

where

$$n = \frac{1}{\Gamma-1}, \qquad T_p = \frac{k_B T}{\tilde{\mu} m_p c^2} = \frac{T}{\tilde{\mu}\times 1.09\times 10^{13}\mathrm{K}}. \tag{4.188}$$

For the polytropic equation of state, we have [351]

$$p + \epsilon = (n + 1)p. \tag{4.189}$$

With the use of the equation of state of the gas, Eq. (4.179) can be written as

$$V^2 = \frac{(n + 1)T_p}{n\left[1 + (n + 1)T_p\right]}. \tag{4.190}$$

By estimating this equation at the sonic point, where the gas temperature is T_{pc}, and comparing with Eq. (4.183) yields

$$T_{pc} = \frac{n}{n + 1} \frac{e^{\lambda(r_c)} U_c^2}{1 + (1 - n)e^{\lambda(r_c)} U_c^2}. \tag{4.191}$$

The equations of motion of the inflowing gas with a polytropic equation of state are given by

$$c^2 \left(\frac{c^2}{K}\right)^n T_p^n U r^2 e^{(\nu+\lambda)/2} = B_1, \tag{4.192}$$

and

$$\left[1 + (n + 1)\, T_p\right]^2 \left(e^\nu + e^{\nu+\lambda} U^2\right) = B_3, \tag{4.193}$$

respectively, where B_1 and B_3 are constants of integration. Evaluating Eqs. (4.192) and (4.193) at $r = r_\infty$, where $e^{\nu(r_\infty)} = e^{\nu_\infty}$, $e^{\lambda(r_\infty)} = e^{\lambda_\infty}$, and the velocity and temperature are $U(r_\infty) = U_\infty$ and $T_p(r_\infty) = T_{p\infty}$, respectively, gives

$$B_1 = c^2 \left(\frac{c^2}{K}\right)^n T_{p\infty}^n U_\infty r_\infty^2 e^{(\nu_\infty+\lambda_\infty)/2}, \tag{4.194}$$

and

$$B_3 = \left[1 + (n + 1)\, T_{p\infty}\right]^2 \left(e^{\nu_\infty} + e^{\nu_\infty+\lambda_\infty} U_\infty^2\right), \tag{4.195}$$

respectively.

Therefore, the two equations in the unknowns T_p and U describing the motion of the polytropic gas in an arbitrary static and spherically symmetric metric are

$$T_p^n U r^2 e^{(\nu+\lambda)/2} = T_{p\infty}^n U_\infty r_\infty^2 e^{(\nu_\infty+\lambda_\infty)/2}, \tag{4.196}$$

and

$$\left[1 + (n + 1)\, T_p\right]^2 \left(e^\nu + e^{\nu+\lambda} U^2\right) = \left[1 + (n + 1)\, T_{p\infty}\right]^2 \left(1 + e^{\nu_\infty+\lambda_\infty} U_\infty^2\right). \tag{4.197}$$

Evaluating Eq. (4.196) at the sonic point $r = r_c$ gives an algebraic equation for the determination of r_c,

$$T_{pc}^n r_c^{5/2} e^{\nu(r_c)} \sqrt{\nu'}\,\big|_{r=r_c} = 2T_{p\infty}^n U_\infty r_\infty^2 e^{(\nu_\infty+\lambda_\infty)/2}. \tag{4.198}$$

By assuming that $T_p \ll 1$ and $e^{\nu_\infty+\lambda_\infty}u_\infty^2 \ll 1$, Eq. (4.197) immediately provides

$$U(r) \approx (1-e^\nu)^{1/2} e^{-(\nu+\lambda)/2}. \tag{4.199}$$

By substituting this expression of the velocity in Eq. (4.196), we obtain the temperature and the density profiles of the cold accreting gas as

$$T_p(r) \approx T_{p\infty} \frac{\left[U_\infty r_\infty^2 e^{(\nu_\infty+\lambda_\infty)/2}\right]^{1/n}}{r^{2/n}(1-e^\nu)^{1/2n}}, \tag{4.200}$$

and

$$\rho(r) \approx \left(\frac{c^2}{K}\right) \frac{U_\infty r_\infty^2 e^{(\nu_\infty+\lambda_\infty)/2}}{r^2(1-e^\nu)^{1/2}}, \tag{4.201}$$

respectively. By considering the accretion onto a black hole, at its event horizon we have $r = r_h$. Therefore, we obtain the velocity $U_h = U(r_h)$, the temperature $T_{ph} = T_p(r_h)$, and the density $\rho_h = \rho(r_h)$ of the gas at the event horizon of the black hole as

$$U_h \approx \left[1-e^{\nu(r_h)}\right]^{1/2} e^{-[\nu(r_h)+\lambda(r_h)]/2}, \tag{4.202}$$

$$T_{ph} \approx T_{p\infty} \frac{\left[U_\infty r_\infty^2 e^{(\nu_\infty+\lambda_\infty)/2}\right]^{1/n}}{r_h^{2/n}\left[1-e^{\nu(r_h)}\right]^{1/2n}}, \tag{4.203}$$

and

$$\rho_h \approx \left(\frac{c^2}{K}\right) \frac{U_\infty r_\infty^2 e^{(\nu_\infty+\lambda_\infty)/2}}{r_h^2\left[1-e^{\nu(r_h)}\right]^{1/2}}, \tag{4.204}$$

respectively.

If $\exp[\nu(r_h)+\lambda(r_h)] = 1$, as is the case for the standard Schwarzschild solution of GR, and because at the event horizon $\exp[\nu(r_h)] = 0$, the gas particles reach the "surface" (event horizon) of the black hole with a four-velocity equal to the speed of light, we have $U_h \approx 1$. However, for black holes for which the event horizon is located so that $\exp[\nu(r_h)+\lambda(r_h)] \neq 1$, $\exp[-\lambda(r_h)] = 0$, and $\exp[\nu(r_h)] \neq 0$, we have $U_h \approx 0$.

An important physical quantity in the description of the accretion is the speed of sound $a^2 = \partial p/\partial\rho = \Gamma p/\rho = c^2\Gamma T_p$. Hence, the speed of sound at infinity a_∞ can be related to the temperature at infinity by the simple relation $T_{p\infty} = a_\infty^2/\Gamma c^2$. At the sonic point, the speed of sound is

$$a_c = c\sqrt{\frac{e^{\lambda(r_c)}U_c^2}{1+(1-n)e^{\lambda(r_c)}U_c^2}} \leq c. \tag{4.205}$$

Sometimes, an alternative form of the equation of state of the inflowing gas is used in astrophysical applications, by assuming that the pressure p of the gas is related to the baryon number density n_B by the simple polytropic relation $p = K n_B^{\Gamma}$. On the other hand n_B and ρ can be related by

$$n_B = \exp\left[\int \frac{c^2 d\rho}{\rho c^2 + K\rho^{\Gamma}}\right], \tag{4.206}$$

which can be easily integrated to give

$$n_B = \rho\left(\frac{c^2}{K}\right)^{-n}\left(1 + a^2/\Gamma c^2\right)^{-n}. \tag{4.207}$$

Therefore, in the present model, the baryon number density is given as a function of the radial distance r by

$$n_B \approx \left(\frac{c^2}{K}\right)^{1-n} \frac{U_\infty r_\infty^2 e^{(\nu_\infty + \lambda_\infty)/2}}{r^2\left(1 - e^{\nu}\right)^{1/2}\left(1 + a^2/\Gamma c^2\right)^{n}}. \tag{4.208}$$

4.5 Electromagnetic Emission Properties of Thin Accretion Disks

As mentioned in the previous section, most astrophysical objects grow via mass accretion, which is the principal factor determining the mass increase of stars and black holes. It is expected that, due to the presence of interstellar matter, accretion disks generally form around compact objects. Observationally the presence of the accretion disks is well documented. They can be described as flattened cosmic structures, made of rapidly rotating hot matter that slowly spirals onto a central very dense and massive object. The internal stresses and the dynamical friction in the hot gas represent a source of heat, produced from the gravitational energy of the motion of the gas. A small fraction of the heat generated in the accretion disk is converted into electromagnetic radiation, which is partially emitted from the disk surface, so that the accretion disk cools down due to electromagnetic radiation emission. Thus, important information about the physical and astrophysical properties of the accretion disks comes from the study of the radiation emitted by them. The radiation allows astronomers to analyze the electromagnetic spectrum of the accretion disks in the radio, optical or X-ray frequency bands, as well as its time variability. Hence, astronomical observations can provide essential data about the physical processes taking place in the accretion disks.

The electromagnetic radiation emission from the surface of the disk leads to its cooling, and thus it represents an efficient physical process that averts its extreme heating. Moreover, the thermodynamic equilibrium established through radiative cooling allows the stabilization of the thin vertical size of the disk. The inner edge of the thin disk is usually located at the marginally stable orbit of the gravitational potential of the massive central object. Therefore it follows that the hot gas has a Keplerian motion in higher orbits [404, 482]. On the other hand, the electromagnetic radiation emission properties from the thin disks are determined by the external gravitational potentials, which, if the massive central object is star-like, are directly determined by the equation of state of the dense neutron or quark matter. Therefore the equation of state of the dense matter inside compact general relativistic objects could be directly tested, at least in principle, by astrophysical observations of the emission spectra from accretion disks.

The electromagnetic emissivity features of the accretion disks have been used to investigate theoretically large classes of compact astrophysical objects, including naked singularities [298], gravastars [226], and wormholes [225, 228], and to obtain distinctive astrophysical signatures that can be used to discriminate between these classes of objects. Different modified gravity theories, such as $f(R)$ gravity, brane world models, Chern–Simons, models and the Horava–Lifshitz theory generate a number of specific electromagnetic disk signatures, which were carefully investigated in [227, 229, 231, 425, 426]. On the other hand the properties of accretion disks around rotating and nonrotating fermion, neutron, boson, or quark stars have been studied in [295, 296, 484, 515].

In order to study the accretion disk properties around compact objects of arbitrary mass, we first describe the general theoretical formalism that permits us to investigate in the general relativistic framework of stationary axisymmetric space-times the electromagnetic radiation emissivity properties of thin accretion disks. We follow closely in our presentation the approach introduced in [297]. We begin our study by considering the basic kinematic properties of massive test particles traveling in stationary and axially symmetric geometries in stable circular orbits. We present in detail the general theoretical formalism necessary for deriving the constants of the motion in the gravitational field of the massive object (the energy and the angular momentum), the effective gravitational potential, and the radii of the marginally stable orbits, respectively. Then we introduce the basic equations describing the electromagnetic radiation emission from accretion disks, and we present the general expressions for the flux, temperature distribution, and luminosity of the disk.

4.5.1 The Motion of Massive Particles in Stationary and Axially Symmetric Space-times

In the following, we analyze first the physical and geometrical properties of the motion of massive test particles, forming a thin accretion disk, and evolving in circular stable orbits around general relativistic compact objects. We assume that the exterior geometry created by the central dense object is stationary and axially symmetric, and it is given, in full generality, by the following metric

$$ds^2 = g_{tt}\, dt^2 + 2g_{t\phi}\, dtd\phi + g_{rr}\, dr^2 + g_{\theta\theta}\, d\theta^2 + g_{\phi\phi}\, d\phi^2 . \qquad (4.209)$$

In our study we adopt the equatorial approximation, which requires $|\theta - \pi| \ll 1$. In this approximation all the components g_{tt}, $g_{t\phi}$, g_{rr}, $g_{\theta\theta}$, and $g_{\phi\phi}$ of the metric tensor are functions of the radial coordinate r only [404, 482]. As before, the square root of the determinant of the metric tensor is denoted in the following by $\sqrt{-g}$.

To obtain the electromagnetic emissivity properties of the thin accretion disk, we first obtain for particles moving in circular orbits around compact objects in the geometry given by Eq. (4.209) the radial dependence of the angular velocity Ω, of the specific energy $\widetilde{E}$, and of the specific angular momentum $\widetilde{L}$, respectively. The expressions of all these physical parameters, characterizing particle motion, can be derived from the geodesic equations, which for the considered stationary and axially symmetric metric take the following form [298, 425, 426]

$$\frac{dt}{ds} = \frac{\widetilde{E} g_{\phi\phi} + \widetilde{L} g_{t\phi}}{g_{t\phi}^2 - g_{tt} g_{\phi\phi}}, \tag{4.210}$$

$$\frac{d\phi}{ds} = -\frac{\widetilde{E} g_{t\phi} + \widetilde{L} g_{tt}}{g_{t\phi}^2 - g_{tt} g_{\phi\phi}}, \tag{4.211}$$

$$g_{rr}\left(\frac{dr}{ds}\right)^2 = -1 + \frac{\widetilde{E}^2 g_{\phi\phi} + 2\widetilde{E}\widetilde{L} g_{t\phi} + \widetilde{L}^2 g_{tt}}{g_{t\phi}^2 - g_{tt} g_{\phi\phi}}. \tag{4.212}$$

The effective potential term $V_{eff}(r)$ is defined as

$$V_{\text{eff}}(r) = -1 + \frac{\widetilde{E}^2 g_{\phi\phi} + 2\widetilde{E}\widetilde{L} g_{t\phi} + \widetilde{L}^2 g_{tt}}{g_{t\phi}^2 - g_{tt} g_{\phi\phi}}. \tag{4.213}$$

For massive test particles moving in the equatorial plane in stable circular orbits the effective potential $V_{\text{eff}}(r)$ must satisfy the following two important conditions,

$$V_{\text{eff}}(r) = 0, \tag{4.214}$$

and

$$V_{\text{eff},\, r}(r) = 0, \tag{4.215}$$

respectively, where we have denoted by a comma the ordinary derivative with respect to the radial coordinate r. For the case of a spinning central general relativistic compact object the specific energy, the specific angular momentum and the angular velocity of particles moving in circular orbits can be obtained from these conditions in the form [297]

$$\widetilde{E} = -\frac{g_{tt} + g_{t\phi}\Omega}{\sqrt{-g_{tt} - 2g_{t\phi}\Omega - g_{\phi\phi}\Omega^2}}, \tag{4.216}$$

$$\widetilde{L} = \frac{g_{t\phi} + g_{\phi\phi}\Omega}{\sqrt{-g_{tt} - 2g_{t\phi}\Omega - g_{\phi\phi}\Omega^2}}, \tag{4.217}$$

$$\Omega = \frac{d\phi}{dt} = \frac{-g_{t\phi,r} + \sqrt{(g_{t\phi,r})^2 - g_{tt,r} g_{\phi\phi,r}}}{g_{\phi\phi,r}}. \tag{4.218}$$

The supplementary condition

$$V_{\text{eff},\,rr}(r) = 0, \tag{4.219}$$

allows us to determine the marginally stable orbits around the massive central object from the following important equation

$$\widetilde{E}^2 g_{\phi\phi,rr} + 2\widetilde{E}\widetilde{L} g_{t\phi,rr} + \widetilde{L}^2 g_{tt,rr} - (g_{t\phi}^2 - g_{tt}g_{\phi\phi})_{,rr} = 0. \tag{4.220}$$

Once the metric coefficients g_{tt}, $g_{t\phi}$, and $g_{\phi\phi}$ are known explicitly, we obtain the radii of the marginally stable orbits of massive particles in circular motion around a rotating high density general relativistic compact object by substituting Eqs. (4.216)–(4.218) into Eq. (4.220), and solving numerically for the radial coordinate r the resulting equation [225, 226, 227, 228, 229, 231, 296, 297, 298].

4.5.2 Electromagnetic Emissivity Properties of Thin Accretion Disks

In the following, we concentrate on thin accretion disks only, that is, accretion disks having their vertical extents negligible small as compared to their horizontal stretching. This implies that the disk height H, defined as the maximum half thickness in the vertical direction of the disk, is much smaller than the characteristic horizontal radius R of the disk, $H \ll R$. A further assumption in modeling the disk properties is that it is in hydrodynamical equilibrium, and as such no macroscopic motions occur in it. Moreover, the possible effects of the vertical entropy gradient and of the pressure gradient in the disk are fully neglected.

The internal stresses and the dynamical friction generate a large amount of heat, which is emitted over the entire disk surface. This radiation emission mechanism prevents the heating up of the disk to extremely high temperatures. Moreover, this equilibrium allows the disk to stabilize its narrow vertical extent. The inner edge of the thin disk, representing an important astrophysical parameter, is positioned at the marginally stable orbit of the gravitational potential of the massive central general relativistic object. Therefore, in higher orbits the accreting matter around the central object has a Keplerian motion.

The mass accretion rate $\dot{M}_0$ is a constant in the steady state accretion disk model, and it does not change in time. By averaging all the physical parameters describing the dust matter orbiting around the massive object over a characteristic time scale Δt, over the azimuthal angle $\Delta\phi = 2\pi$, and over the height H of the disk, respectively, an effective physical description of the disk properties can be obtained [404].

The particles moving in Keplerian orbit around the dense compact object with the four-velocity u^μ have a rotational velocity $\Omega = d\phi/dt$, a specific energy $\widetilde{E}$, and a specific angular momentum $\widetilde{L}$, respectively. All these quantities depend in the steady state thin disk model only on the radii r of the particle trajectories.

The matter particles gravitating around the central object form a disk having an averaged surface density Σ, which is obtained as the average in the vertical direction of the rest mass density ρ_0 of the gas. In the following, we describe the matter in the disk by an anisotropic, fluid-type energy-momentum tensor $T^{\mu\nu}$. We measure in the averaged rest-frame the density ρ_0 of the disk, the energy flow vector q^μ, as well as the dissipative part $t^{\mu\nu}$ of the total energy-momentum tensor $T^{\mu\nu}$. The specific heat and heat transfer processes are neglected in this specific frame of reference. An important physical parameter characterizing the disk structure is the surface density of the disk, given by [404],

$$\Sigma(r) = \int_{-H}^{H} \langle \rho_0 \rangle dz, \tag{4.221}$$

which is obtained as the integral over the infinitesimal vertical direction dz of the rest mass density $\langle \rho_0 \rangle$, averaged over Δt and 2π. The torque W, defined as

$$W_\phi{}^r = \int_{-H}^{H} \langle t_\phi{}^r \rangle dz, \tag{4.222}$$

is another important disk parameter. It represents the average of the component $\langle t^r_\phi \rangle$ of the energy-momentum tensor over the specific time interval Δt, and 2π. The radiation flux $F(r)$ from the disk surface is given by the time and orbital average of the energy flux vector q^z, and it is defined as

$$F(r) = \langle q^z \rangle. \tag{4.223}$$

We represent the energy-momentum tensor of the matter in the disk in its standard form, according to the relation

$$T^{\mu\nu} = \rho_0 u^\mu u^\nu + 2u^{(\mu} q^{\nu)} + t^{\mu\nu}, \tag{4.224}$$

where the four-velocity u^μ of the disk matter satisfies the orthogonality conditions $u_\mu q^\mu = 0$, $u_\mu t^{\mu\nu} = 0$. The energy and angular momentum flux four-vectors are defined by

$$-E^\mu \equiv T^\mu_\nu \left(\frac{\partial}{\partial t}\right)^\nu, \tag{4.225}$$

and

$$J^\mu \equiv T^\mu_\nu \left(\frac{\partial}{\partial \phi}\right)^\nu, \tag{4.226}$$

respectively. It is important to note at this point that we can derive the structure equations of the thin disk by integrating the conservation laws of the angular momentum, of the energy, and of the rest mass of the matter forming the thin disk, respectively [404]. The important result according to which the time averaged mass accretion rate is independent of the disk radius

$$\dot{M}_0 \equiv -2\pi\sqrt{-g}\Sigma u^r = \text{constant}, \tag{4.227}$$

follows from the equation of the rest mass conservation

$$\nabla_\mu(\rho_0 u^\mu) = 0. \tag{4.228}$$

The conservation law of the energy,

$$\nabla_\mu E^\mu = 0, \tag{4.229}$$

can be rewritten in an integral form, which is more relevant from a physical point of view, as [297]

$$[\dot{M}_0 \widetilde{E} - 2\pi\sqrt{-g}\Omega W_\phi{}^r]_{,r} = 4\pi\sqrt{-g}F\widetilde{E}. \tag{4.230}$$

This equation shows that the energy carried away by the mass flow in the disk, $\dot{M}_0\widetilde{E}$, and the energy transmitted by the dynamical stresses, $2\pi\sqrt{-g}\Omega W_\phi{}^r$, are compensated exactly by the energy released away from the disk surface, $4\pi\sqrt{-g}F\widetilde{E}$. The angular momentum conservation law

$$\nabla_\mu J^\mu = 0, \tag{4.231}$$

implies the equilibrium condition between the three forms of the angular momentum transport

$$[\dot{M}_0 \widetilde{L} - 2\pi r W_\phi{}^r]_{,r} = 4\pi\sqrt{-g}F\widetilde{L}\ . \tag{4.232}$$

With the use of the universal energy-angular momentum relation for circular geodesic orbits

$$dE = \Omega dJ, \tag{4.233}$$

written in the form

$$\widetilde{E}_{,r} = \Omega \widetilde{L}_{,r}, \tag{4.234}$$

and after eliminating $W_\phi{}^r$ from Eqs. (4.230) and (4.232), the electromagnetic energy flux F of the radiation emitted by the disk surface is obtained as [404, 296]

$$F(r) = -\frac{\dot{M}_0}{4\pi\sqrt{-g}}\frac{\Omega_{,r}}{(\widetilde{E}-\Omega\widetilde{L})^2}\int_{r_{in}}^{r}(\widetilde{E}-\Omega\widetilde{L})\widetilde{L}_{,r}dr\ , \tag{4.235}$$

where by r_{in} we have denoted the inner edge of the disk. In our presentation we assume always that $r_{in} = r_{ms}$. It is important to note that the radiation flux depends on angular velocity, on the angular momentum, and on the specific energy of the gas in stable motion around the general relativistic central compact object.

In the steady-state thin disk model we are considering the gas forming the accretion disk is supposed to be in thermodynamical equilibrium. Hence, we may consider the electromagnetic radiation emitted by the disk surface as a perfect black body radiation, with properties depending on the disk temperature only, and with the energy flux represented as

$$F(r) = \sigma_{SB} T^4(r), \tag{4.236}$$

where σ_{SB} denotes the Stefan–Boltzmann constant. Thus, the temperature distribution on the disk surface can be obtained from Eq. (4.236), once the radiative flux is determined.

The luminosity $L(\nu)$ of the disk surface has a redshifted black body spectrum, which can be obtained as [484],

$$L(\nu) = 4\pi d^2 I(\nu) = \frac{8\pi h \cos\gamma}{c^2} \int_{r_{in}}^{r_f} \int_0^{2\pi} \frac{\nu_e^3 r d\phi dr}{\exp(h\nu_e / k_B T) - 1}, \tag{4.237}$$

where h is the Planck constant, k_B is the Boltzmann constant, ν_e is the frequency of the emitted electromagnetic radiation, and $I(\nu)$ is the Planck distribution function, while d is the distance from the observer to the source (disk), r_{in} and r_f denote the positions of the inner and outer edges of the accretion disk, and γ is the disk inclination angle, respectively. In the natural system of units, with $\hbar = c = k_B = 1$, for the disk luminosity $L(\nu)$ we obtain the expression

$$L(\nu) = 16\pi^2 \cos\gamma \int_{r_{in}}^{r_f} \int_0^{2\pi} \frac{\nu_e^3 r d\phi dr}{\exp(2\pi\nu_e / T) - 1}. \tag{4.238}$$

As for the upper limit of integration in Eq. (4.237), in the following we take it as infinity, that is, we assume $r_f \to \infty$. Furthermore, we suppose that at $r \to \infty$ the energy flux from the surface of the disk vanishes. This condition does not depend on the local geometry of the disk, as well as of the physical properties of the general relativistic central compact object. The radiation from the disk is emitted at a frequency given by the relation

$$\nu_e = \nu(1 + z), \tag{4.239}$$

where the redshift factor z can be obtained as [297]

$$1 + z = \frac{1 + \Omega r \sin\phi \sin\gamma}{\sqrt{-g_{tt} - 2\Omega g_{t\phi} - \Omega^2 g_{\phi\phi}}}. \tag{4.240}$$

In the above equation we have neglected the bending of the light. For small inclination angles this approximation works well, but for large inclination angles (edge-on disks), it is not particularly good.

The efficiency ϵ with which the central object converts the mass of the hot gas into electromagnetic radiation is another important astrophysical characteristic of the mass accretion processes. The parameter ϵ can be defined as the ratio of two rates, the first being the rate of the energy of the photons emitted from the surface of the disk, while the second is the rate at which mass-energy is transported to the high density compact general relativistic object [404]. The energies are both measured at infinity. By assuming that all photons reach the observer located at infinity, it turns out that the efficiency ϵ depends only on the specific energy of the particles in the disk evaluated at the marginally stable orbit r_{ms}

$$\epsilon = 1 - \widetilde{E}_{ms}. \tag{4.241}$$

The efficiency ϵ is about 6% for Schwarzschild black holes. This numerical value of ϵ is independent on whether we consider the photon capture by the black hole, or not. The efficiency is around 42% for rapidly rotating black holes, for which the capture of radiation is ignored. The efficiency is 40% for a Kerr black hole for which the photon capture is explicitly considered [482].

The emission spectra and fluxes from the accretion disks around high density compact general relativistic objects satisfy some scaling relations, which are simple, but play an important role in many applications. One can find such scaling relations by considering the scaling transformation $r \to \widetilde{r} = r/M$ of the radial coordinate, where M is the mass of the compact object. Generally the metric tensor components are invariant with respect to such scaling transformation, while the specific energy, the angular momentum and the angular velocity transform as

$$\widetilde{E} \to \widetilde{E}, \qquad \widetilde{L} \to M\widetilde{L}, \qquad \Omega \to \frac{\widetilde{\Omega}}{M}. \tag{4.242}$$

The electromagnetic flux scales as

$$F(r) \to \frac{F(\widetilde{r})}{M^4}, \tag{4.243}$$

giving for the temperature the simple scaling law

$$T(r) \to \frac{T(\widetilde{r})}{M}. \tag{4.244}$$

The transformation of the frequency of the emitted radiation under coordinate rescaling is

$$\nu \to \widetilde{\nu} = \frac{\nu}{M}. \tag{4.245}$$

The luminosity of the disk scales under this rescaling of the radiation frequency as [297]

$$L(\nu) \to \frac{L(\widetilde{\nu})}{M}. \tag{4.246}$$

Since the energy flux from the disk is proportional to the accretion rate $\dot{M}_0$, we find the important result that an increase in the mass accretion rate results in a linear increase of the electromagnetic radiation flux from the hot accretion disk.

The employment of the thermal electromagnetic radiation from thin accretion disks to investigate the nature of a compact general relativistic object is called the *continuum-fitting method*. It was introduced in [520] to study the observational significance of the spin of the black holes in X-ray binaries. To analyze the emissivity of accretion disks around black holes the standard thin accretion disk model introduced in [404] was applied. However, it is important to note that several important corrections to the simple formula given by Eq. (4.235) must be performed in order to describe the observed disk spectrum in the case of Kerr black holes. Since the X-rays are emitted from the hot inner disk, in this region the free-free absorption (the absorption of light when scattering a free ion, also called *bremsstrahlung*) may be dominated by electron scattering. Hence, as a direct consequence, the obtained color temperature could be greater than the effective temperature of the disk. In this case the inner disk radiates like a diluted black body. Near the event horizon one must take into account the general relativistic corrections (gravitational redshift and focusing), which can make the observed integrated flux and the color temperature to be different from the true local physical values. The continuum-fitting method was intensively used for the observational investigation of the spin of the black holes [38, 208].

As an application of the present formalism we will consider the electromagnetic emission properties from thin disks formed around compact neutron stars described by the Relativistic Mean Field equations of state with isovector scalar in the presence of the δ-meson field (for the details of this equation of state of high density nuclear matter see [300]). The δ-meson mean field vanishes in symmetric nuclear matter, but even so, the field can still have a significant influence on the properties of asymmetric nuclear matter in neutron stars. The Relativistic Mean Field contribution to the nuclear symmetry energy due to the δ-field is smaller than zero. At high densities the energy per particle of neutron matter is larger, as compared to the case when no δ-field is included. The pressure dependence on the density of the neutron matter for the case of the Relativistic Mean Field equation of state is presented in Fig. 4.1.

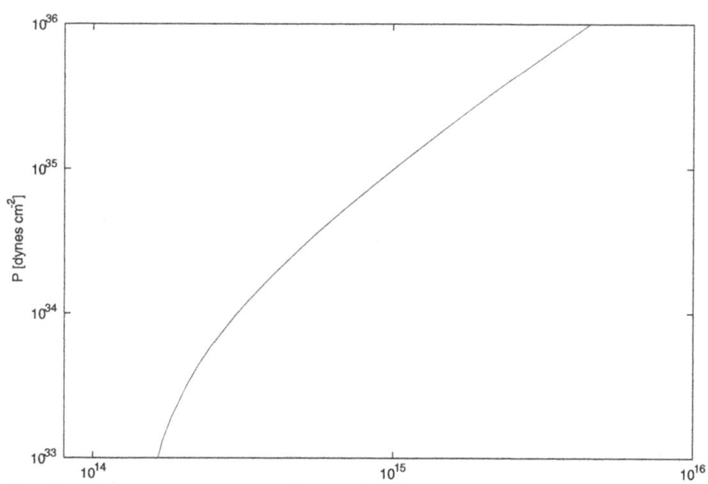

Figure 4.1 Variation of the pressure as a function of the density for the Relativistic Mean Field equation of state [300] of the dense neutron matter.

In the following we will present the physical properties of the electromagnetic radiation from the disk around Relativistic Mean Field stars in three cases. The first case corresponds to the case of stellar models with fixed mass and angular velocity. In our example we have chosen as the central object a star with mass $M = 1.8M_{\odot}$, which rotates at an angular frequency of $\Omega = 5 \times 10^3$ s^{-1}. The second class of models describe accretion disk properties around neutron stars with fixed central density, and fixed ratio of the polar and equatorial radii r_p and r_e of the star. For this model we have chosen for the central density of the star the value $\rho_c = 10^{15}$ g/cm^3, while $r_p/r_e = 0.85$. Finally, the third class of stellar models are represented by stars rotating at Keplerian frequencies. The variations of the electromagnetic fluxes from the thin disk formed around these three classes of rotating neutron star are plotted in Figs. 4.2–4.4, respectively. The electromagnetic luminosities of the disk for these three models are depicted in Figs. 4.5–4.7, respectively.

The study of the electromagnetic emission from the disks around compact objects represents an indirect method that could help distinguishing observationally between different classes of compact objects, and their equations of state. All the astrophysical quantities related to the observable properties of the accretion disks are dependent, and can be directly obtained from the metric of the central compact object. Due to the major differences in their exterior space-time geometry, different types of stars (neutron, quark, exotic, etc.) exhibit, at least on the theoretical level, a number of distinct signatures with respect to the electromagnetic emission properties of the disk. Hence, the observational

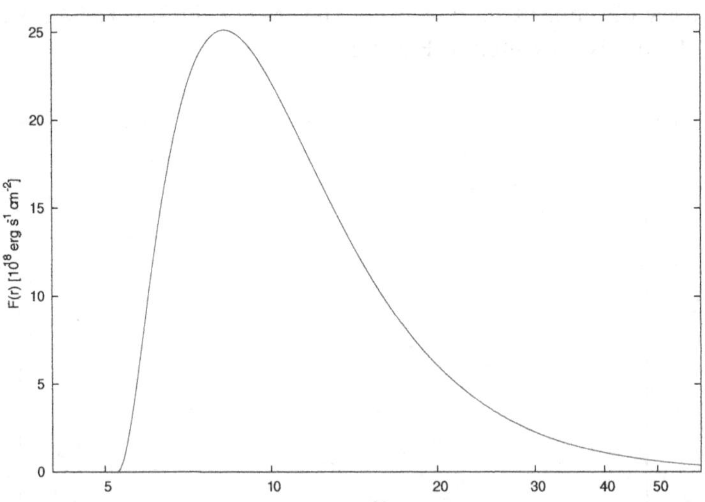

Figure 4.2 Electromagnetic flux from an accretion disk gravitating around a compact general relativistic object, described by the Relativistic Mean Field equation of state [300], having a mass $M = 1.8M_{\odot}$, and rotating at an angular velocity of $\Omega = 5 \times 10^3$ s^{-1}.

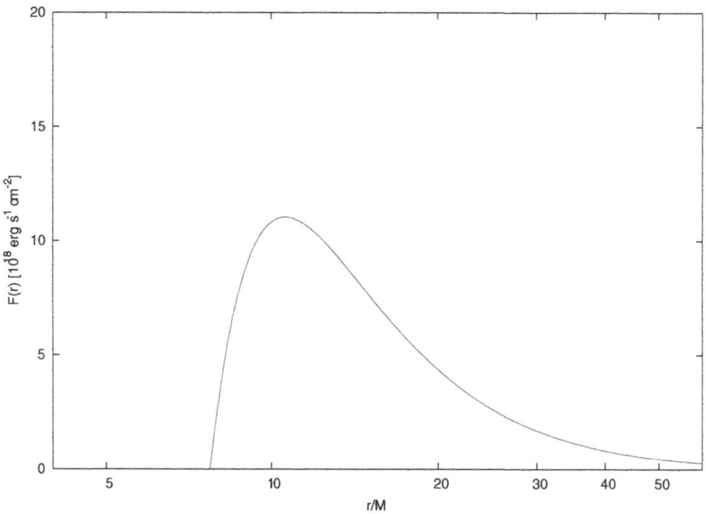

Figure 4.3 Electromagnetic flux from an accretion disk gravitating around a compact general relativistic object, described by the Relativistic Mean Field equation of state [300], with fixed central density $\rho_c = 10^15$ g/cm^3 and polar to the equatorial radius ratio $r_p/r_e = 0.85$.

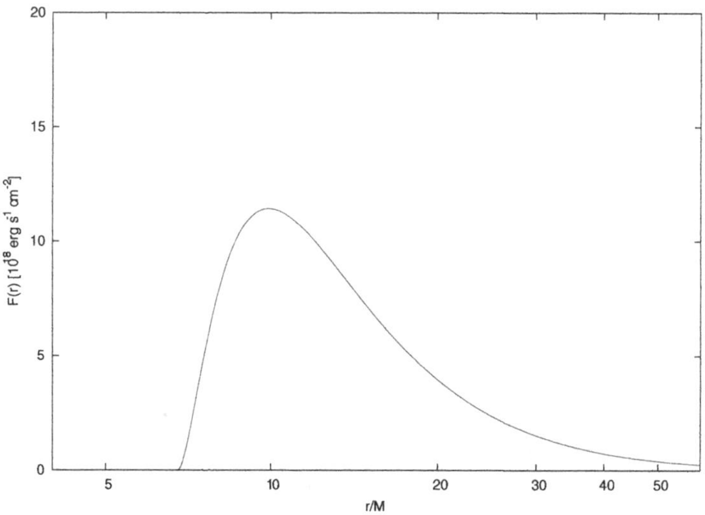

Figure 4.4 Electromagnetic flux from an accretion disk gravitating around a compact general relativistic object, described by the Relativistic Mean Field equation of state [300], and rotating at the Keplerian frequency.

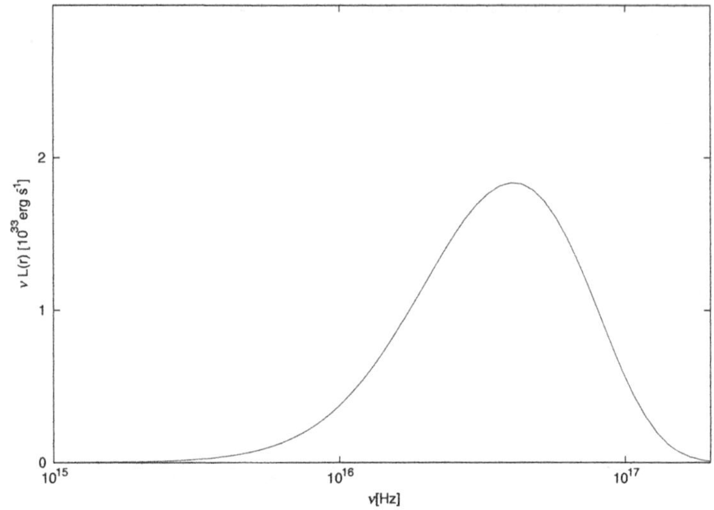

Figure 4.5 Luminosity of an accretion disk gravitating around a compact general relativistic object, described by the Relativistic Mean Field equation of state [300], having a mass $M = 1.8M_{\odot}$, and rotating at an angular velocity of $\Omega = 5 \times 10^3$ s^{-1}.

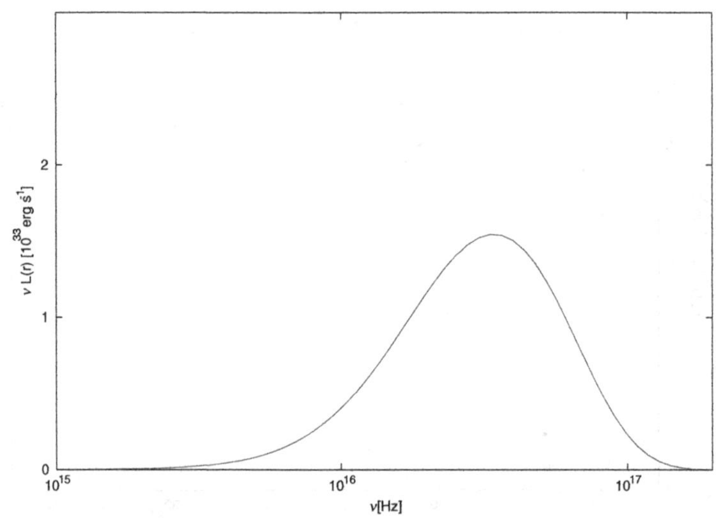

Figure 4.6 Luminosity of an accretion disk gravitating around a compact general relativistic object, described by the Relativistic Mean Field equation of state [300], with fixed central density $\rho_c = 10^15$ g/cm^3 and polar to the equatorial radius ratio $r_p/r_e = 0.85$.

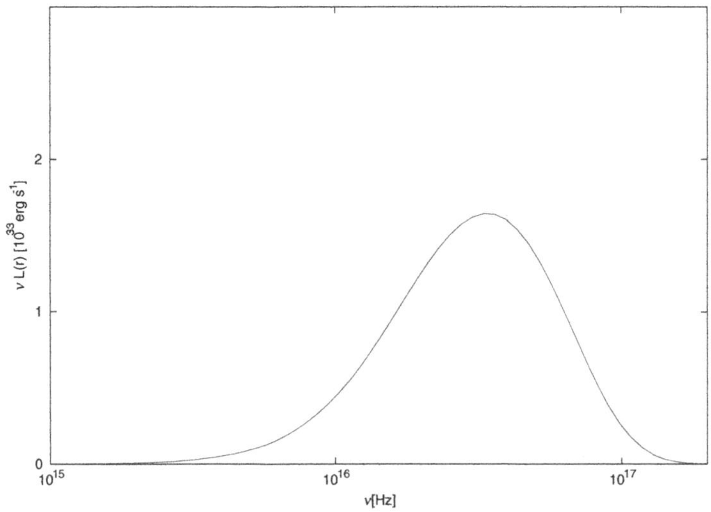

Figure 4.7 Luminosity of an accretion disk gravitating around a compact general relativistic object, described by the Relativistic Mean Field equation of state [300], and rotating at the Keplerian frequency.

analysis of the electromagnetic radiation from accretion disks, which is associated with some specific signatures induced by the presence of the strong gravitational field, may allow to discriminate between different types of compact astrophysical objects.

4.6 Summary and Discussion

In the present chapter, we have investigated some of the astrophysical applications of GR, ranging from the three fundamental solar system tests to the matter accretion by compact objects. General Relativity has a very large area of applications, and the behavior of compact objects, be they neutron stars, or black holes, can be understood only in the framework of a geometric theory of gravity. In (almost) all astrophysical tests performed up to now, GR has been fully confirmed as providing a high precision description of natural phenomena. This description is valid for both static systems, like compact objects, whose global parameters fit well into the formalism of GR, and for dynamical phenomena, like accretion processes, involving the flow of gas streams around massive stars. Moreover, a number of other specific gravitational signatures, like the gravitational redshift, and gravitational waves [2, 3, 4, 5, 6, 7], have not been discussed in this chapter. These have also been given strong experimental and observational support for the correctness of GR up to very high gravitational field intensities. However, the astrophysical phenomena analyzed in this chapter, even involving strong fields, are restricted to a small scale, of the order of the radius of compact objects/black holes, or of the solar system. But the gravitational force dominates not only in our planet or solar system, but extends far beyond in the realm of galaxies, and of the entire Universe. Can GR describe the dynamics of such very far away objects? This is the question we will try to answer in the next chapters.

5

Cosmology

In this chapter, we will review some of the basic results and models in present-day cosmology. We will start our investigation in the framework of Newtonian cosmology, whose results, interestingly enough, coincide, in some particular situations, with those of the best gravitational theory we have today, namely, General Relativity. We will then proceed to a brief presentation of general relativistic cosmology, and of its more significant results. Since the problem of the dark matter and of dark energy are the central questions of present-day cosmology, we will also discuss them.

5.1 Newtonian Cosmology

For a complete and deep understanding of the structure and evolution of the Universe one must resort to the basic theoretical tools provided by GR. However, to facilitate the understanding of the overall dynamics of the Universe, it is useful to develop first a more intuitive approach by analyzing the expansion of the Universe from a Newtonian perspective. In the framework of the inverse square law of gravity, Hubble's law is a natural result of a developing Universe that is both *isotropic and homogeneous, emerging and looking the same in all spatial directions and at all locations.* This key hypothesis and basic assumption of an isotropic and homogeneous Universe is called *the cosmological principle.* Although in the initial formulation of Hubble's law it was assumed that the value of the Hubble constant is the same for all observers (that is, it is a true constant), presently we know that it is actually a function of time, $H = H(t)$. If today's age of the Universe is t_0, then $H(t_0) = H_0$, meaning that H_0 is the present-day value of the Hubble function.

There are several approaches to Newtonian cosmology, and in the following we will present three such approaches, called the energy conservation approach, the hydrodynamic approach, and the Neo-Newtonian approach, respectively.

5.1.1 The Energy Conservation Approach

Let's consider a physical model in which the matter content of the expanding Universe consists of a single component, namely, a pressureless dust having *uniform density*, which we denote by $\rho(t)$. As for the origin of our system of coordinates we choose an arbitrary point P located somewhere in the Universe. Let us choose in this Universe model a thin spherical shell of mass m, and consider that in our coordinate system r be the radius of the shell. This shell expands together with the whole of the Universe, and since the number of particles is conserved, the shell will always contain a conserved number of matter particles, in the form of dust. Denote the expansionary velocity of the shell by v. The total mechanical energy of the thin shell is then $E = K + U$, where K denotes the kinetic energy, while U represents the potential energy. Due to the expansion of the shell in space, the gravitational attraction from the masses within radius r leads to a decrease of the kinetic energy, while, simultaneously, the potential energy of the shell must increase. However, due to the conservation of the energy, the total energy E of the expanding shell remains constant as the shell expands in the cosmic medium.

To simplify our mathematical formalism, we write the total energy E of the expanding shell by introducing two arbitrary constants, denoted k and R_0, respectively, as

$$E = -\frac{1}{2}mkc^2R_0^2. \tag{5.1}$$

The physical units of the constant k are $(\text{length})^{-2}$. The second constant R_0 designates a given shell of mass m. From a geometrical point of view it can be interpreted as the present-day radius of the shell, given by $R_0 = r\,(t_0)$. Then the conservation of the energy of this mass shell can be written as

$$\frac{1}{2}mv^2 - \frac{GmM_r}{r} = -\frac{1}{2}mkc^2R_0^2. \tag{5.2}$$

Here M_r is the mass interior to the shell

$$M_r = \frac{4}{3}\pi\rho(t)r^3. \tag{5.3}$$

Due to the expansion of the Universe, the radius r of the shell, as well as the matter density inside it, are rapidly changing in time. However, due to mass conservation, the quantity $\rho(t)r^3$ does not vary as the Universe expands. Substituting the expression of M_r in Eq. (5.2) gives

$$v^2 - \frac{8}{3}\pi G\rho r^2 = -kc^2R_0^2. \tag{5.4}$$

The constant k determines the overall large time dynamics and, as we will see later, the global geometry of the Universe. Thus:

a. if $k > 0$, it follows that the total energy of the shell is smaller than zero, $E < 0$. This Universe model is called *bounded*, or *closed*. For such a closed Universe its expansionary evolution will *stop* sometime in the future, and reverse itself.
b. if $k = 0$, it immediately follows that the total energy of the shell is zero, $E = 0$. The corresponding Universe is called *flat*, and it is *neither open nor closed*. In this cosmological model the cosmological expansion will continue to slow down, stopping in the far future for $t \to \infty$, when the matter density of the Universe is negligibly small, with $\rho \to 0$.
c. if $k < 0$, the total energy of the shell is positive, $E > 0$, and the Universe is *unbounded*. We call this Universe model as *open*. For an open Universe the cosmological expansion *will continue forever*, and it will never stop.

The cosmological principle, the fundamental concept in cosmology, imposes that for all shells the expansion takes place according to the same dynamics (in the same way). This implies that the radius $r(t)$ of a given shell (labeled by R_0) can be written at any arbitrary time as

$$r(t) = R(t)R_0. \tag{5.5}$$

In the above expression $r(t)$ is designated as the *coordinate distance*, while $R(t)$ is the scale factor, a dimensionless quantity that describes the time dynamics of the expansionary evolution. It is the same for all shells. Since R_0 designates a particular shell, and describes its evolution as it expands with the Universe, it is called the *comoving coordinate*. In the following we will adopt the convention that at $t = t_0$, $R(t_0) = 1$, and hence the present radius of the shell is $R_0 = r(t_0) = 1$. Moreover, since for a specific shell $\rho(t)r^3$ remains constant in time, it follows that the quantity $R^3(t)\rho(t)$ also remains constant for all shells. Hence

$$R^3(t)\rho(t) = R^3(t_0)\rho(t_0) = \rho_0, \tag{5.6}$$

where ρ_0 denotes the present-day density of the pressureless Universe.

The dynamics of the Newtonian Universe can now be characterized by the time evolution of the dimensionless scale factor $R(t)$. As a function of $R(t)$, the basic law of cosmology, the Hubble law, takes the equivalent form

$$v(t) = H(t)r = H(t)R(t)R_0, \tag{5.7}$$

where $H(t)$ denotes the Hubble function that describes the rate of change of the velocity of the cosmic expansion. By taking the time derivative of Eq. (5.5), we obtain $v = (dR(t)/dt)\,R_0$, and comparing this result with Eq. (5.7) we obtain the Hubble function as expressed with the help of the scale factor, and of its time derivative, as

$$H(t) = \frac{1}{R(t)}\frac{dR(t)}{dt}. \tag{5.8}$$

Substituting Eqs. (5.5) and (5.8) into (5.4) gives

$$3H^2 + \frac{3kc^2}{R^2} = 8\pi G\rho, \tag{5.9}$$

or, equivalently, in terms of the scale factor,

$$3\left(\frac{1}{R}\frac{dR}{dt}\right)^2 + \frac{3kc^2}{R^2} = 8\pi G\rho. \tag{5.10}$$

Equations (5.9) and (5.10) appear in the same form in general relativistic cosmology, where they are known as the first Friedmann equation. Using the relation (5.6), then Eq. (5.10) can be reformulated as

$$\left(\frac{dR}{dt}\right)^2 = \frac{8\pi G\rho_0}{3R} - kc^2. \tag{5.11}$$

Taking the time derivative of this equation we obtain

$$\frac{d^2R}{dt^2} = -\frac{4}{3}\pi G\rho R. \tag{5.12}$$

The above equation is a corollary of the important Birkhoff theorem. In 1923 the American mathematician George D. Birkhoff proved, in the framework of GR, that for a spherically symmetric matter distribution, Einstein's gravitational field equations have a unique solution. As an important corollary of the Birkhoff theorem it follows that the acceleration of a mass shell in a pressureless Universe is controlled only by the matter located inside the given shell. Therefore, as one can easily see, the acceleration of the Universe depends only on the scale factor R and the energy density ρ.

Now we can investigate in more detail the time evolution of the Universe. Estimating the first Friedmann equation (5.9) at the present time t_0 we can relate the present-day values of H_0, ρ_0, and k as

$$H_0^2 + kc^2 = \frac{8\pi G}{3}\rho_0. \tag{5.13}$$

For simplicity, in the following we will concentrate only to the specific case of $k = 0$. From a cosmological point of view this case corresponds to each shell moving with a velocity exactly equal to the escape velocity. We define the value of the density corresponding to $k = 0$ as the *critical density* ρ_c of the Universe. With the use of Eq. (5.9), we immediately obtain

$$\rho_c = \frac{3H^2}{8\pi G}. \tag{5.14}$$

The ratio of the average density of the Universe to its critical density is called the density parameter Ω. It is an important cosmological parameter, and is mathematically defined as

$$\Omega(t) = \frac{\rho(t)}{\rho_c(t)}. \tag{5.15}$$

Its present-day value is

$$\Omega_0 = \frac{\rho_0}{\rho_{c,0}}. \tag{5.16}$$

Generally, the density parameter can be written as

$$\Omega(t) = 1 + \frac{kc^2}{(dR/dt)^2}. \tag{5.17}$$

Expressing the density parameter with the help of Eqs. (5.13) and (5.14) we obtain

$$H_0^2 (\Omega_0 - 1) = kc^2. \tag{5.18}$$

If $\Omega_0 = 1$, then the Universe is flat, $k = 0$. The time evolution of the scale factor $R(t)$, describing the full expansionary dynamics of a flat Universe can be obtained easily by solving the simple differential equation Eq. (5.11) with $k = 0$,

$$\sqrt{R}\frac{dR}{dt} = \sqrt{\frac{8\pi G\rho_{c,0}}{3}}, \tag{5.19}$$

with the general solution given by

$$R(t) = (6\pi G\rho_{c,0})^{1/3} t^{2/3} = \left(\frac{3}{2}\right)^{2/3} \left(\frac{t}{t_H}\right)^{2/3}, \tag{5.20}$$

where we have defined the Hubble time as $t_H = 1/H_0$.

Finally, we consider again the energy conservation, which in the general case can be formulated as

$$dE + pdV = TdS, \tag{5.21}$$

where p is the pressure, V is the cosmological volume, and S is the entropy of the Universe. For a unit comoving radius $R_0 = 1$, the volume is $V = 4\pi R^3/3$, while the energy $E = mc^2$ contained within the volume V is

$$E = \frac{4}{3}\pi R^3 \rho c^2. \tag{5.22}$$

By taking the time derivative we obtain

$$\frac{dE}{dt} = 4\pi R^3 \rho c^2 \left(\frac{1}{R}\frac{dR}{dt} + \frac{1}{3}\frac{1}{\rho}\frac{d\rho}{dt}\right), \qquad \frac{dV}{dt} = 4\pi R^2 \frac{dR}{dt}. \tag{5.23}$$

By assuming that the evolution of the Universe is adiabatic, we have $dS = 0$, and then with the use of the results in Eq. (5.23) the energy conservation equation takes the form

$$\frac{d\rho}{dt} + 3H\left(\rho + \frac{p}{c^2}\right) = 0. \tag{5.24}$$

In general relativistic cosmology it can be shown that the energy conservation equation (5.24) follows directly from the gravitational field equations.

5.1.2 The Hydrodynamic Model

In the hydrodynamic approach of Newtonian cosmology [171, 294, 349, 353, 409] it is assumed that the motion of the galaxies can be described by the basic equations of the fluid dynamics, the continuity equation, and the Euler equations, respectively. Hence, to study the evolution of the Universe we start with the following system of partial differential equations,

$$\frac{\partial \rho}{\partial t}+\nabla \cdot (\rho \vec{v})=0, \tag{5.25}$$

$$\frac{d\vec{v}}{dt}=\frac{\partial \vec{v}}{\partial t}+(\vec{v}\cdot \nabla)\,\vec{v}=-\frac{1}{\rho}\nabla p-\nabla \Phi, \tag{5.26}$$

$$\Delta \Phi = 4\pi G\rho, \tag{5.27}$$

where the first equation is the equation of continuity of the cosmological fluid, the second equation is the Euler equation, while Eq. (5.27) is the Poisson equation for the gravitational potential Φ. As usual, $\vec{v}$, ρ, and p denote the velocity, density, and pressure of the cosmological fluid. We assume again the cosmological principle, which requires that the density and pressure are functions of time only, so that $\rho=\rho(t)$, and $p=p(t)$, respectively. By introducing the velocity of the cosmological expansion of the fluid by means of the definition $\vec{v}(t)=H(t)\vec{r}=\left(\dot{R}/R\right)\vec{r}$, where $\vec{r}$ is the comoving radius vector, and by taking into account that $\nabla \cdot \vec{r}=3$, the continuity equation (5.25) immediately yields

$$\rho\,(t)=\frac{\rho_0}{R^3(t)}, \tag{5.28}$$

where by ρ_0 we have denoted an arbitrary constant of integration. Then the Euler equation (5.26) takes the form

$$\left(\dot{H}+H^2\right)\vec{r}=-\nabla\Phi, \tag{5.29}$$

and after applying the ∇ operator to both sides of the equation, with the use of the Poisson equation, we obtain

$$\frac{d^2R}{dt^2}=-\frac{4}{3}\pi G\rho R=-\frac{4\pi G\rho_0}{3R^2}. \tag{5.30}$$

Multiplication with $\dot{R}$ gives

$$\frac{1}{2}\frac{d}{dt}\dot{R}^2=\frac{d}{dt}\frac{4\pi G\rho_0}{3R}, \tag{5.31}$$

or

$$3\frac{\dot{R}^2}{R^2}=8\pi G\rho+\frac{C}{R^2}, \tag{5.32}$$

where C is an arbitrary constant of integration. By taking $C=-3kc^2$, we recover the first Friedmann equation as given by (5.10).

Perturbations of the Hydrodynamic Equations

The main advantage of the Newtonian hydrodynamic approach is that it offers the possibility of investigating the cosmological perturbations that have a fundamental role in the formation of the large scale structure in the Universe. The Newtonian approach gives an excellent approximation for structure formation on cosmological scales that are small as compared to the Hubble length, and also situated very far away from black holes, where the gravitational field can be considered as weak. To obtain the perturbation equations of the cosmological quantities we introduce first the coordinate $\vec{x}$ comoving with the mean rate of expansion, and defined as

$$\vec{x} = \frac{\vec{r}}{R(t)}. \tag{5.33}$$

In these expanding coordinates the velocity field $\vec{v}$ can be represented as

$$\vec{v} = \dot{R}\vec{x} + \vec{u}\,(\vec{x}, t) = H\vec{r} + \vec{u}\,(\vec{x}, t)\,, \tag{5.34}$$

where by $\vec{u}\,(\vec{x}, t) = \delta\vec{v}\,(\vec{x}, t)$ we denote the perturbation of the Hubble flow. We write the perturbation of the energy density in the form

$$\rho\,(\vec{x}, t) = \rho_b(t)\,(1 + \delta\,(\vec{x}, t))\,, \tag{5.35}$$

where for the background density $\rho_b(t)$ we adopt the expression (5.28), $\rho_b(t) = \rho_0/R^3(t)$. The time derivative at fixed $\vec{r}$ of a function $f = f\,(t, \vec{x} = \vec{r}R)$ is obtained as [409]

$$\left(\frac{\partial f}{\partial t}\right)_{\vec{r}} = \left(\frac{\partial f}{\partial t}\right)_{\vec{x}} - \frac{\dot{R}}{R}\vec{x}\cdot\nabla f, \tag{5.36}$$

where the gradients at fixed time in $\vec{x}$, ∇, and the gradient at $\vec{r}$, $\nabla_{\vec{r}}$, are related by the relation

$$\nabla = R\nabla_{\vec{r}}. \tag{5.37}$$

Then the continuity equation of the cosmic matter in the adopted comoving coordinates becomes

$$\left[\left(\frac{\partial}{\partial t}\right)_{\vec{x}} - \frac{\dot{R}}{R}\vec{x}\cdot\nabla\right]\rho_b(t)\,(1 + \delta\,(\vec{x}, t)) + \frac{\rho_b}{R}\nabla\cdot\left[(1 + \delta\,(\vec{x}, t))\left(\dot{R}\vec{x} + \vec{u}\,(\vec{x}, t)\right)\right] = 0, \tag{5.38}$$

which after some simple calculations gives

$$\frac{\partial\delta\,(\vec{x}, t)}{\partial t} + \frac{1}{R}\nabla\cdot\left[(1 + \delta\,(\vec{x}, t))\,\vec{u}\,(\vec{x}, t)\right] = 0. \tag{5.39}$$

The perturbed Poisson equation takes the form

$$\frac{1}{R^2}\Delta\Phi\,(\vec{x}, t) = 4\pi G\rho_b\,(1 + \delta\,(\vec{x}, t)) - \Lambda, \tag{5.40}$$

where for the sake of completeness we have also added the cosmological constant Λ. We assume that the gravitational potential Φ is given by the sum of the perturbed part $\phi(\vec{x},t)$ and of the unperturbed part $\phi_b(\vec{x},t)$, so that

$$\Phi(\vec{x},t) = \phi(\vec{x},t) + \phi_b(\vec{x},t) = \phi(\vec{x},t) + \frac{2}{3}\pi G\rho_b(t)R^2\vec{x}^2 - \frac{1}{6}\Lambda R^2\vec{x}^2. \tag{5.41}$$

Thus, Eq. (5.40) takes the form

$$\Delta\phi(\vec{x},t) = 4\pi G\rho_b(t)R^2\delta(\vec{x},t). \tag{5.42}$$

It remains now to consider the perturbed Euler equation (5.26). After a simple calculation we find

$$\left[\left(\frac{\partial}{\partial t}\right)_{\vec{x}} - \frac{\dot{R}}{R}\vec{x}\cdot\nabla\right]\left(\dot{R}\vec{x} + \vec{u}(\vec{x},t)\right) + \frac{1}{R}\left(\dot{R}\vec{x} + \vec{u}(\vec{x},t)\right)\nabla\cdot\left(\dot{R}\vec{x} + \vec{u}(\vec{x},t)\right)$$
$$= -\frac{1}{R}\nabla\phi - \frac{4}{3}\pi G\rho_b(t)R\vec{x} + \frac{1}{3}\Lambda R\vec{x}. \tag{5.43}$$

By performing further calculations, and taking into account the background solution, the Euler equation reduces to

$$\frac{\partial\vec{u}}{\partial t} + \frac{\dot{R}}{R}\vec{u} + \frac{1}{R}(\vec{u}\cdot\nabla)\vec{u} = -\frac{1}{R}\nabla\phi. \tag{5.44}$$

Equations (5.39), (5.42), and (5.44) provides the evolution of the density fluctuations in a dust universe. By averaging over the concentration of matter on a small scale we obtain

$$\frac{\partial\delta}{\partial t} + \frac{1}{R}\nabla\cdot\vec{u} = 0, \tag{5.45}$$

$$\frac{\partial\vec{u}}{\partial t} + \frac{\dot{R}}{R}\vec{u} = -\frac{1}{R}\nabla\phi. \tag{5.46}$$

Now we take the time derivative of the first equation, and the divergence of the second equation. Then, after subtracting the obtained results, and with the use of the Poisson equation, we arrive at the following equation describing the time variation of the *density contrast* $\delta = \delta\rho/\rho$ of the pressureless dust

$$\frac{\partial^2\delta}{\partial t^2} + 2\frac{\dot{R}}{R}\frac{\partial\delta}{\partial t} = 4\pi G\rho_b\delta. \tag{5.47}$$

As a simple application of the density perturbation equation we adopt as the background solution of the field equation (5.20), according to which $R \propto t^{2/3}$. Then the perturbation equation (5.47) takes the form

$$\frac{\partial^2\delta}{\partial t^2} + \frac{4}{3t}\frac{\partial\delta}{\partial t} = \frac{2}{3t^2}\delta, \tag{5.48}$$

with the general solution given by

$$\delta(t) = At^{2/3} + \frac{B}{t}. \tag{5.49}$$

Thus, in the large time limit there will be an increasing perturbation mode of the matter density.

5.1.3 The Neo-Newtonian Approach

In order to investigate realistically the gravitational effects in the Universe, a full general relativistic treatment is required. In particular, the general relativistic approach naturally includes the effect of the matter pressure, which plays an important role in many astrophysical and cosmological situations. On the other hand, the pressure does not appear naturally in the Newtonian approach. However, it is possible to include the pressure in the dynamical equations of motion, and to reobtain both Friedmann equations in the framework of the so-called Neo-Newtonian approach, which we will present in this section.

The equations of motion of any material system can be obtained from the equation of conservation of the energy-momentum tensor, $\nabla_\mu T^\mu_\nu = 0$, with T^μ_ν given by

$$T^\mu_\nu = \left(\rho_m c^2 + P\right) u_\nu u^\mu + P\delta^\mu_\nu, \tag{5.50}$$

where by ρ_m and P we denote the matter density and the thermodynamic pressure, respectively, and u^μ is the four-velocity of the fluid, satisfying the normalization condition $u_\mu u^\mu = -1$, as well as the condition $\nabla_\nu u^\mu u_\mu = 0$. By taking, with the help of the operator ∇, the covariant derivative of T^μ_ν, we immediately obtain the equation

$$\begin{aligned}\left(\rho_m c^2 + P\right)_{,\mu} u_\nu u^\mu + \left(\rho_m c^2 + P\right) \nabla_\mu u_\nu u^\mu \\ + \left(\rho_m c^2 + P\right) \left(\nabla_\mu u_\nu\right) u^\mu = P_{,\nu},\end{aligned} \tag{5.51}$$

where by a comma we have denoted the ordinary derivative with respect to the coordinate four-vector x^μ. By contracting Eq. (5.51) with u^ν, we find

$$\left(\rho_m c^2 + P\right)_{,\mu} u^\mu + \left(\rho_m c^2 + P\right) \nabla_\mu u^\mu = P_{,\nu} u^\nu. \tag{5.52}$$

In the following we adopt the Newtonian limit of small particle velocities, in which the matter four-velocity is given by $u^\mu = (1, \vec{v}/c)$, where $\vec{v}$ is the three-velocity of the cosmological fluid. As is well known, the four-divergence of the four-velocity is given by the expression

$$\nabla_\mu u^\mu = \frac{1}{\sqrt{-g}} \frac{\partial\left(\sqrt{-g} u^\mu\right)}{\partial x^\mu}, \tag{5.53}$$

where by $(-g)$ we have denoted the determinant of the metric tensor $g_{\mu\nu}$. In the adopted Newtonian limit of small velocities, we can safely assume that $\sqrt{-g} \rightarrow 1$, that is, the deviations of the cosmic geometry from the Minkowski-type metric are negligibly small. With the use of these approximations, from Eq. (5.52) we obtain the Neo-Newtonian equation of continuity of the cosmological flow as

$$\left(\frac{\partial \rho_m}{\partial t}\right)_r + \nabla_r \cdot (\rho_m \vec{v}) + \frac{P}{c^2}\nabla_r \cdot \vec{v} = 0, \tag{5.54}$$

where all differential operations involving the operator ∇ must now be considered with respect to the physical coordinate $\vec{r}$. After contraction of Eq. (5.51) with the projection operator

$$h_\alpha^\nu = \delta_\alpha^\nu + u_\alpha u^\nu, \tag{5.55}$$

having the property $h_\alpha^\nu u_\nu \equiv 0$, we obtain the relativistic Euler equation of motion of the cosmological fluid as

$$(\rho_m c^2 + P)(\nabla_\mu u_\alpha) u^\mu = P_{,\alpha} - P_{,\nu} u^\nu u_\alpha. \tag{5.56}$$

In the Neo-Newtonian approximation of small velocities and negligible distortions of the space-time geometry, the generalized Euler equation of motion of the cosmic fluid becomes

$$\left(\frac{\partial \vec{v}}{\partial t}\right)_r + (\vec{v} \cdot \nabla_r)\vec{v} = -\nabla_r V - \frac{c^2 \nabla_r P + \dot{P}\vec{v}}{\rho_m c^2 + P}. \tag{5.57}$$

The gravitational potential of the fluid V satisfies, in the same approximation, the generalized Poisson equation,

$$\nabla_r^2 V = 4\pi G\left(\rho_m + 3\frac{P}{c^2}\right). \tag{5.58}$$

Equations (5.54), (5.57), and (5.58) provide the fundamental equations describing the cosmological dynamics of a gravitationally confined fluid in the Neo-Newtonian (or first Post-Newtonian) approximation [9, 219, 248, 315, 348, 402, 433, 489].

The Background Evolution

The system of equations Eqs. (5.54), (5.57), and (5.58) admit a homogeneous and isotropic cosmological background solution in which the physical parameters of the cosmic fluid are functions of time only, so that $\rho_m = \rho_b(t)$ and $P = P_b(t)$. For the background solution the fluid's three-velocity velocity is given by

$$\vec{v}_b = \frac{\dot{R}}{R}\vec{r}. \tag{5.59}$$

It is then easy to check that the evolution of the scale factor R is described by the Friedmann equations,

$$3\frac{\dot{R}^2}{R^2} = 3H^2 = 8\pi G \rho_b, \tag{5.60}$$

and

$$\frac{\ddot{R}}{R} = -\frac{4\pi G}{3}\left(\rho_b + \frac{P_b}{c^2}\right), \tag{5.61}$$

respectively, where we have introduced the Hubble function H, defined as $H = \dot{R}/R$. The continuity equation (5.54) reduces to the energy conservation equation of the cosmic fluid,

$$\frac{d\rho_b}{dt} + 3H\left(\rho_b + \frac{P_b}{c^2}\right) = 0. \tag{5.62}$$

Cosmological Perturbations in the Neo-Newtonian Cosmology

Now we assume the existence of small perturbations of the physical quantities in the gravitationally bounded cosmological fluid, with the perturbations generated around the homogeneous background. These perturbations are of the form

$$\rho_m(\vec{r},t) = \rho_b(t) + \delta\rho(\vec{r},t), \tag{5.63}$$

$$P(\vec{r},t) = P_b(t) + \delta P(\vec{r},t), \tag{5.64}$$

$$V(\vec{r},t) = V_b + \varphi(\vec{r},t), \tag{5.65}$$

$$\vec{v}(\vec{r},t) = \vec{v}_b + \vec{u}(\vec{r},t), \tag{5.66}$$

where the index b corresponds to the background cosmological quantities. Substituting first the perturbed cosmological quantities into the continuity equation (5.54), we find

$$\left(\frac{\partial\delta\rho}{\partial t}\right)_r + \left(\rho_b + \frac{P_b}{c^2}\right)\nabla_r\cdot\vec{u} + \nabla_r\cdot(\delta\rho\vec{v}_b) + \frac{\delta P}{c^2}\nabla_r\cdot\vec{v}_b = 0. \tag{5.67}$$

Next we take the variation of the equation of motion (5.57), which gives

$$\left(\frac{\partial\vec{u}}{\partial t}\right)_r + (\vec{v}_b\cdot\nabla_r)\,\vec{u} + (\vec{u}\cdot\nabla_r)\,\vec{v}_b = -\nabla_r\varphi - \frac{\nabla_r\delta P + \dot{P}_b\vec{u}/c^2}{\rho_b + P_b/c^2}. \tag{5.68}$$

The Poisson equation describing the perturbations of the gravitational potential, which can be found by perturbing Eq. (5.58), can be written in the first-order approximation as

$$\nabla_r^2\varphi = 4\pi G\left(\delta\rho + 3\frac{\delta P}{c^2}\right). \tag{5.69}$$

In order to investigate the cosmological evolution of the matter perturbations we perform a change of variables to the comoving coordinate system, so that $\vec{r} = a\vec{q}$, $\nabla_q = \nabla = a\nabla_r$, and

$$\left(\frac{\partial}{\partial t}\right)_q = \frac{\partial}{\partial t} = \left(\frac{\partial}{\partial t}\right)_r + \frac{\dot{R}}{R}(\vec{q}\cdot\nabla_q), \tag{5.70}$$

respectively. To simplify the mathematical notation we introduce two new parameters w and c_{eff}^2, defined as $w = P_b/\rho_b c^2$ and $c_{eff}^2 = \delta P/\delta\rho$, respectively, which are generally functions of the cosmological time only. We also introduce the density contrast of the cosmic fluid by means of the definition $\delta = \delta\rho/\rho_b$. The first time derivative of the background (non-perturbed) fluid pressure is related to the speed of sound of the background matter $c_s^2 = \partial P_b/\partial\rho_b$ by the relation

$$\dot{P}_b = -3\left(\frac{\dot{R}}{R}\right)c_s^2\rho_b(1+w). \tag{5.71}$$

With the use of the above results the perturbation equations Eqs. (5.64)–(5.66) can be reformulated as

$$\dot{\delta} + 3H\left(\frac{c_{eff}^2}{c^2} - w\right)\delta + \frac{1+w}{R}\nabla\cdot\vec{u} = 0, \tag{5.72}$$

$$\frac{d\vec{u}}{dt} + \left(1 - 3\frac{c_s^2}{c^2}\right)\frac{\dot{R}}{R}\vec{u} + \frac{1}{a}\nabla\varphi + \frac{c_{eff}^2}{c^2}\frac{1}{R}\frac{1}{1+w}\nabla\cdot\delta = 0, \tag{5.73}$$

$$\nabla^2\varphi = 4\pi G R^2\rho_b\left(1 + 3\frac{c_{eff}^2}{c^2}\right)\delta. \tag{5.74}$$

To simplify the mathematical formalism, in the following we introduce the following notation

$$\alpha_{eff} = \frac{c_{eff}^2}{c^2} - w, \tag{5.75}$$

and

$$\alpha_s = \frac{c_s^2}{c^2} - w, \tag{5.76}$$

respectively. The quantity α_s is related to the time derivative of w by means of the simple relation

$$\frac{\dot{w}}{1+w} = -3H\alpha_s. \tag{5.77}$$

Now we take the time derivative of Eq. (5.72), and the divergence of Eq. (5.73). Finally, by eliminating $\nabla\cdot\vec{u}$ by using the perturbed equation of continuity, and with the use of Eq. (5.74), we obtain the basic equation giving the time evolution of the density contrast δ in the Neo-Newtonian approach to the cosmological perturbations as

$$\ddot{\delta} + 3H\left(\frac{c_{eff}^2}{c^2} - 2w + \frac{2}{3}\right)\dot{\delta} + \frac{3}{2}H^2 \times$$

$$\left[9w^2 - 2w - 2(1+6w)\frac{c_{eff}^2}{c^2} + \frac{2}{H}\frac{d}{dt}\alpha_{eff} - 1\right]\delta = \frac{c_{eff}^2}{c^2}\frac{1}{a^2}\Delta\delta. \tag{5.78}$$

A simpler form of the perturbation equation for the density contrast can be obtained by a change of the independent variable from the cosmic time t to the scale factor R with the use of the relations $\partial/\partial t = RH(R)\partial/\partial R$, and

$$\frac{\partial^2}{\partial t^2} = R^2H^2\frac{\partial^2}{\partial R^2} - \left[\frac{1}{2}(1+3w)RH^2\right]\frac{\partial}{\partial R}, \tag{5.79}$$

respectively. Then we obtain the evolution equation for the density contrast in the alternative form

$$R^2\frac{\partial^2\delta}{\partial R^2} + 3R\left(\frac{c_{eff}^2}{c^2} - \frac{5}{2}w + \frac{1}{2}\right)\frac{\partial\delta}{\partial R} + \frac{3}{2}\times$$
$$\left[9w^2 - 2w - 2\left(1+6w\right)\frac{c_{eff}^2}{c^2} + 2R\frac{d}{dR}\alpha_{eff} - 1\right]\delta = \frac{c_{eff}^2}{c^2}\frac{1}{R^2H^2}\Delta\delta. \tag{5.80}$$

Equation (5.78) differs from the density perturbation equation derived in the Newtonian cosmology with the pressure [9, 315, 433]. The main reason is that in the present analysis we have also included the term $\dot{P}_b\vec{u}/c^2$, which was neglected in the previous studies. The term $\dot{P}_b\vec{u}/c^2$ gives rise to the new term $-\left(3c_s^2/c^2\right)H\vec{u}$ in the left-hand side of Eq. (5.73), which contributes to the final perturbation equation of the density contrast. On the other hand, it is important to note that in the present approach the term $\delta P\left(\vec{r},t\right)/c^2$ was systematically neglected.

Generally, Eqs. (5.78) and (5.80) cannot be solved analytically, and therefore to investigate their behavior numerical methods are required. To integrate the density perturbation evolution equation we must choose some appropriate cosmological and astrophysical initial conditions. In the present-day standard model for large scale structure formation in the Universe, it is assumed that the quantum fluctuations that are responsible for the development of the local gravitational instabilities were produced during the initial period of the cosmological inflation that reshaped the entire matter structure of the Universe. Due to the accelerated de Sitter-type expansion during inflation, these initially small-scale fluctuations inflated up exponentially to super-horizon scales, where they generated a set of primordial fluctuations of the gravitational potential that were near scale-invariant, and near Gaussian. When inflation ended, the Universe was reheated, and during the reheating phase the present particle and radiation content was produced. In this very hot early phase of the Universe's evolution, a large number of cold thermal relics (dark matter) may have also been produced [403, 410]. The interaction cross section of the dark matter particles with ordinary matter is extremely small, with dark matter particles interacting (mainly) gravitationally, and perhaps through the weak interaction.

Therefore, in order to find some cosmologically realistic initial conditions for the state of the Universe during the period of large scale structure formation, one must evolve back in time to the cosmological perturbations to find out the initial conditions of the matter perturbations. Hence one must go to the early stages of the radiation epoch, and far away from the Hubble radius. In the framework of the standard ΛCDM cosmological model, the initial conditions for cold dark matter particles, neutrinos, photons, and baryons have been obtained, in the synchronous as well as Newtonian gauges [335]. In the standard present-day approach, to compute the power spectrum of the matter density fluctuations in the Universe, one must solve numerically the Boltzmann equation. The power spectrum of the density fluctuations is usually obtained first in the linear theory, and then extrapolated to the present phase of the evolution of the Universe. In the first-order Eulerian perturbation theory, and in the standard cosmological approach, all perturbation modes evolve independently of each other. Therefore the power spectrum can be scaled back in time to the initial cosmological eras with the use of the so-called growth function [335]. In the standard procedures of generating initial cosmological conditions, the possible role of the dark matter pressure is usually ignored. Therefore, finding a rigorous and cosmologically well-motivated set of initial conditions for the density contrast δ, and for its time derivative, imposes the necessity of the full investigation of the cosmological dynamics from the reheating era up to the beginning of the large scale formation phase.

5.1.4 Gravitational Condensation of Dark Energy

As a final example of Newtonian gravitational mechanics in the investigation of the cosmological processes we consider the evolution of the Newtonian perturbations with a *dark energy* fluid, with an initial density $\rho_0 = \Lambda c^2/8\pi G$, where Λ denotes the cosmological constant, and an initial pressure p_0. The initial thermodynamic parameters of this cosmological fluid satisfy the equation of state [73]

$$\rho_0 c^2 + p_0 = 0. \tag{5.81}$$

This equation of state is specific for the dark energy, interpreted as a physical fluid with negative pressure, or for the cosmological constant Λ. We will assume that as a result of some small density and pressure fluctuations, the dark energy fluid undergoes a phase transition into a nonrelativistic, dissipationless material cosmological fluid. This dark energy generated fluid can be described, as usual, by a matter density ρ, a thermodynamic pressure p, a Newtonian velocity $\vec{v}$, and a gravitational acceleration $\vec{g}$, respectively. In the Newtonian approximation, the time evolution of the perturbed dark energy system is described by a set of three differential equations, given by the continuity equation, the hydrodynamical Euler equation, and the Poisson equation, respectively, which can be written down as

$$\frac{\partial \rho}{\partial t} + \nabla \cdot (\rho \vec{v}) = 0, \qquad \frac{\partial \vec{v}}{\partial t} + (\vec{v} \cdot \nabla)\, \vec{v} = -\frac{1}{\rho} \nabla p + \vec{g}, \tag{5.82}$$

$$\nabla \times \vec{g} = 0, \qquad \nabla \cdot \vec{g} = -4\pi G \rho. \tag{5.83}$$

We consider as the initial (unperturbed) state of the fluid system the dark energy phase, which is characterized by the absence of the hydrodynamical flow, $\vec{v} = \vec{v}_0 = 0$, of any "real" gravitational forces acting in the dark energy fluid, $\vec{g} = \vec{g}_0 = 0$, and by constant values of the density and pressure, $\rho = \rho_0$ and $p = p_0$, respectively, satisfying the dark energy equation of state $\rho_0 c^2 + p_0 = 0$. The phase transition process, associated to the change of the equation of state of the initial dark energy fluid, determines the generation of the gravitational interactions in the system, which are associated to the presence of small perturbations of the hydrodynamical quantities, so that

$$\rho = \rho_0 + \rho_1, \qquad p = p_0 + p_1, \qquad \vec{v} = \vec{v}_0 + \vec{v}_1, \qquad \vec{g} = \vec{g}_0 + \vec{g}_1, \tag{5.84}$$

where the small perturbations ρ_1 and p_1 satisfy the conditions $-1 \ll \rho_1/\rho_0 \ll 1$ and $-1 \ll p_1/p_0 \ll 1$, respectively. By keeping only first-order terms, Eqs. (5.82) and (5.83) take the form

$$\frac{\partial \rho_1}{\partial t} + \rho_0 \nabla \cdot \vec{v}_1 = 0, \qquad \frac{\partial \vec{v}_1}{\partial t} = -\frac{v_s^2}{\rho_0} \nabla \rho_1 + \vec{g}_1, \tag{5.85}$$

$$\nabla \times \vec{g}_1 = 0, \qquad \nabla \cdot \vec{g}_1 = -4\pi G \rho_1, \tag{5.86}$$

where in order to simplify the mathematical form of the equations we have defined the adiabatic speed of sound v_s in the perturbed fluid as $v_s = \sqrt{p_1/\rho_1} = \sqrt{\partial p/\partial \rho}$. By applying now the partial derivative operator with respect to the time of the perturbed continuity equation in Eqs. (5.85), we find the propagation equation of the density perturbations in the perturbed dark matter fluid as

$$\frac{\partial^2 \rho_1}{\partial t^2} = v_s^2 \nabla^2 \rho_1 + \frac{\Lambda c^2}{2} \rho_1. \tag{5.87}$$

We look for a simple solution of the form $\rho_1 \propto \exp\left[i\left(\vec{k}\cdot\vec{r} - \omega t\right)\right]$ of Eq. (5.87). After substituting this functional form for ρ_1, we obtain the following dispersion relation for the angular frequency ω

$$\omega^2 = v_s^2 \vec{k}^2 - \frac{\Lambda c^2}{2}. \tag{5.88}$$

From the above dispersion equation we can immediately see that for $k < k_J$, where

$$k_J = \sqrt{\frac{\Lambda c^2}{2 v_s^2}}, \tag{5.89}$$

is the Jeans wave number for the dark energy fluid, the angular frequency ω becomes a complex valued, imaginary physical quantity. This situation corresponds to an instability of the dark energy fluid – ρ_1 can either increase or decrease by following an exponential law, thus leading to a gravitational condensation (or rarefaction) of the fluid. Therefore, for $k < k_J$, we have

$\omega = \pm v_s\sqrt{k^2 - k_J^2} = i\mathrm{Im}\,\omega$, where $\mathrm{Im}\,\omega = \pm v_s\sqrt{k_J^2 - k^2}$, and consequently the density perturbation behaves like $\rho_1 \propto \exp\left[\pm\left|\mathrm{Im}\,\omega\right| t\right]$.

When the mass of the dark energy fluid becomes larger than the mass of a sphere with radius $2\pi/k_J$, a gravitational instability occurs in the system, and the fluid eventually collapses. The critical mass describing this physical situation is called the Jeans mass, and it is defined according to the relation $M_J = (4\pi/3)\left(2\pi/k_J\right)^3 \rho_0$. For the perturbed dark energy fluid generated after the phase transition to the normal fluid state, it is given by

$$M_J = \frac{8\sqrt{2}}{3}\pi^3\left(\frac{v_s}{c}\right)^3\frac{c^2}{G}\Lambda^{-1/2} \approx 1.6\times 10^{30}\times\left(\Lambda\,\mathrm{cm}^{-2}\right)^{-1/2}\left(\frac{v_s}{c}\right)^3\,\mathrm{g}. \qquad (5.90)$$

For $\Lambda = 3\times 10^{-56}\mathrm{cm}^{-2}$ we obtain for the Jeans mass of the perturbed dark energy fluid the expression $M_J = 9.24\times 10^{57}\left(v_s/c\right)^3$ g. The effective radius R_J of the post-phase transition stable dark energy fluid system can be obtained as

$$R_J = 2^{3/2}\pi\frac{v_s}{c}\Lambda^{-1/2}. \qquad (5.91)$$

5.2 Relativistic Cosmology

In building general relativistic theoretical models of the Universe we adopt the fundamental assumption that *there are no privileged galactic observers.* This means that all observers moving with the galactic flow are assumed to experience the same large-scale dynamical processes of evolution of the Universe. This is *the cosmological principle.* From the cosmological principle it follows that the distribution of the matter in Universe must be *isotropic and homogeneous*, as mentioned before. This assumption is certainly not valid for the matter distribution near the Earth (for example, in the Milky Way, and in the associated galactic systems), where we can observe a significant degree of inhomogeneity in the mass distribution of the stars, galaxies, clusters of galaxies, etc. However, in this section, we assume that on the global scale of the Universe these inhomogeneities are small, and can be ignored. Therefore, in studying the large scale structure, the dynamics and the evolution of the Universe we approximate the distribution of cosmological matter by a smoothed-out average density.

If the cosmological principle is adopted, it is possible *to define an absolute cosmological time*, that is, we have the possibility of introducing a time coordinate for cosmic events that is independent of the observer. This is because all cosmic observers will observe the same universal process of evolution, and one can assign some time coordinates according to a common scale for different phases of the cosmological expansion.

5.2.1 The Friedmann–Lemaître–Robertson–Walker Metric

As a first step in developing relativistic cosmology we need to find an appropriate metric of the Universe. This task is very much simplified if we adopt as our basic mathematical and physical starting point the cosmological principle, requiring the isotropy and homogeneity of the Universe. In a Universe with an isotropic and homogeneous geometry, the scalar curvature of the space must have the same value for all points of the Universe at a given time after the big bang "explosion." Therefore, as anticipated by Friedmann and Lemaître in the early 1920s, and rigorously shown by Robertson [439] and Walker [494], the only metric that can describe such a Universe is given, in spherical coordinates (r, θ, φ), by

$$ds^2 = -c^2 dt^2 + a^2(t)\left[\frac{dr^2}{1-kr^2} + r^2\left(d\theta^2 + \sin^2\theta d\varphi^2\right)\right], \tag{5.92}$$

where $a(t)$ is the scale factor, and k is a constant. This metric is known as the Friedmann–Lemaître–Robertson–Walker (FLRW) metric. The FLRW metric gives us the space-time interval ds between two distinct events occurring independently in the expanding Universe. The coordinates in the metric are defined in such a way that the corresponding symmetries of the space-time can be easily obtained. The coordinate system in which the FLRW metric is written down in the form (5.92) is called *the comoving coordinate system.* In these coordinates the world lines with x^α = constant are geodesics of the space-time. The observers that travel along these world lines are called *comoving* (or sometimes *fundamental*) observers. The spatial hypersurfaces $x^0 = 0$ of the FLRW metric have *negative, zero, and positive curvatures* for $k = -1, 0$ and $+1$, respectively. The corresponding geometries of these universes are called open, flat, and closed, respectively.

The study of the geometrical properties of these spaces can be performed in several coordinate systems. For example, in many applications it is advantageous to introduce a new radial-type coordinate χ, defined as

$$\chi = \int \frac{dr}{\sqrt{1-kr^2}} = \begin{cases} \arcsin r, & k=+1, \\ r, & k=0, \\ \text{arcsinh} r, & k=-1. \end{cases} \tag{5.93}$$

In the new coordinate system with coordinates $(ct, \chi, \theta, \varphi)$ the FLRW metric becomes

$$ds^2 = -c^2 dt^2 + a^2(t)\left[d\chi^2 + S_k^2(\chi)\left(d\theta^2 + \sin^2\theta d\varphi^2\right)\right], \tag{5.94}$$

where

$$S_k(\chi) = \begin{cases} \sin\chi, & k=+1, \\ \chi, & k=0, \\ \sinh\chi, & k=-1. \end{cases} \tag{5.95}$$

The Geometry of the FLRW Metric

The FLRW metric has a number of interesting geometric properties. Thus, for $k = +1$, it can be interpreted geometrically as a three-sphere of radius a

embedded in a flat, four-dimensional fictitious Euclidian space. In Cartesian coordinates (x_1, x_2, x_3, x_4) such a three-sphere is defined by the equation

$$x_1^2 + x_2^2 + x_3^2 + x_4^2 = a^2. \tag{5.96}$$

Now we introduce a set of new angular coordinates (χ, θ, φ) on the three-sphere with the help of the definitions

$$x_1 = a \cos\chi \sin\theta \sin\varphi, \qquad x_2 = a \cos\chi \sin\theta \cos\varphi, \tag{5.97}$$

$$x_3 = a \cos\chi \cos\theta, \qquad x_4 = a \sin\chi. \tag{5.98}$$

After calculating the differentials of the dx_i's, and substituting them into the Euclidean four-dimensional metric,

$$dL^2 = \sum_{i=1}^{4} dx_i^2 = dx_1^2 + dx_2^2 + dx_3^2 + dx_4^2, \tag{5.99}$$

we find the metric of the three-sphere as

$$dL^2_{3-\text{sphere}} = a^2 \left[d\chi^2 + \sin^2\chi \left(d\theta^2 + \sin^2\theta d\varphi^2\right)\right], \tag{5.100}$$

which is exactly the same as the spatial part of the FLRW metric for $k = +1$. For this geometry the entire space is covered by the coordinate ranges $0 \le \chi \le \pi$, $0 \le \theta \le \pi$, and $0 \le \varphi \le 2\pi$, respectively. An interesting property of this Universe model is that it has a finite three-dimensional volume V, given by

$$V = \int_0^{2\pi} d\varphi \int_0^{\pi} d\theta \int_0^{\pi} \sqrt{g} d\chi = a^3 \int_0^{2\pi} d\varphi \int_0^{\pi} \sin\theta d\theta \int_0^{\pi} \sin^2\chi d\chi = 2\pi a^3. \tag{5.101}$$

On the other hand, the surface area of a two-sphere located in this geometry, defined by the condition $\chi =$ constant, can be obtained as $S = 4\pi a^2 \sin^2\chi$. As the coordinate χ increases, the surface S first increases with χ, then it reaches a maximal value given by $4\pi a^2$ at $\chi = \pi/2$, and then it decreases. Hence in this geometry the Universe is characterized by a closed three-space which. However, there are no boundaries in the geometry.

For $k = -1$, the spatial part of the FLRW metric is described equivalently by the geometry of a hyperboloid embedded in a fictitious four-dimensional geometric space with Lorentzian metric signature (note that this is not the physical space-time of special relativity). Such a geometry is realized by using the line element

$$dL^2 = dx_1^2 + dx_2^2 + dx_3^2 - dx_4^2. \tag{5.102}$$

A three-dimensional hyperboloid, which is embedded in this space, is described analytically by the equation

$$x_4^2 - x_1^2 - x_2^2 - x_3^2 = a^2. \tag{5.103}$$

By introducing a set of coordinates (χ, θ, φ) defined as

$$x_1 = a \sinh \chi \sin \theta \sin \varphi, \qquad x_2 = a \cos \chi \sin \theta \cos \varphi, \tag{5.104}$$

$$x_3 = a \sinh \chi \cos \theta, \qquad x_4 = a \cosh \chi, \tag{5.105}$$

we obtain the metric of the hyperboloid as

$$dL^2_{3-\mathrm{sphre}} = a^2 \left[d\chi^2 + \sinh^2 \chi \left(d\theta^2 + \sin^2 \theta d\varphi^2 \right) \right]. \tag{5.106}$$

In this geometry the range of the coordinates is $0 \leq \chi \leq \infty$, $0 \leq \theta \leq \pi$, and $0 \leq \varphi \leq 2\pi$, respectively. The hyperboloidal open space has an infinite volume, similarly to the ordinary flat three-space of the Euclidean geometry. A two-sphere, defined by the condition χ =constant, has a surface area given by $S = 4\pi a^2 \sinh^2 \chi$. The surface area of the two sphere increases monotonically with χ.

In some cosmological applications it is useful to introduce a different cosmological time coordinate τ, related to t by the definition $d\tau = a^{-1}(t)dt$. In this time coordinate the FLRW metric becomes

$$ds^2 = a^2 (\tau) \left\{ -d\tau^2 + \left[d\chi^2 + S_k^2 (\chi) \left(d\theta^2 + \sin^2 \theta d\varphi^2 \right) \right] \right\}. \tag{5.107}$$

In this representation of the metric all the time dependence of the evolution of the Universe is described by an overall multiplicative factor.

The Gravitational Redshift

Let's consider now a photon traveling in the FLRW metric from an initial point r to the observer, located at the origin of the coordinate system. Since the Universe is isotropic and homogeneous, we can fix freely the origin of the coordinates at any point. The photon is emitted at a time t_{emit}, and reaches the observer at the time $t = t_{obs}$. In the following, for simplicity, we will limit our analysis to the case of the flat geometries only, and therefore we take $k = 0$. Photons travel along geodesic lines $ds = 0$, and in the considered flat geometry, for radial motion from the FLRW metric, we obtain

$$cdt = \pm a(t)dr. \tag{5.108}$$

Since the photon we are considering is traveling toward the origin, we choose the negative sign in Eq. (5.108). After integration we find the comoving coordinate r as given by

$$r = c \int_{t_{emit}}^{t_{obs}} \frac{dt}{a(t)}. \tag{5.109}$$

Consider now a photon emitted at the same point r at an instant $t_{emit} + \delta t_{emit}$, and reaching the observer at a time $t_{obs} + \delta t_{obs}$. Hence we find the relation

$$\int_{t_{emit}}^{t_{obs}} \frac{dt}{a(t)} = \int_{t_{emit}+\delta t_{emit}}^{t_{obs}+\delta t_{obs}} \frac{dt}{a(t)}. \tag{5.110}$$

By assuming that the frequency ν^{-1} of the photon is much smaller than the difference Δ between the observation and emission times, defined as $\Delta t = t_{obs} - t_{emit}$, we can approximate the time variation δt as $\delta t = \nu^{-1}$. With this value of δt Eq. (5.110) leads immediately to

$$a\left(t_{emit}\right)\nu_{emit} = a\left(t_{obs}\right)\nu_{obs}. \tag{5.111}$$

By introducing the wavelength λ of the photon defined as $\lambda = c/\nu$, and defining the redshift z of the radiation as

$$z = \frac{\lambda_{obs} - \lambda_{emit}}{\lambda_{emit}}, \tag{5.112}$$

we obtain the important cosmological relation

$$\frac{a\left(t_{emit}\right)}{a\left(t_{obs}\right)} = \frac{1}{1+z}. \tag{5.113}$$

Once the redshift of the radiation is obtained from observations, by using the above relation we can find the numerical value of the scale factor of the Universe at the moment of photon emission. On the other hand, an important observational result is that the frequency of the electromagnetic radiation emitted by astrophysical objects varies in time, due to the expansion of the Universe, according to the simple law $\nu \propto a^{-1}$.

5.2.2 The Friedmann Equations

We will now investigate the behavior of general relativistic cosmological models for different types of matter sources, described by different energy-momentum tensors. The FLRW metric (5.92) contains an arbitrary constant k, describing the geometry of the Universe, and an unknown time-dependent function $a(t)$, the scale factor, which can be determined from Einstein's gravitational field equations (3.43), which we rewrite here

$$G^{\mu}_{\nu} = R^{\mu}_{\nu} - \frac{1}{2}\delta^{\mu}_{\nu}R = \frac{8\pi G}{c^4}T^{\mu}_{\nu}. \tag{5.114}$$

The assumption of the isotropy of the Universe requires that the components T^{i}_{0}, $i = 1, 2, 3$, of the energy-momentum tensor T^{μ}_{ν} must be equal to zero, and that its spatial components must have a simple diagonal form, with $T^{1}_{1} = T^{2}_{2} = T^{3}_{3}$. Moreover, the homogeneity of space implies that all the energy-momentum tensor components T^{μ}_{ν} must not depend on the spatial radial coordinate r. In the comoving reference frame in which the FLRW metric is written down, the components of the four velocity of the cosmological matter flow become $U^0 = -1$, and $U^i = 0$, $i = 1, 2, 3$. Therefore, it turns out that for a Universe obeying the cosmological principle, the energy-momentum tensor has the simple diagonal form

$$T^{\mu}_{\nu} = \mathrm{diag}\left[-\rho\left(t\right)c^2, p(t), p(t), p(t)\right], \tag{5.115}$$

where ρ denotes, as usual, the mass density of the cosmic fluid, while p is its thermodynamic pressure. In writing down the energy-momentum tensor in this form we have assumed that the matter in the Universe consists of an ideal fluid, in which dissipative or heat transfer processes are ignored. Of course, in a realistic astrophysical approach the matter content of the Universe may not be describable by an ideal cosmological fluid, but such an idealized physical description can give at least a qualitative understanding of the cosmological processes. For the FLRW metric the Einstein tensor G^{μ}_{ν} can be calculated easily. The only nonzero components are

$$G^0_0 = -\frac{3}{a^2}\left(\dot{a}^2 + kc^2\right), \qquad G^j_i = -\frac{1}{a^2}\left(2a\ddot{a} + \dot{a}^2 + kc^2\right)\delta^j_i. \tag{5.116}$$

Thus, the Einstein field equations reduce to only two independent equations, given by

$$3\frac{\dot{a}^2}{a^2} + \frac{3kc^2}{a^2} = 8\pi G\rho, \tag{5.117}$$

$$2\frac{\ddot{a}}{a} + \frac{\dot{a}^2}{a^2} + \frac{kc^2}{a^2} = -\frac{8\pi G}{c^2}p. \tag{5.118}$$

These two equations are called the Friedmann equations, and they are the basic equations of cosmology. In order to close the system of cosmological evolution equations we must also specify the *equation of state* of the cosmic matter, which is given by the relation

$$p = p(\rho). \tag{5.119}$$

Once this is known, the two Friedmann equations completely determine the three unknown cosmological functions $a(t)$, $\rho(t)$, and $p(t)$. Eliminating the term $\left(\dot{a}^2 + kc^2\right)/a^2$ between the Friedmann equations gives

$$\frac{\ddot{a}}{a} = -\frac{4\pi G}{3}\left(\rho + 3\frac{p}{c^2}\right). \tag{5.120}$$

For normal cosmological matter, the condition $\rho + 3p/c^2 > 0$ always holds (this satisfies the strong energy condition), thus implying that $\ddot{a} < 0$. From Eq. (5.117) we obtain that

$$\rho a^3 = \frac{3}{8\pi G}a\left(\dot{a}^2 + kc^2\right). \tag{5.121}$$

By differentiating the above relation, with the use of the second Friedmann equation (5.118), we obtain immediately

$$\frac{d}{dt}\left(\rho a^3\right) = -3a^2\dot{a}\frac{p}{c^2} = -\frac{p}{c^2}\frac{d}{dt}a^3. \tag{5.122}$$

This equation can be reformulated in two equivalent forms as

$$\frac{d\left(\rho a^3\right)}{da} = -3a^2\frac{p}{c^2}, \tag{5.123}$$

or

$$\frac{d\rho}{dt} + 3H\left(\rho + \frac{p}{c^2}\right) = 0, \tag{5.124}$$

where $H = \dot{a}/a$. Thus, once the equation of state $p = p(\rho)$ of the cosmological matter is given, we can immediately integrate Eq. (5.123) to find the density as a function of the scale factor, $\rho = \rho(a)$. After the density-scale factor relation is known, by substituting this relation into Eq. (5.117) we can obtain the time evolution of the scale factor a. From Eq. (5.117) it follows that

$$\frac{k}{a^2} = \frac{8\pi G}{3}\rho - \frac{\dot{a}^2}{a^2} = \frac{\dot{a}^2}{a^2}\left(\frac{\rho}{3H^2/8\pi G} - 1\right). \tag{5.125}$$

We can introduce now the critical density $\rho_c = 3H^2/8\pi G$, and the density parameter $\Omega(t) = \rho(t)/\rho_c$. Hence we obtain

$$\frac{k}{a^2} = H^2\left[\Omega(t) - 1\right]. \tag{5.126}$$

Evaluating the above equation at $t = t_0$ gives

$$\frac{k}{a_0^2} = H_0^2\left[\Omega(t_0) - 1\right]. \tag{5.127}$$

Thus, in the Friedmann cosmological models the density parameter Ω fully establishes the spatial three-geometry of the Universe. In particular, for $\Omega(t) \equiv 1$, the geometry of the Universe is flat.

The Friedmann equations can be reformulated in terms of the Hubble function as

$$3H^2 + \frac{3kc^2}{a^2} = 8\pi G\rho, \tag{5.128}$$

$$2\dot{H} + 3H^2 + \frac{kc^2}{a^2} = -\frac{8\pi G}{c^2}p. \tag{5.129}$$

An important cosmological quantity is the deceleration parameter q, defined as

$$q = \frac{d}{dt}\frac{1}{H} - 1 = -\frac{\dot{H}}{H^2} - 1 = -\frac{\ddot{a}}{aH^2}. \tag{5.130}$$

After dividing Eq. (5.129) by H^2, replacing the term $\dot{H}/H^2$ in the definition of the deceleration parameter, and by using Eq. (5.128) to replace H^2, we obtain for the deceleration parameter the expression

$$q = \frac{1}{2} + \frac{kc^2}{2\dot{a}^2} + \frac{4\pi G}{c^2}\frac{p}{8\pi G\rho/3 - kc^2/a^2}. \tag{5.131}$$

In the case of the flat geometry with $k = 0$, the deceleration parameter can be written in the simple form

$$q = \frac{1}{2}(1 + 3w), \tag{5.132}$$

where we have introduced the parameter of the matter equation of state w, defined as

$$w = \frac{p}{\rho c^2}. \tag{5.133}$$

For a cosmic object having an intrinsic luminosity L, we can obtain the luminosity distance d_L to the object, once its energy flux F is measured, by using the fact that the intensity of the radiation follows a Newtonian-type inverse square law, decreasing as it travels away from the source. The luminosity distance d_L is defined as

$$d_L(z) = \sqrt{\frac{L}{4\pi F}} = (1+z)\, r(z), \tag{5.134}$$

where by $r(z)$ we have denoted the comoving distance to the emitting object, located at the cosmological redshift z and defined, for a flat geometry with $k = 0$, by the relation

$$r(z) = \int_0^z \frac{dz'}{H(z')} = \int_{1/(1+z)}^1 \frac{da}{a^2 H(a)}, \tag{5.135}$$

where, by convention, we assume that $a = 1$ today. By Taylor expanding the scale factor around its present-day value, we have

$$a(t) = 1 + H_0(t-t_0) - \frac{1}{2} q_0 H_0^2 (t-t_0)^2 + ..., \tag{5.136}$$

which allows to write the distance–redshift relation as

$$H_0 d_L = z + \frac{1}{2}(1-q_0)\, z^2 + ... \tag{5.137}$$

This relation is valid only for $z \ll 1$.

Exact Cosmological Models

We now turn back to the Friedmann equations, and obtain some of their simple solutions. We assume that the matter content of the Universe is represented by an ideal fluid that can be characterized by a linear polytropic equation of state, with

$$p = w\rho c^2, \tag{5.138}$$

where $w =$ constant, is called the parameter of the equation of state. For the linear polytropic equation of state we can easily integrate Eq. (5.123), to obtain

$$\rho(a) = \frac{\rho_0}{a^{3(1+w)}}, \tag{5.139}$$

where ρ_0 is an arbitrary integration constant. From a physical point of view there are several important models that can be described by the relation in Eq. (5.139). The case $w = 0$ corresponds to nonrelativistic dust matter, whose energy density scales as $\rho_{NR} \propto a^{-3}$. For electromagnetic radiation, the parameter w is $w = 1/3$, and the radiation energy density evolves as $\rho_R \propto a^{-4}$ in the expanding Universe. If $w = 1$, then $p = \rho c^2$, and $\rho_S \propto a^{-6}$. A fluid satisfying an equation of state with $w = 1$ is called a stiff fluid, and it has the property that the speed of sound $c_s^2 = \partial p/\partial \rho$, describing the propagation of compressional matter perturbations, has the same value as the speed of light, $c_s = c$.

Of particular importance is the case $w = -1$, for which we find

$$\rho_V = \text{constant}. \tag{5.140}$$

In this case the pressure

$$p = -\rho c^2, \tag{5.141}$$

should be negative, since we must have $\rho > 0$ in order to keep $\dot{a}^2/a^2 > 0$. The negative pressure makes the energy density inside a given volume to remain constant, even when the volume increases due to the expansion of the Universe.

For $k = 0$ we can easily integrate the first Friedmann equation. Taking into account the expression of the density we find

$$\frac{\dot{a}}{a} \propto a^{3(1+w)/2}, \tag{5.142}$$

and

$$a(t) \propto \begin{cases} t^{2/3(1+w)}, & w \neq -1, \\ e^{\lambda t}, & w = -1. \end{cases}, \tag{5.143}$$

where λ is an arbitrary integration constant. The total mass density of the Universe ρ_{tot} can be represented as

$$\begin{aligned} \rho_{tot} &= \rho_R(a) + \rho_{NR}(a) + \rho_V(a) \\ &= \rho_c \left[\Omega_R \left(\frac{a_0}{a}\right)^4 + \Omega_{NR} \left(\frac{a_0}{a}\right)^3 + \Omega_V\right], \end{aligned} \tag{5.144}$$

where the critical density ρ_c and the various Ω correspond to their values evaluated at $a = a_0$. The nonrelativistic density parameter Ω_{NR} can generally be written as $\Omega_{NR} = \Omega_B + \Omega_{DM} + ...$, where Ω_B denotes the density parameter of the baryonic matter, while Ω_{DM} is the density parameter of the dark matter, respectively. By substituting the expression of the total energy density of the Universe into the first Friedmann equation gives

$$\frac{\dot{a}^2}{a^2} + \frac{kc^2}{a^2} = H_0^2 \left[\Omega_R \left(\frac{a_0}{a}\right)^4 + \Omega_{NR} \left(\frac{a_0}{a}\right)^3 + \Omega_V\right]. \tag{5.145}$$

The solution of this differential equation provides the full picture of the evolution of the Universe in its different phases.

5.2.3 Cosmological Constant, Vacuum Energy, or Dark Energy?

The equation of state $p = -\rho c^2$ is of particular importance in cosmology, since it also naturally arises if the Einstein field equations are generalized by adding a constant term $-\Lambda\delta^\mu_\nu$ on their left-hand side. This term was introduced by Einstein in 1917 in order to build the first static general relativistic cosmological model, and for this reason the constant Λ is called the *cosmological constant.* However, adding a cosmological constant to the geometric left-hand side of the Einstein equations is formally equivalent to adding a new term $+\Lambda\delta^\mu_\nu$ to the right-hand side of the gravitational field equations, where it acts as a source term. Therefore one can also adopt the alternative point of view according to which the cosmological constant represents a special kind of "matter" source, satisfying a peculiar equation of state.

In the presence of a cosmological constant the Friedmann equations can be written as

$$3H^2 + \frac{3kc^2}{a^2} = 8\pi G\rho + \Lambda c^2, \tag{5.146}$$

$$2\dot{H} + 3H^2 + \frac{kc^2}{a^2} = -\frac{8\pi G}{c^2}p + \Lambda c^2. \tag{5.147}$$

By a simple calculation one can check that for the case of a flat vacuum Universe, with $k = 0$, and $\rho = p = 0$, in the presence of a cosmological constant the Friedmann equations have the expanding solution

$$H \equiv H_0 = \sqrt{\frac{\Lambda c^2}{3}} = \text{constant}, \tag{5.148}$$

which gives an exponentially increasing scale factor of the form

$$a(t) = a_0 \exp\left(\sqrt{\frac{\Lambda c^2}{3}}t\right) = a_0 \exp\left(H_0 t\right). \tag{5.149}$$

This solution of the Einstein field equations is called the de Sitter solution, and it plays a fundamental role in modern cosmology. In the presence of a cosmological constant the deceleration parameter is given by

$$q = \frac{1}{2} + \frac{kc^2}{2\dot{a}^2} + \frac{4\pi G}{c^2}\frac{p}{8\pi G\rho/3 + \Lambda c^2/3 - kc^2/a^2} - \frac{\Lambda c^2}{2H^2}. \tag{5.150}$$

In the limit of a flat vacuum Universe with $k = 0$ and $\rho = p = 0$, we obtain the deceleration parameter of the de Sitter Universe as $q = -1$.

The cosmological constant is sometimes called the vacuum energy. Its interpretation as a vacuum energy density can be acquired through the following simple line of reasoning that uses the basic principles of quantum mechanics. In quantum field theory, it is customary to define the vacuum as the ground state of the energy operator. By using the Lorentz invariance principle of the quantum fields one can prove that the mean value of the energy momentum tensor of the vacuum must have the following expression,

$$\langle T_{\mu\nu}\rangle_{vac} = \langle \varepsilon \rangle g_{\mu\nu}, \tag{5.151}$$

where $\langle \varepsilon \rangle$ is the mean value of the energy density operator of the vacuum, and $g_{\mu\nu}$ is the metric tensor, respectively. In a general quantum framework $\langle \varepsilon \rangle$ is different from zero, and it can be obtained by summing the zero-point energies of all the normal modes of the considered physical fields. From a physical point of view the zero-point energy of a quantum field can also be interpreted as the result of the spontaneous production in the quantum vacuum of a large number of virtual particles. Since the law of the conservation of energy is a key result of physics, the creation of such virtual particles is restricted by the limits assessed by the quantum energy-time uncertainty relations. Then the energy conservation together with the energy-time uncertainty principle immediately tells us that the virtual particle must necessarily have a very short existence.

Consider a virtual quantum particle that lives for a very short time interval $\sim \tau$. Hence, according to the energy-time uncertainty principle, its energy E_τ must satisfy the constraint $E_\tau \geq \hbar/\tau$. While they exist, virtual quantum particles produced from the vacuum can travel only up to a distance $\sim c\tau$. Hence, we can construct a small cell of volume $\sim (c\tau)^3$ encompassing each of them. Therefore the energy density associated with the virtual particles can be written as $\sim \hbar c/(c\tau)^4$. We assume now that the smallest possible length scale in nature is the Planck length $l_P = \sqrt{\hbar G/c^3} = 1.6 \times 10^{-33}$ cm. Below the Planck length the very concept of space-time continuum breaks down. Adopting for τ the value $c\tau = l_P$, it turns out that the energy density associated to the virtual particles created from the vacuum must be larger than

$$\langle \varepsilon \rangle \geq \frac{\hbar c}{l_P^4} = 4.8 \times 10^{114}\ \mathrm{erg/cm^3}. \tag{5.152}$$

Hence, the above arguments indicate that the numerical value of $\langle \varepsilon \rangle$ must be at least of this order of magnitude. As a consequence, it follows that due to virtual particle creation the energy density associated with the vacuum has a non-zero, and very large value. The mean value of the vacuum energy density cannot be taken to be equal to zero, and therefore it should be included as a matter source for the gravitational field in the Einstein field equations, where it acts as a cosmological constant. The observational determination of the actual value of the energy density associated to the cosmological constant gives an energy density of around 10^{-6} erg/cm^3. Therefore there is a difference of the order of 10^{120} between the theoretical prediction of the cosmological constant as a form of vacuum energy, and the astronomical and astrophysical observations.

To explain the astronomical observations that revealed the recent acceleration of the Universe, a mysterious and dominant component is needed, which is generically called dark energy. Even if we do not really know what dark energy is, or even if it really exists, and it is not only a modification of gravity at cosmological scales, in the standard ΛCDM cosmological paradigm it is assumed that the late acceleration of the Universe is produced by the cosmological constant Λ. However, as we have just seen, the numerical value of Λ that we find from cosmological observations is many orders of magnitude smaller than the theoretically expected contribution from the vacuum energy, which also conducts itself as a cosmological constant. This important problem in theoretical physics and cosmology is known as the old cosmological constant problem. But there is another problem with the cosmological constant. In the ΛCDM model, the Universe is dominated by matter in the past, up to almost the present time, while its future dynamics will be dominated by dark energy. The question then is why we are living in the short cosmological time interval when the two densities of the matter and of the dark energy are approximately equal, having the same order of magnitude. This intriguing question is known presently as the new cosmological constant problem.

As we have seen, vacuum energy, and/or the cosmological constant, do not change in space and time, and hence they are not dynamical variables. However, if we insert in the cosmological gravitational theory a new degree of freedom in the form of a scalar field ϕ, we can transform vacuum energy in an effective dynamical quantity varying in space and time [197, 430, 505, 521]. For this reason, the study of scalar fields may shed some light on the dark energy mystery. In the next section we will briefly review some basic properties of scalar field models.

5.2.4 Scalar Field Cosmologies with Arbitrary Self-Interaction Potential

In the following we will consider a general class of scalar field models, with the scalar field minimally coupled to the gravitational field. The Lagrangian density for such a system in the Einstein frame reads [470]

$$L = \frac{1}{2\kappa}\sqrt{|g|}\left\{R + \kappa\left[g^{\mu\nu}\left(\partial_\mu\phi\right)\left(\partial_\nu\phi\right) - 2V\left(\phi\right)\right]\right\}, \tag{5.153}$$

where ϕ is the scalar field, $V(\phi)$ is its self-interaction potential, and $\kappa = 8\pi G/c^4$ is the gravitational coupling constant, respectively. The first term in L is the Einstein–Hilbert gravitational Lagrangian, the second term corresponds to the kinetic energy of the scalar field, while the third term is its potential energy. In the following we will use natural units with $c = 8\pi G = \hbar = 1$.

For a flat FLRW Universe in the presence of a scalar field, the cosmological evolution is described by the system of the two Friedmann field equations, given by

$$3H^2 = \rho_\phi = \frac{\dot{\phi}^2}{2} + V\left(\phi\right), \tag{5.154}$$

$$2\dot{H} + 3H^2 = -p_\phi = -\frac{\dot{\phi}^2}{2} + V\left(\phi\right), \tag{5.155}$$

together with the Klein–Gordon time evolution equation of the scalar field

$$\ddot{\phi}+3H\dot{\phi}+V'(\phi)=0, \tag{5.156}$$

where the prime in the expression of the potential V denotes the derivative with respect to the scalar field ϕ. In the following, we will limit our investigations to the case of the expansionary cosmological models, which satisfy the basic requirement that the scale factor a of the Universe is a monotonically increasing function of the cosmological time. Note that for such cosmological models with a continuous expansion the important condition $H>0$ is always satisfied. On the other hand cosmological models with $H<0$ describe collapsing scalar field structures, in which the scale factor a is a monotonically decreasing function of time.

After the addition of Eqs. (5.154) and (5.155), it follows that H satisfies a Riccati-type equation of the form

$$\dot{H}=V-3H^2. \tag{5.157}$$

We substitute now the Hubble function as given by Eq. (5.154) into Eq. (5.156), thus obtaining the generalized Klein–Gordon equation describing the dynamics of the scalar field, and its time evolution, as [238]

$$\ddot{\phi}+\sqrt{3}\sqrt{\frac{\dot{\phi}^2}{2}+V(\phi)}\,\dot{\phi}+\frac{dV}{d\phi}=0. \tag{5.158}$$

To further simplify the above Klein–Gordon-type equation describing the dynamics of the scalar fields in the flat FLRW geometry, we perform several transformations of the dynamical variables. First, we introduce a new function $f(\phi)$ so that the time derivative of the scalar field is given by $\dot{\phi}=\sqrt{f(\phi)}$. As the next step we change the independent time variable t to ϕ. Thus Eq. (5.158) takes the form

$$\frac{1}{2}\frac{df(\phi)}{d\phi}+\sqrt{3}\sqrt{\frac{f(\phi)}{2}+V(\phi)}\sqrt{f(\phi)}+V'(\phi)=0, \tag{5.159}$$

and it can be rewritten as

$$\frac{\frac{1}{2}\frac{df(\phi)}{d\phi}+V'(\phi)}{2\sqrt{\frac{f(\phi)}{2}+V(\phi)}}+\frac{\sqrt{3}}{2}\sqrt{f(\phi)}=0. \tag{5.160}$$

Now, we define a new function

$$F(\phi)=\sqrt{\frac{f(\phi)}{2}+V(\phi)}, \tag{5.161}$$

which gives $f(\phi) = 2\left[F^2(\phi) - V(\phi)\right]$. Hence Eq. (5.160) becomes

$$\frac{dF(\phi)}{d\phi} + \sqrt{\frac{3}{2}}\sqrt{V(\phi)}\sqrt{\left[\frac{F(\phi)}{\sqrt{V(\phi)}}\right]^2 - 1} = 0. \tag{5.162}$$

We introduce at this moment a new function $u(\phi)$, given by $F(\phi) = u(\phi)\sqrt{V(\phi)}$. Hence Eq. (5.162) takes the form

$$\frac{1}{\sqrt{u^2-1}}\frac{du}{d\phi} + \frac{1}{2V}\frac{dV}{d\phi}\frac{u}{\sqrt{u^2-1}} + \sqrt{\frac{3}{2}} = 0. \tag{5.163}$$

Finally, we define a new function G, which is related to $u(\phi)$ by $u(\phi) = \cosh G(\phi)$. Then, by expressing u in terms of G, we find the fundamental dynamical equation describing the time evolution of the scalar fields with self-interaction potential V in the flat FLRW Universe as

$$\frac{dG}{d\phi} + \frac{1}{2V}\frac{dV}{d\phi}\coth G + \sqrt{\frac{3}{2}} = 0. \tag{5.164}$$

Equation (5.164) is a first-order strongly nonlinear differential equation. For the function f we find

$$f(\phi) = 2V(\phi)\sinh^2 G(\phi), \tag{5.165}$$

giving for $\dot{\phi}$ the expression

$$\dot{\phi} = \sqrt{2V(\phi)}\sinh G(\phi). \tag{5.166}$$

The time evolution of G is obtained from the first-order differential equation

$$\frac{dG}{dt} = -\sqrt{2V(\phi)}\sinh G\left[\sqrt{\frac{3}{2}} + \frac{1}{2V(\phi)}\frac{dV}{d\phi}\coth G\right]. \tag{5.167}$$

The function $G(\phi)$ can be calculated, once the scalar field ϕ is known, from the equation

$$G(\phi) = \operatorname{arccosh}\sqrt{1 + \frac{\dot{\phi}^2}{2V(\phi)}}. \tag{5.168}$$

Moreover, as a function of the scalar field ϕ, the scale factor a of the Universe can be obtained by solving the differential equation

$$\frac{1}{a(\phi)}\frac{da(\phi)}{d\phi} = \frac{1}{\sqrt{6}}\coth G(\phi). \tag{5.169}$$

As a function of G, the scale factor can be obtained alternatively as a solution of the following differential equation

$$\frac{1}{a}\frac{da}{dG} = -\frac{1}{\sqrt{6}}\frac{\coth G}{\sqrt{\frac{3}{2} + \frac{1}{2V}\frac{dV}{d\phi}\coth G}}. \tag{5.170}$$

The deceleration parameter q can be represented as a function of F and V in the form

$$q(\phi) = \frac{d}{dt}\left(\frac{1}{H}\right) - 1 = \sqrt{6\left[F^2(\phi) - V(\phi)\right]}\frac{d}{d\phi}\left[F(\phi)\right]^{-1} - 1. \tag{5.171}$$

Alternatively, as a function of the potential V and of the function G, the deceleration parameter is obtained as

$$q(\phi) = \sqrt{6V(\phi)}\sinh G(\phi)\frac{d}{d\phi}\left[\frac{1}{\sqrt{V(\phi)}\cosh G(\phi)}\right] - 1. \tag{5.172}$$

With the use of Eqs. (5.164) and Eq. (5.172), the latter gives the compact result

$$q(\phi) = 3\tanh^2 G(\phi) - 1. \tag{5.173}$$

By substituting Eq. (5.168) into Eq. (5.173), we find for the deceleration parameter the following form

$$q(\phi) = 2 - 3\left(1 + \frac{\dot{\phi}^2}{2V(\phi)}\right)^{-1}. \tag{5.174}$$

If, during a specific phase of cosmological evolution the potential energy dominates the kinetic energy of the scalar field, $V(\phi) \gg \dot{\phi}^2/2$, then from Eq. (5.174) it follows that $q \to -1$, with the Universe experiencing an exponential, de Sitter-type expansion. It is important to mention that this property of the deceleration parameter is independent of the functional form of the scalar field self-interaction potential $V(\phi)$, and depends only on the numerical values of the ratio of the kinetic and potential energy of the field.

With the use of the new functional variables $f(\phi)$ and $F(\phi)$, from Eqs. (5.154) and (5.155) we obtain

$$\dot{H} = -\frac{\dot{\phi}^2}{2} = -\frac{f(\phi)}{2} \tag{5.175}$$

and

$$3H^2 = F^2(\phi), \tag{5.176}$$

respectively. Eqs. (5.175) and (5.176) tells us that the functions $f(\phi)$ and $F(\phi)$ are closely related to the Hubble function, and its time derivative, respectively.

A similar approach, in which the Hubble function was taken to be a function of the scalar field ϕ, was developed for the study of the cosmological scalar field models in [443].

Cosmological Application: The Case of the Exponential Potential

As we have already mentioned in the Introduction, scalar fields are considered to play a central role in cosmological models of the early and late Universe. The self-interaction potential energy density V of the scalar fields is not influenced directly by the cosmological expansion of the Universe, and hence it can play the role of an effective cosmological constant, inducing a period of inflationary, de Sitter-type expansion in the very early Universe, or driving the late-time acceleration of the Universe. However, the evolution of the Universe strongly depends on the specific functional form of the scalar field potential $V(\phi)$, with different forms of the potential generating very different cosmological scenarios.

A simple form of the scalar field self-interaction potential, which has been intensively investigated in the recent physical literature, is the exponential-type potential. As one can see immediately, Eq. (5.164) can be integrated easily in the case of scalar field potentials satisfying the general mathematical condition V'/V = constant. Therefore, for potentials belonging to this class the gravitational field equations can be solved exactly, and their solution can be represented in an exact analytical form. Classes of exact solutions of the field equations can be obtained by assuming that V'/V is some given function of G, i.e., $V'/V = f(G)$. For many choices of the function $f(G)$, the first-order evolution equation for the scalar field, described by Eq. (5.164), can be solved exactly by using standard mathematical techniques, and the solution corresponding to a given scalar field potential can be represented in a closed analytical form, which can easily be studied. In the following, we will restrict our analysis to the case of the exponential potential.

If V'/V has a constant value for all ϕ, that is, $V'/V = \sqrt{6}\alpha_0$ =constant, then the scalar field self-interaction potential can be written as an exponential function of the field,

$$V = V_0 \exp\left(\sqrt{6}\alpha_0\phi\right), \tag{5.177}$$

where V_0 is an arbitrary constant. The conditions for the exact integrability of the gravitational field equations for exponential-type scalar field potentials were considered in [133, 206, 370, 451].

By taking into account Eq. (5.177), then Eq. (5.164) can be rewritten in the equivalent form

$$\frac{dG}{d\phi} + \sqrt{\frac{3}{2}}\left(\alpha_0 \coth G + 1\right) = 0. \tag{5.178}$$

Next we investigate several solutions of cosmological interest of this equation.

1. The case $\alpha_0 \neq \pm 1$: For this case, Eq. (5.178) can be integrated immediately to give

$$\sqrt{\frac{3}{2}}\left[\phi(G)-\phi_0\right]=\frac{G-\alpha_0 \ln\left|\sinh G+\alpha_0 \cosh G\right|}{\alpha_0^2-1}, \tag{5.179}$$

where ϕ_0 is an arbitrary constant of integration. The dependence of the cosmological time on the scalar field parameters, contained in the function G, can be calculated from Eq. (5.167) as

$$t(G)-t_0=-\frac{1}{\sqrt{3V_0}}\int \frac{dG}{e^{\sqrt{3/2}\alpha_0\phi}\left(\sinh G+\alpha_0 \cosh G\right)}, \tag{5.180}$$

where t_0 is an arbitrary constant of integration. With the use of Eq. (5.179), the above equation can be transformed to give the following integral representation of the cosmological time t

$$t(G)-t_0=-\frac{e^{-\sqrt{3/2}\alpha_0\phi_0}}{\sqrt{3V_0}}\int e^{\left[\alpha_0/\left(1-\alpha_0^2\right)\right]G}\left(\sinh G+\alpha_0 \cosh G\right)^{1/\left(\alpha_0^2-1\right)} dG. \tag{5.181}$$

Thus, Eqs. (5.180) and (5.181) give a closed-form parametric representation of the time evolution of the scalar field ϕ with an exponential-type self-interaction potential in a flat FLRW Universe, with G taken as a parameter. As for the scale factor, it can also be obtained in a parametric form as

$$a(G)=a_0 e^{\left[\alpha_0/3\left(1-\alpha_0^2\right)\right]G}\left(\sinh G+\alpha_0 \cosh G\right)^{1/3\left(\alpha_0^2-1\right)}, \tag{5.182}$$

where a_0 denotes an arbitrary constant of integration. The deceleration parameter is given parametrically as a function of G by the simple expression

$$q(G)=3\tanh^2 G-1. \tag{5.183}$$

For some particular values of α_0 the expression of the time integral given by Eq. (5.181) can be obtained in an exact form. For example, if $\alpha_0=\pm\sqrt{2}$, we find

$$t_{\pm}(G)-t_0^{\pm}=-\frac{e^{\mp\left(\sqrt{2}G+\sqrt{3}\phi_0\right)}}{\sqrt{3V_0}}\left[3\cosh G\pm 2\sqrt{2}\sinh G\right]. \tag{5.184}$$

For $\alpha_0=\pm\sqrt{3/2}$ we obtain the time dependence of t on G as

$$t_{\pm}(G)-t_0^{\pm}=\pm\frac{1}{24\sqrt{V_0}}\mathrm{e}^{\mp\left(\sqrt{6}G+\frac{3\phi_0}{2}\right)}\left[\sqrt{2}+27\sqrt{2}\cosh(2G)\pm 22\sqrt{3}\sinh(2G)\right], \tag{5.185}$$

while for $\alpha_0 = \pm 2/\sqrt{3}$ we deduce the result

$$t_{\pm}(G) - t_0^{\pm} = \frac{1}{396\sqrt{3V_0}} e^{\mp\left(2\sqrt{3}G + \sqrt{2}\phi_0\right)} \times$$
$$\left\{45\cosh G + 1067\cosh(3G) \pm 8\sqrt{3}\left[3\sinh G + 77\sinh(3G)\right]\right\}. \quad (5.186)$$

Therefore, the exact solution of the gravitational field equations in the presence of a scalar field with exponential potential can be obtained in an exact parametric form in several cases, and consequently there is no need to resort to numerical integration for an in-depth study of the cosmological properties of the solutions.

2. The case $G(\phi) = G_0 =$ constant: Another case in which the gravitational field equations can be integrated exactly corresponds to the choice $G(\phi) = G_0 =$ constant, for which a particular solution can be obtained. With this form of G, Eq. (5.178) is identically satisfied, with G_0 given by

$$G_0 = \operatorname{arccoth}\left(-\frac{1}{\alpha_0}\right) = \frac{1}{2}\ln\left|\frac{1-\alpha_0}{1+\alpha_0}\right|, \qquad 0 < |\alpha_0| < 1. \quad (5.187)$$

From Eq. (5.169) it immediately follows that the scale factor a of the flat Universe can be obtained as a function of the scalar field ϕ as

$$a = a_0 e^{-\phi/\sqrt{6}\alpha_0}. \quad (5.188)$$

The time variation of the scalar field is determined by the equation

$$\dot{\phi} = \pm\sqrt{2V_0}\sinh\left(G_0\right) e^{\sqrt{3/2}\alpha_0\phi}, \quad (5.189)$$

which follows from Eq. (5.168). The general solution of Eq. (5.189) is given by

$$e^{-\sqrt{3/2}\alpha_0\phi} = \pm\sqrt{3V_0}\sinh\left(G_0\right)\alpha_0\left(t_0 - t\right), \quad (5.190)$$

where t_0 is an arbitrary constant of integration. By using Eq. (5.188) we obtain the scale factor of the scalar field filled Universe in the form

$$a(t) = a_0\left[\pm\sqrt{3V_0}\sinh\left(G_0\right)\alpha_0\left(t_0 - t\right)\right]^{\frac{1}{3\alpha_0^2}}. \quad (5.191)$$

The scalar field potential has a time variation given by the expression

$$V(t) = \frac{V_0}{3V_0\alpha_0^2\sinh^2\left(G_0\right)}\frac{1}{\left(t - t_0\right)^2}$$
$$= \left(\frac{1-\alpha_0^2}{3\alpha_0^4}\right)\frac{1}{\left(t - t_0\right)^2} = \frac{V_0}{\left(t - t_0\right)^2}, \quad (5.192)$$

with the constants V_0, α_0 and G_0 satisfying the consistency condition

$$3V_0\alpha_0^2 \sinh^2(G_0) = 1, \tag{5.193}$$

which can be obtained immediately by comparing the second and the last terms in Eq. (5.192).

Several simple power-law solutions describing inflationary cosmological models containing scalar fields with exponential potentials have been obtained and studied in [206].

5.2.5 The Dark Matter Problem

The assumption of the existence of dark matter, a component of the Universe as mysterious as dark energy, is one of the cornerstones of present-day cosmology and galactic astrophysics, and it is an essential part of the standard ΛCDM cosmological paradigm [21, 29, 141, 476, 522]. The first evidence for its presence in the Universe was provided by the study of galactic rotation curves – specifically, the finding that these curves decayed far more slowly than one would expect when one included only the effects of baryonic matter (gas and stars) in the analysis of their motion. Consider a massive test particle in rotational motion far away from the baryonic matter distribution M_B of a galaxy. Then its tangential velocity v_{tg} should decay according to the simple Keplerian law

$$v_{tg}^2 = \frac{GM_B}{r}. \tag{5.194}$$

This means that with the increase of the distance from the galactic center, the velocity of the test particles must be smaller and smaller. However, a large body of astronomical observations has shown that the rotational velocities do not decay, but keep a constant value that extends to very large distances from the galactic center.

This behavior can be explained by assuming the existence of a supplementary (and exotic) galactic mass component, forming a large halo around the baryonic matter distribution, and most likely consisting of new particle(s) lying beyond the standard model of particle physics. The behavior of the rotation curves still represents the most powerful and convincing evidence for dark matter [213, 276, 414, 431]. But various other astrophysical and cosmological observations have also provided compelling evidence for the existence of dark matter, such as the recent determination of the cosmological parameters from the Planck satellite observations of the Cosmic Microwave Background (CMB) radiation [11]. These observations have also shown that the mass content of the Universe cannot be explained by baryonic matter only, thus confirming again the standard ΛCold Dark Matter (ΛCDM) cosmological paradigm. Other types of observations, such as gravitational lensing also require the existence of dark matter for a consistent interpretation of the data [136, 372, 500]. A particularly strong evidence for the existence of dark matter is provided by the observations of a galaxy cluster called the Bullet Cluster. In this cluster the baryonic matter and the dark matter components are separated due to a long-ago collision of its two components [346]. Measurements of the cosmological parameters by using the Planck data on the (CMB) radiation indicate that the Universe is composed of ~4–5% baryons, ~25% non-baryonic dark matter, and ~70% dark energy [11].

Based on the energy of the particles composing them, dark matter particle models can be divided into three types: cold, warm, and hot dark matter models, respectively [400]. The main candidates for dark matter are WIMPs (Weakly Interacting Massive Particles) and axions [400]. WIMPs are heavy particles that interact via the weak force [142, 347]. Axions are bosons that were first proposed to solve the strong CP problem [124, 458]. If the axions form the dark matter, then at low temperature the axion gas may form a Bose–Einstein condensate.

There are also other theories that try to explain the observations without introducing dark matter. These theories are based on a modification of the law of gravity at the galactic scales. The earliest of them is the MOND theory (Modified Newtonian Dynamics) [352]. Modified gravity theories [44, 191, 211, 233, 242, 243, 390] have also been used extensively as an alternatives to dark matter, as we will see in the next chapters of our book.

Even though dark matter models provide a good explanation of the qualitative behavior and constancy of the rotation curves, an important contradiction arises as a result of the in-depth comparison of simulation results with the observations. Data on almost all observed rotation curves show that they rise less steeply than cosmological simulations of structure formation in the standard ΛCDM model in the presence of a single pressureless dark matter component predict. The simulations indicate a central dark matter density profile that behaves as $\rho \sim 1/r$ (a cusp) [375], while the observed rotation curves indicate the presence of constant density cores [391]. This is the so-called core-cusp problem in dark matter physics. Another important open question that dark matter models have to face is the "too big to fail" problem [86, 87]. By using the Aquarius simulations it was shown that the most massive subhalos in the dark matter halos predicted in the ΛCDM model are obviously inconsistent with the dynamics of the brightest Milky Way dwarf spheroidal galaxies [87]. This is related to the fact that while the best-fitting galaxies related to dwarf spheroidal galaxies have velocities in the range $12\,\mathrm{km/s} < V_{max} < 25$ km/s, the ΛCDM simulations have found at least ten subhalos with velocities $V_{max} > 25$ km/s. These results cannot be explained in the framework of the ΛCDM-based models of the satellite populations of the Milky Way galaxy. The main problem emerging here is related to the densities of the satellites, with the dwarf spheroidals required to have more massive halos (by a factor of ~5) than observed in the astronomical investigations.

These problems related to the physics of dark matter may be solvable if one goes one step beyond the standard ΛCDM model, and assumes that the dark matter particles may possess some forms of self-interaction. Such a possibility has gained some observational support after the study of the data provided by the observations of 72 cluster collisions, including both "major" and "minor" mergers, with the observations done by using the Chandra and Hubble Space Telescopes [249]. Collisions between galaxy clusters can provide an important test of the nongravitational forces acting on dark matter, and the analysis done in [249] gives an upper limit of the ratio of the self-interaction cross-section σ_{DM} and of the mass m of the dark matter particle as $\sigma_{DM}/m < 0.47$ cm^2/g (95% Confidence Level). A new upper limit on the self-interaction crosssection of dark matter of $\sigma_{DM} < 1.28$ cm^2/g (68% Confidence Level), was obtained in [269]. From a theoretical point of view different self-interacting dark matter models were investigated in [118, 159, 448, 449].

However, after almost 80 years since the dark matter concept was introduced, the dark matter particle is still missing from the table of the elementary particles of nature.

5.3 Open Questions and Going Beyond General Relativity

In the present chapter we have briefly reviewed some of the fundamental theoretical issues of cosmology, the science devoted to the understanding of the Universe as a whole. Modern cosmology was born in 1917 as a direct application of the physical and mathematical principles of GR to the study of the entire Universe. In the past century it has undergone amazing changes and led to a consistent picture of nature that is both amazing and intriguing. Even though a full understanding of the structure of the Universe, dominated by the gravitational force, requires the use of GR, the Newtonian description can still give a reasonable approximation in many cosmological situations. It is due to this fact that we have presented in detail the Newtonian and Neo-Newtonian approaches to cosmology, and briefly outlined some of their applications. The Neo-Newtonian approach can obtain both Friedmann equations, with the pressure effects included, and the perturbation equation for the matter density fluctuations is identical to the general relativistic one. However, there is a huge conceptual difference between all forms of the Newtonian approach, with the corresponding dynamics taking place in absolute space and time of the Euclidian universe, and the geometric approach of GR, in which the arena of the physical phenomena is represented by the curved Riemannian geometry of the space-time continuum.

We have briefly introduced the foundations of relativistic cosmology, starting from the FLRW metric that encompasses the cosmological principle requiring an isotropic and homogeneous universe. The basic equations describing the time evolution of the basic cosmological parameters, the Friedmann equations, were also introduced, and some of their simple solutions were presented. A mathematical approach to the investigation of the scalar field cosmologies was also briefly outlined.

When dealing with modern cosmology two essential problems naturally arise: the dark energy and the dark matter problems, respectively. The dark energy (cosmological constant Λ) problem was born in 1917, and ever since it continues to play a central role in GR and cosmology. The best fit of the cosmological observations can be done assuming the existence of a simple Λ, constant in space and time. But this approach naturally raises some questions about the nature of the cosmological constant: why is it so small and why is it so fine-tuned? Is it part of the geometry, or is it a kind of special matter source? Or in fact is it just an artifact, and a particular case of a more complex structure, called dark energy, is the dominant energy component in the Universe, which in some limiting cases takes a nearly constant value? We also have the dark matter problem. What is its nature, and why have the many experiments performed up to now failed to detect it? We know that the dark matter particle cannot be a known particle of the standard particle physics model, a very successful theoretical model. Then is it a very light particle, like the axion? Or is it a WIMP?

However, there is another possible explanation of the mysteries of cosmology. What if no dark energy or dark matter exist? What if what we observe is in fact just a modification of the laws of gravity on cosmological scales? Einstein's gravity is very successful on the scale of the solar system, but can it be extended automatically to galactic or cosmological scales? What if other laws of the gravitational interaction come in to dominate the Universe far beyond the solar system? Then dark energy and dark matter would be just some consequences of these new laws of gravity we may call *dark gravity.* But are there such extensions of Einstein's theory that could explain dark energy and dark matter from the new perspective of dark gravity? If so, how can these theories be constructed, what is their mathematical structure, how are they tested, and what are their cosmological implications? These, and many other related questions, are the subject of the present book. In the next chapters we will try to formulate the appropriate answers to them, and see if indeed dark gravity is a viable alternative to Einstein's General Relativity, and its two dark components.

6

$f(R)$ Gravity

6.1 Introduction

The discovery in 1998 of the recent acceleration of the Universe [413, 436] has triggered a tremendous shift in our understanding of the cosmological dynamics, composition, and evolution of the cosmic structures. The combination of a wide range of astronomical observations have led to the formulation of the so-called ΛCDM cosmological paradigm, according to which our Universe is spatially flat, homogeneous, and isotropic on large scales [473]. What is more astonishing is that it is composed of ordinary matter, radiation, dark matter, plus a mysterious component, usually called dark energy, which is best described by a cosmological constant Λ. The cosmological constant, or its dark energy equivalent, is responsible for the acceleration of the distant type Ia supernovae [410]. Hence, based on these observations it turns out that around 95–96% of the content of the Universe is in the form of dark matter + a cosmological constant (dark energy), with only about 4–5% being represented by baryonic matter. More intriguingly, around 70% of the energy-density of the Universe should be in the form of the energy density associated to the cosmological constant.

The ΛCDM model is one of the simplest possible cosmological models. It fits very well and is compatible with all the observational data, with no other theoretical model providing a better fit or explanation. However, it faces a deep theoretical problem, namely that the existing present-day fundamental theories of physics cannot explain it. The inclusion of a simple cosmological constant in the theoretical framework of the gravitational interaction raises at least two fundamental questions: why is it so small and so fine-tuned? Why is it so different from other forms of matter? In a Universe with scale factor a, the density of the ordinary matter ρ_m decreases according to the law $\rho_m \sim a^{-3}$, while the energy density ρ_Λ associated to the cosmological constant does not vary. Moreover, we may also ask why the Universe started to accelerate only recently? In the previous chapter we also saw that the popular interpretation of the cosmological constant as vacuum energy faces some extraordinary challenges, due to the huge difference in the theoretical predictions and observations. Hence, today's theories of gravitation and cosmology face two fundamental problems, that can be epitomized as the dark energy problem, and the dark matter problem, respectively.

These theoretical problems are too severe to simply ignore, and we await for their possible solution sometime in the future. Hence, searching for alternatives to the ΛCDM cosmological model allows us not only to provide a better explanation/fit of the observational data, but also to test the very theoretical foundations of our present-day gravitational theory. As a basic feature these alternatives and extensions of Einstein gravity require that vacuum energy does not gravitate, and that $\rho_{vac} \equiv 0$.

There are two possibilities of formulating alternative descriptions of the cosmological and gravitational dynamics. The first one is to assume the existence of a universal dynamical component, filling the Universe, and which is called *dark energy.* In the presence of dark energy the generalized Einstein gravitational field equations take the form

$$G_{\mu\nu} = \frac{8\pi G}{c^4} T_{\mu\nu} + \frac{8\pi G}{c^4} T_{\mu\nu}^{\text{dark}}, \tag{6.1}$$

where $T_{\mu\nu}^{\text{dark}}$ is the energy-momentum tensor of the time varying dark energy field, which is described by two thermodynamic parameters, its pressure p_{DE}, and its energy density ρ_{DE}. In order to obtain consistency with observations we need to have $w = p_{DE}/\rho_{DE} < -1/3$. There are many candidates for dark energy, including quintessence fields, effective dark energy via nonlinear effects of structure formation, higher dimensional brane world models, etc. (For recent reviews of dark energy, see [25, 313, 331, 368].) The second alternative for the explanation of the cosmic dynamics is to assume that at large scales Einstein's General Relativity breaks down, and a more general theory describes the gravitational field. We tentatively refer to this type of modification of GR as *dark gravity*, and generally the geometric modification/extension of GR can be schematically represented as

$$G_{\mu\nu} + G_{\mu\nu}^{\text{dark}} = \frac{8\pi G}{c^4} T_{\mu\nu}, \tag{6.2}$$

where $G_{\mu\nu}^{\text{dark}}$ is a new geometric degree of freedom included in the gravitational theory, and which is responsible for the modification of the cosmological dynamics. It is the main goal of this book to present some of the recently proposed dark gravity theories, and to investigate their astrophysical and cosmological implications.

Almost immediately after its formulation, geometric extensions of GR were proposed, either to unify gravity and electromagnetism [507], to include the spin in the field equations [121, 122], or to formulate its teleparallel equivalent [184]. Hence GR proved to be a very fertile ground for mathematical and theoretical "experimentation," with many of these extensions also having profound physical implications.

An interesting theoretical model of gravity, in which the standard Einstein–Hilbert action is replaced by an arbitrary function of the Ricci scalar R, was propsed in 1970 by Buchdahl [95]. This theory is called today $f(R)$ gravity. It became an established field of theoretical gravity and cosmology after the influential work by Starobinsky [474]. For other early research in this field, see [39, 177, 278].

In the present chapter we will briefly review the basic mathematical formalism of $f(R)$ gravity, and we will discuss some of its cosmological and astrophysical applications. There are a number of good reviews on this subject, such as [103, 157, 384, 385, 471], which we recommend to the reader for further study.

6.2 The Gravitational Field Equations

One of the simplest extensions of standard GR, based on the Einstein–Hilbert action, is the $f(R)$ modified theory of gravity, whose action takes the form [39, 95, 474]

$$S = \int \left[\frac{1}{16\pi G} f(R) + L_m\right] \sqrt{-g}\, d^4x = S_g[g] + S_m\left[g, \psi, ...\right], \tag{6.3}$$

where $f(R)$ is an arbitrary analytical function of the Ricci scalar R, and L_m is the Lagrangian density corresponding to matter. The matter Lagrangian may be a function of the metric tensor, and of different matter fields ψ. The only restriction imposed on the function f is to be analytic, that is, it must possess a Taylor series expansion about any point. In order to obtain the field equations of $f(R)$ gravity we will discus two methods of considering the variation of the action with respect to the metric tensor: as performed in a locally geodesic frame [103], and in an arbitrary frame, respectively.

6.2.1 Variation of the $f(R)$-Gravity Action in a Locally Geodesic Frame

We begin our analysis of $f(R)$ gravity by considering the variation with respect to the metric tensor of the gravitational part of Eq. (6.3). The corresponding Lagrangian density is given by $\sqrt{-g}\,\mathcal{L} = \sqrt{-g}\, f(R)$, and in the absence of matter it must satisfy the variational principle $\delta \int d^4x \sqrt{-g}\, f(R) = 0$. By taking into account that

$$R = R_{\mu\nu} g^{\mu\nu}, \tag{6.4}$$

and Eq. (3.7), which we rewrite here

$$\delta\sqrt{-g} = -\frac{1}{2}\sqrt{-g} g_{\mu\nu}\delta g^{\mu\nu}, \tag{6.5}$$

we obtain as a first step in our variational procedure

$$\delta S_g[g] = \delta \int d^4x \sqrt{-g}\, f(R) = \int d^4x \left[\delta\left(\sqrt{-g}\, f(R)\right) + \sqrt{-g}\, \delta\left(f(R)\right)\right]$$
$$= \int d^4x \sqrt{-g}\left[f'(R)R_{\mu\nu} - \tfrac{1}{2}g_{\mu\nu}f(R)\right]\delta g^{\mu\nu} + \int d^4x \sqrt{-g}\, f'^{\mu\nu}\delta R_{\mu\nu}, \quad (6.6)$$

where the prime denotes differentiation with respect to R. To evaluate the second integral in Eq. (6.6), we introduce a locally geodesic frame, in which at a fixed point P all the Christoffel symbols vanish, so that $\Gamma^{\lambda}_{\mu\nu}\big|_P = 0$. Hence, as we have already seen in our discussion of the derivation of the Einstein field equations, in the locally geodesic frame P the Ricci tensor can be represented in P as

$$R_{\mu\nu}|_P = \frac{\partial \Gamma^{\sigma}_{\mu\nu}}{\partial x^{\sigma}} - \frac{\partial \Gamma^{\sigma}_{\nu\sigma}}{\partial x^{\mu}}, \quad (6.7)$$

while for $g^{\mu\nu}\delta R_{\mu\nu}$ we find

$$[g^{\mu\nu}(\delta R_{\mu\nu})]|_P = \left[g^{\mu\nu}\left(\frac{\partial \delta\Gamma^{\sigma}_{\mu\nu}}{\partial x^{\sigma}} - \frac{\partial \delta\Gamma^{\lambda}_{\nu\lambda}}{\partial x^{\mu}}\right)\right]\Bigg|_P = \frac{\partial w^{\sigma}}{\partial x^{\sigma}}\Bigg|_P,$$

where we have denoted

$$w^{\sigma} = \left(g^{\mu\nu}\delta\Gamma^{\sigma}_{\mu\nu} - g^{\sigma\nu}\Gamma^{\lambda}_{\nu\lambda}\right)\big|_P. \quad (6.8)$$

We can now transform the second integral in Eq. (6.6) to the locally geodesic frame as

$$\int d^4x \sqrt{-g}\, f'^{\mu\nu}\delta R_{\mu\nu}\Bigg|_P = \int d^4x \sqrt{-g}\, f'(R)\frac{\partial}{\partial x^{\sigma}} w^{\sigma}\Bigg|_P. \quad (6.9)$$

Integrating by parts Eq. (6.9) provides

$$\int d^4x \sqrt{-g}\, f'^{\mu\nu}\delta R_{\mu\nu}\Bigg|_P = \left\{\int d^4x \frac{\partial}{\partial x^{\sigma}}\left[\sqrt{-g}\, f'^{\sigma}\right]\right.$$
$$\left.- \int d^4x \frac{\partial}{\partial x^{\sigma}}\left[\sqrt{-g}f'(R)\right] W^{\sigma}\right\}\Bigg|_P. \quad (6.10)$$

By assuming that the gravitational field vanishes at infinity, we can discard the first integrand as being a total divergence, thus obtaining

$$\int d^4x \sqrt{-g}\, f'^{\mu\nu}\delta R_{\mu\nu}\Bigg|_P = -\int d^4x \frac{\partial}{\partial x^{\sigma}}\left[\sqrt{-g}\, f'(R)\right] w^{\sigma}\Bigg|_P. \quad (6.11)$$

To proceed further we now need to calculate the term w^σ appearing in Eq. (6.11). Since in the locally geodesic frame introduced here we have $\partial_\alpha g_{\mu\nu} = \nabla_\alpha g_{\mu\nu} = 0$, for the variation of the Christoffel symbol we obtain

$$\begin{aligned}\delta\Gamma^\sigma_{\mu\nu}\Big|_P &= \delta\left[\frac{1}{2}g^{\sigma\alpha}\left(\frac{\partial}{\partial x^\mu}g_{\alpha\nu} + \frac{\partial}{\partial x^\nu}g_{\mu\alpha} - \frac{\partial}{\partial x^\alpha}g_{\mu\nu}\right)\right]\Bigg|_P \\ &= \frac{1}{2}g^{\sigma\alpha}\left[\frac{\partial}{\partial x^\mu}\left(\delta g_{\alpha\nu}\right) + \frac{\partial}{\partial x^\nu}\left(\delta g_{\mu\alpha}\right) - \frac{\partial}{\partial x^\alpha}\left(\delta g_{\mu\nu}\right)\right]\Bigg|_P . \end{aligned} \tag{6.12}$$

Similarly we obtain

$$\delta\Gamma^\nu_{\mu\nu}\Big|_P = \frac{1}{2}\,g^{\nu\alpha}\frac{\partial}{\partial x^\mu}\left(\delta g_{\nu\alpha}\right)\Big|_P .$$

Thus, we have

$$\begin{aligned} g^{\mu\nu}\delta\Gamma^\sigma_{\mu\nu}\Big|_P &= \left\{\frac{1}{2}\,g^{\mu\nu}\left[-\frac{\partial}{\partial x^\mu}\left(g_{\alpha\nu}\delta g^{\alpha\sigma}\right) - \frac{\partial}{\partial x^\nu}\left(g_{\mu\alpha}\delta g^{\sigma\alpha}\right) - g^{\sigma\alpha}\frac{\partial}{\partial x^\alpha}\left(\delta g_{\mu\nu}\right)\right]\right\}\Bigg|_P \\ &= \left\{\frac{1}{2}\frac{\partial}{\partial x_\sigma}\left(g_{\mu\nu}\delta g^{\mu\nu}\right) - \frac{\partial}{\partial x_\mu}\left(g_{\alpha\mu}\delta g^{\nu\alpha}\right)\right\}\Bigg|_P , \end{aligned} \tag{6.13}$$

$$g^{\mu\sigma}\delta\Gamma^\nu_{\mu\nu}\Big|_P = -\frac{1}{2}\frac{\partial}{\partial x_\sigma}\left(g_{\nu\alpha}\delta g^{\nu\alpha}\right)\Big|_P , \tag{6.14}$$

from which it immediately follows that

$$w^\sigma = \frac{\partial}{\partial x_\sigma}\left(g_{\mu\nu}\delta g^{\mu\nu}\right) - \frac{\partial}{\partial x_\mu}\left(g_{\mu\nu}\delta g^{\sigma\nu}\right).$$

Using the above equation we obtain

$$\begin{aligned}\int d^4x\sqrt{-g}\,f'^{\mu\nu}\delta R_{\mu\nu}\Big|_P &= \int d^4x\frac{\partial}{\partial x^\sigma}\left[\sqrt{-g}\,f'(R)\right] \\ &\times\left[\frac{\partial}{\partial x_\mu}\left(g_{\mu\nu}\delta g^{\sigma\nu}\right) - \frac{\partial}{\partial x_\sigma}\left(g_{\mu\nu}\delta g^{\mu\nu}\right)\right]\Bigg|_P .\end{aligned}$$

After an integration by parts, and after discarding the total divergences, we find

$$\begin{aligned}\int d^4x\,\sqrt{-g}\,f'^{\mu\nu}\delta R_{\mu\nu}\Big|_P = &\left\{\int d^4x\,g_{\mu\nu}\frac{\partial}{\partial x_\sigma}\frac{\partial}{\partial x^\sigma}\left[\sqrt{-g}f'(R)\right]\delta g^{\mu\nu}\right. \\ &\left.-\int d^4x\,g_{\mu\nu}\frac{\partial}{\partial x_\mu}\frac{\partial}{\partial x^\sigma}\left[\sqrt{-g}\,f'(R)\right]\delta g^{\sigma\nu}\right\}\Bigg|_P . \end{aligned} \tag{6.15}$$

The variation of the gravitational part of the action in the locally geodesic frame is then

$$\delta S_g[g] = \delta \int d^4x\sqrt{-g}\, f(R) = \int d^4x\sqrt{-g}\, \left[f\prime(R)R_{\mu\nu} - \tfrac{1}{2}f(R)g_{\mu\nu}\right]\delta g^{\mu\nu}$$
$$+ \int d^4x \left[g_{\mu\nu}\frac{\partial}{\partial x_\sigma}\frac{\partial}{\partial x^\sigma}\left(\sqrt{-g}\, f'(R)\right) - g_{\sigma\nu}\frac{\partial}{\partial x_\mu}\frac{\partial}{\partial x^\sigma}\left(\sqrt{-g}f'(R)\right)\right]\delta g^{\mu\nu}. \quad (6.16)$$

We go now to an arbitrary frame of reference by substituting the ordinary derivatives with the covariant ones. Hence the vanishing of the variation of the gravitational part of the action of the $f(R)$ theory gives the vacuum field equations

$$f'(R)R_{\mu\nu} - \frac{f(R)}{2}g_{\mu\nu} - \left[\nabla_\mu\nabla_\nu) - g_{\mu\nu}\Box\right]f'(R) = 0.$$

By varying simultaneously in the action (6.3) both the gravitational and the matter part, so that

$$\delta S = \delta S_g + \delta S_m = 0, \quad (6.17)$$

we obtain the full system of the field equations of the $f(R)$ gravity theory as given by

$$F(R)R_{\mu\nu} - \frac{1}{2}f(R)g_{\mu\nu} - \left(\nabla_\mu\nabla_\nu - g_{\mu\nu}\Box\right)F(R) = 8\pi G T_{\mu\nu}, \quad (6.18)$$

where we have denoted

$$F(R) = \frac{df(R)}{dR}, \quad (6.19)$$

while $T_{\mu\nu}$ represents the standard, minimally coupled matter energy-momentum tensor defined, as usual, by

$$T_{\mu\nu} = -\frac{2}{\sqrt{-g}}\frac{\delta\left(\sqrt{-g}L_m\right)}{\delta g^{\mu\nu}}. \quad (6.20)$$

6.2.2 The Derivation of the Field Equations in an Arbitrary Frame

In order to derive the field equations of $f(R)$ gravity in an arbitrary frame we start with the variation of the Ricci scalar, which can be obtained from Eq. (6.4) in the form

$$\delta R = \delta\left(R_{\mu\nu}\delta g^{\mu\nu}\right) = R_{\mu\nu}\delta g^{\mu\nu} + g^{\mu\nu}\delta R_{\mu\nu}$$
$$= R_{\mu\nu}\delta g^{\mu\nu} + g^{\mu\nu}\left(\nabla_\rho\delta\Gamma^\rho_{\nu\mu} - \nabla_\nu\delta\Gamma^\rho_{\rho\mu}\right), \quad (6.21)$$

where we have substituted $\delta R_{\mu\nu}$ by its expression given by the Palatini identity (3.15), which we have discussed in Chapter 3. As a next step in our derivation we have to evaluate the variations of the Christoffel symbols, which in an arbitrary frame can be written as

$$\delta\Gamma^{\lambda}_{\mu\nu} = \frac{1}{2}g^{\lambda\alpha}\left(\nabla_{\mu}\delta g_{\alpha\nu} + \nabla_{\nu}\delta g_{\alpha\mu} - \nabla_{\alpha}\delta g_{\mu\nu}\right). \tag{6.22}$$

To obtain this relation we have taken into account the fact that $\delta\Gamma^{\lambda}_{\mu\nu}$, being a difference of two connections, transforms as a tensor. By substituting the variation of the Christoffel symbols into Eq. (6.21), we obtain for the variation of the Ricci scalar the expression

$$\delta R = R_{\mu\nu}\delta g^{\mu\nu} + g_{\mu\nu}\Box\delta g^{\mu\nu} - \nabla_{\mu}\nabla_{\nu}\delta g^{\mu\nu}, \tag{6.23}$$

where by $\Box$ we have denoted the d'Alembert operator, $\Box = \nabla_{\nu}\nabla^{\nu}$. Then for the gravitational part of the action, we obtain

$$\begin{aligned}
\delta S_g[g] &= \frac{1}{16\pi G}\int\left[\delta f(R)\sqrt{-g} + f(R)\,\delta\sqrt{-g}\right]d^4x \\
&= \frac{1}{16\pi G}\int\left[f'(R)\delta R\sqrt{-g} - \frac{1}{2}\sqrt{-g}g_{\mu\nu}\delta g^{\mu\nu}f(R)\right]d^4x \\
&= \frac{1}{16\pi G}\int\Bigg\{\left[R_{\mu\nu}\delta g^{\mu\nu} + g_{\mu\nu}\Box\delta g^{\mu\nu} - \nabla_{\mu}\nabla_{\nu}\delta g^{\mu\nu}\right]f'(R) \\
&\quad -\frac{1}{2}g_{\mu\nu}\delta g^{\mu\nu}f(R)\Bigg\}\sqrt{-g}d^4x.
\end{aligned} \tag{6.24}$$

Now with the use of the expressions

$$g_{\mu\nu}f'(R)\Box\delta g^{\mu\nu} = \nabla_{\alpha}\left[g_{\mu\nu}f'(R)\nabla^{\alpha}\delta g^{\mu\nu}\right] - \left[g_{\mu\nu}\nabla_{\alpha}f'(R)\right]\nabla^{\alpha}\delta g^{\mu\nu}, \tag{6.25}$$

$$\left[g_{\mu\nu}\nabla_{\alpha}f'(R)\right]\nabla^{\alpha}\delta g^{\mu\nu} = \nabla^{\alpha}\left[g_{\mu\nu}\nabla_{\alpha}f'(R)\delta g^{\mu\nu}\right] - \left[g_{\mu\nu}\nabla_{\alpha}\nabla^{\alpha}f'(R)\right]\delta g^{\mu\nu}, \tag{6.26}$$

$$f'(R)\nabla_{\mu}\nabla_{\nu}\delta g^{\mu\nu} = \nabla_{\mu}\left[f'(R)\nabla_{\nu}\delta g^{\mu\nu}\right] - \left[\nabla_{\mu}f'(R)\right]\left[\nabla_{\nu}\delta g^{\mu\nu}\right], \tag{6.27}$$

$$\left[\nabla_{\mu}f'(R)\right]\left[\nabla_{\nu}\delta g^{\mu\nu}\right] = \nabla_{\nu}\left[\delta g^{\mu\nu}\nabla_{\mu}f'(R)\right] - \left[\nabla_{\nu}\nabla_{\mu}f'(R)\right]\delta g^{\mu\nu}, \tag{6.28}$$

and after discarding the total derivatives, we obtain for the variation of the action the expression

$$\delta S_g[g] = \frac{1}{16\pi G}\int\left[f'(R)R_{\mu\nu} - \frac{1}{2}g_{\mu\nu}f(R) + \left(g_{\mu\nu}\Box - \nabla_{\mu}\nabla_{\nu}\right)f'(R)\right]\sqrt{-g}d^4x. \tag{6.29}$$

By taking the variation of the total action (6.3), we reobtain again the gravitational field equations of $f(R)$ gravity as given by Eqs. (6.18).

6.2.3 The Palatini Approach in $f(R)$ Gravity

As mentioned in Section 3.1.1, the Einstein equations can also be derived using the Palatini formalism, which consists on an independent variation with respect to the metric and an independent connection (Palatini variation). More specifically, the Riemann tensor and the Ricci tensor are constructed in terms of the independent connection. The action $f(R)$ gravity is given by the following form

$$S_P = \frac{1}{2\kappa}\int d^4x\sqrt{-g}\, f(\mathcal{R}) + S_m(g_{\mu\nu}, \psi), \tag{6.30}$$

where the Ricci scalar is defined as $\mathcal{R} = g^{\mu\nu}\mathcal{R}_{\mu\nu}$, and the matter action S_m only depends on the metric and the matter fields, and not on the independent connection.

Following a similar analysis as in Section 3.1.1, and varying the action (6.30) with respect to the metric and the connection, respectively, yields the following equations of motion [471]

$$f'(\mathcal{R})\mathcal{R}_{(\mu\nu)} - \frac{1}{2}f(\mathcal{R})g_{\mu\nu} = \kappa G T_{\mu\nu}, \tag{6.31}$$

$$\bar{\nabla}_\lambda\left(\sqrt{-g}f'(\mathcal{R})g^{\mu\nu}\right) = 0, \tag{6.32}$$

Consider Eq. (6.32), where we define a conformal metric to $g_{\mu\nu}$ given by

$$h_{\mu\nu} \equiv f'(\mathcal{R})g_{\mu\nu}, \tag{6.33}$$

so we have

$$\sqrt{-h}\, h^{\mu\nu} = \sqrt{-g}\, f'(\mathcal{R})g^{\mu\nu}. \tag{6.34}$$

Thus, Eq. (6.32) translates that the Levi–Civita connection of $h_{\mu\nu}$ is given as

$$\Gamma^\lambda{}_{\mu\nu} = \frac{1}{2}h^{\lambda\sigma}\left(\partial_\mu h_{\nu\sigma} + \partial_\nu h_{\mu\sigma} - \partial_\sigma h_{\mu\nu}\right), \tag{6.35}$$

or, equivalently, in terms of $g_{\mu\nu}$,

$$\Gamma^\lambda{}_{\mu\nu} = \frac{1}{2}\frac{1}{f'(\mathcal{R})}g^{\lambda\sigma}\Big[\partial_\mu\left(f'(\mathcal{R})g_{\nu\sigma}\right) + \partial_\nu\left(f'(\mathcal{R})g_{\mu\sigma}\right) - \partial_\sigma\left(f'(\mathcal{R})g_{\mu\nu}\right)\Big], \tag{6.36}$$

We can then easily show that the Ricci tensor transforms under conformal transformations of $f(R)$ gravity and is given by

$$\begin{aligned}\mathcal{R}_{\mu\nu} = R_{\mu\nu} &+ \frac{3}{2}\frac{1}{(f'(\mathcal{R}))^2}\left(\nabla_\mu f'(\mathcal{R})\right)\left(\nabla_\nu f'(\mathcal{R})\right)\\ &- \frac{1}{f'(\mathcal{R})}\left(\nabla_\mu\nabla_\nu - \frac{1}{2}g_{\mu\nu}\Box\right)f'(\mathcal{R}).\end{aligned} \tag{6.37}$$

Contracting with $g^{\mu\nu}$ provides the relation

$$\mathcal{R} = R + \frac{3}{2(f'(\mathcal{R}))^2}\left(\nabla_\mu f'(\mathcal{R})\right)\left(\nabla^\mu f'(\mathcal{R})\right) + \frac{3}{f'(\mathcal{R})}\Box f'(\mathcal{R}). \tag{6.38}$$

Finally, substituting Eqs. (6.37) and (6.38) in Eq. (6.31) yields the following field equation

$$G_{\mu\nu} = \frac{\kappa}{f'} T_{\mu\nu} - \frac{1}{2} g_{\mu\nu} \left(\mathcal{R} - \frac{f}{f'} \right) + \frac{1}{f'} \left(\nabla_\mu \nabla_\nu - g_{\mu\nu} \Box \right) f' \\ - \frac{3}{2} \frac{1}{f'^2} \left[(\nabla_\mu f')(\nabla_\nu f') - \frac{1}{2} g_{\mu\nu} (\nabla f')^2 \right]. \qquad (6.39)$$

We refer the reader to Ref. [396] for an excellent review on the Palatini formalism. We will return to the Palatini approach in Chapter 17.

6.2.4 The Effective Einstein Equation

In order to obtain a simpler form of the field equations of the $f(R)$ gravity theory we introduce the operator

$$\hat{P}_{\mu\nu} = \nabla_\mu \nabla_\nu - g_{\mu\nu} \Box. \qquad (6.40)$$

Then the field equation (6.18) take the form

$$F(R) R_{\mu\nu} - \frac{1}{2} f(R) g_{\mu\nu} - \hat{P}_{\mu\nu} F(R) = 8\pi G T_{\mu\nu}. \qquad (6.41)$$

These equations can be rewritten in a form similar to the standard Einstein field equations by adding and subtracting a term of the form $f'(R)\, g_{\mu\nu} R/2$, so that

$$F(R) R_{\mu\nu} - \frac{F(R)}{2} g_{\mu\nu} R + \frac{F(R)}{2} g_{\mu\nu} R - \frac{f(R)}{2} g_{\mu\nu} - \hat{P}_{\mu\nu} F(R) = 8\pi G T_{\mu\nu}. \qquad (6.42)$$

Then the field equation of $f(R)$ gravity takes the form

$$G_{\mu\nu} + G_{\mu\nu}^{\text{dark}} = \frac{8\pi G}{F(R)} T_{\mu\nu}, \qquad (6.43)$$

where

$$G_{\mu\nu}^{\text{dark}} = \frac{1}{F(R)} \left\{ \frac{1}{2} \left[F(R) R - f(R) \right] g_{\mu\nu} - \hat{P}_{\mu\nu} F(R) \right\}. \qquad (6.44)$$

The field equation (6.43) can be interpreted as either describing a gravitating system with an effective energy-momentum tensor

$$T_{\mu\nu}^{\text{eff}} = 8\pi G \left[\frac{1}{F(R)} T_{\mu\nu} - \frac{1}{8\pi G} G_{\mu\nu}^{\text{dark}} \right], \qquad (6.45)$$

which sources the effective Einstein equations, or as an example of dark gravity, in which the gravitational force is effectively modified at large astrophysical and cosmological scales.

By contracting the field equation (6.18), we obtain for the trace T of the energy-momentum tensor the relation

$$3\Box F(R) + F(R)R - 2f(R) = 8\pi G T, \tag{6.46}$$

which indicates that in the present theory the Ricci scalar has become a fully dynamical quantity.

We consider now the problem of the conservation of the matter energy-momentum tensor in $f(R)$ gravity [284]. Taking the covariant divergence on both sides of Eq. (6.18) yields the following equations

$$(\nabla^\mu F)R_{\mu\nu} + F\nabla^\mu R_{\mu\nu} - \frac{1}{2}F\nabla_\nu R = (\Box\nabla_\nu - \nabla_\nu\Box)F + \nabla^\mu T_{\mu\nu}. \tag{6.47}$$

By using the definition $G_{\mu\nu} \equiv R_{\mu\nu} - \frac{1}{2}g_{\mu\nu}R$ we can rewrite this equation as

$$(\nabla^\mu F)R_{\mu\nu} + F\nabla^\mu G_{\mu\nu} = (\Box\nabla_\nu - \nabla_\nu\Box)F + \nabla^\mu T_{\mu\nu}. \tag{6.48}$$

On purely geometrical grounds we obtain immediately $\nabla^\mu G_{\mu\nu} = 0$, and

$$(\Box\nabla_\nu - \nabla_\nu\Box)F = R_{\mu\nu}\nabla^\mu F, \tag{6.49}$$

respectively. These identities follow from the definitions of the tensors $G_{\mu\nu}$ and $R_{\mu\nu}$, as introduced in the previous chapters. Therefore from Eq. (6.48), we find

$$\nabla^\mu T_{\mu\nu} = 0, \tag{6.50}$$

and thus it follows that the energy-momentum tensor in $f(R)$ gravity in the metric formulation is independently conserved [284].

6.2.5 The Scalar–Tensor Representation

The $f(R)$ theory in the metric formalism can be reformulated in a form similar to the Brans–Dicke theory [88], in the presence of a potential for the effective scalar-field degree of freedom. To investigate this analogy we start with the following action with a new scalar field χ [157]

$$S = \frac{1}{16\pi G}\int [f(\chi) + f_{,\chi}(\chi)(R-\chi)]\sqrt{-g}d^4x + \int L_m(g_{\mu\nu},\psi)\, d^4x, \tag{6.51}$$

where a comma denotes the derivative with respect of the argument of the function. After variation of this action with respect to the scalar field χ, we find the relation

$$f_{,\chi\chi}(\chi)(R-\chi) = 0\,. \tag{6.52}$$

If the condition

$$f_{,\chi\chi}(\chi) \neq 0 \tag{6.53}$$

is satisfied, it follows that

$$\chi = R, \tag{6.54}$$

and the action (6.51) takes the same form as the action (6.3) in $f(R)$ gravity. If we now define a new field φ as

$$\varphi \equiv f_{,\chi}(\chi)\,, \tag{6.55}$$

the action (6.51) can be reformulated as

$$S = \int \left[\frac{1}{16\pi G}\varphi R - U(\varphi)\right]\sqrt{-g}\,d^4x + \int L_m(g_{\mu\nu}, \psi)\,d^4x, \tag{6.56}$$

where $U(\varphi)$ is the potential associated to the field φ, and given by

$$U(\varphi) = \frac{\chi(\varphi)\,\varphi - f(\chi(\varphi))}{2\kappa^2}\,. \tag{6.57}$$

On the other hand in the Brans–Dicke theory [88] the total action with a potential $U(\varphi)$ for the scalar field is given by

$$S = \int \left[\frac{1}{16\pi G}\varphi R - \frac{\omega_{\rm BD}}{2\varphi}g^{\mu\nu}\partial_\mu\varphi\partial_\nu\varphi - U(\varphi)\right]\sqrt{-g}d^4x + \int L_m(g_{\mu\nu}, \psi)\,d^4x\,, \tag{6.58}$$

where $\omega_{\rm BD}$ is the Brans–Dicke parameter. Comparing Eq. (6.58) with Eq. (6.56), it follows that $f(R)$ theory in the metric formalism is equivalent to the Brans–Dicke theory with the parameter $\omega_{\rm BD} = 0$.

At the level of the field equations the transition to a scalar–tensor representation of the field equations can be performed by introducing the Legendre transformation [103]

$$\{R, f\} \to \{\phi, V\}\,, \tag{6.59}$$

defined as

$$\phi \equiv F(R)\,, \quad V(\phi) \equiv R(\phi)\,F - f\,(R(\phi))\,. \tag{6.60}$$

Then the field equations of $f(R)$ gravity can be reformulated as [103]

$$R_{\mu\nu} - \frac{1}{2}g_{\mu\nu}R = 8\pi\frac{G}{\phi}T_{\mu\nu} + \theta_{\mu\nu}, \tag{6.61}$$

where

$$\theta_{\mu\nu} = -\frac{1}{2}V(\phi)\,g_{\mu\nu} + \frac{1}{\phi}\left(\nabla_\mu\nabla_\nu - g_{\mu\nu}\Box\right)\phi. \tag{6.62}$$

In the new variable the trace of the field equations gives the relation

$$3\Box\phi + 2V(\phi) - \phi\frac{dV}{d\phi} = 8\pi GT. \tag{6.63}$$

The field equations of $f(R)$ gravity can be derived, in this scalar–tensor representation, from a Brans–Dicke-type gravitational action, with parameter $\omega_{\rm BD} = 0$, which can be written down as

$$S = \frac{1}{16\pi G} \int \left[\phi R - V(\phi) + L_m\right] \sqrt{-g}\, d^4x. \tag{6.64}$$

In this approach the only requirement for the $f(R)$ gravity theory equations to be expressible in the form of a Brans–Dicke theory is that the function $F(R)$ be invertible, that is, $R(F)$ exists [394]. This condition is necessary for the construction of the potential $V(\phi)$.

In the scalar–tensor representation the $f(R)$-type modification of the standard Einstein–Hilbert action leads to the occurrence in the gravitational field equations of an effective gravitational constant $G_{\rm eff} = G/\phi$, which is a function of the scalar field. Secondly, a new source term for the gravitational field, given by the tensor $\theta_{\mu\nu}$, also appears. The tensor $\theta_{\mu\nu}$ can be obtained from the trace of the energy-momentum tensor with the use of Eq. (6.63), which thus becomes an independent physical quantity determining the metric of the space-time.

6.3 Late-Time Cosmic Acceleration in $f(R)$ Gravity

It is a simple matter to show that $f(R)$ gravity may play the role of an effective dark energy, and drive the late-time cosmic acceleration. Consider the FLRW metric

$$ds^2 = -dt^2 + a^2(t)\left[\frac{dr^2}{1-kr^2} + r^2(d\theta^2 + \sin^2\theta\, d\phi^2)\right]. \tag{6.65}$$

The gravitational field equation (6.18), provides the generalized Friedmann equations in the following form [102, 468]

$$\left(\frac{\dot{a}}{a}\right)^2 - \frac{1}{3F(R)}\left\{\frac{1}{2}\left[f(R) - RF(R)\right] - 3\left(\frac{\dot{a}}{a}\right)\dot{R}F'(R)\right\} = \frac{\kappa}{3}\rho, \tag{6.66}$$

$$\left(\frac{\ddot{a}}{a}\right) + \frac{1}{2F(R)}\left\{\frac{\dot{a}}{a}\dot{R}F'(R) + \ddot{R}F'(R) + \dot{R}^2F''(R) - \frac{1}{3}\left[f(R) - RF(R)\right]\right\} = -\frac{\kappa}{6}(\rho + 3p). \tag{6.67}$$

We have taken into account the perfect fluid description for matter given by

$$T^\mu_\nu = \text{diag}\left[-\rho(t)\,c^2, p(t), p(t), p(t)\right], \tag{6.68}$$

where, as usual, ρ denotes the mass density of the cosmic fluid, and p the thermodynamic pressure.

Equations (6.66) and (6.67) may be rewritten in the following more familiar manner

$$\left(\frac{\dot{a}}{a}\right)^2 = \frac{\kappa}{3}\left(\rho + \rho_{(c)}\right), \tag{6.69}$$

$$\left(\frac{\ddot{a}}{a}\right) = -\frac{\kappa}{6}\left[\rho + \rho_{(c)} + 3\left(p + p_{(c)}\right)\right], \tag{6.70}$$

where the curvature energy-momentum components, $\rho_{(c)}$ and $p_{(c)}$, are defined as

$$\rho_{(c)} = \frac{1}{\kappa F(R)}\left\{\frac{1}{2}\left[f(R) - RF(R)\right] - 3\left(\frac{\dot{a}}{a}\right)\dot{R}F'(R)\right\}, \tag{6.71}$$

$$p_{(c)} = \frac{1}{\kappa F(R)}\left\{2\left(\frac{\dot{a}}{a}\right)\dot{R}F'(R) + \ddot{R}F'(R) + \dot{R}^2F''(R) - \frac{1}{2}\left[f(R) - RF(R)\right]\right\}, \tag{6.72}$$

respectively.

For simplicity, consider the absence of matter, $\rho = p = 0$. From Eq. (6.70), one may obtain the late-time cosmic acceleration if the condition $\rho_{(c)} + 3p_{(c)} < 0$ is obeyed. Taking into account the equation of state $\omega_{\text{eff}} = p_{(c)}/\rho_{(c)}$, with the specific choice of $f(R) \propto R^n$ and a generic power law for the scale factor $a(t) = a_0(t/t_0)^\alpha$ [102], it can be shown that the parameters of the model ω_{eff} and α are given by

$$\omega_{\text{eff}} = -\frac{6n^2 - 7n - 1}{6n^2 - 9n + 3}, \qquad \alpha = \frac{-2n^2 + 3n - 1}{n - 2}, \tag{6.73}$$

respectively, for $n \neq 1$. A suitable choice of n can lead to the value of $\omega_{\text{eff}} < -1/3$, and thus achieve the late-time cosmic acceleration.

One is at liberty to consider other interesting forms of $f(R)$. For instance, consider $f(R) = R - \mu^{2(n+1)}/R^n$ [120], and choosing once again a generic power law for the scale factor, the parameter can be written as

$$\omega_{\text{eff}} = -1 + \frac{2(n+2)}{3(2n+1)(n+1)}. \tag{6.74}$$

The desired value of $\omega_{\text{eff}} < -1/3$ may be attained, by appropriately choosing the value of the parameter n. Note that as $n \to \infty$, the space-time is approximately de Sitter.

Other forms of $f(R)$ have also been considered in the literature, for instance those involving logarithmic terms, such as $f(R) = R + \alpha \ln(R/\mu^2) + \beta R^m$ or $f(R) = R + \gamma R^{-n}[\ln(R/\mu^2)]^m$ [382, 381]. These models also yield acceptable values for the effective equation of state parameter, resulting in the late-time cosmic acceleration.

6.4 Vacuum Solutions of the Gravitational Field Equations

6.4.1 Metric and Field Equations

In this section we will consider static and spherically symmetric $f(R)$ gravity models, with the metric given by

$$ds^2 = -e^{\nu(r)}dt^2 + e^{\lambda(r)}dr^2 + r^2\left(d\theta^2 + \sin^2\theta d\phi^2\right). \tag{6.75}$$

Then, for a metric of the form given by Eq. (6.75), the vacuum field equations of $f(R)$ gravity can be written down as [371]

$$\frac{F''}{F} - \frac{1}{2}(\nu' + \lambda')\frac{F'}{F} - \frac{\nu' + \lambda'}{r} = 0, \tag{6.76}$$

$$\nu'' + \nu'^2 - \frac{1}{2}(\nu' + \lambda')\left(\nu' + \frac{2}{r}\right) - \frac{2}{r^2}(1 - e^{\lambda}) = -2\frac{F''}{F} + \left(\lambda' + \frac{2}{r}\right)\frac{F'}{F}, \tag{6.77}$$

$$\frac{f}{F}e^{\lambda} = -\nu'' - \frac{1}{2}(\nu' - \lambda')\nu' + \frac{2}{r}\lambda' + \left(\nu' + \frac{4}{r}\right)\frac{F'}{F}, \tag{6.78}$$

$$R = 2\frac{f}{F} - 3e^{-\lambda}\left[\frac{F''}{F} + \frac{1}{2}\left(\nu' - \lambda' + \frac{4}{r}\right)\frac{F'}{F}\right], \tag{6.79}$$

where the prime denotes differentiation with respect to r. By introducing a new independent variable ξ by means of the transformation $\xi = \ln r$, the field equations (6.76) and (6.77) take the form

$$\frac{F_{,\,\xi\xi}}{F} - \left[1 + \frac{1}{2}(\nu_{,\,\xi} + \lambda_{,\,\xi})\right]\frac{F_{,\,\xi}}{F} - (\nu_{,\,\xi} + \lambda_{,\,\xi})F = 0\,, \tag{6.80}$$

$$\nu_{,\,\xi\xi} - \nu_{,\,\xi} + \nu_{,\,\xi}^2 - \frac{1}{2}(\nu_{,\,\xi} + \lambda_{,\,\xi})(\nu_{,\,\xi} + 2) + 2(1 - e^{\lambda}) = -2\frac{F_{,\,\xi\xi}}{F} + (\lambda_{,\,\xi} + 4)\frac{F_{,\,\xi}}{F}\,, \tag{6.81}$$

where a comma denotes differentiation with respect to the variable ξ. The basic field equations Eqs. (6.76) and (6.77) are now, as a result of introducing the new independent variable, independent of the coordinate ξ. We introduce the following formal representation of the function F as

$$\frac{F_{,\,\xi}}{F} = u, \quad \frac{F_{,\,\xi\xi}}{F} = u_{,\,\xi} + u^2, \tag{6.82}$$

where u is a new function of ξ. Then Eq. (6.76) can be written as

$$u_{,\,\xi} + u^2 - \left[1 + \frac{1}{2}(\nu_{,\,\xi} + \lambda_{,\,\xi})\right]u - (\nu_{,\,\xi} + \lambda_{,\,\xi}) = 0. \tag{6.83}$$

Equation (6.83) is a Riccati-type first order ordinary differential equation. In terms of the newly defined function u, Eq. (6.77) becomes

$$\nu_{,\,\xi\xi} - \nu_{,\,\xi} + \nu_{,\,\xi}^2 - \frac{1}{2}(\nu_{,\,\xi} + \lambda_{,\,\xi})(\nu_{,\,\xi} + 2) + 2(1 - e^{\lambda}) = -2u_{,\,\xi} - 2u^2 + (\lambda_{,\,\xi} + 4)u. \tag{6.84}$$

Substitution, with the use of Eq. (6.83), of the term $u_{,\,\xi} + u^2$ into Eq. (6.84), yields

$$2(1 - e^{\lambda}) + \nu_{,\,\xi\xi} - \nu_{,\,\xi} + \nu^2_{,\,\xi} + (\nu_{,\,\xi} + \lambda_{,\,\xi})(1 - \frac{1}{2}\nu_{,\,\xi}) - (2 - \nu_{,\,\xi})u = 0. \quad (6.85)$$

Hence the general solution of the field equations for the static vacuum case of $f(R)$ gravity is given by the general solution of Eqs. (6.83) and (6.85). Once $\nu(\xi)$ and $\lambda(\xi)$ are known, we can obtain immediately u, and then F (by integration), as well as all the other relevant geometrical and physical parameters. On the other hand, if the metric tensor coefficients and the function F are specified, f can be obtained as a function of R from Eqs. (6.78) and (6.79), respectively, in a parametric form, as $f = f(\xi)$, $R = R(\xi)$.

6.4.2 Schwarzschild-Type Solutions of the Field Equations

We consider now general solutions of the field equations in which the metric tensor components satisfy the condition

$$\nu_{,\,\xi} + \lambda_{,\,\xi} = 0. \quad (6.86)$$

We call the corresponding solutions of the static spherically symmetric equations of $f(R)$ gravity as *Schwarzschild-type solutions*. The constant of integration that appears after the integration in Eq. (6.86) may be set to zero, without any loss of generality, by re scaling the time coordinate. Hence we may consider the Schwarzschild-type solution as satisfying the condition

$$\nu + \lambda = 0. \quad (6.87)$$

Therefore, Eq. (6.83) reduces to the form

$$u_{,\,\xi} + u^2 - u = 0, \quad (6.88)$$

which has the general solution

$$u = \frac{e^{\xi}}{e^{\xi} + C} = \frac{r}{r + C}, \quad (6.89)$$

where C is an arbitrary constant of integration. Next, by reverting back to the r coordinate, from the definition of u we obtain

$$F(r) = Ar + B, \quad (6.90)$$

where A and B are arbitrary constants of integration. After solving Eq. (6.85), we find that the Schwarzschild-type general solution in $f(R)$ gravity is given by

$$e^{\nu} = e^{-\lambda} = 1 - \frac{AC_2}{2B^2} + \frac{C_2}{3Br} - \frac{A(B^2 - AC_2)}{B^3} r + \frac{1}{B^2}\left(A^2 - \frac{C_1}{6B^2}\right) r^2$$
$$+ \frac{A^2(B^2 - AC_2)r^2}{B^4} \ln\left(A + \frac{B}{r}\right), \tag{6.91}$$

where C_1 and C_2 are arbitrary constants of integration.

There is an interesting difference between the vacuum solutions in standard GR and in $f(R)$ gravity, and this is related to the presence in the vacuum solution of the $f(R)$ gravity of a linearly increasing term in the radial coordinate r, as well as of a term with a logarithmic dependence of r. The Schwarzschild-type exact vacuum solution in $f(R)$ gravity depends, in the general case, on four arbitrary integration constants A, B, C_1, and C_2, respectively. The limiting case of the Schwarzschild–de Sitter solution of standard GR is obtained by setting for the constants in the solution the following values

$$A = 0, \qquad C_2 = -6BM, \qquad C_1 = 2B^4\Lambda, \tag{6.92}$$

where M is the mass of the central gravitating massive object [371].

In our next discussion we assume that, in the general Schwarzschild-type solution of the $f(R)$ gravity, C_2 also takes the value $C_2 = -6BM$, a value which is necessary to recover the standard Schwarzschild solution. Moreover, by taking $C_1 = 0$, we neglect the possible effect of a cosmological constant. Hence, by denoting the ratio A/B as $A/B = 1/\eta$, it follows that we can represent the Schwarzschild-type static spherically symmetric vacuum metric in $f(R)$ gravity in the form [425]

$$e^{\nu} = e^{-\lambda} = 1 + \frac{3M}{\eta} - \frac{2M}{r} - \left(1 + \frac{6M}{\eta}\right)\left(\frac{r}{\eta}\right)$$
$$+ \left(\frac{r}{\eta}\right)^2 \left\{1 + \left(1 + \frac{6M}{\eta}\right) \ln\left[A\left(1 + \frac{\eta}{r}\right)\right]\right\}. \tag{6.93}$$

With the use of the Schwarzschild-type solution of the field equations in $f(R)$ gravity, given by Eqs. (6.93), we obtain for the functions $f(r)$, $R(r)$, and $dF/dR = d^2f/dR^2$ the following expressions

$$f(r) = \frac{1}{\eta^2 (r+\eta) r^2} \{2A\eta \left[3\left(6M - r\right) r^2 + \eta r \left(9M + r\right)\right) - 3\eta^2 \left(M - r\right)]$$
$$-6Ar^2 \left(6M + \eta\right)\left(r + \eta\right) \ln\left[A\left(1 + \eta r^{-1}\right)\right]\}, \tag{6.94}$$

$$R(r) = \frac{1}{\eta^3 (r+\eta)^2 r^2} \Big\{\eta \big[6M \left(2r + \eta\right)\left(6r^2 + 6\eta r - \eta^2\right)$$
$$+ r\left(-12r^3 - 12\eta r^2 + 7\eta^2 r + 6\eta^3\right)\big]$$
$$-12r^2 \left(6M + \eta\right)\left(r + \eta\right) \ln\left[A\left(1 + \eta r^{-1}\right)\right] \Big\} \tag{6.95}$$

and

$$\frac{dF}{dR} = \frac{d^2 f}{dR^2} = -\frac{Ar^3 (r+\eta)^3}{2r^3 + 6\eta r^2 + 6\eta^2 (r - 2M)}, \tag{6.96}$$

respectively. Generally, the functional dependence of f on R cannot be obtained in an explicit analytical form. However, in the limiting cases of small and large r, respectively, some approximate representations are possible.

The Limit of Small r

In the limit of small r, corresponding to values of r close to a few M, the logarithmic correction $\ln(A + B/r)$ becomes the dominant term in the metric. Thus we immediately obtain

$$f(r) \approx -\chi_f^{-1} \frac{\ln B}{r}, \tag{6.97}$$

where

$$\chi_f = \frac{\eta^2}{6A(6M+\eta)}, \tag{6.98}$$

$$R(r) \approx -\chi_R (r+\eta)^{-1} \frac{\ln B}{r}, \tag{6.99}$$

and we have denoted

$$\chi_R = 12 \frac{(6M+\eta)}{\eta^3}. \tag{6.100}$$

After eliminating r with the use of the relation

$$r = B \exp(-\chi_f f), \tag{6.101}$$

we find

$$R = \frac{\chi_f \chi_R}{B \exp(-\chi_f f) + \eta} f. \tag{6.102}$$

If f is small we reobtain the Lagrangian of GR, $f \sim R$. By performing a series expansion of the exponential factor we obtain

$$f(R) \approx \frac{B+\eta}{\chi_f} \frac{R}{BR + \chi_R} \approx \frac{B+\eta}{\chi_f \chi_R} R \left(1 - \frac{B}{\chi_R} R\right). \tag{6.103}$$

The Limit of Large r

In the limit of large r, when the conditions $r \gg M$ and $r \gg \eta$ are satisfied, by keeping only the terms in $1/r$, we immediately obtain

$$f(r) \approx \chi_{f\infty} + \frac{1}{\eta r}, \tag{6.104}$$

$$R(r) \approx \frac{\chi_{R\infty}}{r} - \frac{12}{\eta^2}, \tag{6.105}$$

where

$$\chi_{f\infty} = -6\left[6\eta + (6M+\eta)\ln A\right]A, \tag{6.106}$$

and

$$\chi_{R\infty} = 12\frac{\left[6M\eta - \eta^2 - (6M+\eta)\ln A\right]}{\eta^3}. \tag{6.107}$$

Eliminating $1/r$ we obtain

$$f(R) = \frac{1}{\eta\chi_{R\infty}}R + \Lambda, \tag{6.108}$$

where

$$\Lambda = \frac{12}{\eta^3\chi_{R\infty}} + \chi_{f\infty}. \tag{6.109}$$

Therefore the modified $f(R)$ gravity model induces at large distances a constant term, which plays the role of a cosmological constant, and which is responsible for the accelerated expansion of the Universe.

The Viability of the $f(R)$ Gravity Model

Equations (6.94)–(6.96) can be used to discuss the conditions under which the generalized Schwarzschild-type solution, as well as the corresponding $f(R)$ generalized gravity theory, represents a viable model. The conditions under which $f(R)$ theories can represent viable models of cosmic acceleration have been summarized in [416]. First of all, the requirement of the existence of a stable, high curvature regime imposes the constraint

$$\frac{d^2f}{dR^2} > 0, \tag{6.110}$$

for $R \gg d^2f/dR^2$. From Eq. (6.96) it follows that this condition can be satisfied for all r if $A < 0$ and $\eta > 0$. Moreover, the condition

$$\left(\frac{r}{\eta}\right)^3 + 3\left(\frac{r}{\eta}\right)^2 + 3\left(\frac{r}{\eta}\right) > \frac{6M}{\eta}, \tag{6.111}$$

must also hold for all r, η, and M. The Big Bang nucleosynthesis and the CMB radiation pattern provide tight constraints on the model parameters, and based on them we require that $F = df/dR$ must be a negative, monotonically increasing function of R, which approaches zero asymptotically from below. Moreover, the condition $F = -Ar + B < 0$ can be used to constrain the admissible range of the radial coordinate as $r > \eta$, where we have assumed that $B > 0$.

From the requirement that the effective Newtononian constant $G_{\text{eff}} = G/(1+F)$ is not allowed to change sign we obtain the condition $1 + F > 0$, at all finite R. This condition gives $r < \eta + 1/A$. Therefore in the present model the range of the radial variable r is restricted to $\eta < r < \eta + 1/A$. A must have a very small numerical value in order for the model to be viable on very large distances, corresponding to cosmic and galactic scales. Finally, F must be a small quantity, a condition which is required for the model to pass the solar and galactic scale constraints. The conditions of viability of $f(R)$ gravity with the Schwarzschild-like vacuum solution can be formulated, in terms of the scalar curvature R, and in the limit of small r, as [425]

$$-\frac{2B(B+\eta)}{\chi_f \chi_R^2} < 0, \tag{6.112}$$

a condition that follows from the requirement $d^2f/dR^2 < 0$, and

$$\frac{\chi_R}{2B} < R < \frac{\chi_R}{2B} + \frac{\chi_f \chi_R^2}{2B(B+\eta)}, \tag{6.113}$$

respectively, which follows from the conditions $df/dR < 0$ and $1 + df/dR > 0$, respectively.

Since A must be a very small (dimensionless) quantity, $A \ll 1$, the condition of the smallness of F is automatically satisfied by also assuming a small numerical value for B. By assuming that the value of $|F|$ should not exceed today the numerical value of 10^{-6} at any point in the space-time [416], the requirement that the modified gravity model is viable on a scale of $r = 1000M$, for example, where M is the mass of the central black hole, and by representing B as $B = \beta M$, where β is a constant, gives the condition $\beta - 1000A = 10^{-6}/M$. For a black hole with a mass of around three solar masses we obtain $\beta - 1000A \approx 6.74 \times 10^{-12}$. For $\beta = 1$ it follows that A is very close to $A \approx 10^{-3}$ [425].

The Schwarzschild-type solution of $f(R)$ gravity can be used to test observationally its possible realization in nature, by using the study of the electromagnetic radiation emitted by accretion disks gravitating around massive astrophysical objects.

The effective potential which determines the geodesic motion of the test particles in the equatorial plane of the metric given by Eq. (6.93) can be written as [425]

$$V_{\text{eff}}(r) = \left(1 + \frac{\widetilde{L}^2}{r^2}\right)\left\{1 + \frac{3M}{\eta} - \frac{2M}{r} - \left(1 + \frac{6M}{\eta}\right)\left(\frac{r}{\eta}\right) + \left(\frac{r}{\eta}\right)^2 + \left(\frac{r}{\eta}\right)^2\left(1 + \frac{6M}{\eta}\right)\ln\left[A\left(1 + \frac{\eta}{r}\right)\right]\right\}. \tag{6.114}$$

All astrophysical quantities, related to the observable properties of the accretion disk, can be obtained from the vacuum metric. Due to the differences in the space-time structure, the modified $f(R)$ gravity black holes present some very important differences with respect to the disk properties as compared to the standard general relativistic Schwarzschild case. For a detailed discussion of these differences see [425].

6.5 Dark Matter in $f(R)$ Gravity

The problem of the dark matter is one of the central problems in present-day astrophysics. Two important observations, namely, the behavior of the galactic rotation curves and the mass discrepancy in clusters of galaxies, suggest the necessity of considering the existence of a special form of matter, called dark matter, at a galactic and extra-galactic scale. The determination of the cosmological matter composition of the Universe also indicates the presence of the dark matter component. The rotation curves of all known galaxies, independently of their morphological type, show that the rotational velocities of test particles, moving around the galactic center, increase from the center of the galaxy, and then reach an approximately constant value, $v_{tg\infty} \sim 200 - 300$ km/s, within a distance r from the center of the galaxy [65, 85, 414, 444]. In these regions the mass increases linearly with the radius, even where very little luminous matter can be detected. A second important astrophysical evidence for dark matter is provided by the study of the velocity and mass distribution of galaxies in galaxy clusters, i.e., cosmic structures formed of a large number of galaxies bound together by the gravitational interaction. The total mass of a cluster can be estimated in two ways. First, by taking into account the motions of its member galaxies, the virial theorem provides an estimate, M_V. Secondly, the total baryonic mass M may be estimated by considering the total sum of each individual member galaxy's mass. The problem of the mass discrepancy in galaxy clusters arises as one generally finds that M_V is considerably greater than M, with typical values of $M_V/M \sim 2030$ [65, 85, 414, 444]. This interesting behavior of the rotation curves, as well as the mass discrepancy in galactic clusters, is usually explained by postulating the existence of dark matter, assumed to be a cold and pressureless form of matter, distributed in a spherical halo around the galaxies.

In the following section we will consider the problem of the dark matter in the framework of the $f(R)$ modified theory of gravity. We will investigate both the problems mentioned in the previous paragraph, namely, the problem of the galactic rotation curves, and the problem of the virial mass in galactic clusters. As a result of these investigations an alternative interpretation of dark matter emerges, according to which dark matter is nothing but a manifestation of dark gravity, or more exactly, a geometric effect induced by the modification of gravity at large astrophysical scales.

6.5.1 The Galactic Rotation Curves

One of the most convincing evidences for the possible failure of Newtonian gravity and of the General Theory of Relativity on galactic and intergalactic scales is provided by the behavior of the rotational velocity of test particles around the galactic center. In many galaxies, neutral hydrogen clouds can be observed at large distances from the galactic centers, much beyond the extent of the luminous matter. These clouds are usually moving in circular orbits with nearly constant tangential velocities v_{tg}. The orbits of the particles are determined by the balance between the centrifugal acceleration v_{tg}^2/r and the gravitational attraction $GM(r)/r^2$ exerted by the total mass $M(r)$ contained within radius r. The dynamical equilibrium condition yields for the galactic mass profile a relation of the form

$$M(r) = rv_{tg}^2/G, \tag{6.115}$$

indicating that the galactic mass increases linearly with r, even at distances where very little luminous matter can be observed.

A possible explanation for the behavior of the galactic rotation curves in the $f(R)$ modified gravity theory was considered in [100, 101], using R^n gravity model. As a first step in this analysis a modified Newtonian potential of the form

$$\Phi(r) = -\frac{Gm}{2r}\left[1 + \left(\frac{r}{r_c}\right)^{\beta}\right], \tag{6.116}$$

was obtained, where m is the mass of the particle, r_c a constant, and the coefficient β depends on the "slope" parameter n in the modified R^n-type action. In the weak gravitational field limit, and in the slow motion approximation, the coefficient β can be expressed as

$$\beta = \frac{12n^2 - 7n - 1 - \sqrt{36n^4 + 12n^3 - 83n^2 + 50n + 1}}{6n^2 - 4n + 2}. \tag{6.117}$$

Using the modified Newtonian potential, given by Eq. (6.116), it was found that the best fit to 15 low-luminosity rotation curves in R^n gravity is obtained for $n = 3.5$ [100, 101]. These results seem to suggest that a strong modification, i.e., a rather large value of n as opposed to $n = 1$, of standard GR is required to explain the observed behavior of the galactic rotation curves. As one can see from Eq. (6.116), the potential obtained in this approach is still asymptotically decreasing, but the corrected rotation curve, although not flat, is higher than the Newtonian one, thus offering the possibility of fitting the rotation curves without dark matter.

An alternative approach to the dark matter problem in $f(R)$ gravity was proposed in [74]. In this study, the starting point for finding an exact analytic description of the galactic dynamics of test particles in $f(R)$ gravity models was the general relativistic expression of the tangential velocity $v_{\rm tg}$ of massive test particles in static and spherically symmetric space-times, moving in stable circular orbits around the galactic center. By assuming a general form of the metric of the form (6.75), it turns out that the rotational velocity is determined by the $g_{tt} = e^{\nu}$ component of the metric tensor, and by the radial coordinate r trough the relation [74]

$$v_{\rm tg}^2 = \frac{r\nu'}{2}. \tag{6.118}$$

This simple expression, which relates the tangential velocity of test particles to one of the metric tensor components, has three important implications. First of all, it is an exact general relativistic expression, valid for all static and spherically symmetric space-times. Secondly, it turns out that the tangential velocity depends on only one of the two metric functions g_{tt} and g_{rr}, or, more exactly, it does not depend on the functional form of the metric tensor coefficient g_{rr}. Finally, since the motion of test particles is described by the geodesic equations, the expression of the tangential velocity of test particles is independent of the form of the gravitational action. Even if one considers modified theories of gravity containing arbitrary contractions of the Ricci and Riemann tensors, Eq. (6.118) would still be valid.

The metric tensor component ν is determined in regions with constant tangential velocity by the condition $v_{\rm tg} \approx$ constant. Hence, after the integration of Eq. (6.118), we find

$$\nu = 2v_{\rm tg}^2 \ln\left(\frac{r}{r_0}\right), \tag{6.119}$$

where r_0 is a constant of integration. Therefore in the constant tangential velocity regions in the galactic halo, the most general static and spherically symmetric metric has the form

$$ds^2 = -\left(\frac{r}{r_0}\right)^{2v_{\rm tg}^2} dt^2 + e^{\lambda(r)}dr^2 + r^2 d\Omega^2. \tag{6.120}$$

As a next step in the analysis of the geometry in the constant velocity region one must consider the spherically symmetric vacuum solutions of the gravitational field equations in $f(R)$ modified theories of gravity. By introducing several coordinate and functional transformations, the field equations can be reduced to an autonomous system of differential equations [74]. By using the general form of $g_{tt} = e^{\nu}$ as given by Eq. (6.119), one can obtain a second order differential equation, fixing the functional form of g_{rr} in the constant velocity region, and which is given by [74]

$$\begin{aligned}\frac{1}{2}\frac{d^2\lambda}{d\eta^2} - \frac{1}{1-m}\left(\frac{3}{2}e^{\lambda} - m^2 + 1 - m\right)\frac{d\lambda}{d\eta} + \frac{1}{(1-m)^2}\left(1-e^{\lambda}\right)^2 \\ + \frac{3m^2-1}{(1-m)^2}\left(1-e^{\lambda}\right) + \frac{m^2\left(2m^2-1\right)}{(1-m)^2} - 2m = 0,\end{aligned} \tag{6.121}$$

where $\eta = \ln r$ and $m = v_{\rm tg}^2 \approx 10^{-6}$. Eq. (6.121) is a first order approximation in m for e^{λ}.

Equation (6.121) admits an exact solution of the form λ = constant. By denoting $1 - e^{\lambda} = \delta$, and considering again a first order approximation in m, it follows that δ satisfies the second order algebraic equation

$$\delta^2 - \delta - 2m = 0, \tag{6.122}$$

which has only one physical solution, namely, $\delta = -2m$ (the other solution contradicts the condition $e^{\lambda} \geq 1$, which essentially represents the condition of the positivity of the mass). Therefore, in the first order of approximation the metric tensor component e^{λ} takes in the "dark matter" region the approximate form

$$e^{\lambda} \approx 1 + 2v_{\rm tg}^2, \qquad e^{-\lambda} \approx 1 - 2v_{\rm tg}^2. \tag{6.123}$$

Equation (6.123) has a simple physical interpretation. In the Newtonian approximation we have $v_{\rm tg}^2 = GM(r)/r$, where $M(r)$ is the total mass of the galaxy inside radius r. Therefore we obtain the metric tensor component e^{λ} in a form that is very similar to the Schwarzschild solution of GR, that is,

$$e^{-\lambda} \approx 1 - \frac{2GM(r)}{r}. \tag{6.124}$$

Moreover, since $v_{\rm tg}^2 =$ constant, the mass within the radius r must increase so that $M(r) \sim r$. This "mass," which is linearly increasing with the radial distance r from the galactic center, is usually interpreted in astrophysical and cosmological observations as being due to the presence of the dark matter.

Once the metric coefficients are known, one can obtain the form of the gravitational Lagrangian of the $f(R)$ gravity model as given by [74]

$$f(R) = f_0 R^{1+m}, \tag{6.125}$$

where f_0 is a constant.

It is also interesting to investigate the Newtonian limit of this approach. In the Newtonian limit of the weak gravitational fields the e^{ν} component of the metric tensor is given by

$$e^{\nu} \approx 1 + 2\Phi_N, \tag{6.126}$$

where Φ_N is the Newtonian gravitational potential, satisfying the Poisson equation

$$\Delta\Phi_N = 4\pi\rho. \tag{6.127}$$

In the constant velocity region the mass $M(r)$ of the dark matter varies as $M(r) = v_{\rm tg}^2 r$, while the energy density ρ of the matter changes with the distance according to the relation

$$\rho = \frac{v_{\rm tg}^2}{4\pi r^2}, \tag{6.128}$$

respectively. Therefore in the dark matter dominated region of the galactic halo the Poisson equation for the gravitational potential is obtained as [74]

$$\frac{1}{r^2}\frac{d}{dr}\left(r^2 \frac{d\Phi_N}{dr}\right) = \frac{v_{\rm tg}^2}{r^2}, \tag{6.129}$$

and it has the general solution given by

$$\Phi_N(r) = v_{\rm tg}^2 \ln \frac{r}{r_0} - \frac{C_N}{r}, \tag{6.130}$$

where C_N and r_0 are arbitrary constants of integration. The Newtonian potential can be represented in the limit of large r, corresponding to the constant velocity regions in the galactic halo, as

$$\Phi_N(r) \approx v_{\rm tg}^2 \ln \frac{r}{r_0}, \tag{6.131}$$

indicating a logarithmic dependence on the radial distance r. Moreover, in the constant velocity "dark matter" dominated halo the e^{ν} component of the metric tensor can be represented as

$$\begin{aligned} e^{\nu} &\approx \left(\frac{r}{r_0}\right)^{2v_{\rm tg}^2} = \exp\left[\ln\left(\frac{r}{r_0}\right)^{2v_{\rm tg}^2}\right] \\ &\approx 1 + 2v_{\rm tg}^2 \ln\left(\frac{r}{r_0}\right) = 1 + 2\Phi_N(r). \end{aligned} \tag{6.132}$$

Therefore, since the model has a well-defined Newtonian limit, to describe the geometry of the space-time in the dark matter-dominated regions the metric given by Eq. (6.120) can indeed be used.

As one can see from Eq. (6.131), in the "dark-matter"-dominated region of a galaxy, where the rotation curves are strictly flat, the correction term to the Newtonian potential must have a logarithmic dependence on the radial coordinate r. A correction term of this form does not appear in Eq. (6.116), where to describe the observed behavior of the galactic rotation curves a power-law-modified Newtonian potential is assumed. The differences in the two models are also related to the asymptotic behavior of the metric tensor components. These differences in the Newtonian limit of the two models lead to different values of the parameter n in the power-law-modified action of gravity. Moreover, we have to point out that in the analysis performed in [74], the possible effect of the luminous galactic baryonic matter on the space-time geometry was completely neglected.

Hence, in the $f(R)$ gravity approach to dark matter, the appearance of a dark "mass" is an effect of a purely geometric origin, resulting from the modification of the basic equations of the gravitational field. However, one can formally introduce a mass $M(r)$ for the dark matter, even that the origin of this "mass" cannot be related to the existence of some physical particles. Hence, the galactic rotation curves can be explained naturally in $f(R)$ gravity theories, both qualitatively and quantitatively, without introducing any additional hypothesis about the existence of an exotic form of matter. In this geometric interpretation, the galaxy is embedded in a modified spherically symmetric geometry, generated by the non-zero contributions of the modified gravitational action through the dark gravity term. The extra-terms in the gravitational field equations, of purely geometric origin, act as a "matter" distribution outside the galaxy.

Another important conclusion of this study is that only small deviations from standard GR are needed to explain the flat galactic rotation curves. A gravitational Lagrangian having the form $f(R) \propto R^{1+v_{\rm tg}^2}$ can fully explain the observed galatic dynamics, a somewhat natural result in the context of modified gravity theories. Hence, to explain dark matter it is enough for the deviations from standard GR to be very small, with the exponent of the power-law-modified gravity being given by $n = 1 + v_{\rm tg}^2$. Then it follows that this gravitational model is defined by the action

$$S = \int f_0 \, R^{1+v_{\rm tg}^2} \sqrt{-g} \, d^4x, \tag{6.133}$$

where f_0 is a positive constant, which can be obtained in terms of the tangential velocity. Usually $v_{\rm tg}^2$ takes numerical values of the order of 10^{-6}. For this modified gravity model the function $f(R)$ can also be approximated as

$$f(R) = f_0 R^{1+v_{\rm tg}^2} \approx f_0 R \left(1 + v_{\rm tg}^2 \ln R\right). \tag{6.134}$$

Hence in this model of $f(R)$ gravity the standard Einstein–Hilbert action is corrected by a logarithmic term.

An important problem the $f(R)$ theories of gravity must face is the problem of their stability. When the condition $f''(R) < 0$ is satisfied, a "fatal instability" quickly develops on an extremely short time scale of the order of $\tau \approx 10^{-26}$ s [190]. For the Lagrangian given by Eq. (6.133) one obtains easily

$$f''(R) = f_0 \left(\frac{v_{tg}^2}{2}\right) \left(\frac{v_{tg}^2}{2} + 1\right) R^{v_{tg}^2/2-1} > 0. \tag{6.135}$$

Therefore in the present model this type of instability does not develop.

6.5.2 The Generalized Virial Theorem

An important role in astrophysics is played by the virial theorem plays, a role that can be explained by its generality, as well as its wide range of applications. Under the assumption of a steady state, with the use of the virial theorem, one can derive the mean density of astrophysical objects such as galaxies, clusters, and superclusters. This important result can be tested observationally through the comparison of the theoretical predictions with the determinations of the velocities of the test particles rotating in or around cosmic structures. Hence, the virial theorem can be used to predict the total mass of the galaxies and clusters of galaxies, and it is also a powerful tool for stability studies. In a general relativistic framework several versions of the virial theorem have been proposed [202, 209]. The inclusion of the effects of the cosmological constant has been done in [31, 32, 267], while its generalization to brane world models has been considered in [222].

In its simplest version the virial theorem can be derived as follows. Consider a system of n particles of masses m_i, $i = 1, 2, ..., n, ...$, bound by gravitation, with only the gravitational force acting in the system, so that the force $\vec{F}_{ij} = -\vec{F}_{ji}$ between two particles is given by

$$\vec{F}_{ij} = -\frac{Gm_i m_j \left(\vec{r}_i - \vec{r}_j\right)}{\left|\vec{r}_i - \vec{r}_j\right|^3}. \tag{6.136}$$

The equation of motion of each particle, having a momentum $\vec{P}_i$, is

$$\frac{d}{dt}\vec{P}_i = \vec{F}_i = m_i \frac{d^2}{dt^2}\vec{r}_i, i = 1, 2, .., n. \tag{6.137}$$

The time derivative of the quantity $\sum_{i=1}^{n} \vec{P}_i \cdot \vec{r}_i$ is given by

$$\frac{d}{dt}\left(\sum_{i=1}^{n} \vec{P}_i \cdot \vec{r}_i\right) = \sum_{i=1}^{n} \frac{d}{dt}\vec{P}_i \cdot \vec{r}_i + \sum_{i=1}^{n} \vec{P}_i \cdot \frac{d}{dt}\vec{r}_i. \tag{6.138}$$

The left hand side of Eq. (6.138) can also be written as

$$\frac{d}{dt}\left(\sum_{i=1}^{n} \vec{P}_i \cdot \vec{r}_i\right) = \frac{d}{dt}\left(\sum_{i=1}^{n} m_i \vec{r}_i \cdot \frac{d}{dt}\vec{r}_i\right) = \frac{1}{2}\frac{d^2}{dt^2}\left(\sum_{i=1}^{n} m_i \vec{r}_i^{\,2}\right) = \frac{1}{2}\frac{d^2 I}{dt^2}, \tag{6.139}$$

where $I = \sum_{i=1}^{n} m_i \vec{r}_i^{\,2}$ is the moment of inertia of the system. The first term on the right hand side of Eq. (6.138) can be rewritten as

$$\begin{aligned}\sum_{i=1}^{n} \frac{d}{dt}\vec{P}_i \cdot \vec{r}_i &= \sum_{i=1}^{n} \vec{F}_i \cdot \vec{r}_i = \frac{1}{2} \sum_{i,j,\, i\neq j} \left(\vec{F}_{ij} \cdot \vec{r}_i + \vec{F}_{ji} \cdot \vec{r}_j\right) \\ &= \frac{1}{2} \sum_{i,j,\, i\neq j} \vec{F}_{ij} \cdot \left(\vec{r}_i - \vec{r}_j\right) = \frac{1}{2}\sum_{i,j} \Phi_{ij} = \Phi,\end{aligned} \tag{6.140}$$

where Φ is the total gravitational potential energy, and

$$\Phi_{ij} = -\frac{Gm_i m_j}{\left|\vec{r}_i - \vec{r}_j\right|}. \tag{6.141}$$

The second term in the right-hand side of Eq. (6.138) is

$$\sum_{i=1}^{n} \vec{P}_i \cdot \frac{d}{dt}\vec{r}_i = 2\sum_{i=1}^{n} \frac{\vec{P}_i}{2m_i} = 2T, \tag{6.142}$$

where T is the total kinetic energy of the system. Therefore Eq. (6.138) becomes

$$\frac{1}{2}\frac{d^2 I}{dt^2} = \Phi + 2T. \tag{6.143}$$

In equilibrium, we obtain

$$\Phi + 2T = 0, \tag{6.144}$$

and this relation represents the virial theorem in classical mechanics.

The mass and extension (radius) of a galaxy is defined in astronomy in terms of its virial mass and its virial radius, respectively. Cosmic structures like galaxies and clusters of galaxies are highly extended objects, and therefore in practice it is hard to introduce well-defined, finite descriptions of their masses and sizes. A convenient way to describe these basic properties of galaxies is provided by the virial theorem.

The mass of a galaxy is determined by measuring the rotation velocity of its gas and stars. Under the assumption that particles move on circular Keplerian orbits, since the kinetic energy (per particle) of the system is $T = (1/2)\, mv^2 = (3/2)\, \sigma^2$, where σ is the dispersion velocity, and by taking for the potential energy (per particle) the expression $U = (3/5)(GM/R)$, where by R we have denoted the radius at which the velocity dispersion is measured, and M is the mass within that radius, with the use of the virial theorem we obtain

$$\sigma^2 = \frac{GM}{R}. \tag{6.145}$$

The virial mass M_V and virial radius R_V are defined as the values of the mass and radius corresponding to the maximum of the velocity dispersion

$$\sigma^2_{\rm max} = \frac{GM_V}{R_V}. \tag{6.146}$$

In the following discussion, we consider the virial theorem, and some of its applications, in the framework of $f(R)$ modified theories of gravity [75]. To derive a generalized virial identity that could describe the mass content of complex galactic structures without the need of introducing some exotic forms of matter we use the covariant form of the collisionless Boltzmann equation [319] and the field equations of $f(R)$ gravity. Due to the modification of the gravitational action, and of the basic field equations, supplementary terms, geometric in their nature, do also appear in the Boltzmann equation. These new terms give an effective contribution to the total energy-momentum tensor of the theory, and can be interpreted as generating an effective mass. Moreover, from the virial theorem it follows that the total virial mass becomes proportional to the effective geometric mass generated by the geometric effects induced by the modification of the gravitational action. Hence, the virial mass discrepancy in clusters of galaxies can be interpreted as being a direct consequence of the geometric modification of the Einstein field equations. An important physical result that follows from the use of the virial theorem, together with the gravitational field equations of the $f(R)$ gravity theory, is that the metric inside the cluster of galaxies can be obtained as a function of astrophysical parameters that can obtained from the astronomical observations.

The Field Equations for a System of Identical and Collisionless Point Particles

In this section we consider an isolated and spherically symmetric cluster of galaxies described by the static and spherically symmetric metric (6.75). A galaxy cluster, or cluster of galaxies, is a cosmic structure consisting of a large number of galaxies, ranging from hundreds to thousands, which are bound together by the force of gravity. The typical masses of the galaxy clusters are of the order of $10^{14} - 10^{15}$ solar masses. Together with the superclusters they are the largest known gravitationally bound structures in the Universe. For the galaxies in the cluster we adopt a simplified description by assuming that they can be described as identical and collisionless point particles. Their distribution in the cluster is described by a distribution function f_B, which gives the probability that at a time t the galaxies are located in the interval $(\vec{r}, \vec{r} + d\vec{r})$, and have velocities in the range $(\vec{v}, \vec{v} + d\vec{v})$. For the distribution function of the galaxies in the cluster we assume that it the general relativistic Boltzmann equation.

The energy-momentum tensor of the matter can be written, in terms of the distribution function of the particles, as [319]

$$T_{\mu\nu} = \int f_B \, m \, u_\mu u_\nu \, du, \tag{6.147}$$

where m is the mass of the particle (galaxy), u_μ is the four-velocity of the galaxy, and $du = du_r du_\theta du_\varphi / u_t$ is the invariant volume element in the velocity space. Hence, the energy-momentum tensor of the matter in a cluster of galaxies can be obtained in terms of an effective density ρ_{eff} and of an effective anisotropic pressure, with radial $p_{\text{eff}}^{(r)}$ and tangential $p_{\text{eff}}^{(\perp)}$ components, given by

$$\rho_{\text{eff}} = \rho \left\langle u_t^2 \right\rangle, \qquad p_{\text{eff}}^{(r)} = \rho \left\langle u_r^2 \right\rangle, \qquad p_{\text{eff}}^{(\perp)} = \rho \left\langle u_\theta^2 \right\rangle = \rho \left\langle u_\varphi^2 \right\rangle, \tag{6.148}$$

where at each point $\langle u_i^2 \rangle$, $i = t, r, \theta, \varphi$ is the average value of u_i^2, and ρ is the mass density [267].

In the scalar–tensor representation of $f(R)$ gravity, given by Eqs. (6.61), with the use of this form of the energy-momentum tensor, the gravitational field equations, describing a cluster of galaxies, can be written as [75]

$$-e^{-\lambda}\left(\frac{1}{r^2} - \frac{\lambda'}{r}\right) + \frac{1}{r^2} = 8\pi \frac{G}{\phi} \rho \left\langle u_t^2 \right\rangle - \frac{1}{2\phi} V(\phi) + \frac{1}{\phi}\left(\nabla_t \nabla^t - \Box\right)\phi, \tag{6.149}$$

$$e^{-\lambda}\left(\frac{\nu'}{r} + \frac{1}{r^2}\right) - \frac{1}{r^2} = 8\pi \frac{G}{\phi} \rho \left\langle u_r^2 \right\rangle + \frac{1}{2\phi} V(\phi) - \frac{1}{\phi}\left(\nabla_r \nabla^r - \Box\right)\phi, \tag{6.150}$$

$$\begin{aligned}\frac{1}{2} e^{-\lambda}\left(\nu'' + \frac{\nu'^2}{2} + \frac{\nu' - \lambda'}{r} - \frac{\nu'\lambda'}{2}\right) &= 8\pi \frac{G}{\phi} \rho \left\langle u_\theta^2 \right\rangle + \frac{1}{2\phi} V(\phi) - \frac{1}{\phi}\left(\nabla_\theta \nabla^\theta - \Box\right)\phi \\ &= 8\pi \frac{G}{\phi} \rho \left\langle u_\varphi^2 \right\rangle + \frac{1}{2\phi} V(\phi) - \frac{1}{\phi}\left(\nabla_\varphi \nabla^\varphi - \Box\right)\phi.\end{aligned} \tag{6.151}$$

By adding the gravitational field equations (6.149)–(6.151) we obtain a useful relationship, which can be expressed as the following equation

$$e^{-\lambda}\left(\frac{\nu''}{2}+\frac{\nu'^2}{4}+\frac{\nu'}{r}-\frac{\nu'\lambda'}{4}\right)=4\pi\frac{G}{\phi}\rho\left\langle u^2\right\rangle+\frac{1}{\phi}V\left(\phi\right)+\frac{1}{\phi}\left(2\nabla_t\nabla^t+\Box\right)\phi, \tag{6.152}$$

where $\langle u^2\rangle = \langle u_t^2\rangle + \langle u_r^2\rangle + \langle u_\theta^2\rangle + \langle u_\varphi^2\rangle$.

For astrophysical applications at the extragalactic level we may safely assume that the deviations from standard GR (which is recovered in the limit of the background value $\phi = 1$) are very small. Hence, we can represent ϕ as

$$\phi = 1 + \epsilon g'(R), \tag{6.153}$$

where ϵ is a small quantity, and $g'(R)$ describes the modifications of the geometry due to the contribution of the tensor $\theta_{\mu\nu}$ in the gravitational field equations [394]. Consequently

$$\frac{1}{\phi} \simeq 1 - \epsilon g'(R), \tag{6.154}$$

and Eq. (6.152) can be rewritten as

$$e^{-\lambda}\left(\frac{\nu''}{2}+\frac{\nu'^2}{4}+\frac{\nu'}{r}-\frac{\nu'\lambda'}{4}\right)\simeq 4\pi G\rho\left\langle u^2\right\rangle+4\pi G\rho_\phi, \tag{6.155}$$

where

$$4\pi G\rho_\phi \simeq -4\pi G\epsilon\rho\left\langle u^2\right\rangle g'(R)+\left.\left[\frac{1}{\phi}V\left(\phi\right)+\frac{1}{\phi}\left(2\nabla_t\nabla^t+\Box\right)\phi\right]\right|_{\phi=1+\epsilon g'(R)}. \tag{6.156}$$

The Virial Theorem in f(R) Generalized Gravity Models

To derive the virial theorem for galaxy clusters we have first to write down the Boltzmann equation giving the evolution of the distribution function f_B of the galaxies in the cluster. Then this equation can be integrated over the velocity space, to give an equation from which, used together with the gravitational field equations of $f(R)$ gravity theory, we obtain the desired generalization of the virial theorem. The Boltzmann equation for the distribution function of the particles in a galaxy cluster and in spherical coordinates is given, under the assumption that the only coordinate dependence of the distribution function is through the radial coordinate r, by [75, 267]

$$\begin{aligned}u_1\frac{\partial f_B}{\partial r}-\left(\frac{1}{2}u_0^2\frac{\partial\nu}{\partial r}-\frac{u_2^2+u_3^2}{r}\right)\frac{\partial f_B}{\partial u_1}-\frac{1}{r}u_1\left(u_2\frac{\partial f_B}{\partial u_2}+u_3\frac{\partial f_B}{\partial u_3}\right)\\-\frac{1}{r}e^{\lambda/2}u_3\cot\theta\left(u_2\frac{\partial f_B}{\partial u_3}-u_3\frac{\partial f_B}{\partial u_2}\right)=0.\end{aligned} \tag{6.157}$$

After multiplication of Eq. (6.157) by $mu_r du$, then integrating it over the velocity space, and by assuming that f_B vanishes sufficiently rapidly as the velocities tend to $\pm\infty$, we find the result

$$r\frac{\partial}{\partial r}\left[\rho\left\langle u_1^2\right\rangle\right]+\frac{1}{2}\rho\left[\left\langle u_0^2\right\rangle+\left\langle u_1^2\right\rangle\right]r\frac{\partial \nu}{\partial r}-\rho\left[\left\langle u_2^2\right\rangle+\left\langle u_3^2\right\rangle-2\left\langle u_1^2\right\rangle\right]=0. \tag{6.158}$$

We now multiply Eq. (6.158) by $4\pi r^2$, and we integrate it over the entire cluster. Thus we obtain the following relationship [267, 75]

$$\int_0^R 4\pi\rho\left[\left\langle u_1^2\right\rangle+\left\langle u_2^2\right\rangle+\left\langle u_3^2\right\rangle\right]r^2dr-\frac{1}{2}\int_0^R 4\pi r^3\rho\left[\left\langle u_0^2\right\rangle+\left\langle u_1^2\right\rangle\right]\frac{\partial \nu}{\partial r}dr=0. \tag{6.159}$$

At this moment it is useful to introduce some approximations that apply to the galactic clusters, as well as to test particles in stable circular motion around galaxies. First of all, we assume that ν and λ are slowly varying functions (which implies that ν' and λ' are small), so that in Eq. (6.155) the quadratic terms can be safely neglected. Moreover, based on the observational evidence, we can assume that the galaxies in the cluster have velocities much smaller than the velocity of the light, so that $\left\langle u_1^2\right\rangle \approx \left\langle u_2^2\right\rangle \approx \left\langle u_3^2\right\rangle \ll \left\langle u_0^2\right\rangle \approx 1$. Thus, Eqs. (6.155) and (6.159) become

$$\frac{1}{2r^2}\frac{\partial}{\partial r}\left(r^2\frac{\partial \nu}{\partial r}\right)=4\pi G\rho+4\pi G\rho_\phi, \tag{6.160}$$

and

$$2K-\frac{1}{2}\int_0^R 4\pi r^3\rho\frac{\partial \nu}{\partial r}dr=0, \tag{6.161}$$

respectively, where

$$K=\int_0^R 2\pi\rho\left[\left\langle u_1^2\right\rangle+\left\langle u_2^2\right\rangle+\left\langle u_3^2\right\rangle\right]r^2dr, \tag{6.162}$$

is the total kinetic energy of the galaxies. The total mass of the cluster is defined as $M=\int_0^R dM(r)=\int_0^R 4\pi\rho r^2dr$. The main contribution to M is due to the baryonic mass of the intra-cluster gas and of the mass of the stars. Other particles, such as, for example, massive neutrinos, may also give a significant contribution to M.

Multiplying Eq. (6.160) by r^2 and integrating from 0 to r gives

$$GM(r)=\frac{1}{2}r^2\frac{\partial \nu}{\partial r}-GM_\phi\left(r\right), \tag{6.163}$$

where we have denoted

$$M_\phi\left(r\right)=4\pi\int_0^r \rho_\phi(r')r'^2dr'. \tag{6.164}$$

This useful astrophysical quantity is called the *geometric mass* of the cluster. After multiplication of Eq. (6.163) with $dM(r)$, followed by an integration from 0 to R, and by defining

$$\Omega = -\int_0^R \frac{GM(r)}{r}\, dM(r), \tag{6.165}$$

and

$$\Omega_\phi = \int_0^R \frac{GM_\phi(r)}{r}\, dM(r), \tag{6.166}$$

we arrive at the relation

$$\Omega = \Omega_\phi - \frac{1}{2}\int_0^R 4\pi r^3 \rho \frac{\partial \nu}{\partial r}\, dr, \tag{6.167}$$

where Ω represents the usual gravitational potential energy of the cluster.

With the use of Eq. (6.161) we obtain finally the generalization of the virial theorem in $f(R)$ gravity, which can be written down as

$$2K + \Omega - \Omega_\phi = 0. \tag{6.168}$$

The generalized virial theorem, given by Eq. (6.168), can be reformulated in an alternative representation by introducing the radii R_V and R_ϕ, defined by

$$R_V = M^2 \Big/ \int_0^R \frac{M(r)}{r}\, dM(r), \tag{6.169}$$

and

$$R_\phi = M_\phi^2 \Big/ \int_0^R \frac{M_\phi(r)}{r}\, dM(r), \tag{6.170}$$

respectively. We call in the following R_ϕ as the *geometric radius* of the cluster of galaxies. Thus, the quantities Ω and Ω_ϕ are finally obtained as

$$\Omega = -\frac{GM^2}{R_V}, \tag{6.171}$$

and

$$\Omega_\phi = \frac{GM_\phi^2}{R_\phi}, \tag{6.172}$$

respectively.

We define the virial mass M_V of the galaxy cluster as [267]

$$2K = \frac{GMM_V}{R_V}. \tag{6.173}$$

After substituting this definition into the virial theorem, given by Eq. (6.168), we find

$$\frac{M_V}{M} = 1 + \frac{M_\phi^2 R_V}{M^2 R_\phi}. \tag{6.174}$$

For $M_V/M > 3$, a condition that is valid for most of the observed galactic clusters, the generalized virial mass in $f(R)$ gravity is provided by Eq. (6.174) in the approximate form

$$M_V \approx \frac{M_\phi^2}{M}\frac{R_V}{R_\phi}. \tag{6.175}$$

In the framework of Newtonian gravity and in the presence of a cosmological constant the virial theorem can also be derived, and it can be formulated as [31, 32, 65]

$$\frac{d^2 I_{jk}}{dt^2} = 4K_{jk} + 2W_{jk} + \frac{2}{3}\Lambda I_{jk} + \ldots, \tag{6.176}$$

where I_{jk} denotes the moment of inertia tensor of the gravitating system, K_{jk} is the kinetic energy tensor, W_{jk} is the gravitational potential energy tensor, and Λ is the cosmological constant, respectively. In the presence of an external force with potential $\Phi_{\rm ext}$, acting on the system, a new term of the form

$$V_{jk} = -\frac{1}{2}\int \rho \left(x_k \frac{\partial \Phi_{\rm ext}}{\partial x_j} + x_j \frac{\partial \Phi_{\rm ext}}{\partial x_k} \right) dV, \tag{6.177}$$

must be added to the right-hand side of Eq. (6.176). When $d^2 I_{jk}/dt^2 = 0$, that is, in gravitational equilibrium, and with the assumption $\Lambda = 0$, the virial theorem gives $2K - |W| = 0$, where K and W are the traces of K_{jk} and W_{jk}, respectively. In the case of a constant density matter distribution we have $|W| = (3/5)\, GM_V^2/R_V$, and we obtain that $K = (3/10)\, GM_V^2/R_V$.

It is important to note that in the "geometric mass" approach to the dark matter problem, we must make a clear distinction between matter, in the usual sense, and geometric effects. In the geometric approach only the baryonic matter has kinetic energy, and this feature requires a modification of the definition of the virial mass. On the other hand, the geometric effects induced by the $f(R)$ gravity theory cannot be investigated systematically in a purely Newtonian framework.

The virial mass M_V is determined, from the observational point of view, from the study of the velocity dispersion σ_r^2 of the stars and of the galaxies in the clusters. Based on our interpretation, most of the mass in a cluster with mass $M_{\rm tot}$ must be in the form of the geometric mass M_ϕ, and therefore $M_\phi \approx M_{\rm tot}$. The presence of the geometric mass and of the astrophysical effects of the $f(R)$ extensions of GR could be detected through gravitational lensing. This is an observational method that could provide direct evidence of the mass distribution and of the gravitational effects induced by the modification of the action, even at distances extending far beyond of the virial radius of the galaxy cluster.

Astrophysical Applications

Astrophysical observations, together theoretical modeling and cosmological simulations, have convincingly shown that the virialized part of a cluster corresponds roughly to a fixed density contrast $\delta \sim 200$. Hence the virial density ρ_V of the cluster is given by

$$\rho_V = \frac{3M_V}{4\pi R_V^3} = \delta\rho_c(z), \tag{6.178}$$

where M_V and R_V are the virial mass and radius, and the critical density of the Univere at a given redshift z is given by

$$\rho_c(z) = \frac{3H_0^2}{8\pi G} h^2(z), \tag{6.179}$$

where H_0 is the Hubble constant, and $h(z)$ is the Hubble parameter normalized to its local value, defined as

$$h^2(z) = \Omega_m (1+z)^3 + \Omega_\Lambda, \tag{6.180}$$

where Ω_m is the mass density parameter and Ω_Λ is the dark energy density parameter [24].

A physically meaningful fiducial radius for galaxy clusters can be defined once the integrated mass as a function of the radius is determined. Two such radii are commonly used, and they are either r_{200} or r_{500}. These radii are defined as the radii corresponding to the mean gravitational mass density of the matter $\langle\rho_{\text{tot}}\rangle = 200\rho_c$ or $500\rho_c$, respectively. A current approach to the virial mass of a cluster is to use r_{200} to define the outer boundary [432]. The numerical values of the radius r_{200} are in the range $r_{200} = 0.85$ Mpc (for the cluster NGC 4636) and $r_{200} = 4.49$ Mpc (for the cluster A2163). A typical value for r_{200} is approximately 2 Mpc. The masses corresponding to r_{200} and r_{500} are denoted by M_{200} and M_{500}, respectively, and it is usually assumed that $M_V = M_{200}$ and $R_V = r_{200}$ [432].

Geometric Mass and Geometric Radius from Galactic Cluster Observations

In the clusters of galaxies most of the baryonic mass is in the form of the intra-cluster gas. The profile of the gas mass density ρ_g in the cluster can be fitted with the observational data by using the following representation for the radial baryonic mass (gas) distribution [432]

$$\rho_g(r) = \rho_0 \left(1 + \frac{r^2}{r_c^2}\right)^{-3\beta/2}, \tag{6.181}$$

where r_c is the core radius, and ρ_0 and β are constants, whose numerical values are cluster dependent.

A standard assumption in astrophysics is that the hot, ionized intra-cluster gas emitting the X-ray radiation is in isothermal equilibrium. Therefore, it follows that the pressure P_g of the gas satisfies the equation of state

$$P_g = \left(\frac{k_B T_g}{\mu m_p}\right)\rho_g, \tag{6.182}$$

where $\mu \approx 0.61$ is the mean atomic weight of the particles in the gas, m_p is the proton' mass, k_B is Boltzmann's constant, and T_g is the gas temperature, respectively [432]. Therefore, with the use of the Jeans equation [65] the total mass distribution in the cluster can be obtained as [432]

$$M_{\rm tot}(r) = -\frac{k_B T_g}{\mu m_p G} r^2 \frac{d}{dr}\ln\rho_g. \tag{6.183}$$

By adopting for the density distribution of the gas the expression given by Eq. (6.181), for the total mass profile inside the cluster we obtain the following relationship [432]

$$M_{\rm tot}(r) = \frac{3k_B\beta T_g}{\mu m_p G}\frac{r^3}{r_c^2 + r^2}. \tag{6.184}$$

According to the virial theorem in $f(R)$ gravity, the total mass of the cluster consists of the sum of the geometric mass, and of the baryonic mass (given mainly the intra-cluster gas), so that

$$M_{\rm tot}(r) = 4\pi\int_0^r (\rho_g + \rho_\phi)\, r^2 dr. \tag{6.185}$$

Therefore it follows that $M_{\rm tot}(r)$ satisfies the following mass continuity equation,

$$\frac{dM_{\rm tot}(r)}{dr} = 4\pi r^2\rho_g(r) + 4\pi r^2\rho_\phi(r). \tag{6.186}$$

Since the total mass profile inside the cluster and the gas density distribution are given by Eqs. (6.184) and (6.181), respectively, we can obtain immediately the expression of the geometric density distribution inside the cluster as

$$4\pi\rho_\phi(r) = \frac{3k_B\beta T_g\left(r^2 + 3r_c^2\right)}{\mu m_p\left(r_c^2 + r^2\right)^2} - \frac{4\pi G\rho_0}{\left(1 + r^2/r_c^2\right)^{3\beta/2}}. \tag{6.187}$$

In the limit $r \gg r_c$ for ρ_ϕ we obtain the simple relation

$$4\pi\rho_\phi(r) = \left[\frac{3k_B\beta T_g}{\mu m_p} - 4\pi G\rho_0 r_c^{3\beta} r^{2-3\beta}\right]\frac{1}{r^2}. \tag{6.188}$$

For the geometric mass we find the following general expression

$$\begin{aligned} GM_\phi(r) &= 4\pi\int_0^r r^2\rho_\phi(r)\, dr \\ &= \frac{3k_B\beta T_g}{\mu m_p}\frac{r}{1 + r_c^2/r^2} - 4\pi G\rho_0\int_0^r \frac{r^2 dr}{\left(1 + r^2/r_c^2\right)^{3\beta/2}}. \end{aligned} \tag{6.189}$$

In the limit $r \gg r_c$, the geometric mass can be approximated as

$$GM_\phi(r) \approx \left[\frac{3k_B\beta T_g}{\mu m_p} - \frac{4\pi G\rho_0 r_c^{3\beta} r^{2-3\beta}}{3(1-\beta)}\right] r. \tag{6.190}$$

Next we assume that the contribution of the mass of the gas, and of its density, to the geometric mass and density, respectively, can be neglected. Such an approximation is very well supported by the astrophysical observations, which show that indeed the gas mass represents only a small fraction of the total mass [24, 432, 456]. Therefore, we obtain

$$4\pi G\rho_\phi(r) \approx \left(\frac{3k_B\beta T_g}{\mu m_p}\right) r^{-2}, \tag{6.191}$$

and

$$GM_\phi(r) \approx \left(\frac{3k_B\beta T_g}{\mu m_p}\right) r, \tag{6.192}$$

respectively.

The form of the metric tensor component e^ν inside the cluster can be obtained by using the previously derived mass profile. From Eq. (6.160) it follows that ν' must satisfy the condition

$$r^2\nu' = 2M_{\rm tot}(r) \approx 2M_\phi(r), \tag{6.193}$$

from which it follows that

$$e^\nu \approx C_\nu r^{2s}, \tag{6.194}$$

where for notational simplicity we have defined

$$s = \frac{3k_B\beta T_g}{\mu G m_p}. \tag{6.195}$$

In order to determine the metric tensor coefficient $e^{-\lambda}$, we can assume that in the first approximation it is given by its standard general relativistic form,

$$e^{-\lambda} \approx 1 - \frac{2GM_{\rm tot}}{r}, \tag{6.196}$$

and therefore it is given by

$$e^{-\lambda} \approx 1 - \frac{6k_B\beta T_g}{\mu m_p} = 1 - 2Gs. \tag{6.197}$$

However, it is important to note that this approximation could not be necessarily correct inside the clusters of galaxies.

Hence, the metric inside a cluster of galaxies can be obtained directly in $f(R)$ gravity from astrophysical observations.

An upper bound for the cutoff of the geometric mass can also be estimated for the present model. The basic idea is to consider the radius at which the decaying density profile of the geometric density, associated to the galaxy cluster, becomes equal, or smaller, than the average energy density of the Universe. Let $R_\phi^{(cr)}$ be the value of the coordinate radius at the point where the two densities become equal. Then, at this point, $\rho_\phi(R_\phi^{(cr)}) = \rho_{\text{univ}}$, where ρ_{univ} is the mean energy density of the universe. By assuming

$$\rho_{\text{univ}} = \rho_c = 3H^2/8\pi G = 4.6975 \times 10^{-30} h_{50}^2 \text{ g/cm}^{-3}, \tag{6.198}$$

where $H = 50h_{50}$ km/Mpc/s [432], we obtain

$$R_\phi^{(cr)} = \left(\frac{3k_B \beta T_g}{\mu m_p G \rho_c}\right)^{1/2} = 91.33\sqrt{\beta}\left(\frac{k_B T_g}{5\text{keV}}\right)^{1/2} h_{50}^{-1}\text{Mpc}. \tag{6.199}$$

The total geometric mass corresponding to this value is [75]

$$M_\phi^{(cr)} = M_\phi\left(R_\phi^{(cr)}\right) = 4.83 \times 10^{16} \beta^{3/2} \left(\frac{k_B T_g}{5\text{keV}}\right)^{3/2} h_{50}^{-1} M_\odot. \tag{6.200}$$

This value of the total mass is consistent with the astrophysical observations of the mass values of the clusters of galaxies. However, according to $f(R)$ modified theories of gravity, the geometric mass and its effects could extend far beyond the virial radius of the clusters [75], which is of the order of only a few Mpc, as compared to the geometric radius of the cluster, which has a much larger value.

Radial Velocity Dispersion in Galactic Clusters

The virial mass of a cosmic structure can also be expressed generally in terms of the characteristic velocity dispersion σ_1 as [116]

$$M_V = \frac{3}{G}\sigma_1^2 R_V. \tag{6.201}$$

Under the assumption that the velocity distribution in the cluster is isotropic, we obtain

$$\langle u^2 \rangle = \langle u_1^2 \rangle + \langle u_2^2 \rangle + \langle u_3^2 \rangle = 3\langle u_1^2 \rangle = 3\sigma_r^2, \tag{6.202}$$

where σ_r^2 is the radial velocity dispersion. The characteristic velocity dispersion σ_1 and the radial velocity dispersion σ_r are related by $3\sigma_1^2 = \sigma_r^2$.

To derive the radial velocity dispersion relation for galaxy clusters in $f(R)$ gravity we start from Eq. (6.158). Since the velocity distribution of the galaxies in the cluster is isotropic, we obtain easily

$$\frac{d}{dr}\left(\rho\sigma_r^2\right) + \frac{1}{2}\rho\frac{d\nu}{dr} = 0. \tag{6.203}$$

Now we may assume that inside the cluster the condition $e^{-\lambda} \approx 1$ is still valid. Hence, in the first order of approximation, and in the limit of small velocities the modified field equation Eq. (6.155) can be integrated to give

$$r^2\nu' = 2GM_\phi(r) + 2GM(r) + 2C, \tag{6.204}$$

where C is an arbitrary constant of integration. But from Eq. (6.203) it follows that we have

$$\nu' = -\frac{2}{\rho}\frac{d}{dr}\left(\rho\sigma_r^2\right), \tag{6.205}$$

we obtain the results that in $f(R)$ gravity the radial velocity dispersion of the galactic clusters satisfies the differential equation given by

$$\frac{d}{dr}\left(\rho\sigma_r^2\right) = -\frac{GM_\phi(r)}{r^2}\rho(r) - \frac{GM(r)}{r^2}\rho(r) - \frac{C}{r^2}\rho(r). \tag{6.206}$$

This equation has the general solution

$$\sigma_r^2(r) = -\frac{1}{\rho}\int^r \left[\frac{GM_\phi(r')}{r'^2}\rho(r') + \frac{GM(r')}{r'^2}\rho(r') + \frac{C}{r'^2}\rho(r')\right] dr' + \frac{C_1}{\rho}, \tag{6.207}$$

where C_1 is an integration constant.

We apply now Eq. (6.207) in an astrophysical context, by considering the simple the case in which the density ρ of the normal matter (gas plus stars) inside the cluster has a power law profile, given by

$$\rho(r) = \rho_0 r^{-\gamma}, \tag{6.208}$$

with ρ_0 and $\gamma \neq 1, 3$ positive constants. The corresponding normal matter mass distribution is

$$M(r) = \frac{4\pi\rho_0 r^{3-\gamma}}{(3-\gamma)}. \tag{6.209}$$

The geometric mass, given by

$$GM_\phi \approx \left(\frac{3k_B\beta T_g}{\mu m_p}\right) r = q_0 r, \tag{6.210}$$

where we have denoted

$$q_0 = \left(\frac{3k_B\beta T_g}{\mu m_p}\right), \tag{6.211}$$

is linearly proportional to r, as we have already shown in the previous section.

Therefore, for $\gamma \neq 1, 3$, we obtain the following solution of Eq. (6.207)

$$\sigma_r^2(r) = \frac{q_0}{\gamma} + \frac{2\pi G\rho_0}{(\gamma-1)(3-\gamma)} r^{2-\gamma} + \frac{C}{\gamma+1}\frac{1}{r} + \frac{C_1}{\rho_0} r^\gamma. \tag{6.212}$$

For the particular case $\gamma = 1$, Eq. (6.207) has the solution

$$\sigma_r^2(r) = q_0 + \frac{C}{2r} - 2\pi G\rho_0 r \ln r + \frac{C_1}{\rho_0} r, \tag{6.213}$$

while for $\gamma = 3$, we find

$$\sigma_r^2(r) = \frac{q_0}{3} - \pi G\rho_0\left(\ln r + \frac{1}{4}\right)\frac{1}{r^4} + \frac{C}{4}\frac{1}{r} + \frac{C_1}{\rho_0} r^3. \tag{6.214}$$

In clusters of galaxies the observational data for the velocity dispersion are usually analyzed by assuming for the radial velocity dispersion the simple form

$$\sigma_r^2(r) = \frac{B}{(r+b)}, \tag{6.215}$$

with B and b constants. For the matter density of the galaxies in the clusters the relation

$$\rho(r) = \frac{A}{r(r+a)^2}, \tag{6.216}$$

with A and a constants, is used. By using a nonlinear fitting procedure the observational data are then fitted with these functions [116]. For $r \ll a$, $\rho(r) \approx A/r$, while for $r \gg a$, $\rho(r)$ behaves like $\rho(r) \approx A/r^3$. Therefore the comparison of the observed matter density distributions of the galaxy clusters, as well as of the velocity dispersion profiles predicted by $f(R)$ modified theories of gravity with the observational data, may give a powerful method to test the predictions of this model, and to also discriminate between the different modified gravity theories.

The Lagrangian of the $f(R)$ Gravity Model:

The virial theorem in $f(R)$ gravity models, which leads to the possibility of obtaining the metric tensor components inside the cluster, also opens the possibility of directly obtaining the Lagrangian $f(R)$ of the theory from astrophysical observations. By using the gravitational field equations for a static spherically symmetric metric, we obtain f as a function of R in the form

$$f(R) = f_0 R^{1+q/2}, \tag{6.217}$$

where

$$f_0 = 2C_1 \frac{4(1+s) - 2(2+s)\sqrt{1+10s+s^2}}{\left[(1+5s)\sqrt{1+10s+s^2} - 1 - 34s - 9s^2\right]^{1+q/2}} e^{q\lambda/2}, \tag{6.218}$$

and

$$q = \frac{1}{2}\left[\sqrt{1+10s+s^2} - (1+s)\right] \geq 0. \tag{6.219}$$

Therefore, once the main physical parameters of the gas in the cluster, such as the gas temperature T_g or the gas density profile, described by the parameter β, are known, the action of the modified gravity model can be completely obtained from observations.

It is interesting to note that the modified gravitational action given by Eq. (6.134), obtained from the study of the rotation curves at galactic level, has the same functional form with the action (6.218), derived from the study of the geometric effects of the $f(R)$ modified gravity theory at the cluster of galaxies level. Therefore the $f(R)$ gravity in its $f(R) \sim R^{1+\epsilon}$, $\epsilon \ll 1$, formulation gives a consistent description of gravitational dynamics from galactic to cosmological scales. The presence of only very mild deviations from the standard Einstein–Hilbert action can give a full geometric description of gravitational phenomena at both galactic and extragalactic levels, which are usually explained by postulating the existence of the mysterious dark matter.

6.6 Final Considerations

In this chapter, we have considered an extension of the Einstein–Hilbert action, namely $f(R)$ gravity, which contains several appealing features, as it combines mathematical simplicity and a fair amount of generality. The literature in $f(R)$ is extensive and it is not our aim to present an exhaustive bibliographical list on the subject. However, we do recommend the interested reader to the excellent reviews that are available, such as [103, 157, 384, 385, 471], which were also provided in the Introduction. This chapter also serves essentially as a bridge to the two next parts of the book, where we considered an intensive analysis of two extensions to $f(R)$ gravity considered in the literature, namely, modified theories of gravity with generalized couplings between curvature and matter, and the hybrid approach to $f(R)$ gravity, where one considers a hybrid combination of the metric and the Palatini formalisms. This hybrid metric-Palatini modified theory of gravity contains elements from both of the latter formalisms and turns out to be extremely successful in accounting for the observed astrophysical and cosmological phenomenology.

Part II

Generalized Curvature-Matter Couplings in Modified Gravity

7
Introduction

Recent observations of supernovae, together with the WMAP and SDSS data, lead to the remarkable conclusion that our universe is not just expanding, but has begun to accelerate [413, 436, 438]. The resolution of this fundamental question is extremely important for theoretical cosmology, looking beyond the standard theory of gravity. The standard model of cosmology has favored the dark energy models as fundamental candidates responsible for the cosmic expansion. However, it is clear that these questions involve not only gravity, but also particle physics. String theory provides a synthesis of these two branches of physics and is widely believed to be moving toward a viable quantum gravity theory. One of the key predictions of string theory is the existence of extra spatial dimensions. In the brane-world scenario, motivated by recent developments in string theory, the observed 3-dimensional universe is embedded in a higher-dimensional spacetime [334]. The new degrees of freedom belong to the gravitational sector, and can be responsible for the late-time cosmic acceleration.

Indeed, detailed theoretical and phenomenological analysis of the relation between the effective field yielding dark energy and noncanonical Lagrangians and nonlinear gauge kinetic functions have been extensively explored in the literature. Note that generalizations of the action functional can be approached in several ways. For instance, prescriptions consist of replacing the linear scalar curvature term in the Einstein–Hilbert action by a function of the scalar curvature, $f(R)$, or by more general scalar invariants of the theory. This class of theories is often termed higher-order gravity theories [324, 383, 386]. In this context, infrared modifications of GR have been extensively explored, including 4-dimensional modifications to the Einstein–Hilbert action, and the consistency of various candidate models has been extensively analyzed. All modified gravity theories induce observational signatures at the post-Newtonian level, which are translated by the parameterized post-Newtonian (PPN) metric coefficients arising from these extensions of GR. Thus, generalizations of the Einstein–Hilbert Lagrangian, including quadratic Lagrangians that involve second order curvature invariants, have also been extensively explored [324]. While these modified theories of gravity offer an alternative explanation to the standard cosmological model [120], it offers a paradigm for nature fundamentally distinct from dark energy models of cosmic acceleration, even those that perfectly mimic the same expansion history. It is a fundamental question to understand how one may observationally and theoretically differentiate these modified theories of gravity from dark energy models. Thus, one should test these models against large-scale structure and lensing, astrophysical and laboratory measurements, as well as laboratory and space-based equivalence principle experiments. These solar system tests, large-scale structure and lensing essentially restrict the range of allowed modified gravity theories, and thus offer a window into understanding the perplexing nature of the cosmic acceleration, and of gravity itself.

In this Part II, we consider an interesting possibility, which includes nonminimal couplings between the scalar curvature and the matter Lagrangian density, introduced in [44], and its extensions and generalizations, which we follow closely [237]. This nonminimal coupling between matter and the gravitational field was first studied by Goenner, in 1984 [204], who considered a departure from GR by breaking the principle of minimal coupling. Recall that the latter is incorporated into Einstein's theory and states that in the matter Lagrangian density of a dynamical variational principle, the dynamical variables describing matter do not couple to the curvature tensor, which represents the gravitational field. An explicit curvature-dark matter coupling was also considered in [379]. The specific cases of a curvature coupling to dark energy [380], the Maxwell field [36], and an explicit curvature-Yang–Mills coupling [35] was also explored in the context of inflation and of the late-time cosmic acceleration. Indeed, in the context of $f(R)$-modified theories of gravity, it was shown that an explicit coupling between an arbitrary function of the scalar curvature R and the Lagrangian density of matter induces a nonvanishing covariant derivative of the energy-momentum tensor, implying non-geodesic motion and consequently leading to the appearance of an extra force [44]. The extra force is orthogonal to the four-velocity and the corresponding acceleration law was obtained in the weak field limit. Connections with MOND and with the Pioneer anomaly were further discussed. These curvature-matter coupling theories include the more evolved generalizations of $f(R, L_m)$ [233] and $f(R,T)$ gravities [243]. Among other features, these models allow for an explicit breaking of the Equivalence Principle (EP), which is highly constrained by solar system experimental tests [62, 186], by imposing a matter-dependent deviation from geodesic motion.

Note that the weak equivalence principle is considered one of the pillars of GR and, in fact, even a sizable part of the modified gravity community considers this principle as truly fundamental [509]. This fact has to be stressed because it demonstrates somehow a limitation for this class of theories. However, it has been recently reported, from data of the Abell Cluster A586, that interaction of dark matter and dark energy does imply the violation of the equivalence principle [50]. Notice that the violation of the EP is also found as a low-energy feature of some compactified version of higher-dimensional theories. Indeed, as emphasized by Thibault Damour, it is important to note that the EP is not one of the "universal" principles of physics [144]. It is a heuristic hypothesis which was introduced by Einstein in 1907, and used by him to construct his theory of GR. In modern language, the (Einsteinian) EP consists of assuming that the only long-range field with gravitational-strength couplings to matter is a massless spin-2 field. Modern unification theories, and notably string theory, suggest the existence of new fields (in particular, scalar fields: "dilaton" and "moduli") with gravitational-strength couplings. In most cases the couplings of these new fields violate the EP. If the field is long ranged, these EP violations lead to many observable consequences, such as the variation of fundamental "constants," the nonuniversality of free fall, and the relative drift of atomic clocks, among others. The best experimental probe of a possible violation of the EP is to compare the free-fall acceleration of different materials. Further tests of this principle remain important and relevant for new physics, and do indeed strongly restrict the parameters of the considered theory [143, 146]. However, it is important to note that, in this context, the violations of the EP does not in principle rule out the specific theory.

The $f(R, L_m)$ gravitational theory further generalizes $f(R)$ gravity by assuming that the gravitational Lagrangian is given by an arbitrary function of the Ricci scalar R and of the matter Lagrangian L_m [233]. This may also be considered a maximal extension of the Einstein–Hilbert action. Here we use the term "maximal extension" in a strict mathematical sense, and we define it as "an element in an ordered set that is followed by no other." Thus, the $f(R, L_m)$ gravity theory represents the maximal extension of the set of all gravitational theories constructed in a Riemann space, and with the action depending on the Ricci scalar and the matter Lagrangian only. In [233] the gravitational field equations of $f(R, L_m)$ gravity theory in the metric formalism were obtained, as well as the equations of motion for test particles, which follow from the covariant divergence of the energy-momentum tensor. The equations of motion for test particles can also be derived from a variational principle in the particular case in which the Lagrangian density of the matter is an arbitrary function of the energy density of the matter only. In [233], the Newtonian limit of the equation of motion was also considered, and a procedure for obtaining the energy-momentum tensor of the matter was presented. In particular, the gravitational field equations and the equations of motion for a particular model in which the action of the gravitational field has an exponential dependence on the standard general relativistic Hilbert–Einstein Lagrangian density were also derived. The $f(R, L_m)$ gravitational theory was further generalized by considering the novel inclusion of a scalar field and a kinetic term constructed from the gradients of the scalar field, respectively [242]. Specific models with a nonminimal coupling between the scalar field and the matter Lagrangian were further explored. We emphasize that these models are extremely useful for describing an interaction between dark energy and dark matter, and for explaining the late-time cosmic acceleration.

In the context of $f(R,T)$-modified theories of gravity, the gravitational Lagrangian is given by an arbitrary function of the Ricci scalar R and of the trace of the energy-momentum tensor T [243]. The gravitational field equations in the metric formalism were obtained, and it was shown that these field equations explicitly depend on the nature of the matter source. The field equations of several particular models, corresponding to some explicit forms of the function $f(R,T)$, were also presented. Furthermore, in [243] the Newtonian limit of the equation of motion was also analyzed, and constraints on the magnitude of the extra-acceleration were obtained by analyzing the perihelion precession of the planet Mercury in the framework of the particular model considered. An interesting specific case, namely, the $f(R,T^{\phi})$ model, was also analyzed in detail, where T^{ϕ} is the trace of the stress-energy tensor of a self-interacting scalar field. The cosmological implications of the model were briefly considered. We refer the reader to [243] for more details.

The nonminimally curvature-matter-coupled $f(R,T)$ gravitational theory was further generalized by considering the inclusion of a contraction of the Ricci tensor with the matter energy-momentum tensor [211, 212, 390]. We emphasize that examples of such couplings can be found in the Einstein–Born–Infeld theories [164] when one expands the square root in the Lagrangian. An interesting feature of this theory is that in considering a traceless energy-momentum tensor, i.e., $T = 0$, the field equations of $f(R,T)$ gravity reduce to those of $f(R)$ gravity theories, while the presence of the $R_{\mu\nu}T^{\mu\nu}$ coupling term still entails a nonminimal coupling to matter. In [211] the Newtonian limit of the $f(R,T,R_{\mu\nu}T^{\mu\nu})$ gravitational theory was considered, and an explicit expression for the extra-acceleration, which depends on the matter density, was obtained in the small velocity limit for dust particles. The so-called Dolgov–Kawasaki instability [172] was also analyzed in detail, and the stability conditions of the model with respect to local perturbations was obtained. A particular class of gravitational field equations can be obtained by imposing the conservation of the energy-momentum tensor [211]. In this context, the corresponding field equations for the conservative case was also derived by using a Lagrange multiplier method, from a gravitational action that explicitly contains an independent parameter multiplying the divergence of the energy-momentum tensor [211]. The cosmological implications of the model were extensively investigated for both the conservative and nonconservative cases, and several classes of analytical solutions were obtained. In [390], the FRLW cosmological dynamics for several versions of the $f(R,T,R_{\mu\nu}T^{\mu\nu})$ gravity theory was also considered. The reconstruction of the above action was explicitly analyzed, including the numerical reconstruction for the occurrence of the ΛCDM universe. De Sitter universe solutions in the presence of nonconstant fluids were also presented, and the problem of matter instability was further discussed.

All of the gravitational modifications listed here are based on the curvature description of gravity. However, an interesting and rich class of modified gravity arises by modifying the action of the equivalent torsional formulation of GR. The latter approach has been denoted the Teleparallel Equivalent of General Relativity (TEGR) [12, 184, 252, 340, 361, 363], and consists by replacing the torsionless Levi–Civita connection by the curvatureless Weitzenböck connection, and using the vierbein instead of the metric as the fundamental field. In this formulation, the Lagrangian of the theory is constructed by contractions of the torsion tensor. Thus, in analogy to $f(R)$ gravity, if one wishes to modify gravity in this formulation, the simplest approach would be in generalizing the torsion scalar $\mathcal{T}$ to an arbitrary function $f(\mathcal{T})$ [191, 318].

Now, in the context of TEGR models, one may also construct an extension of $f(\mathcal{T})$ gravity with the inclusion of a nonminimal torsion-matter coupling in the action [245]. The resulting theory is a novel gravitational modification, since it is different from both $f(\mathcal{T})$ gravity, as well as from the nonminimal curvature-matter-coupled theory. The cosmological applications of this new theory proves to be very interesting. In particular, an effective dark energy sector was obtained, where the equation-of-state parameter can be quintessence or phantom-like, or cross the phantom-divide, while for a large range of the model parameters the Universe results in a de Sitter, dark-energy-dominated, accelerating phase. Additionally, early-time inflationary solutions were also obtained, and thus one can provide a unified description of the cosmological history. We refer the reader to [245] for more details.

In addition to the latter non-minimal torsion-matter coupling, an alternative extension of $f(\mathcal{T})$ gravity was explored by considering a general coupling of the torsion scalar $\mathcal{T}$ with the trace of the matter energy-momentum tensor T [244]. The resulting $f(\mathcal{T},T)$ theory is a new modified gravity, since it is different from all the existing torsion or curvature based constructions. Applied to a cosmological framework, it also leads to interesting phenomenology. In particular, one can obtain a unified description of the initial inflationary phase, the subsequent non-accelerating, matter-dominated expansion, and then the transition to a late-time accelerating phase. In the far future, the Universe results either in a de Sitter exponential expansion, or in eternal power-law-accelerated expansions. A similar analysis was investigated in [281], where, using a perturbational approach, the stability of the solutions and in particular of the de Sitter phase, was explored. Furthermore, the constraints imposed by the energy conditions was also considered. We refer the reader to [244, 281] for more details.

The possibility that the behavior of the rotational velocities of test particles gravitating around galaxies can be explained in the framework of modified gravity with a nonminimal curvature-matter coupling was considered in [217]. Generally, the dynamics of test particles around galaxies, as well as the corresponding mass deficit, is explained by postulating the existence of dark matter. The extra-terms in the gravitational field equations with a curvature-matter coupling modify the equations of motion of test particles, and induce a supplementary gravitational interaction. Starting from the variational principle describing the particle motion in the presence of the nonminimal coupling, the expression of the tangential velocity of a test particle, moving in the vacuum on a stable circular orbit in a spherically symmetric geometry, was derived. The tangential velocity depends on the metric tensor components, and on the coupling function between matter and geometry. The Doppler velocity shifts were also obtained in terms of the coupling function. If the tangential velocity profile is known, the coupling term between matter and curvature can be obtained explicitly in an analytical form. The functional form of this function was obtained in two cases: for a constant tangential velocity, and for an empirical velocity profile obtained from astronomical observations, respectively. Therefore, these results open the possibility of directly testing the modified gravity theories with a nonminimal curvature-matter coupling by using direct astronomical and astrophysical observations at the galactic or extragalactic scale. These issues will be reviewed in detail in the chapters that follow.

It is important to discuss the theoretical motivations for these curvature-matter coupling theories more carefully. For instance, one may wonder if there is a fundamental theory or model that reduces to one of these theories in some limit. Indeed, for the specific case of the linear nonminimal curvature-matter coupling, one can show that this theory can be expressed as a scalar–tensor theory with two scalar fields [470]. These are reminiscent of the extensions of scalar–tensor gravity which include similar couplings, such as the theories considered by Damour and Polyakov [151], where it was shown that string-loop modifications of the low-energy matter couplings of the dilaton may provide a mechanism for fixing the vacuum expectation value of a massless dilaton. In addition to this, the results presented in [151] provide a new motivation for trying to improve by several orders of magnitude the various experimental tests of the Einstein Equivalence Principle, such as the universality of free fall and constancy of the fundamental constants, among others. In addition to this, note that the modern revival of the Kaluza–Klein theory often leads to the introduction of several scalar fields, i.e., the "compacton" [147]. Superstring theory also lead to several scalar fields coupled to the macroscopic distribution of energy; indeed, the "dilaton" is already present in ten dimensions, and the "compacton" comes from a dimensional reduction [147]. Quantum-motivated higher-order generalizations of Einstein's GR, under some conditions, are also equivalent to adding several scalar fields to the Einstein–Hilbert action [207]. In addition to this, we note that the nonminimal curvature-matter coupling also arises from one-loop vacuum-polarization effects in the formulation of quantum electrodynamics in curved space-times [173]. All of these considerations make it natural to consider modified fields of gravity containing several scalar fields, which in some cases are equivalent, in the present context, to the presence of a nonminimal curvature-matter couplings. These considerations further motivate the analysis of the coupling between matter and curvature.

8

Gravity Theories with Linear Curvature-Matter Coupling

8.1 Linear Nonminimal Curvature-Matter Coupling

In the present chapter, we review some of the basic features and properties of the simplest class of models involving a nonminimal coupling between curvature and matter, which is linear in the matter Lagrangian.

The action of $f(R)$ gravity can be generalized by introducing in the action the linear nonminimal coupling between curvature-matter. The action of this modified theory of gravity is given by [44]

$$S = \int \left\{ \frac{1}{2} f_1(R) + [1 + \lambda f_2(R)] \, L_m \right\} \sqrt{-g} \, d^4x \, , \tag{8.1}$$

where the factors $f_i(R)$ (with $i = 1, 2$) are arbitrary functions of the Ricci scalar R, L_m is the matter Lagrangian density, and the coupling constant λ determines the strength of the interaction between $f_2(R)$ and the matter Lagrangian.

Now, varying the action with respect to the metric $g_{\mu\nu}$ provides the following field equations

$$\begin{aligned} F_1(R) R_{\mu\nu} - \frac{1}{2} f_1(R) g_{\mu\nu} - \nabla_\mu \nabla_\nu \, F_1(R) + g_{\mu\nu} \Box \, F_1(R) = -2\lambda F_2(R) L_m R_{\mu\nu} \\ +2\lambda(\nabla_\mu \nabla_\nu - g_{\mu\nu} \Box) L_m F_2(R) + [1 + \lambda f_2(R)] T_{\mu\nu}^{(m)} \, , \end{aligned} \tag{8.2}$$

where $F_i(R) = f_i'(R)$, and the prime represents the derivative with respect to the scalar curvature. The matter energy-momentum tensor is defined as

$$T_{\mu\nu}^{(m)} = -\frac{2}{\sqrt{-g}} \frac{\delta(\sqrt{-g} \, L_m)}{\delta(g^{\mu\nu})} \, . \tag{8.3}$$

Throughout this section, we use the metric formalism, while the Palatini approach will be analyzed in Section 8.3 [223].

A general property of these nonminimal curvature-matter coupling theories is the nonconservation of the energy-momentum tensor. This can be easily verified by taking into account the covariant derivative of the field equation (8.2), the

Bianchi identity, $\nabla^\mu G_{\mu\nu} = 0$, and the following identity $(\Box\nabla_\nu - \nabla_\nu\Box)F_i = R_{\mu\nu}\nabla^\mu F_i$, which then imply the relationship

$$\nabla^\mu T^{(m)}_{\mu\nu} = \frac{\lambda F_2}{1+\lambda f_2}\left[g_{\mu\nu}L_m - T^{(m)}_{\mu\nu}\right]\nabla^\mu R\,. \tag{8.4}$$

Thus, the coupling between the matter and the higher derivative curvature terms may be interpreted as an exchange of energy and momentum between both. Analogous couplings arise after a conformal transformation in the context of scalar–tensor theories of gravity, and also in string theory. In the absence of the coupling, one verifies the conservation of the energy-momentum tensor [284], which can also be verified from the diffeomorphism invariance of the matter part of the action. It is also interesting to note that, from Eq. (8.4), the conservation of the energy-momentum tensor is verified if $f_2(R)$ is a constant or the matter Lagrangian is not an explicit function of the metric.

In order to test the motion in our model, we consider for the energy-momentum tensor of matter a perfect fluid $T^{(m)}_{\mu\nu} = (\rho+p)\,U_\mu U_\nu + p g_{\mu\nu}$, where ρ is the overall energy density and p is the pressure, respectively. The four-velocity, U_μ, satisfies the conditions $U_\mu U^\mu = -1$ and $\nabla_\nu U^\mu U_\mu = 0$. We also introduce the projection operator $h_{\mu\lambda} = g_{\mu\lambda} + U_\mu U_\lambda$ from which one obtains $h_{\mu\lambda}U^\mu = 0$.

By contracting Eq. (8.4) with the projection operator $h_{\mu\lambda}$, one deduces the following expression

$$(\rho+p)\,g_{\mu\lambda}U^\nu\nabla_\nu U^\mu - (\nabla_\nu p)(\delta^\nu_\lambda - U^\nu U_\lambda) \\ - \frac{\lambda F_2}{1+\lambda f_2}(L_m - p)(\nabla_\nu R)(\delta^\nu_\lambda - U^\nu U_\lambda) = 0\,. \tag{8.5}$$

Finally, contraction with $g^{\alpha\lambda}$ gives rise to the equation of motion for a fluid element

$$\frac{DU^\alpha}{ds} \equiv \frac{dU^\alpha}{ds} + \Gamma^\alpha_{\mu\nu}U^\mu U^\nu = f^\alpha\,, \tag{8.6}$$

where the extra force is given by

$$f^\alpha = \frac{1}{\rho+p}\left[\frac{\lambda F_2}{1+\lambda f_2}(L_m - p)\,\nabla_\nu R + \nabla_\nu p\right]h^{\alpha\nu}\,. \tag{8.7}$$

As one can immediately verify, the extra force f^α is orthogonal to the four-velocity of the particle, $f^\alpha U_\alpha = 0$, which can be seen directly from the properties of the projection operator. This is consistent with the usual interpretation of the force, according to which only the component of the four-force that is orthogonal to the particle's four-velocity can influence its trajectory.

It has also been shown that an $f(G)$-modified Gauss–Bonnet gravity with a nonminimal coupling to matter also induces an extra force, which is normal to their four-velocities and as a result move along nongeodesic world lines [360].

A particularly intriguing feature is that the extra force depends on the form of the Lagrangian density. Note that considering the Lagrangian density $L_m = p$, where p is the pressure, the extra force vanishes [53]. It has been argued that this is not the unique choice for the matter Lagrangian density, and that more natural forms for L_m, such as $L_m = -\rho$, do not imply the vanishing of the extra force. Indeed, in the presence of the nonminimal coupling they give rise to two distinct theories with different predictions [189], and this issue has been further investigated in different contexts [61, 356], including phantom energy [67]. In this context, a matter Lagrangian density as an arbitrary function of the energy-density of the matter was explored [215] (this possibility has also been explored in five dimensions [512] and in the brane world context [398]).

It was also argued that the corresponding energy-momentum tensor of matter in modified gravity models with a nonminimal coupling is more general than the usual general-relativistic energy-momentum tensor for perfect fluids [218], and it contains a supplementary equation of state dependent term, which could be related to the elastic stresses in the body, or to other forms of internal energy. Therefore, the extra force induced by the coupling between matter and geometry never vanishes as a consequence of the thermodynamic properties of the system, or for a specific choice of the matter Lagrangian, and it is non-zero in the case of a fluid of dust particles.

In the following subsection we discuss in detail the problem of the matter Lagrangian and of the energy-momentum tensor in modified gravity theories with a linear curvature-matter coupling [218].

8.2 The Matter Lagrangian and the Energy-Momentum Tensor

It is an interesting and novel feature of $f(R)$ gravity with a nonminimal curvature-matter coupling that the matter Lagrangian, and the energy-momentum tensor obtained from it, are not model-independent quantities, but they are completely and uniquely determined by the coupling function between matter and geometry. This important result can be obtained by deriving first the equations of motion of test particles (or test fluid) in the modified gravity model from a variational principle, and then considering the Newtonian limit of the particle action for a fluid obeying a barotropic equation of state [218]. The energy-momentum tensor of the matter obtained in this way is more general than the usual general-relativistic energy-momentum tensor for perfect fluids, and it contains a supplementary term that may be related to the elastic stresses in the test fluid, or to other sources of internal energy, as mentioned in the previous section. Since we assume that the matter obeys a barotropic equation of state, the matter Lagrangian can be expressed either in terms of the energy density, or in terms of the pressure, and in both representations the physical description of the system is equivalent. Therefore the presence (or absence) of the extra force is independent of the specific form of the matter Lagrangian, and it never vanishes, except in the particular case of (un)physical matter systems with zero sound speed. In particular, in the case of dust particles, the extra force is always non-zero.

We define the energy-momentum tensor of matter by Eq. (8.3), which can take the following form

$$T_{\mu\nu} = L_m g_{\mu\nu} - 2\frac{\partial L_m}{\partial g^{\mu\nu}}. \tag{8.8}$$

By taking into account the explicit form of the field equations (8.2), one obtains for the covariant divergence of the energy-momentum tensor the equation, given by Eq. (8.4), which can be rewritten as

$$\nabla^\mu T_{\mu\nu} = 2\left\{\nabla^\mu \ln\left[1+\lambda f_2(R)\right]\right\}\frac{\partial L_m}{\partial g^{\mu\nu}}. \tag{8.9}$$

As a specific example of generalized gravity models with a linear curvature-matter coupling, we consider the case in which matter, assumed to be a thermodynamic perfect fluid, obeys a barotropic equation of state, with the thermodynamic pressure p being a function of the *rest mass density of the matter* (for short: *matter density*) ρ only, so that $p = p(\rho)$. In this case, the matter Lagrangian density, which in the general case could be a function of both density and pressure, $L_m = L_m(\rho, p)$, or of only one of the thermodynamic parameters, becomes an arbitrary function of the matter density ρ only, so that $L_m = L_m(\rho)$. Then the matter energy-momentum tensor is obtained as [218]

$$T^{\mu\nu} = \rho\frac{dL_m}{d\rho}U^\mu U^\nu + \left(L_m - \rho\frac{dL_m}{d\rho}\right)g^{\mu\nu}, \tag{8.10}$$

where the four-velocity $U^\mu = dx^\mu/ds$ satisfies the condition $g^{\mu\nu}U_\mu U_\nu = -1$. To obtain Eq. (8.10) we have imposed the condition of the conservation of the matter current,

$$\nabla_\nu\left(\rho U^\nu\right) = 0, \tag{8.11}$$

and used the relation

$$\delta\rho = \frac{1}{2}\rho\left(g_{\mu\nu} - U_\mu U_\nu\right)\delta g^{\mu\nu}, \tag{8.12}$$

whose proof is given in the appendix of [218].

With the use of the identity $U^\nu\nabla_\nu U^\mu = d^2x^\mu/ds^2 + \Gamma^\mu_{\nu\alpha}U^\nu U^\alpha$, from Eqs. (8.9) and (8.10) we obtain the equation of motion of a massive test particle, or of a test fluid in the modified gravity model with a linear curvature-matter coupling as

$$\frac{dU^\mu}{ds} + \Gamma^\mu_{\nu\alpha}U^\nu U^\alpha = f^\mu, \tag{8.13}$$

where

$$f^\mu = -\nabla_\nu \ln\left\{\left[1+\lambda f_2(R)\right]\frac{dL_m(\rho)}{d\rho}\right\}\left(U^\mu U^\nu + g^{\mu\nu}\right). \tag{8.14}$$

The extra force f^μ, generated due to the presence of the coupling between matter and geometry, is perpendicular to the four-velocity, $f^\mu U_\mu = 0$.

The equation of motion (8.13) can also be obtained from the variational principle [218]

$$\delta S_p = \delta \int L_p ds = \delta \int \sqrt{Q}\sqrt{g_{\mu\nu}U^\mu U^\nu} ds = 0, \tag{8.15}$$

where S_p and $L_p = \sqrt{Q}\sqrt{g_{\mu\nu}U^\mu U^\nu}$ are the action and the Lagrangian density for test particles (test fluid), respectively, and

$$\sqrt{Q} = [1 + \lambda f_2(R)] \frac{dL_m(\rho)}{d\rho}. \tag{8.16}$$

To prove this result we start with the Lagrange equations corresponding to the action (8.15)

$$\frac{d}{ds}\left(\frac{\partial L_p}{\partial U^\alpha}\right) - \frac{\partial L_p}{\partial x^\alpha} = 0. \tag{8.17}$$

Since

$$\frac{\partial L_p}{\partial U^\alpha} = \sqrt{Q} U_\alpha, \tag{8.18}$$

and

$$\frac{\partial L_p}{\partial x^\alpha} = \frac{1}{2}\sqrt{Q} g_{\mu\nu,\alpha} U^\mu U^\nu + \frac{1}{2}\frac{Q_{,\alpha}}{Q}, \tag{8.19}$$

where a comma indicates the derivative with respect to x^λ, a straightforward calculation provides the following equation of motion of the particle

$$\frac{d^2 x^\mu}{ds^2} + \Gamma^\mu{}_{\nu\alpha} U^\nu U^\alpha + \left(U^\mu U^\nu - g^{\mu\nu}\right) \nabla_\nu \ln \sqrt{Q} = 0. \tag{8.20}$$

By simple identification with the equation of motion of the modified gravity model with linear curvature-matter coupling, given by Eq. (8.13), we obtain the explicit form of $\sqrt{Q}$, as given by Eq. (8.16). When $\sqrt{Q} \to 1$ we reobtain the standard general relativistic equation for geodesic motion.

8.2.1 The Newtonian Limit

One can also study the Newtonian limit of the model by using the variational principle (8.15). In the limit of the weak gravitational fields

$$ds \approx \sqrt{1 + 2\phi - \vec{v}^2} dt \approx \left(1 + \phi - \frac{\vec{v}^2}{2}\right) dt, \tag{8.21}$$

where ϕ is the Newtonian potential and $\vec{v}$ is the usual tridimensional velocity of the fluid.

By taking into account Eq. (8.16) in the first order of approximation, the equation of motion of the fluid can be obtained from the variational principle

$$\delta \int \left[\frac{dL_m(\rho)}{d\rho} + \lambda f_2(R) \frac{dL_m(\rho)}{d\rho} + \phi - \frac{\vec{v}^2}{2} \right] dt = 0, \tag{8.22}$$

and is given by

$$\vec{a} = -\nabla\phi - \nabla \frac{dL_m(\rho)}{d\rho} - \nabla U_E = \vec{a}_N + \vec{a}_H + \vec{a}_E, \tag{8.23}$$

where $\vec{a}$ is the total acceleration of the system. The term $\vec{a}_N$

$$\vec{a}_N = -\nabla\phi, \tag{8.24}$$

is the Newtonian gravitational acceleration, and $\vec{a}_E$ given by

$$\vec{a}_E = -\nabla U_E = -\lambda \nabla \left[f_2(R) \frac{dL_m(\rho)}{d\rho} \right], \tag{8.25}$$

is a supplementary acceleration induced due to the curvature-matter coupling. The term $\vec{a}_H$

$$\vec{a}_H = -\nabla \left[\frac{dL_m(\rho)}{d\rho} \right], \tag{8.26}$$

can be identified with the hydrodynamic acceleration term in the perfect fluid Euler equation,

$$\vec{a}_H = -\nabla \frac{dL_m(\rho)}{d\rho} = -\nabla \int_{\rho_0}^{\rho} \frac{dp}{d\rho} \frac{d\rho}{\rho}, \tag{8.27}$$

where ρ_0 is an integration constant and plays the role of a limiting density.

As an application, we consider the solar system tests in subsection 9.2.

8.2.2 The Matter Lagrangian

With the use of Eq. (8.27), the matter Lagrangian in modified theories of gravity with a curvature-matter coupling can be obtained by a simple integration as [218]

$$L_m(\rho) = \rho \left[1 + \Pi(\rho)\right] - \int_{p_0}^{p} dp, \tag{8.28}$$

where

$$\Pi(\rho) = \int_{p_0}^{p} \frac{dp}{\rho}, \tag{8.29}$$

and we have normalized an arbitrary integration constant to 1; p_0 is an integration constant, or a limiting pressure.

Thus, the corresponding energy-momentum tensor of matter is given by

$$T^{\mu\nu} = \left[\rho \left(1 + \Phi(\rho)\right) + p(\rho)\right] U^\mu U^\nu + p(\rho) g^{\mu\nu}, \tag{8.30}$$

respectively, where

$$\Phi(\rho) = \int_{\rho_0}^{\rho} \frac{p}{\rho^2} d\rho = \Pi(\rho) - \frac{p(\rho)}{\rho}, \tag{8.31}$$

and with all the constant terms included in the definition of p. By introducing the energy density of the body according to the definition

$$\varepsilon = \rho\left[1 + \Phi(\rho)\right], \tag{8.32}$$

the energy-momentum tensor of a test fluid can be written in the modified gravity theories with a curvature-matter coupling in a form similar to the standard general relativistic case

$$T^{\mu\nu} = \left[\varepsilon(\rho) + p(\rho)\right] U^{\mu}U^{\nu} + p(\rho)\, g^{\mu\nu}. \tag{8.33}$$

From a physical point of view $\Phi(\rho)$ can be interpreted as the elastic (deformation) potential energy of the body, and therefore Eq. (8.30) corresponds to the energy-momentum tensor of a compressible elastic isotropic system [193]. The matter Lagrangian can also be written in the simpler form

$$L_m(\rho) = \rho\Phi(\rho). \tag{8.34}$$

If the pressure does not have a thermodynamic or radiative component, one can take $p_0 = 0$. If the pressure is a constant background quantity, independent of the density, so that $p = p_0$, then $L_m(\rho) = \rho$, and the matter energy-momentum tensor takes the form corresponding to dust,

$$T^{\mu\nu} = \rho U^{\mu}U^{\nu}. \tag{8.35}$$

Since matter is supposed to obey a barotropic equation of state, these results are independent of the concrete representation of the matter Lagrangian in terms of the thermodynamic quantities [218]. The same results are obtained by assuming $L_m = L_m(p)$ – due to the equation of state ρ and p are freely interchangeable thermodynamic quantities, and the Lagrangians expressed in terms of ρ and p only are completely equivalent. More general situations, in which the density and pressure are functions of the particle number and temperature, respectively, and the equation of state is given in a parametric form, can be analyzed in a similar way.

The form of the matter Lagrangian, and the energy-momentum tensor, are strongly dependent on the equation of state of the test fluid. For example, if the barotropic equation of state is linear, $p = (\gamma - 1)\rho$, $\gamma =$ constant, $1 \leq \gamma \leq 2$, then

$$L_m(\rho) = \rho\left\{1 + (\gamma - 1)\left[\ln\left(\frac{\rho}{\rho_0}\right) - 1\right]\right\}, \tag{8.36}$$

and $\Phi(\rho) = (\gamma - 1)\ln(\rho/\rho_0)$, respectively.

For the case of a polytropic equation of state $p = K\rho^{1+1/n}$, where K, n =constant, we have

$$L_m(\rho) = \rho + K\left(\frac{n^2}{n+1} - 1\right)\rho^{1+1/n}, \tag{8.37}$$

and $\Phi(\rho) = Kn\rho^{1+1/n} = np(\rho)$, respectively, where we have taken for simplicity $\rho_0 = p_0 = 0$.

For a test fluid satisfying the ideal gas equation of state

$$p = \frac{k_B}{\mu}\rho T, \tag{8.38}$$

where k_B is the Boltzmann constant, T is the temperature, and μ is the mean molecular weight, we obtain

$$L_m(\rho) = \rho\left\{1 + \frac{k_B T}{\mu}\left[\ln\left(\frac{\rho}{\rho_0}\right) - 1\right]\right\} + p_0. \tag{8.39}$$

In the case of a physical system satisfying the ideal gas equation of state, the extra-acceleration induced by the presence of the nonminimal coupling between curvature and matter is given by

$$\vec{a}_E \approx -\lambda\frac{k_B T}{\mu}\nabla\left[f_2(R)\ln\frac{\rho}{\rho_0}\right], \tag{8.40}$$

and it is proportional to the temperature of the fluid. It is also interesting to note that the limiting density and pressure ρ_0 and p_0 generate in the energy-momentum tensor some extra constant terms, which may be interpreted as dark energy.

In conclusion, the extra force induced by the coupling between matter and geometry does not vanish for any specific choices of the matter Lagrangian. In the case of the dust, with $p = 0$, the extra force is given by

$$f^\mu = -\nabla_\nu \ln\left[1 + \lambda f_2(R)\right]\left(U^\mu U^\nu - g^{\mu\nu}\right), \tag{8.41}$$

and it is independent on the thermodynamic properties of the system, being completely determined by the geometry, kinematics and coupling. In the limit of small velocities and weak gravitational fields, the extra-acceleration of a dust fluid is given by

$$\vec{a}_E = -\lambda\nabla\left[f_2(R)\right]. \tag{8.42}$$

The thermodynamic condition for the vanishing of the extra force is given by

$$\frac{\partial L_m}{\partial g^{\mu\nu}} = \frac{1}{2}\left(\frac{\partial L_m}{\partial \rho}\right)\rho\left(g_{\mu\nu} - U_\mu U_\nu\right) = 0. \tag{8.43}$$

If the matter Lagrangian is written as a function of the pressure, then

$$\frac{\partial L_m}{\partial \rho} = \left(\frac{\partial L_m}{\partial p}\right)\left(\frac{\partial p}{\partial \rho}\right), \tag{8.44}$$

and for all physical systems satisfying an equation of state (or, equivalently, for all systems with a non-zero sound velocity), the extra force is non-zero. Therefore, in this context, the curvature-matter coupling is introduced in a consistent way. The coupling determines all the physical properties of the system, including the extra force, the matter Lagrangian, and the energy-momentum tensor, respectively.

Hence, we have shown that in $f(R)$ gravity with a nonminimal coupling, the matter Lagrangian – and the corresponding energy-momentum tensor, and thermodynamic parameters – are not model-independent quantities, but are completely and uniquely determined by the nature of the coupling between matter and geometry, which, in the present model, is given by the function $f_2(R)$. We have obtained this result by deriving first the equations of motion in the modified gravity model with curvature-matter coupling from a variational principle, and then by taking the Newtonian limit of the particle action for a fluid obeying a barotropic equation of state. The energy-momentum tensor of matter obtained in this way is more general than the usual general-relativistic energy-momentum tensor for a perfect fluid, and it contains a supplementary term that may be related to the elastic stresses in the body, or to other sources of internal energy. The matter Lagrangian can be expressed either in terms of the energy density, or in terms of the pressure, and in both representations the physical description of the system is equivalent. Therefore, the presence (or absence) of the extra force is independent of the specific form of the matter Lagrangian, and it never vanishes, except in the case of (un)physical systems with zero sound speed. In particular, in the case of dust particles, the extra force is always non-zero.

8.3 Curvature-Matter Coupling in the Palatini Formalism

8.3.1 Gravitational Field Equations

In this section, we will analyze the curvature-matter coupling in the Palatini formalism. By assuming an arbitrary coupling between curvature and matter, consider the following action [44, 215]

$$S = \int \left\{\frac{1}{2}f_1\left[R(g,\tilde{\Gamma})\right] + f_2\left[R(g,\tilde{\Gamma})\right] G\left[L_m\left(g,\psi\right)\right]\right\} \times \sqrt{-g}\, d^4x\,, \tag{8.45}$$

where $f_i(R)$ (with $i = 1, 2$) are arbitrary functions of the Ricci scalar $R = g^{\mu\nu}\tilde{R}_{\mu\nu}$; the Ricci tensor $\tilde{R}_{\mu\nu}$ is expressed solely in terms of the connection $\tilde{\Gamma}$. The term $G\left(L_m\right)$ is an arbitrary function of the matter Lagrangian density L_m, which is a function of the metric tensor g and of the physical fields ψ. The only requirement for the functions $f_i(R)$, $i = 1, 2$, and G, is that they be analytical functions of the Ricci scalar R and of the matter Lagrangian density L_m, respectively, i.e., they must possess a Taylor-series expansion about any point [215, 425].

The Ricci tensor is defined in the following manner [284, 302]

$$\tilde{R}_{\mu\nu} = \partial_\lambda \tilde{\Gamma}^\lambda_{\mu\nu} - \partial_\nu \tilde{\Gamma}^\lambda_{\mu\lambda} + \tilde{\Gamma}^\lambda_{\mu\nu}\tilde{\Gamma}^\alpha_{\lambda\alpha} - \tilde{\Gamma}^\alpha_{\mu\lambda}\tilde{\Gamma}^\lambda_{\nu\alpha}. \tag{8.46}$$

The connection $\tilde{\Gamma}^\lambda_{\mu\nu}$ is obtained through the independent variation of the gravitational action given by Eq. (8.45), and not directly constructed from the metric by using the Levi–Civita prescription. The energy-momentum tensor of the matter is given by Eq. (8.3).

The Palatini formalism consists in taking variations, separately and independently, of the metric and the connection, respectively. Note that the action is formally the same as in the metric formalism, but the Riemann tensor and the Ricci tensor are now constructed in terms of the independent connection. The variation of the action (8.45) with respect to the metric g yields the following

$$\begin{aligned} \left[f_1'(R) + 2f_2'(R)G\left(L_m\right)\right]\tilde{R}_{\mu\nu} - \Bigg\{ & \frac{1}{2}f_1(R) + f_2(R)G\left(L_m\right) \\ & \times \left[1 - L_m\frac{G'\left(L_m\right)}{G\left(L_m\right)}\right]\Bigg\} g_{\mu\nu} = f_2(R)G'\left(L_m\right)T_{\mu\nu}\,, \end{aligned} \tag{8.47}$$

where the prime denotes a derivative with respect to the argument, i.e., $f_i'(R) = df_i(R)/dR$ and $G'\left(L_m\right) = dG\left(L_m\right)/dL_m$. For the specific case of $f_2(R) = 1$, $G\left(L_m\right) = L_m$, one obtains the Palatini formalism field equations of $f(R)$ gravity [284, 470]. For the case $f_1(R) = R$, $f_2(R) = 1$, and $G\left(L_m\right) = L_m$, the field equations of standard GR are recovered.

We now take the variation of the action with respect to the connection Γ. Note that the variation is simplified by using the Palatini identity, given by Eq. (2.138), which we rewrite here

$$\delta\tilde{R}_{\mu\nu} = \tilde{\nabla}_\lambda\left(\delta\tilde{\Gamma}^\lambda_{\mu\nu}\right) - \tilde{\nabla}_\nu\left(\delta\tilde{\Gamma}^\lambda_{\mu\lambda}\right), \tag{8.48}$$

where $\tilde{\nabla}_\lambda$ is the covariant derivative associated with the connection $\tilde{\Gamma}$.

By taking the variation of the action (8.45) with respect to the connection $\tilde{\Gamma}$ we obtain

$$\frac{\delta S}{\delta\tilde{\Gamma}} = \frac{1}{2}\int A^{\mu\nu}\left[\tilde{\nabla}_\lambda\left(\delta\tilde{\Gamma}^\lambda_{\mu\nu}\right) - \tilde{\nabla}_\mu\left(\delta\tilde{\Gamma}^\lambda_{\nu\lambda}\right)\right]\sqrt{-g}d^4x, \tag{8.49}$$

where, for simplicity, we have defined $A^{\mu\nu}$ by

$$A^{\mu\nu} = \left[\frac{1}{2}f_1'(R) + f_2'(R)G\left(L_m\right)\right]g^{\mu\nu}. \tag{8.50}$$

Integration by parts yields the following relation

$$\frac{\delta S}{\delta\tilde{\Gamma}}=\frac{1}{2}\int\tilde{\nabla}_\lambda\left[\sqrt{-g}\left(A^{\mu\nu}\delta\tilde{\Gamma}^\lambda_{\mu\nu}-A^{\lambda\nu}\delta\tilde{\Gamma}^\alpha_{\nu\alpha}\right)\right]d^4x$$
$$+\frac{1}{2}\int\tilde{\nabla}_\mu\left[\sqrt{-g}\left(A^{\mu\nu}\delta^\lambda_\alpha-A^{\lambda\nu}\delta^\mu_\alpha\right)\right]\delta\tilde{\Gamma}^\alpha_{\lambda\nu}d^4x. \tag{8.51}$$

Note that the first term in the right-hand side is a total derivative, and can be discarded. Therefore, the variation of the action with respect to the connection provides

$$\tilde{\nabla}_\mu\left[\sqrt{-g}\left(A^{\mu\nu}\delta^\lambda_\alpha-A^{\lambda\nu}\delta^\mu_\alpha\right)\right]=0, \tag{8.52}$$

which for $\alpha=\lambda$ is identically zero. For the case $\lambda\neq\alpha$, we obtain

$$\tilde{\nabla}_\alpha\left\{\sqrt{-g}\left[f_1'(R)+2f_2'(R)G\left(L_m\right)\right]g^{\mu\nu}\right\}=0, \tag{8.53}$$

which shows that the connection is compatible with a conformal metric. Defining a new metric $h_{\mu\nu}$ [284, 470], conformal to $g_{\mu\nu}$, which is given by

$$h_{\mu\nu}\equiv\left[f_1'(R)+2f_2'(R)G\left(L_m\right)\right]g_{\mu\nu}, \tag{8.54}$$

one obtains

$$\sqrt{-h}h^{\mu\nu}=\sqrt{-g}\left[f_1'(R)+2f_2'(R)G\left(L_m\right)\right]g^{\mu\nu}, \tag{8.55}$$

where h is the determinant of the metric $h_{\mu\nu}$.

Therefore, Eq. (8.53) becomes the definition of the Levi–Civita connection $\tilde{\Gamma}$ of $h_{\mu\nu}$, given by

$$\tilde{\Gamma}^\lambda_{\mu\nu}=\frac{1}{2}h^{\lambda\rho}\left(\partial_\nu h_{\mu\rho}+\partial_\mu h_{\nu\rho}-\partial_\rho h_{\mu\nu}\right), \tag{8.56}$$

and using the explicit form of $h_{\mu\nu}$, takes the following form

$$\tilde{\Gamma}^\lambda_{\mu\nu}=\frac{1}{2}\frac{g^{\lambda\rho}}{F}\left[\partial_\nu\left(Fg_{\mu\rho}\right)+\partial_\mu\left(Fg_{\nu\rho}\right)-\partial_\rho\left(Fg_{\mu\nu}\right)\right], \tag{8.57}$$

where we have defined

$$F=F\left(R,L_m\right)=f_1'(R)+2f_2'(R)G\left(L_m\right). \tag{8.58}$$

Now, taking into account the Levi–Civita connection $\Gamma^\lambda_{\mu\nu}$ associated to the metric g

$$\Gamma^\lambda_{\mu\nu}=\frac{1}{2}g^{\lambda\rho}\left(\partial_\nu g_{\mu\rho}+\partial_\mu g_{\nu\rho}-\partial_\rho g_{\mu\nu}\right), \tag{8.59}$$

the independent connection $\tilde{\Gamma}^\lambda_{\mu\nu}$ is expressed as

$$\tilde{\Gamma}^\lambda_{\mu\nu}=\Gamma^\lambda_{\mu\nu}+\partial_\nu\ln\sqrt{F}\delta^\lambda_\mu+\partial_\mu\ln\sqrt{F}\delta^\lambda_\nu-g_{\mu\nu}g^{\lambda\rho}\partial_\rho\ln\sqrt{F}. \tag{8.60}$$

Using these relations, we can write the Ricci tensor $\tilde{R}_{\mu\nu}$ in terms of $R_{\mu\nu}$, which is constructed from the metric by using the Levi–Civita connection Eq. (8.59) in the following manner [470]

$$\tilde{R}_{\mu\nu} = R_{\mu\nu}(g) + \frac{3}{2}\frac{1}{F^2}\left(\nabla_\mu F\right)\left(\nabla_\nu F\right) - \frac{1}{F}\left(\nabla_\mu\nabla_\nu - \frac{1}{2}g_{\mu\nu}\Box\right)F. \tag{8.61}$$

In an analogous manner, the Ricci scalar and the Einstein tensor are given by

$$\tilde{R} = R\left(g\right) + 3\frac{1}{F}\Box F + \frac{3}{2}\frac{1}{F^2}\left(\nabla_\mu F\right)\left(\nabla^\mu F\right), \tag{8.62}$$

and

$$\begin{aligned}\tilde{G}_{\mu\nu} = \tilde{R}_{\mu\nu} - \frac{1}{2}g_{\mu\nu}\tilde{R} = G_{\mu\nu}(g) + \frac{3}{2}\frac{1}{F^2}\left(\nabla_\mu F\right)\left(\nabla_\nu F\right)\\ -\frac{1}{F}\left(\nabla_\mu\nabla_\nu + g_{\mu\nu}\Box\right)F - \frac{3}{4}\frac{1}{F^2}g_{\mu\nu}\left(\nabla_\lambda F\right)\left(\nabla^\lambda F\right),\end{aligned} \tag{8.63}$$

respectively, where all the covariant derivatives are taken with respect to the metric $g_{\mu\nu}$.

Now, using the Einstein tensor Eq. (8.63), the gravitational field equation (8.47) are written as

$$\tilde{G}_{\mu\nu} = \left(\frac{2K}{F} - \frac{1}{2}R - \frac{3}{2}\frac{1}{F}\Box F - \frac{3}{4}\frac{1}{F^2}\nabla_\lambda F\nabla^\lambda F\right)g_{\mu\nu} + \frac{f_2\left(R\right)G\left(L_m\right)}{F}T_{\mu\nu}, \tag{8.64}$$

where, for simplicity, we have defined K by

$$K = K\left(R, L_m\right) = \left\{f_1(R) + 2f_2(R)G\left(L_m\right)\left[1 - L_m\frac{G'\left(L_m\right)}{G\left(L_m\right)}\right]\right\}. \tag{8.65}$$

Thus, substituting Eq. (8.63) into the field equation Eq. (8.64), we obtain the following gravitational field equation of $f(R)$ gravity with a nonminimal coupling between matter and geometry, in the Palatini formalism:

$$\begin{aligned}G_{\mu\nu} + \frac{3}{2}\frac{1}{F^2}\left(\nabla_\mu F\right)\left(\nabla_\nu F\right) - \frac{1}{F}\nabla_\mu\nabla_\nu F =\\ \left(\frac{2K}{F} - \frac{1}{2}\frac{1}{F}\Box F - \frac{R}{2}\right)g_{\mu\nu} + \frac{f_2\left(R\right)G\left(L_m\right)}{F}T_{\mu\nu}.\end{aligned} \tag{8.66}$$

Note that the field equation Eq. (8.47) can be written as

$$F\tilde{R}_{\mu\nu} - \frac{1}{2}K\left(R, L_m\right)g_{\mu\nu} = f_2(R)G'\left(L_m\right)T_{\mu\nu}, \tag{8.67}$$

which provides the following trace equation

$$F\tilde{R} - 2K\left(R, L_m\right) = f_2(R)G'\left(L_m\right)T, \tag{8.68}$$

where $T = T^{\mu}_{\mu}$ is the trace of the energy-momentum tensor. Taking into account Eq. (8.61), the equation determining R as a function of T and the matter Lagrangian L_m yields

$$R(g) - 3\frac{1}{F}\Box F - \frac{3}{2}\frac{1}{F^2}(\nabla_\mu F)(\nabla^\mu F) - 2\frac{1}{F}K(R, L_m) = \frac{f_2(R)G'(L_m)}{F}T. \tag{8.69}$$

8.3.2 Equation of Motion and the Extra Force

In order to write the equation of motion for a fluid element, we use the covariant divergence of Eq. (8.67), which provides

$$\begin{aligned}(\nabla^\mu F)\tilde{R}_{\mu\nu} + F\nabla^\mu\tilde{R}_{\mu\nu} - \frac{1}{2}g_{\mu\nu}\nabla^\mu K(R, L_m) = \\ = [\nabla^\mu f_2(R)G'(L_m)]T_{\mu\nu} + f_2(R)G'(L_m)\nabla^\mu T_{\mu\nu},\end{aligned} \tag{8.70}$$

Now, taking into account the useful identity [284]

$$\nabla^\mu\tilde{G}_{\mu\nu} = -\frac{\nabla^\mu F}{F}\tilde{R}_{\mu\nu}, \tag{8.71}$$

which takes the form

$$F\nabla^\mu\tilde{R}_{\mu\nu} + (\nabla^\mu F)\tilde{R}_{\mu\nu} = \frac{1}{2}g_{\mu\nu}F\nabla^\mu\tilde{R}, \tag{8.72}$$

and substituting this relation this relation in Eq. (8.70), we then obtain

$$\begin{aligned}\nabla_\mu T^\mu{}_\nu = \frac{1}{f_2(R)G'(L_m)}\Big\{ &- [\nabla_\mu f_2(R)G'(L_m)]T^\mu{}_\nu \\ &+ L_m G'(L_m)\nabla_\nu f_2(R) - f_2(R)\nabla_\nu\mathcal{G}\Big\}.\end{aligned} \tag{8.73}$$

For notational convenience we defined $\mathcal{G} = G(L_m) - L_m G'(L_m)$, which vanishes when the action is linear in the matter Lagrangian.

For a perfect fluid, where given by the following matter energy-momentum tensor

$$T^{(m)}_{\mu\nu} = (\epsilon + p)\,u_\mu u_\nu + p g_{\mu\nu}, \tag{8.74}$$

where ϵ is the energy density, p is the thermodynamic pressure, and the four-velocity, u_μ, satisfies the conditions $u_\mu u^\mu = -1$ and $u^\mu u_{\mu;\nu} = 0$. Note that the trace of the energy-momentum tensor is given by $T = 3p - \epsilon$.

We introduce, as before, the projection operator $h_{\mu\lambda} = g_{\mu\lambda} + u_\mu u_\lambda$, from which one obtains $h_{\mu\lambda}u^\mu = 0$ and $T^{\mu\nu}h_{\mu\lambda} = ph^\nu_\lambda$, respectively. Now, contracting Eq. (8.70) with the projection operator $h_{\mu\lambda}$, one deduces the following expression

$$\begin{aligned}(\epsilon + p)\,g_{\mu\lambda}u^\nu\nabla_\nu u^\mu - (\nabla_\nu p)(\delta^\nu_\lambda - u^\nu u_\lambda) = \frac{1}{f_2(R)G'(L_m)}(\delta^\nu_\lambda - u^\nu u_\lambda)\times \\ \times\Big\{p\nabla_\nu\left(f_2(R)G'(L_m)\right)L_m G'(L_m)\nabla_\nu f_2(R) - f_2(R)\nabla_\nu\mathcal{G}\Big\}.\end{aligned} \tag{8.75}$$

A final contraction with the metric tensor $g^{\alpha\lambda}$ yields the equation of motion for a fluid element:

$$\frac{Du^\alpha}{ds} \equiv \frac{du^\alpha}{ds} + \Gamma^\alpha_{\mu\nu} u^\mu u^\nu = f^\alpha . \tag{8.76}$$

Here the space-time connection $\Gamma^\alpha_{\mu\nu}$ is expressed in terms of the Christoffel symbols, constructed from the metric $g_{\mu\nu}$. The extra force is defined as:

$$f^\alpha = \frac{h^{\nu\alpha}}{\epsilon + p}\Big[- \nabla_\nu p - (p - L_m)\, \nabla_\nu \log f_2(R) \\ + p\nabla_\nu \log G' + \frac{1}{G'} \nabla_\nu \left(G - L_m G'\right)\Big] , \tag{8.77}$$

where we have dropped the explicit dependence of G on L_m, for notational simplicity.

As is transparent from Eq. (8.77), the extra force f^α is orthogonal to the four-velocity of the particle,

$$f^\alpha u_\alpha = 0, \tag{8.78}$$

which follows from the properties of the projection operator. This property is clearly consistent with the usual interpretation of the force, according to which only the component of the four-force that is orthogonal to the particle's four-velocity can influence its trajectory. Thus, the presence of the extra force f^α implies that the motion of the particle is non-geodesic (the geodesic equation of motion is recovered for $f^\alpha \equiv 0$). Note also that the usual gravitational effects, due to the presence of an arbitrary mass distribution, are assumed to be contained in the term $a^\mu_N = \Gamma^\mu_{\alpha\beta} u^\alpha u^\beta$.

We can now identify three contributions to the force: (i) The usual general relativistic pressure gradient is the first term in (8.77). (ii) The second term is due to the nonminimal curvature-matter coupling to gravity and of course vanishes in GR (for $L_m = p$, the result that the extra force also disappears, as in the metric formalism, also applies here). Interestingly, the choice of $L_m = p$ is the case in the usual Lagrangian description of scalar fields and holds for more general perfect fluids as well, depending upon the chosen prescription. (iii) Furthermore, the two following terms, consisting of the last line in (8.77), are due to the new self-interactions of matter and disappear when the action is linear in the matter Lagrangian L_m.

All of the three terms can be considered as *contact forces*, so that for a test particle, these vanish in vacuum. This behavior reflects the fact that the Palatini approach of gravity can be understood as a modified response to matter sources whereas it reduces to GR (with a cosmological constant) in vacuum. In other words, though the geodesics of particles are modified within a matter distribution, there are no new propagating degrees of freedom that would mediate interactions in vacuum, and the extra force is absent there.

8.4 Equivalence with an Anomalous Scalar–Tensor Theory

As we have shown in Chapter 6, pure $f(R)$ gravity, with the action given by Eq. (6.3), is equivalent to a scalar–tensor theory (6.64). The equivalence between the modified gravity theories with a linear curvature-matter coupling was established in [188]. In the following argument we show that the action (8.1) is equivalent with a scalar–tensor Brans–Dicke-type theory, with a single scalar field, vanishing Brans–Dicke parameter ω, and an unusual coupling of the potential $U(\psi)$ of the theory to matter.

By introducing a new field ϕ, the action (8.1) becomes

$$S = \int d^4x\sqrt{-g}\left\{\frac{f_1(\phi)}{2} + \frac{1}{2}\frac{df_1}{d\phi}(R-\phi) + [1+\lambda f_2(\phi)]\,\mathcal{L}_m\right\}. \tag{8.79}$$

Next, we further introduce the field $\psi(\phi) \equiv f_1'(\phi)$ (with a prime denoting the differentiation with respect to ϕ), and we obtain for the action the expression

$$S = \int d^4x\sqrt{-g}\left[\frac{\psi R}{2} - V(\psi) + U(\psi)\mathcal{L}_m\right], \tag{8.80}$$

where

$$V(\psi) = \frac{\phi(\psi) f_1'\,[\phi(\psi)] - f_1\,[\phi(\psi)]}{2}, \tag{8.81}$$

$$U(\psi) = 1 + \lambda f_2\,[\phi(\psi)]. \tag{8.82}$$

The field $\phi(\psi)$ must be obtained by inverting $\psi(\phi) \equiv f_1'(\phi)$. The actions (8.1) and (8.80) are equivalent when $f_1''(R) \neq 0$ [188]. We can see this by setting $\phi = R$, and then Eq. (8.79) reduces trivially to Eq. (8.1). On the other hand, the variation of Eq. (8.79) with respect to ϕ gives

$$(R-\phi)\,f_1''(\phi) + 2\lambda f_2'(\phi)\mathcal{L}_m = 0. \tag{8.83}$$

In vacuum, we have $\mathcal{L}_m = 0$, and Eq. (8.83) gives $\phi = R$ whenever $f_1'' \neq 0$ [187, 479, 495, 508]. However, in the presence of matter there seem to be other possibilities, which can be excluded as follows.

When $\mathcal{L}_m \neq 0$, the actions (8.1) and (8.79) are equivalent if

$$(R-\phi)\,f_1''(\phi) + 2\lambda f_2''(\phi)\mathcal{L}_m \neq 0. \tag{8.84}$$

When Eq. (8.83) is satisfied, we have a pathological case. It corresponds to

$$\lambda f_2(\phi)\mathcal{L}_m = \frac{f_1'(\phi)}{2}(\phi - R) - \frac{f_1(\phi)}{2}. \tag{8.85}$$

But if Eq. (8.85) holds, then the action (8.79) reduces to the trivial case of pure matter without the gravity sector. Then, it follows that in the modified theories of gravity with a linear curvature-matter coupling the actions (8.1) and (8.80) are equivalent when $f_1''(R) \neq 0$, similarly to the case of pure $f(R)$ gravity [187, 479, 495, 508].

8.5 Further Theoretical Developments

The linear nonminimal curvature-matter coupling has been extensively explored in a plethora of contexts, in the literature. For instance, the equations of motion of test bodies for a theory with nonminimal coupling by means of a multipole method [422] was studied, and it was shown that the propagation equations for pole-dipole particles allow for a systematic comparison with the equations of motion of GR and other gravity theories.

The consequences that a nonminimal coupling between curvature and matter can have on the dynamics of perfect fluids has also been investigated [55]. It was argued that the presence of a static, axially symmetric pressureless fluid does not imply an asymptotically Minkowski space-time, such as in GR. This feature can be attributed to a pressure mimicking mechanism related to the nonminimal coupling. The case of a spherically symmetric black hole surrounded by fluid matter was analyzed, and it was shown that under equilibrium conditions the total fluid mass is about twice that of the black hole [55].

The consequences of the curvature-matter coupling on stellar equilibrium and constraints on the nonminimal coupling was considered, where particular attention was paid to the validity of the Newtonian regime, and on the boundary and exterior matching conditions [56, 57]. This explicit "anomalous" coupling of the Ricci curvature to matter has also raised the question of curvature instabilities, and in [63, 188, 497, 498] constraints imposed by the energy conditions and the conditions in order to avoid the notorious Dolgov–Kawasaki instability [172] were obtained. It has also been claimed that the curvature-matter action leads to a theory of gravity that includes higher-order derivatives of the matter fields without introducing more dynamics in the gravity sector and, therefore, cannot be a viable theory for gravitation [469]. However, we emphasize that the results of [469] only apply to the specific case of $f_2(R) = R$ and not to action (8.1) in general. For more generic functional forms, the theory propagates extra degrees of freedom and the conclusions of [469] do not apply. The relation between these theories and ordinary scalar–tensor gravity was also analyzed, and its implications for the equivalence principle [470].

In fact, the theoretical consistency of a nonminimal curvature-matter coupling was studied using a scalar field Lagrangian to model the matter content [482]. The conditions that the coupling does not introduce ghosts, classical instabilities, or superluminal propagation of perturbations were derived. These consistency conditions were then employed to rule out or severely restrict the forms of the nonminimal coupling functions [381]. Lagrange–Noether methods have been used to derive the conservation laws for models in which matter interacts nonminimally with the gravitational field [389]. Furthermore, a covariant derivation of the equations of motion for test bodies for a wide class of gravitational theories with nonminimal coupling was presented in [423], encompassing a general interaction via the complete set of nine parity-even curvature invariants. The equations of motion for spinning test bodies in such theories were explicitly derived by means of Synge's expansion technique. The authors' findings generalize previous results in the literature and allowed for a direct comparison to the general relativistic equations of motion of pole-dipole test bodies. In [424], the authors derived multipolar equations of motion for gravitational theories with general nonminimal coupling in space-times admitting that admit torsion. Their general findings allow for a systematic testing of whole classes of theories by means of extended test bodies. One peculiar feature of certain subclasses of nonminimal theories turns out to be their sensitivity to post-Riemannian space-time structures even in experiments without microstructured test matter. Weak field constraints have also been studied in detail [54, 125], and wormhole solutions were further explored, where it is the higher curvature coupling terms that support these exotic geometries [45, 364, 365].

In a cosmological context, the perturbation equation of matter on subhorizon scales was deduced, and specific bounds on the theory from weak lensing observations and the primordial nucleosynthesis were obtained in order to constrain the parameters of the model [376]. It was also shown that a nonminimal coupling between the scalar curvature and the matter Lagrangian density may account for the accelerated expansion of the Universe [46, 66, 481] and provide, through mimicking, for a viable unification of dark energy and dark matter [46]. It was shown that a generalized nonminimal coupling between curvature and matter is compatible with Starobinsky inflation and leads to a successful process of preheating [47], and the problem of the cosmological constant was further explored [60].

The effects of the nonminimal curvature-matter coupling on the evolution of cosmological perturbations around a homogeneous and isotropic Universe and hence the formation of large-scale structure has also been analyzed [49]. This framework places constraints on the terms that arise due to the coupling with matter and, in particular, on the modification in the growth of matter density perturbations. Approximate analytical solutions were obtained for the evolution of matter overdensities during the matter dominated era and it was shown that these favor the presence of a coupling function that is compatible with the late-time cosmic acceleration.

The observations related to the growth of matter has also shown that there is a small but finite window where one can distinguish the nonminimally coupled $f(R)$ models with the concordance ΛCDM [480]. The possibility that the behavior of the rotational velocities of test particles gravitating around galaxies can be explained in the framework of modified gravity models with nonminimal curvature-matter coupling has also been extensively explored [48, 58, 217].

In fact, the literature is extremely vast, and rather than enumerate all the features of these models, we refer the reader to [52, 59, 234, 237] for a review on the topic of the linear nonminimal curvature-matter coupling.

9

$f(R, L_m)$ Gravity

9.1 Action and Field Equations

In this chapter, we generalize the $f(R)$-type gravity models considered in the previous chapters by assuming a maximal extension of the Einstein–Hilbert action, so that the action takes the following form [233]

$$S = \int f(R, L_m)\sqrt{-g}\, d^4x \;, \tag{9.1}$$

where $f(R, L_m)$ is an arbitrary function of the Ricci scalar R, and of the Lagrangian density corresponding to matter, L_m. The energy-momentum tensor of the matter is defined by Eq. (8.3). Thus, by assuming that the Lagrangian density L_m of the matter depends only on the metric tensor components $g_{\mu\nu}$, and not on its derivatives, we obtain $T_{\mu\nu} = g_{\mu\nu}L_m - 2\,\partial L_m/\partial g^{\mu\nu}$, which will be useful in Section 9.5.

Now, varying the action with respect to the metric, we obtain the following field equation

$$\begin{aligned} &f_R(R, L_m)\, R_{\mu\nu} + (g_{\mu\nu}\Box - \nabla_\mu\nabla_\nu)\, f_R(R, L_m) \\ &-\frac{1}{2}\left[f(R, L_m) - f_{L_m}(R, L_m)\, L_m\right] g_{\mu\nu} = \frac{1}{2} f_{L_m}(R, L_m)\, T_{\mu\nu}. \end{aligned} \tag{9.2}$$

Note that if $f(R, L_m) = R/2 + L_m$ (the Hilbert–Einstein Lagrangian), we recover the standard Einstein field equation of GR. For $f(R, L_m) = f_1(R) + f_2(R)G(L_m)$, where f_1 and f_2 are arbitrary functions of the Ricci scalar, and G a function of the matter Lagrangian density, respectively, we reobtain the field equations of the modified gravity with a curvature-matter coupling [215], considered in the previous chapter.

The contraction of Eq. (9.2) provides the following relation between R, L_m, and the trace $T = T^\mu_\mu$ of the energy-momentum tensor,

$$f_R(R,L_m)R+3\Box f_R(R,L_m)-2\left[f(R,L_m)-f_{Lm}(R,L_m)L_m\right] = \frac{1}{2}f_{Lm}(R,L_m)T. \qquad (9.3)$$

By eliminating the term $\Box f_R(R,L_m)$ between Eqs. (9.2) and (9.3), we obtain another form of the gravitational field equations as

$$f_R(R,L_m)\left(R_{\mu\nu}-\frac{1}{3}Rg_{\mu\nu}\right)+\frac{1}{6}\left[f(R,L_m)-f_{Lm}(R,L_m)L_m\right]g_{\mu\nu} = \frac{1}{2}f_{Lm}(R,L_m)\left(T_{\mu\nu}-\frac{1}{3}Tg_{\mu\nu}\right)+\nabla_\mu\nabla_\nu f_R(R,L_m). \qquad (9.4)$$

Taking the covariant divergence of Eq. (9.2), with the use of the following mathematical identity [284]

$$\nabla^\mu\left[f_R(R,L_m)R_{\mu\nu}-\frac{1}{2}f(R,L_m)g_{\mu\nu}+ (g_{\mu\nu}\Box-\nabla_\mu\nabla_\nu)f_R(R,L_m)\right]\equiv 0\,, \qquad (9.5)$$

we obtain for the divergence of the energy-momentum tensor $T_{\mu\nu}$, the following equation

$$\nabla^\mu T_{\mu\nu} = \nabla^\mu \ln\left[f_{Lm}(R,L_m)\right]\left\{L_m g_{\mu\nu}-T_{\mu\nu}\right\} = 2\nabla^\mu \ln\left[f_{Lm}(R,L_m)\right]\frac{\partial L_m}{\partial g^{\mu\nu}}\,. \qquad (9.6)$$

The requirement of the conservation of the energy-momentum tensor of matter, i.e., $\nabla^\mu T_{\mu\nu}=0$, gives an effective functional relation between the matter Lagrangian density and the function $f_{Lm}(R,L_m)$

$$\nabla^\mu \ln\left[f_{Lm}(R,L_m)\right]\frac{\partial L_m}{\partial g^{\mu\nu}}=0\,. \qquad (9.7)$$

Thus, once the matter Lagrangian density is known, by an appropriate choice of the function $f(R,L_m)$ one can construct, at least in principle, conservative models with arbitrary curvature-matter dependences.

Now, assuming that the matter Lagrangian is a function of the rest mass density ρ of the matter only, from Eq. (9.6) we obtain explicitly the equation of motion of the test particles in $f(R,L_m)$ gravity as

$$\frac{D^2x^\mu}{ds^2}=U^\nu\nabla_\nu U^\mu=\frac{d^2x^\mu}{ds^2}+\Gamma^\mu_{\nu\lambda}U^\nu U^\lambda=f^\mu, \qquad (9.8)$$

where the word line parameter s is taken as the proper time, $U^\mu=dx^\mu/ds$ is the four-velocity of the particle, $\Gamma^\nu_{\sigma\beta}$ are the Christoffel symbols associated to the metric, and the extra force f^μ is defined as

$$f^\mu=-\nabla_\nu \ln\left[f_{Lm}(R,L_m)\frac{dL_m(\rho)}{d\rho}\right](U^\mu U^\nu-g^{\mu\nu})\,. \qquad (9.9)$$

Note that, as in the linear curvature-matter coupling considered in the previous chapter, the extra force f^{μ}, generated by the curvature-matter coupling, is perpendicular to the four-velocity, $f^{\mu}U_{\mu} = 0$. Due to the presence of the extra force f^{μ}, the motion of test particles in modified theories of gravity with an arbitrary coupling between matter and curvature is non-geodesic. From the relation $U_{\mu}\nabla_{\nu}U^{\mu} \equiv 0$ it follows that the force f^{μ} is always perpendicular to the velocity, so that $U_{\mu}f^{\mu} = 0$, as we covered in the previous chapter.

Building on this analysis, the geodesic deviation equation, describing the relative accelerations of nearby particles, and the Raychaudhuri equation, giving the evolution of the kinematical quantities associated with deformations, was extensively considered in this framework of modified theories of gravity with an arbitrary curvature-matter coupling, by taking into account the effects of the extra force [235]. As a physical application of the geodesic deviation equation, the modifications of the tidal forces due to the supplementary curvature-matter coupling were obtained in the weak field approximation. The tidal motion of test particles is directly influenced not only by the gradient of the extra force, which is basically determined by the gradient of the Ricci scalar, but also by an explicit coupling between the velocity and the Riemann tensor. As a specific example, the expression of the Roche limit (the orbital distance at which a satellite will begin to be tidally torn apart by the body it is orbiting) was also obtained for this class of models. These aspects will be presented in the following sections.

Furthermore, the energy conditions and cosmological applications were also explored [496]. In [261] the Wheeler–DeWitt equation of $f(R, L_m)$ gravity was analyzed in a flat FLRW Universe, which is the first step of the study of $f(R, L_m)$ quantum cosmology. In the minisuperspace spanned by the FLRW scale factor and the Ricci scalar, the equivalence of the reduced action was examined, and the canonical quantization of $f(R, L_m)$ gravity was undertaken and the corresponding Wheeler–DeWitt equation derived. This analysis will be extensively considered in later chapters, in a different context. The introduction of the invariant contractions of the Ricci and Riemann tensors were further considered and applications to black hole and wormhole physics were analyzed [483].

9.2 Solar System Tests

One of the basic predictions of the modified theories of gravity with a curvature-matter coupling is the existence of an extra force, which makes the motion of the test particles non-geodesic. The existence of this force can be tested at the level of the solar system, by estimating its effects on the orbital parameters of the motion of the planets around the Sun. The impact on the planetary motion of the extra force can be obtained in a simple way by using the properties of the Runge–Lenz vector, defined as

$$\vec{A} = \vec{v} \times \vec{L} - \alpha \vec{e}_r, \tag{9.10}$$

where $\alpha = GmM_{\odot}$, and $\vec{v}$ is the velocity of the planet of mass m relative to the Sun, with mass $M_{\odot}$, $\vec{r} = r\vec{e}_r$ is the two-body position vector, $\vec{p} = \mu\vec{v}$ is the relative momentum, $\mu = mM_{\odot}/(m + M_{\odot})$ is the reduced mass, and

$$\vec{L} = \vec{r} \times \vec{p} = \mu r^2 \dot{\theta} \vec{k}, \tag{9.11}$$

is the angular momentum (see [215] and references therein). For an elliptical orbit of eccentricity e, major semiaxis a, and period T, the equation of the orbit is given by $(L^2/\mu\alpha) r^{-1} = 1 + e\cos\theta$. The Runge–Lenz vector can be expressed as

$$\vec{A} = \left(\frac{\vec{L}^2}{\mu r} - \alpha \right) \vec{e}_r - \dot{r} L \vec{e}_\theta, \tag{9.12}$$

and its derivative with respect to the polar angle θ is given by

$$\frac{d\vec{A}}{d\theta} = r^2 \left[\frac{dV(r)}{dr} - \frac{\alpha}{r^2} \right] \vec{e}_\theta\,, \tag{9.13}$$

where $V(r)$ is the potential of the central force [215].

The gravitational potential term acting on a planet consists of the post-Newtonian potential

$$V_{PN}(r) = -\frac{\alpha}{r} - 3\frac{\alpha^2}{mr^2}, \tag{9.14}$$

plus the gravitational term induced by the general coupling between curvature and matter. Thus we have

$$\frac{d\vec{A}}{d\theta} = r^2 \left[6\frac{\alpha^2}{mr^3} + m\vec{a}_E(r) \right] \vec{e}_\theta, \tag{9.15}$$

where $\vec{a}_E$ is given by Eq. (8.25), and we have also assumed that $\mu \approx m$. Then we obtain the change of $\Delta\phi$ of the perihelion with a change of θ of 2π as

$$\Delta\phi = \left(\frac{1}{\alpha e} \right) \int_0^{2\pi} \left| \dot{\vec{L}} \times d\frac{\vec{A}}{d\theta} \right| d\theta, \tag{9.16}$$

which can be explicitly calculated as

$$\Delta\phi = 24\pi^3 \left(\frac{a}{T} \right)^2 \frac{1}{1-e^2} + \frac{L}{8\pi^3 me} \frac{\left(1-e^2\right)^{3/2}}{\left(a/T\right)^3}$$
$$\times \int_0^{2\pi} \frac{a_E\left[L^2\left(1+e\cos\theta\right)^{-1}/m\alpha\right]}{\left(1+e\cos\theta\right)^2} \cos\theta d\theta, \tag{9.17}$$

where we have used the relation $\alpha/L = 2\pi\,(a/T)/\sqrt{1-e^2}$. The first term of Eq. (9.17) gives the expression of the standard general relativistic perihelion precession of the planets, while the second term gives the contribution to the perihelion precession due to the presence of the extra force generated by the coupling between matter and curvature.

As an example of the application of Eq. (9.17), we consider the simple case for which the extra force may be considered as a constant, $a_E \approx$ constant, an approximation that could be valid for small regions of space-time. This case may also correspond to a MOND-type acceleration $a_E \approx \sqrt{a_0 a_N} = \sqrt{GM_\odot a_0}/r$, where a_0 is a constant acceleration, which was proposed phenomenologically as a dynamical model for dark matter [42, 352]. In the Newtonian limit the extra-acceleration generated by the curvature-matter coupling can be expressed in a similar form [44]. With the use of Eq. (9.17), one finds for the perihelion precession of a planet in the solar system the expression

$$\Delta\phi = \frac{6\pi GM_\odot}{a\left(1-e^2\right)} + \frac{2\pi a^2\sqrt{1-e^2}}{GM_\odot}a_E, \tag{9.18}$$

where we have also used Kepler's third law, $T^2 = 4\pi^2 a^3/GM_\odot$. For the planet Mercury $a = 57.91 \times 10^{11}$ cm, and $e = 0.205615$, respectively, while $M_\odot = 1.989 \times 10^{33}$ g [509]. With the use of these numerical values, the first term in Eq. (9.18) gives the standard general relativistic value for the precession angle, $(\Delta\phi)_{GR} = 42.962$ arcsec per century. On the other hand, the observed value of the perihelion precession of Mercury is $(\Delta\phi)_{obs} = 43.11 \pm 0.21$ arcsec per century [509]. Therefore the difference $(\Delta\phi)_E = (\Delta\phi)_{obs} - (\Delta\phi)_{GR} = 0.17$ arcsec per century can be attributed to other physical effects. Hence the observational constraint on the perihelion precession of the planet Mercury requires that the value of the constant acceleration a_E induced by the curvature-matter coupling at the scale of the solar system must satisfy the condition $a_E \leq 1.28 \times 10^{-9}$ cm/s^2. This value of a_E, obtained from the high precision solar system observations, is somewhat smaller than the value of the extra-acceleration $a_0 \approx 10^{-8}$ cm/s^2, necessary to explain the "dark matter" properties, as well as the Pioneer anomaly [44]. However, this value does not rule out the possibility of an extra curvature-matter coupling, acting at both the solar system and galactic levels, since the assumption of a constant extra-acceleration may not be correct on larger astronomical scales.

9.3 The Geodesic Deviation Equation and the Raychaudhuri Equation

In this subsection, we present the equation of the geodesic deviation and the Raychaudhuri equation in $f(R, L_m)$ gravity, both of which explicitly contain the effects of the curvature-matter coupling and of the extra force. Some of the physical implications of the geodesic deviation equation, namely, the problem of the tidal forces in $f(R, L_m)$ gravity, are also considered, and the generalization of the Roche limit will be considered in the next subsection. In our presentation we closely follow [235].

Consider a one-parameter congruence of curves $x^\mu(s;\lambda)$, so that for each $\lambda = \lambda_0 =$ constant, $x^\mu(s, \lambda_0)$ satisfies Eq. (9.8). We suppose the parametrization to be smooth, and hence we can introduce the tangent vector fields along the trajectories of the particles as $U^\mu = \partial x^\mu (s;\lambda)/\partial s$ and $n^\mu = \partial x^\mu (s;\lambda)/\partial\lambda$, respectively. We also introduce the four-vector

$$\eta^\mu = \left[\frac{\partial x^\mu (s;\lambda)}{\partial\lambda}\right]\delta\lambda \equiv n^\mu \delta\lambda, \tag{9.19}$$

joining points on infinitely close geodesics, corresponding to parameter values λ and $\lambda+\delta\lambda$, which have the same value of s. By taking into account Eq. (9.8), we obtain the geodesic deviation equation (the Jacobi equation), giving the second-order derivative with respect to the parameter s of the deviation vector η^μ as [303, 250]

$$\frac{D^2\eta^\mu}{ds^2} = R^\mu_{\nu\alpha\beta}\eta^\alpha U^\beta U^\nu + \eta^\alpha \nabla_\alpha f^\mu. \tag{9.20}$$

In the case $f^\mu \equiv 0$ we reobtain the standard Jacobi equation, corresponding to the geodesic motion of test particles in standard GR. The interest in the deviation vector η^μ derives from the fact that if $x_0^\mu(s) = x^\mu (s;\lambda_0)$ is a solution of Eq. (9.20), then to first-order $x_1^\mu(s) = x_0^\mu(s) + \eta^\mu$ is a solution of the geodesic equation as well, since $x^\mu (s;\lambda_1) \approx x^\mu (s;\lambda_0) + n^\mu (s;\lambda_0)\,\delta\lambda \approx x^\mu (s;\lambda_0+\delta\lambda)$.

By taking into account the explicit form of the extra force given by Eq. (9.9) in $f(R, L_m)$ modified theories of gravity with a curvature-matter coupling, the geodesic deviation equation can be written as

$$\begin{aligned}\frac{D^2\eta^\mu}{ds^2} &= R^\mu_{\nu\alpha\beta}\eta^\alpha U^\beta U^\nu \\ &+\eta^\alpha\nabla_\alpha\left\{\nabla_\nu \ln\left[f_{L_m}(R, L_m)\frac{dL_m(\rho)}{d\rho}\right](L_m g^{\mu\nu} - T^{\mu\nu})\right\},\end{aligned} \tag{9.21}$$

which explicitly becomes

$$\begin{aligned}\frac{D^2\eta^\mu}{ds^2} &= R^\mu_{\nu\alpha\beta}\eta^\alpha U^\beta U^\nu + \eta^\alpha\left\{\nabla_\alpha\nabla_\nu \ln\left[f_{L_m}(R, L_m)\frac{dL_m(\rho)}{d\rho}\right]\right\}(L_m g^{\mu\nu} - T^{\mu\nu}) \\ &+\eta^\alpha\nabla_\nu \ln\left[f_{L_m}(R, L_m)\frac{dL_m(\rho)}{d\rho}\right](g^{\mu\nu}\nabla_\alpha L_m - \nabla_\alpha T^{\mu\nu}).\end{aligned} \tag{9.22}$$

In the presence of an extra force the Raychaudhuri equation is obtained as [250]

$$\dot{\theta} + \frac{1}{3}\theta^2 + (\sigma^2 - \omega^2) = \nabla_\mu f^\mu + R_{\mu\nu}U^\mu U^\nu, \tag{9.23}$$

where $\theta = \nabla_\nu U^\nu$ is the expansion of the congruence of particles, and where $\sigma^2 = \sigma_{\mu\nu}\sigma^{\mu\nu}$ and $\omega^2 = \omega_{\mu\nu}\omega^{\mu\nu}$, respectively.

With the explicit use of the field equation (9.2), and the expression of the extra force given by Eq. (9.9), in $f(R, L_m)$ gravity, the Raychaudhuri equation assumes the following generalized form

$$\begin{aligned}\dot{\theta} = &-\frac{1}{3}\theta^2 - \left(\sigma^2 - \omega^2\right) + \Lambda\left(R, L_m\right) \\ &+\nabla_\mu \left\{\nabla_\nu \ln\left[f_{L_m}\left(R, L_m\right)\frac{dL_m\left(\rho\right)}{d\rho}\right]\left(L_m g^{\mu\nu} - T^{\mu\nu}\right)\right\} \\ &+\frac{1}{f_R\left(R, L_m\right)}U^\mu U^\nu \nabla_\mu \nabla_\nu f_R\left(R, L_m\right) \\ &+\Phi\left(R, L_m\right)\left(T_{\mu\nu}U^\mu U^\nu - \frac{1}{3}T\right),\end{aligned} \tag{9.24}$$

where we have denoted

$$\Lambda\left(R, L_m\right) = \frac{2f_R\left(R, L_m\right)R - f\left(R, L_m\right) + f_{L_m}\left(R, L_m\right)L_m}{6f_R\left(R, L_m\right)}, \tag{9.25}$$

and

$$\Phi\left(R, L_m\right) = \frac{f_{L_m}\left(R, L_m\right)}{f_R\left(R, L_m\right)}, \tag{9.26}$$

respectively.

9.4 Tidal Forces and the Roche Limit

Tides are common astrophysical phenomena, and they are due to the presence of a gradient of the gravitational force field induced by a mass above an extended body or a system of particles. In the solar system, tidal perturbations act on compact celestial bodies such as planets, moons, and comets. On larger scales than the solar system, such as in a galactic or cosmological context, one can observe tidal deformations or disruptions of a stellar cluster by a galaxy, or in galaxy encounters [343, 344, 345]. In the relativistic theories of gravitation, as well as in Newtonian gravity, a local system of coordinates can be chosen, which is inertial except for the presence of the tidal forces. In strong gravitational fields, relativistic tidal effects can lead to important physical phenomena, such as the emission of tidal gravitational waves [343, 344, 345].

In the following, we denote by a prime the reference frame in which all the Christoffel symbols vanish. In such a system one can always take the deviation vector component $\eta'^0 = 0$, which means that the particle accelerations are compared at equal times. η'^i is then the displacement of the particle from the origin [391]. Moreover, in the static/stationary case, in which the metric, the Ricci scalar and the thermodynamic parameters of the matter do not depend on time, we have $f'^0 = 0$. With the use of the equation of motion this condition implies $U'^0 = \text{constant} = 1$. Therefore, with these assumptions, in a static or stationary space-time the equation of the geodesic deviation (the Jacobi equation) takes the form [235, 328]

$$\frac{d^2\eta'^i}{dt'^2} = R'^i_{0l0}\eta'^l + R'^i_{jlm}\eta'^l U'^j U'^m + \eta'^l \frac{\partial f'^i}{\partial x'^l}. \tag{9.27}$$

Equation (9.27) can be reformulated as

$$F'^i = \frac{d^2\eta'^i}{dt'^2} = K^i_j \eta'^j, \tag{9.28}$$

where F'^i is the tidal force, and we have also introduced the generalized tidal matrix K^i_l [343, 344, 345], which is defined as

$$K^i_j = R'^i_{0j0} + R'^i_{kjm} U'^k U'^m + \frac{\partial f'^i}{\partial x'^j}. \tag{9.29}$$

The tidal force has the property

$$\frac{\partial F'^i}{\partial \eta'^j} = K^i_j, \tag{9.30}$$

and its divergence is given by

$$\frac{\partial F'^i}{\partial \eta'^i} = K, \tag{9.31}$$

where the trace K of the tidal matrix is

$$K = K^i_i = R'_{00} + R'_{km} U'^k U'^m + \frac{\partial f'^i}{\partial x'^i}. \tag{9.32}$$

With the use of the gravitational field equations (9.2) we can express K as

$$\begin{aligned} K = {} & \Lambda\left(R', L'_m\right)\eta_{00} + \frac{1}{f_R\left(R', L_m\right)}\frac{\partial^2}{\partial t'^2} f_R\left(R', L'_m\right) + \Phi\left(R', L'_m\right)\left(T'_{00} - \frac{1}{3}T'\eta_{00}\right) \\ & + \Lambda\left(R', L'_m\right)\eta_{km}U'^k U'^m + \frac{1}{f_R\left(R', L'_m\right)} U'^k U'^m \frac{\partial^2}{\partial x'^k \partial x'^m} f_R\left(R', L'_m\right) \\ & + \phi\left(R', L'_m\right)\left(T'_{km} - \frac{1}{3}T\eta_{km}\right)U'^k U'^m \\ & + \frac{\partial}{\partial x'^k}\left\{\frac{\partial}{\partial x'^m}\ln\left[f_{Lm}\left(R', L'_m\right)\frac{dL'_m(\rho)}{d\rho}\right]\left(L'_m \eta^{km} - T'^{km}\right)\right\}. \end{aligned} \tag{9.33}$$

Then, since in the Newtonian limit one can omit all time derivatives, we obtain the following expression for R^i_{0l0} [391]

$$R_{i0l0} = \frac{\partial^2 \phi}{\partial x^i \partial x^l}. \tag{9.34}$$

For simplicity, in the following we omit the primes for the geometrical and physical quantities in the Newtonian approximation. Therefore, in $f(R, L_m)$ gravity we obtain for the tidal acceleration of the test particles the following expression

$$\frac{d^2\eta^i}{dt^2} = F^i \approx \frac{\partial^2 \phi}{\partial x^i \partial x^l}\eta^l + R^i_{jlm}\eta^l V^j V^m + \eta^l \frac{\partial f^i}{\partial x^l}, \tag{9.35}$$

where V^j and V^m are the Newtonian three-dimensional velocities. In the Newtonian approximation, in modified theories of gravity with a curvature-matter coupling the tidal force tensor is defined as

$$\frac{\partial F^i}{\partial \eta^l} = \frac{\partial^2 \phi}{\partial x^i \partial x^l} + R^i_{jlm} V^j V^m + \frac{\partial f^i}{\partial x^l}, \tag{9.36}$$

and its trace provides the generalized Poisson equation

$$\frac{\partial F^i}{\partial \eta^i} = \Delta\phi + R_{jm} V^j V^m + \frac{\partial f^i}{\partial x^i}. \tag{9.37}$$

In Newtonian gravity

$$\chi_{il} = -\frac{\partial^2 \phi}{\partial x^i \partial x^l}, \tag{9.38}$$

represents the Newtonian tidal tensor [391]. In the Newtonian approximation, the spherical potential of a given particle with mass M is $\phi(r) = -M/8\pi r$. By choosing a frame of reference so that the x-axis passes through the particle's position, corresponding to the radial spherical coordinate, that is, $(x = r, y = 0, z = 0)$, the Newtonian tidal tensor is diagonal, where the only non-zero components are given by

$$\chi_{ii} = \text{diag}\left(\frac{2M}{8\pi r^3}, -\frac{M}{8\pi r^3}, -\frac{M}{8\pi r^3}\right). \tag{9.39}$$

The Newtonian tidal force $\vec{F}_t$ can be written as

$$F_{tx} = 2M\Delta x/8\pi r^3, \qquad F_{ty} = -GM\Delta y/8\pi r^3, \qquad F_{tz} = -M\Delta z/8\pi r^3, \tag{9.40}$$

respectively [391].

These results can be used to derive the generalization of the Roche limit in modified gravity with an arbitrary curvature-matter coupling between matter and curvature. The Roche limit is the closest distance r_{Roche} that a celestial object with mass m, radius R_m, and density ρ_m, held together only by its own gravity, can come to a massive body of mass M, radius R_M, and density ρ_M, respectively, without being pulled apart by the massive object's tidal (gravitational) force [343, 344, 345]. For simplicity we will consider $M \gg m$, so that the center of mass of the system practically coincides with the geometrical centre of the mass M.

The elementary Newtonian theory of this process is as follows. Consider a small mass Δm located at the surface of the small object of mass m. There are two forces acting on Δm, the gravitational attraction of the mass m, given by

$$F_G = \frac{m\Delta m}{8\pi R_m^2}, \tag{9.41}$$

and the tidal force exerted by the massive object, which is provided by

$$F_{tr} = \frac{M\Delta m R}{8\pi r^3}, \tag{9.42}$$

where r is the distance between the centers of the two celestial bodies. The Roche limit is reached at the distance $r = r_{Roche}$, when the gravitational force and the tidal force exactly balance each other, $F_G = F_{tr}$, thus giving [343, 344, 345]

$$r_{Roche} = R_m \left(\frac{M}{m}\right)^{1/3} = 2^{1/3} R_M \left(\frac{\rho_M}{\rho_m}\right)^{1/3}. \tag{9.43}$$

In modified gravity with a curvature-matter coupling the equilibrium between the gravitational and tidal forces occurs at a distance r_{Roche} given by the equation

$$\left(\frac{M}{8\pi r^3_{Roche}} + R^r_{jrm} V^j V^m + \frac{\partial f^r}{\partial r}\right) R_m = \frac{m}{8\pi R^2_m} + f^r, \tag{9.44}$$

where f^r is the radial component of the extra force, which modifies the Newtonian gravitational force, and the curvature tensor R^r_{jrm} (no summation upon r) must be evaluated in the coordinate system in which the Newtonian tidal tensor is diagonal. Hence we obtain the generalized Roche limit in the presence of arbitrary curvature-matter coupling as [235]

$$r_{Roche} \approx R_m \left(\frac{M}{m}\right)^{1/3} \left[1 + \frac{8\pi R^3_m}{3m}\left(R^r_{jrm} V^j V^m + \frac{\partial f^r}{\partial r}\right) - \frac{8\pi R^2_m}{3m} f^r\right], \tag{9.45}$$

where we have assumed that the gravitational effects due to the coupling between matter and curvature are small as compared to the Newtonian ones. Relativistic corrections to the Newtonian tidal accelerations caused by a massive rotating source, such as, for example, the earth, could be determined experimentally, at least in principle, thus leading to the possibility of testing relativistic theories of gravitation by measuring such effects in a laboratory.

9.5 Extensions with Scalar Field and Kinetic Term Dependencies

9.5.1 Action and Field Equations

We generalize $f(R, L_m)$ gravity, outlined in the previous section, by considering a novel gravitational model given by the following action [242]

$$S = \int f\left(R, L_m, \phi, g^{\mu\nu}\nabla_\mu\phi\nabla_\nu\phi\right)\sqrt{-g}\, d^4x, \tag{9.46}$$

where $\sqrt{-g}$ is the determinant of $g_{\mu\nu}$, and $f\left(R, L_m, \phi, g^{\mu\nu}\nabla_\mu\phi\nabla_\nu\phi\right)$ is an arbitrary function of the Ricci scalar R, the matter Lagrangian density, L_m, a scalar field ϕ, and a kinetic term constructed from the gradients of the scalar field, respectively. The only restriction on the function f is to be an analytical function of R, L_m, ϕ, and of the scalar field kinetic energy, respectively, that is, f must possess a Taylor series expansion about any point.

Note that one may motivate the introduction of Eq. (9.46) by the extensive interest in the literature between couplings between the scalar and the matter fields [16, 23, 26, 123, 145, 149, 150, 151, 152, 153, 155, 201, 355, 506]. Indeed, such couplings generically appear in Kaluza–Klein theories with compactified dimensions [198, 399] or in the low energy effective limit of string theories [150, 151, 152, 153, 201], where the dilaton has been proposed to be a good candidate for quintessence [201] or inflaton [153]. The theory represented by Eq. (9.46) offers a generalization of the earlier theories in a single theoretical framework.

Here, we introduce first the "reduced" energy-momentum tensor $\tau_{\mu\nu}$ of the matter, which is defined as [303],

$$\tau_{\mu\nu} = -\frac{2}{\sqrt{-g}}\frac{\delta\left(\sqrt{-g}L_m\right)}{\delta g^{\mu\nu}} = g_{\mu\nu}L_m - 2\frac{\delta L_m}{\delta g^{\mu\nu}}. \tag{9.47}$$

In addition to this, we assume that the scalar field ϕ is independent of the metric, i.e., $\delta\phi/\delta g^{\mu\nu} \equiv 0$, and in the following we define, for simplicity, $(\nabla\phi)^2 = g^{\mu\nu}\nabla_\mu\phi\nabla_\nu\phi$.

By varying the action S with respect to the metric tensor and to ϕ, we obtain the following field equation

$$\begin{aligned} f_R R_{\mu\nu} + \left(g_{\mu\nu}\nabla_\lambda\nabla^\lambda - \nabla_\mu\nabla_\nu\right) f_R - \frac{1}{2}\left(f - f_{L_m}L_m\right) g_{\mu\nu} \\ = \frac{1}{2} f_{L_m}\tau_{\mu\nu} - f_{(\nabla\phi)^2}\nabla_\mu\phi\nabla_\nu\phi, \end{aligned} \tag{9.48}$$

and the evolution equation for the scalar field

$$\Box_{(\nabla\phi)^2}\phi = \frac{1}{2}f_\phi, \tag{9.49}$$

respectively. The subscript of f denotes a partial derivative with respect to the arguments, i.e., $f_R = \partial f/\partial R$, $f_{L_m} = \partial f/\partial L_m$, $f_{(\nabla\phi)^2} = \partial f/\partial(\nabla\phi)^2$, $f_\phi = \partial f/\partial\phi$, and

$$\Box_{(\nabla\phi)^2} = \frac{1}{\sqrt{-g}}\frac{\partial}{\partial x^\mu}\left[f_{(\nabla\phi)^2}\sqrt{-g}g^{\mu\nu}\frac{\partial}{\partial x^\nu}\right], \tag{9.50}$$

is the generalized D'Alembert operator of $f(R, L_m, \phi, \nabla_\mu\phi\nabla^\mu\phi)$ gravity.

The contraction of Eq. (9.48) provides the following relation between the Ricci scalar R, the matter Lagrangian density L_m, the derivatives of the scalar field, and the trace $\tau = \tau^\mu_\mu$ of the "reduced" energy-momentum tensor,

$$f_R R + 3\Box f_R - 2\left(f - f_{L_m}L_m\right) = \frac{1}{2}f_{L_m}\tau - f_{(\nabla\phi)^2}\nabla_\mu\phi\nabla^\mu\phi. \tag{9.51}$$

By taking the covariant divergence of Eq. (9.48), we obtain for the covariant divergence of the "reduced" energy-momentum tensor the following expression

$$\frac{1}{2}\nabla_\mu \left(f_{L_m} \tau_{\mu\sigma}\right) = \frac{1}{2}\left(L_m \nabla_\mu f_{L_m} - f_\phi \nabla_\mu \phi\right) + f_{(\nabla\phi)^2} \nabla_\mu \phi \,\Box\phi$$
$$+\nabla_\mu \phi \nabla^\sigma \phi \nabla_\sigma f_{(\nabla\phi)^2}. \qquad (9.52)$$

This relationship was deduced by taking into account the following mathematical identities

$$\nabla^\mu R_{\mu\nu} = \frac{1}{2}\nabla_\nu R, \qquad \left(\nabla_\nu \Box - \Box \nabla_\nu\right) f_R = -\left(\nabla^\mu f_R\right) R_{\mu\nu}, \qquad (9.53)$$

and considering torsion-free space-times, such that $\left[\nabla_\sigma \nabla_\epsilon - \nabla_\epsilon \nabla_\sigma\right] \psi = 0$, where ψ is any scalar field. Now, using Eq. (9.50) we get

$$\nabla_\mu \left(f_{L_m} \tau_{\mu\sigma}\right) = L_m \nabla_\mu f_{L_m}. \qquad (9.54)$$

For $\phi \equiv 0$, Eqs. (9.48) reduce to the field equations of the $f\left(R, L_m\right)$ theory considered in the previous section [233]. For $\phi \neq 0$, one recovers the conservation equations for either GR and Brans–Dicke-like scalar–tensor theories (with and without scalar/matter couplings [357]). For instance, the total Lagrangian of the simplest matter-scalar field-gravitational field theory, with a scalar field kinetic term and a self-interacting potential $V(\phi)$, corresponds to the choice

$$f = \frac{R}{2} + L_m + \frac{\lambda}{2} g^{\mu\nu} \nabla_\mu \phi \nabla_\nu \phi + V(\phi), \qquad (9.55)$$

where λ is a constant. The corresponding field equations can be immediately obtained from Eqs. (9.48) as

$$R_{\mu\nu} - \frac{1}{2} R g_{\mu\nu} = \tau_{\mu\nu} - \lambda \nabla_\mu \phi \nabla_\nu \phi + \left[\frac{\lambda}{2} g^{\alpha\beta} \nabla_\alpha \phi \nabla_\beta \phi + V(\phi)\right] g_{\mu\nu}. \qquad (9.56)$$

The scalar field satisfies the evolution equation

$$\frac{1}{\sqrt{-g}} \frac{\partial}{\partial x^\mu}\left[\sqrt{-g} g^{\mu\nu} \frac{\partial \phi}{\partial x^\nu}\right] = \frac{1}{\lambda} \frac{dV(\phi)}{d\phi}, \qquad (9.57)$$

while the energy-momentum tensor obeys the conservation equation $\nabla^\sigma \tau_{\mu\sigma} = 0$.

9.5.2 Models with Nonminimal Matter-Scalar Field Coupling

As an example of the application of the formalism developed in the previous section, we consider a simple phenomenological model, in which a scalar field is non-minimally coupled to pressureless matter with *rest mass density* ρ. For the action of the system we consider

$$S = \int \left[\frac{R}{2} - F(\phi)\rho + \lambda g^{\mu\nu} \nabla_\mu \phi \nabla_\nu \phi\right] \sqrt{-g} d^4 x, \qquad (9.58)$$

where $F(\phi)$ is an arbitrary function of the scalar field that couples nonminimally to ordinary matter. The field equations for this model are given by

$$R_{\mu\nu} - \frac{1}{2}Rg_{\mu\nu} = F(\phi)\rho U_\mu U_\nu + 2\lambda\left[\nabla_\mu\phi\nabla_\nu\phi - \frac{1}{2}g_{\mu\nu}\nabla_\alpha\phi\nabla^\alpha\phi\right], \tag{9.59}$$

where U^μ is the four-velocity of the matter fluid. The scalar field satisfies the evolution equation

$$\Box\phi = -\frac{1}{2\lambda}\frac{dF(\phi)}{d\phi}\rho, \tag{9.60}$$

where $\Box$ is the usual d'Alembert operator defined in a curved space. The total energy-momentum tensor of the scalar field-matter system is given by

$$T_{\mu\nu} = F(\phi)\rho U_\mu U_\nu + 2\lambda\left[\nabla_\mu\phi\nabla_\nu\phi - \frac{1}{2}g_{\mu\nu}\nabla_\alpha\phi\nabla^\alpha\phi\right]. \tag{9.61}$$

The Bianchi identity imply that $\nabla_\nu T^{\mu\nu} = 0$, and in the following, we assume that the mass density current is conserved, i.e., $\nabla_\nu\left(\rho U^\nu\right) = 0$. Using the latter condition, and the mathematical identity given by $\left[\nabla_\sigma\nabla_\epsilon - \nabla_\epsilon\nabla_\sigma\right]\psi = 0$, we have

$$F(\phi)\rho\left(U^\nu\nabla_\nu U^\mu\right) + \rho U^\mu U^\nu\frac{dF(\phi)}{d\phi}\nabla_\nu\phi + 2\lambda\left(\nabla^\mu\phi\right)\Box\phi = 0, \tag{9.62}$$

then by eliminating the term $\Box\phi$ with the help of Eq. (9.60), we obtain

$$U^\nu\nabla_\nu U^\mu + \left[\frac{d}{d\phi}\ln F(\phi)\right]\left(U^\mu U^\nu\nabla_\nu\phi - \nabla^\mu\phi\right) = 0. \tag{9.63}$$

Using the identity $U^\nu\nabla_\nu U^\mu \equiv d^2x^\mu/ds^2 + \Gamma^\mu_{\alpha\beta}u^\alpha u^\beta$, where $\Gamma^\mu_{\alpha\beta}$ are the Christoffel symbols corresponding to the metric, the equation of motion of the test particles nonminimally coupled to an arbitrary scalar field takes the form

$$\frac{d^2x^\mu}{ds^2} + \Gamma^\mu_{\alpha\beta}U^\alpha U^\beta + \left[\frac{d}{d\phi}\ln F(\phi)\right]\left(U^\mu U^\nu\nabla_\nu\phi - \nabla^\mu\phi\right) = 0. \tag{9.64}$$

A particular model can be obtained by assuming that $F(\phi)$ is given by a linear function,

$$F(\phi) = \frac{\Lambda+1}{2}\left[1 + \frac{1}{2}\left(\Lambda - 1\right)\phi\right], \tag{9.65}$$

where Λ is a constant. Then the equation of motion becomes

$$\frac{d^2x^\mu}{ds^2} + \Gamma^\mu_{\alpha\beta}U^\alpha U^\beta + \left(U^\mu U^\nu - g^{\mu\nu}\right)\nabla_\nu\ln\left[1 + \frac{\Lambda-1}{2}\phi\right] = 0. \tag{9.66}$$

In order to simplify the field equations we adopt for λ the value $\lambda = -\left(\Lambda^2 - 1\right)/8$. Then Eq. (9.60), determining the scalar field, takes the simple form $\Box\phi = \rho$.

The gravitational field equations take the form

$$R_{\mu\nu} - \frac{1}{2}g_{\mu\nu}R = \frac{\Lambda+1}{2}T_{\mu\nu}, \tag{9.67}$$

with the total energy-momentum tensor given by

$$T_{\mu\nu} = \left[1 + \frac{\Lambda-1}{2}\phi\right]\rho U_\mu U_\nu - \frac{\Lambda-1}{2}\left[\nabla_\mu\phi\nabla_\nu\phi - \frac{1}{2}g_{\mu\nu}\nabla_\alpha\phi\nabla^\alpha\phi\right]. \tag{9.68}$$

For $\Lambda = 1$ we reobtain the general relativistic model for dust. Other possible choices of the function $F(\phi)$, such as $F(\phi) = \exp(\phi)$, can be discussed in a similar way.

A more general model can be obtained by adopting for the matter Lagrangian the general expression [193, 218, 357]

$$L_m = -\left[\rho + \rho\int\frac{dp(\rho)}{\rho} - p(\rho)\right], \tag{9.69}$$

where ρ is *the rest-mass energy density*, p is the thermodynamic pressure, which, by assumption, satisfies a barotropic equation of state, $p = p(\rho)$. By assuming that the matter Lagrangian does not depend on the derivatives of the metric, and that the particle matter fluid current is conserved $[\nabla_\nu(\rho u^\nu) = 0]$, the Lagrangian given by Eq. (9.69) is the unique matter Lagrangian that can be constructed from the thermodynamic parameters of the fluid, and it is valid for all gravitational theories satisfying the two previously mentioned conditions [357].

The gravitational field equations and the equation describing the matter-scalar field coupling are given by

$$R_{\mu\nu} - \frac{1}{2}g_{\mu\nu}R = F(\phi)\,\epsilon\, U_\mu U_\nu - pg_{\mu\nu} + \lambda Q_{\mu\nu}, \tag{9.70}$$

$$\Box\phi = \frac{1}{2\lambda}\frac{dF(\phi)}{d\phi}\,\epsilon, \tag{9.71}$$

where $Q_{\mu\nu} = \nabla_\mu\phi\nabla_\nu\phi - \frac{1}{2}g_{\mu\nu}\nabla_\lambda\phi\nabla^\lambda\phi$, and the total energy density is $\epsilon = \rho + \rho\int dp/\rho - p$ [193, 357]. With the use of the conservation equation $\nabla_\nu(\rho U^\nu) = 0$, one obtains the equation of motion of massive test particles as

$$\frac{d^2x^\mu}{ds^2} + \Gamma^\mu_{\alpha\beta}U^\alpha U^\beta + (U^\mu U^\nu - g^{\mu\nu})\nabla_\nu \ln\left[1 + \int\frac{dp}{\rho}\right] = 0. \tag{9.72}$$

Models with scalar field-matter coupling were considered in the framework of the Brans–Dicke theory [155], with the action of the model given by

$$S = \int\left[\frac{1}{2}\phi R + \frac{\omega}{\phi}\nabla_\mu\phi\nabla^\mu\phi + F(\phi)L_m\right]\sqrt{-g}d^4x. \tag{9.73}$$

Such models can give rise to a late-time cosmic acceleration of the Universe for very high values of the Brans–Dicke parameter ω. Other models with interacting scalar field and matter have been considered in [16, 123, 149, 506]. We emphasize that the gravitational theory considered in this section generalizes all of these models.

In conclusion, the general formalism outlined in this work can be extremely useful in a variety of scenarios, such as, in describing the interaction between dark energy, modeled as a scalar field, and dark matter, or ordinary matter (neutrinos), with or without pressure, matter-scalar field interactions in inflation, as well as in the study of the interactions of the scalar field (representing dark matter and/or dark energy) and the electromagnetic component in the very early Universe. Moreover they can provide a realistic description of the late cosmic expansion of the Universe, where a possible interaction between ordinary matter and dark energy cannot be excluded a priori.

9.6 Summary and Discussion

In the present chapter, we have considered a generalized gravitational theory consisting of a Lagrangian density of an arbitrary function of the Lagrangian density of the matter and of the Ricci scalar. The proposed action represents the most general extension of the standard Einstein–Hilbert action for the gravitational field. $f(R, L_m)$ modified theories of gravity open the possibility of going beyond the *additive* algebraic structure of the classical Einstein–Hilbert action, as well as most of its generalizations, which are constructed as the *sum* of several terms describing independently geometry, matter, and their coupling. As in the case of the linear curvature-matter coupling, the equations of motion corresponding to $f(R, L_m)$ gravity induce the presence of an extra force acting on test particles, and the motion is generally non-geodesic. On the other hand, the field equations are equivalent to the field equations of $f(R)$ gravity in vacuum, but differ from them, as well as from standard GR, in the presence of matter. Therefore the predictions of the present model could lead to some major differences, as compared to the predictions of GR, in several problems of current interest, such as cosmology, gravitational collapse, or the generation of gravitational waves[51]. The study of these phenomena may also provide some specific signatures and effects, which could distinguish and discriminate between the various theories of modified gravity.

10

$f(R,T)$ Gravity

10.1 Action and Gravitational Field Equations

In this section, we consider another extension of standard General Relativity, namely, $f(R,T)$-modified theories of gravity, where the gravitational Lagrangian is given by an arbitrary function of the Ricci scalar R and of the trace of the energy-momentum tensor T [243]. The dependence from T may be induced by exotic imperfect fluids or quantum effects (conformal anomaly).

The action takes the following form

$$S = \frac{1}{16\pi}\int f\left(R,T\right)\sqrt{-g}\, d^4x + \int L_m\sqrt{-g}\, d^4x\,, \tag{10.1}$$

where $f\left(R,T\right)$ is an arbitrary function of the Ricci scalar R, and of the trace T of the energy-momentum tensor of matter, $T_{\mu\nu}$. For the matter content we assume that it consists of a fluid that can be characterized by two thermodynamic parameters only: the energy density and the pressure, respectively. The trace of the energy-momentum tensor T can be expressed as a function of the matter Lagrangian as

$$T = g^{\mu\nu}T_{\mu\nu} = 4L_m - 2g^{\mu\nu}\frac{\delta L_m}{\delta g^{\mu\nu}}. \tag{10.2}$$

Hence, $f(R,T)$ gravity can be interpreted as an extension of the $f\left(R, L_m\right)$-type gravity theories, with the gravitational action depending not only on the matter Lagrangian only, but also on its variation with respect to the metric.

By varying the action (10.1) with respect to the metric tensor components $g^{\mu\nu}$, we obtain the following gravitational field equations of $f\left(R,T\right)$ gravity

$$\begin{aligned} f_R\left(R,T\right)R_{\mu\nu} - \frac{1}{2}f\left(R,T\right)g_{\mu\nu} + \left(g_{\mu\nu}\Box - \nabla_\mu\nabla_\nu\right)f_R\left(R,T\right) \\ = 8\pi T_{\mu\nu} - f_T\left(R,T\right)T_{\mu\nu} - f_T\left(R,T\right)\Theta_{\mu\nu}\,. \end{aligned} \tag{10.3}$$

We have defined the variation of T with respect to the metric tensor as

$$\frac{\delta\left(g^{\alpha\beta}T_{\alpha\beta}\right)}{\delta g^{\mu\nu}} = T_{\mu\nu} + \Theta_{\mu\nu}\,, \tag{10.4}$$

where

$$\Theta_{\mu\nu} \equiv g^{\alpha\beta}\frac{\delta T_{\alpha\beta}}{\delta g^{\mu\nu}}\,. \tag{10.5}$$

Note that $T = 0$ reduces to $f(R,T) \equiv f(R)$, and from Eq. (10.3) we obtain the field equations of $f(R)$ gravity.

Contracting Eq. (10.3) gives the following relation between the Ricci scalar R and the trace T of the energy-momentum tensor,

$$f_R\left(R,T\right)R + 3\Box f_R\left(R,T\right) - 2f\left(R,T\right) = 8\pi T - f_T\left(R,T\right)T - f_T\left(R,T\right)\Theta\,, \tag{10.6}$$

where we have denoted $\Theta = \Theta_{\mu}^{\ \mu}$.

By eliminating the term $\Box f_R\left(R,T\right)$ between Eqs. (10.3) and (10.6), the gravitational field equations of $f(R,T)$ gravity can be written in the form

$$\begin{aligned} f_R\left(R,T\right)\left(R_{\mu\nu} - \frac{1}{3}Rg_{\mu\nu}\right) + \frac{1}{6}f\left(R,T\right)g_{\mu\nu} &= 8\pi\left(T_{\mu\nu} - \frac{1}{3}Tg_{\mu\nu}\right) - f_T\left(R,T\right) \\ \times\left(T_{\mu\nu} - \frac{1}{3}Tg_{\mu\nu}\right) - f_T\left(R,T\right)\left(\Theta_{\mu\nu} - \frac{1}{3}\Theta g_{\mu\nu}\right) &+ \nabla_\mu\nabla_\nu f_R\left(R,T\right). \end{aligned} \tag{10.7}$$

Taking into account the covariant divergence of Eq. (10.3), with the use of the following mathematical identity [284]

$$\begin{aligned} \nabla^\mu\left[f_R\left(R,T\right)R_{\mu\nu} - \frac{1}{2}f\left(R,T\right)g_{\mu\nu} + \left(g_{\mu\nu}\Box - \nabla_\mu\nabla_\nu\right)f_R\left(R,T\right)\right] \\ \equiv -\frac{1}{2}g_{\mu\nu}f_T(R,T)\nabla^\mu T\,, \end{aligned} \tag{10.8}$$

we obtain for the divergence of the energy-momentum tensor $T_{\mu\nu}$ in $f(R,T)$ gravity the equation

$$\nabla^\mu T_{\mu\nu} = \frac{f_T\left(R,T\right)}{8\pi - f_T\left(R,T\right)}\left[\left(T_{\mu\nu} + \Theta_{\mu\nu}\right)\nabla^\mu \ln f_T\left(R,T\right) + \nabla^\mu\Theta_{\mu\nu} - \frac{1}{2}g_{\mu\nu}\nabla^\mu T\right]. \tag{10.9}$$

We now consider the calculation of the tensor $\Theta_{\mu\nu}$. Once the matter Lagrangian is known, from Eq. (8.8), we first obtain

$$\begin{aligned} \frac{\delta T_{\alpha\beta}}{\delta g^{\mu\nu}} &= \frac{\delta g_{\alpha\beta}}{\delta g^{\mu\nu}}L_m + g_{\alpha\beta}\frac{\partial L_m}{\partial g^{\mu\nu}} - 2\frac{\partial^2 L_m}{\partial g^{\mu\nu}\partial g^{\alpha\beta}} \\ &= \frac{\delta g_{\alpha\beta}}{\delta g^{\mu\nu}}L_m + \frac{1}{2}g_{\alpha\beta}g_{\mu\nu}L_m - \frac{1}{2}g_{\alpha\beta}T_{\mu\nu} - 2\frac{\partial^2 L_m}{\partial g^{\mu\nu}\partial g^{\alpha\beta}}\,. \end{aligned} \tag{10.10}$$

From the condition $g_{\alpha\sigma}g^{\sigma\beta} = \delta^{\beta}_{\alpha}$, we have

$$\frac{\delta g_{\alpha\beta}}{\delta g^{\mu\nu}} = -g_{\alpha\sigma}g_{\beta\gamma}\delta^{\sigma\gamma}_{\mu\nu}\,, \tag{10.11}$$

where $\delta^{\sigma\gamma}_{\mu\nu} = \delta g^{\sigma\gamma}/\delta g^{\mu\nu}$ is the generalized Kronecker symbol. Therefore for $\Theta_{\mu\nu}$ we find

$$\Theta_{\mu\nu} = -2T_{\mu\nu} + g_{\mu\nu}L_m - 2g^{\alpha\beta}\frac{\partial^2 L_m}{\partial g^{\mu\nu}\partial g^{\alpha\beta}}\,. \tag{10.12}$$

More specifically, in the case of the electromagnetic field the matter Lagrangian is given by

$$L_m = -\frac{1}{16\pi}F_{\alpha\beta}F_{\gamma\sigma}g^{\alpha\gamma}g^{\beta\sigma}\,, \tag{10.13}$$

where $F_{\alpha\beta}$ is the electromagnetic field tensor, we obtain $\Theta_{\mu\nu} = -T_{\mu\nu}$. For a massless scalar field ϕ with Lagrangian $L_m = g^{\alpha\beta}\nabla_\alpha\phi\nabla_\beta\phi$, we obtain $\Theta_{\mu\nu} = -T_{\mu\nu} + (1/2)Tg_{\mu\nu}$.

The problem of the perfect fluids, described by an energy density ρ, pressure p and four-velocity U^μ is more subtle, since there is no unique definition of the matter Lagrangian. However, in the present study we *assume* that the energy-momentum tensor of the matter is given by $T_{\mu\nu} = (\rho + p)\,U_\mu U_\nu + pg_{\mu\nu}$, and the matter Lagrangian can be taken as $L_m = p$. The four-velocity U_μ satisfies the conditions $U_\mu U^\mu = -1$ and $U^\mu\nabla_\nu U_\mu = 0$, respectively. Then, with the use of Eq. (10.12), we obtain for the variation of the energy-momentum tensor of a perfect fluid the expression

$$\Theta_{\mu\nu} = -2T_{\mu\nu} + pg_{\mu\nu}\,. \tag{10.14}$$

The choice of the matter Lagrangian given by $L_m = -\rho$ leads to a different form for $\Theta_{\mu\nu}$, $\Theta_{\mu\nu} = -2T_{\mu\nu} + \rho g_{\mu\nu}$, and consequently to different field equations. The nature of the qualitative and quantitative differences between the two different versions of the $f(R,T)$ gravity, corresponding to different matter Lagrangians, is still an open question, although significant contributions were made, as outlined in Section 8.2.

10.2 Specific Cosmological Solution

In the present section, we consider a particular class of $f(R,T)$- modified gravity-model, obtained by explicitly specifying the functional form of f. Generally, the field equations also depend, through the tensor $\Theta_{\mu\nu}$, on the physical nature of the matter field. Hence in the case of $f(R,T)$ gravity, depending on the nature of the matter source, for each choice of f we can obtain several theoretical models, corresponding to different matter models [234].

As a specific case of a $f(R,T)$ model, consider a correction to the Einstein–Hilbert action given by $f\,(R,T) = R{+}2f(T)$, where $f(T)$ is an arbitrary function of the trace of the energy-momentum tensor of matter. The gravitational field equations immediately follow from Eq. (10.3), and are given by

$$R_{\mu\nu} - \frac{1}{2}Rg_{\mu\nu} = 8\pi T_{\mu\nu} - 2f'(T)\,T_{\mu\nu} - 2f'(T)\Theta_{\mu\nu} + f(T)g_{\mu\nu}\,, \tag{10.15}$$

where the prime denotes a derivative with respect to the argument.

Consider a perfect fluid, so that $\Theta_{\mu\nu} = -2T_{\mu\nu} - pg_{\mu\nu}$, and the field equations become

$$R_{\mu\nu} - \frac{1}{2}Rg_{\mu\nu} = 8\pi T_{\mu\nu} + 2f'(T)\,T_{\mu\nu} + \left[2pf'(T) + f(T)\right]g_{\mu\nu}\,. \tag{10.16}$$

In the case of dust with $p = 0$ the gravitational field equations are given by

$$R_{\mu\nu} - \frac{1}{2}Rg_{\mu\nu} = 8\pi T_{\mu\nu} + 2f'(T)T_{\mu\nu} + f(T)g_{\mu\nu}\,. \tag{10.17}$$

These field equations were proposed in [418] to solve the cosmological constant problem. The simplest cosmological model can be obtained by assuming a dust universe ($p = 0$, $T = -\rho$), and by choosing the function $f(T)$ so that $f(T) = \lambda T$, where λ is a constant. By assuming that the metric of the universe is given by the flat Friedmann–Lemaître–Robertson–Walker (FLRW) metric

$$ds^2 = -dt^2 + a^2(t)\left(dx^2 + dy^2 + dz^2\right)\,, \tag{10.18}$$

the modified Friedmann equations are given by

$$3\frac{\dot{a}^2}{a^2} = (8\pi + 3\lambda)\,\rho\,, \tag{10.19}$$

$$2\frac{\ddot{a}}{a} + \frac{\dot{a}^2}{a^2} = \lambda\rho\,, \tag{10.20}$$

respectively. Thus, this $f(R,T)$ gravity model is equivalent to a cosmological model with an effective dark energy density $\Lambda_{\rm eff} \propto H^2$, where $H = \dot{a}/a$ is the Hubble function [418]. It is also interesting to note that generally for this choice of $f(R,T)$ the gravitational coupling becomes an effective and time dependent coupling, of the form $G_{\rm eff} = G \pm 2f'(T)$. Thus the term $2f(T)$ in the gravitational action modifies the gravitational interaction between matter and curvature, replacing G by a running gravitational coupling parameter.

The field equations reduce to a single equation for H

$$2\dot{H} + 3\frac{8\pi + 2\lambda}{8\pi + 3\lambda}H^2 = 0\,, \tag{10.21}$$

with the general solution given by

$$H(t) = \frac{2\,(8\pi + 3\lambda)}{3\,(8\pi + 2\lambda)}\frac{1}{t}\,. \tag{10.22}$$

The scale factor evolves according to $a(t) = t^\alpha$, with $\alpha = 2\,(8\pi + 3\lambda)\,/3\,(8\pi + 2\lambda)$.

We refer the reader to [243] for further examples of cosmological solutions.

10.3 $f(R,T,R_{\mu\nu}T^{\mu\nu})$ Gravity

For the specific case of a traceless energy-momentum tensor, $T = 0$, for instance, when the electromagnetic field is involved, the gravitational field equations for the $f(R,T)$ theory [243] reduce to that of the field equations for $f(R)$ gravity, and all nonminimal couplings of gravity to the matter field vanish. This motivates a further generalization of $f(R,T)$ gravity that consists in including an explicit first-order coupling between the matter energy-momentum $T_{\mu\nu}$ and the Ricci tensor [211, 390]. In contrast to $f(R,T)$ gravity, for $T = 0$, this extra coupling still has a nonminimal coupling to the electromagnetic field via the $R_{\mu\nu}T^{\mu\nu}$ coupling term in the action, which is nonzero in general (see [237] for a review). As in the previous section, we consider the matter content as consisting of a perfect fluid characterized by its energy density and thermodynamic pressure only.

10.3.1 Action and Field Equations

The action for $f(R,T,R_{\mu\nu}T^{\mu\nu})$ gravity, is given by

$$S = \frac{1}{16\pi G}\int d^4x\sqrt{-g}f(R,T,R_{\mu\nu}T^{\mu\nu}) + \int d^4x\sqrt{-g}L_m\,. \tag{10.23}$$

The only requirement imposed on the function $f(R,T,R_{\mu\nu}T^{\mu\nu})$ is that it is an arbitrary analytical function in all arguments. The gravitational field equations takes the following form

$$\begin{aligned}
&(f_R - f_{RT}L_m)G_{\mu\nu} - \nabla_\mu\nabla_\nu f_R + \frac{1}{2}\Box\left(f_{RT}T_{\mu\nu}\right) + 2f_{RT}R_{\alpha(\mu}T_{\nu)}^{\alpha} \\
&\quad + \left[\Box f_R + \frac{1}{2}Rf_R - \frac{1}{2}f + f_T L_m + \frac{1}{2}\nabla_\alpha\nabla_\beta\left(f_{RT}T^{\alpha\beta}\right)\right]g_{\mu\nu} \\
&\quad -\nabla_\alpha\nabla_{(\mu}\left[T^{\alpha}_{\ \nu)}f_{RT}\right] - \left(f_T + \frac{1}{2}f_{RT}R + 8\pi G\right)T_{\mu\nu} \\
&\quad -2\left(f_T g^{\alpha\beta} + f_{RT}R^{\alpha\beta}\right)\frac{\partial^2 L_m}{\partial g^{\mu\nu}\partial g^{\alpha\beta}} = 0.
\end{aligned} \tag{10.24}$$

The trace of the gravitational field equation, (Eq. 10.24), is obtained as

$$\begin{aligned}
&3\Box f_R + \frac{1}{2}\Box\left(f_{RT}T\right) + \nabla_\alpha\nabla_\beta\left(f_{RT}T^{\alpha\beta}\right) + Rf_R - Tf_T - \frac{1}{2}RTf_{RT} \\
&\quad +2R_{\alpha\beta}T^{\alpha\beta}f_{RT} + Rf_{RT}L_m + 4f_T L_m - 2f - 8\pi GT \\
&\quad -2g^{\mu\nu}\left(g^{\alpha\beta}f_T + R^{\alpha\beta}f_{RT}\right)\frac{\partial^2 L_m}{\partial g^{\mu\nu}\partial g^{\alpha\beta}} = 0.
\end{aligned} \tag{10.25}$$

The second derivative of the matter Lagrangian with respect to the metric is nonzero if the matter Lagrangian is of second or higher order in the metric. Thus, for a perfect fluid with $L_m = -\rho$, or a scalar field with $L_m = -\partial_\mu\phi\partial^\mu\phi/2$, this term can be dropped. However, for instance, considering the Maxwell field, if we have $L_m = -F_{\mu\nu}F^{\mu\nu}/4$, this term results in $\partial^2 L_m/\partial g^{\mu\nu}\partial g^{\alpha\beta} = -F_{\mu\alpha}F_{\nu\beta}/2$, thus giving a nonzero contribution to the field equations.

In analogy with the standard Einstein field equation, one can write the field equation (10.24) as

$$G_{\mu\nu} = 8\pi G_{\rm eff} T_{\mu\nu} - \Lambda_{\rm eff}\, g_{\mu\nu} + T^{\rm eff}_{\mu\nu}, \tag{10.26}$$

where we have defined the effective gravitational coupling $G_{\rm eff}$, the effective cosmological constant $\Lambda_{\rm eff}$, and an effective energy-momentum tensor $T^{\rm eff}_{\mu\nu}$ as

$$G_{\rm eff} = \frac{G + \frac{1}{8\pi}\left(f_T + \frac{1}{2} f_{RT} R - \frac{1}{2}\Box\, f_{RT}\right)}{f_R - f_{RT} L_m}, \tag{10.27}$$

$$\Lambda_{\rm eff} = \frac{2\Box\, f_R + R f_R - f + 2 f_T L_m + \nabla_\alpha \nabla_\beta \left(f_{RT} T^{\alpha\beta}\right)}{2\left(f_R - f_{RT} L_m\right)}, \tag{10.28}$$

and

$$\begin{aligned} T^{\rm eff}_{\mu\nu} = \frac{1}{f_R - f_{RT} L_m} \Bigg\{ & \nabla_\mu \nabla_\nu f_R - \nabla_\alpha f_{RT} \nabla^\alpha T_{\mu\nu} - \frac{1}{2} f_{RT} \Box\, T_{\mu\nu} - 2 f_{RT} R_{\alpha(\mu} T_{\nu)}^{\ \alpha} \\ & + \nabla_\alpha \nabla_{(\mu} \left[T^\alpha_{\ \nu)} f_{RT} \right] + 2\left(f_T g^{\alpha\beta} + f_{RT} R^{\alpha\beta} \right) \frac{\partial^2 L_m}{\partial g^{\mu\nu} \partial g^{\alpha\beta}} \Bigg\}, \end{aligned} \tag{10.29}$$

respectively. Note that, in general, $G_{\rm eff}$ and $\Lambda_{\rm eff}$ are not constants, and depend on the specific model considered.

10.3.2 Equation of Motion of Massive Test Particles

The covariant divergence of the energy-momentum tensor can be obtained by taking the divergence of the gravitational field equation (10.24), which takes the following form

$$\begin{aligned} \nabla^\mu T_{\mu\nu} = \frac{2}{(1 + R f_{TR} + 2 f_T)} \Bigg\{ & \nabla_\mu \left(f_{RT} R^{\sigma\mu} T_{\sigma\nu} \right) + \nabla_\nu \left(L_m f_T \right) - G_{\mu\nu} \nabla^\mu \left(f_{RT} L_m \right) \\ & - \frac{1}{2} \Big(f_{RT} R_{\rho\sigma} + f_T g_{\rho\sigma} \Big) \nabla_\nu T^{\rho\sigma} - \frac{1}{2} \left[\nabla^\mu \left(R f_{RT} \right) + 2 \nabla^\mu f_T \right] T_{\mu\nu} \Bigg\}, \end{aligned} \tag{10.30}$$

where we have assumed that $\partial^2 L_m/\partial g^{\mu\nu}\partial g^{\alpha\beta} = 0$, and used the mathematical identities

$$\begin{aligned} & \nabla^\mu \left(f_R R_{\mu\nu} + \Box\, f_R g_{\mu\nu} - \frac{1}{2} f g_{\mu\nu} - \nabla_\mu \nabla_\nu f_R \right) \\ & \qquad = -\frac{1}{2} \left[f_T \nabla_\nu T + f_{RT} \nabla_\nu \left(R_{\rho\sigma} T^{\rho\sigma} \right) \right], \end{aligned} \tag{10.31}$$

$$2 T_{\mu\tau;\delta[;\rho;\sigma]} = T_{\mu\tau;\alpha} R^\alpha_{\ \delta\rho\sigma} + T_{\alpha\tau;\delta} R^\alpha_{\ \mu\rho\sigma} + T_{\mu\alpha;\delta} R^\alpha_{\ \tau\rho\sigma}, \tag{10.32}$$

and $[\Box, \nabla_\nu] T = R_{\mu\nu} \nabla^\mu T$, respectively.

In order to find the equation of motion for a massive test particle, as in the previous chapters, we consider the energy-momentum tensor of a perfect fluid. Following the procedure outlined in Section 8.1, we obtain the equation of motion for a massive test particle, considering the matter Lagrangian $L_m = p$, as

$$\frac{d^2x^\lambda}{ds^2} + \Gamma^\lambda_{\ \mu\nu} U^\mu U^\nu = f^\lambda, \tag{10.33}$$

where the extra force acting on the test particles is given by

$$f^\lambda = \frac{1}{\rho + p}\Big[-(1+3f_T)\nabla_\nu p - (\rho+p) f_{RT} R^{\sigma\rho}\left(\nabla_\nu h_{\sigma\rho} - 2\nabla_\rho h_{\sigma\nu}\right) \\ + (f_T + R f_{RT})\nabla_\nu \rho - f_{RT} R_{\sigma\rho} h^{\sigma\rho} \nabla_\nu (\rho+p)\Big] \frac{h^{\lambda\nu}}{1+2f_T+Rf_{RT}}. \tag{10.34}$$

Contrary to the nonminimal coupling presented in Chapter 8 [44], and as can be seen from the above equations, the extra-force does not vanish even with the Lagrangian $L_m = p$.

The extra-force is perpendicular to the four-velocity, satisfying the relation $f^\mu U_\mu = 0$. In the absence of any coupling between matter and geometry, with $f_T = f_{RT} = 0$, the extra-force takes the usual form of the standard general relativistic fluid motion, i.e., $f^\lambda = -h^{\lambda\nu}\nabla_\nu p/(\rho+p)$. In the case of $f(R,T,R_{\mu\nu}T^{\mu\nu})$ gravity, there is an explicit dependence of the extra-force on the Ricci tensor $R_{\sigma\rho}$, which makes the deviation from the geodesic motion more important for regions with strong curvatures.

10.3.3 Cosmological Applications: Specific Case of $f = R + \alpha R_{\mu\nu}T^{\mu\nu}$

Let us now consider some examples of cosmological solutions of the theory. In order to obtain explicit results and as a first step, one has to fix the functional form of the function $f(R,T,R_{\mu\nu}T^{\mu\nu})$. We analyze the evolution and dynamics of the Universe, assuming that the Universe is isotropic and homogeneous, with the matter content described by the energy density ρ, and thermodynamic pressure p with the matter Lagrangian chosen as $L_m = -\rho$. Consider the FLRW metric, the Hubble parameter $H = \dot{a}/a$, and the deceleration parameter q, defined as

$$q = \frac{d}{dt}\frac{1}{H} - 1. \tag{10.35}$$

Note that if $q < 0$, the expansion of the Universe is accelerating, while positive values of $q > 0$ describe decelerating evolutions.

For simplicity, consider the case in which the interaction between matter and geometry takes place only via the coupling between the energy-momentum and Ricci tensors, i.e.,

$$f = R + \alpha R_{\mu\nu}T^{\mu\nu}. \tag{10.36}$$

In order to pass the solar system and the other astrophysical tests, the correction term in Eq. (10.36) must be small, which implies that α is a small parameter. This simple case serves as an example to show the main differences of the present theory with $f(R,T)$ gravity, considered before [243]. The gravitational field equations for this form of f are given by

$$G_{\mu\nu}+\alpha\Big[2R_{\sigma(\mu}T^{\sigma}{}_{\nu)}-\frac{1}{2}R_{\rho\sigma}T^{\rho\sigma}g_{\mu\nu}-\frac{1}{2}RT_{\mu\nu}-G_{\mu\nu}L_m-2R^{\alpha\beta}\frac{\partial^2 L_m}{\partial g^{\mu\nu}\partial g^{\alpha\beta}}$$
$$-\frac{1}{2}\left(2\nabla_\sigma\nabla_{(\nu}T^{\sigma}{}_{\mu)}-\Box T_{\mu\nu}-\nabla_\alpha\nabla_\beta T^{\alpha\beta}g_{\mu\nu}\right)\Big]-8\pi GT_{\mu\nu}=0. \qquad (10.37)$$

For the case of the FLRW metric, the modified Friedmann equations are

$$3H^2=\frac{\kappa}{1-\alpha\rho}\rho+\frac{3}{2}\frac{\alpha}{1-\alpha\rho}H\left(\dot{\rho}-\dot{p}\right), \qquad (10.38)$$

and

$$2\dot{H}+3H^2=\frac{2\alpha}{1+\alpha p}H\dot{\rho}-\frac{\kappa p}{1+\alpha p}+\frac{1}{2}\frac{\alpha}{1+\alpha p}\left(\ddot{\rho}-\ddot{p}\right), \qquad (10.39)$$

respectively, where we have denoted $\kappa=8\pi G$ for simplicity. When $\alpha=0$ we recover the standard Friedmann equations. To remove the under-determinacy of the field equations, we impose an equation of state for the cosmological matter, $p=p(\rho)$. A standard form of the cosmological matter equation of state is $p=\omega\rho$, where $\omega=$ constant, and $0\leq\omega\leq 1$.

High Cosmological Density Limit of the Field Equations

First, we consider the high energy density limit of the system of modified cosmological equations Eqs. (10.38) and (10.39), and assume that the constant α is small, so that $\alpha\rho\ll 1$, and $\alpha p\ll 1$, respectively. In the high-energy limit, $\rho=p$, then Eqs. (10.38) and (10.39) take the approximate form

$$3H^2=\kappa\rho, \qquad (10.40)$$
$$2\dot{H}+3H^2=-\kappa\rho+2\alpha H\dot{\rho}. \qquad (10.41)$$

The time evolution of the Hubble parameter is described by the equation

$$\left(1-\frac{6\alpha}{\kappa}H^2\right)\dot{H}+3H^2=0, \qquad (10.42)$$

and hence for this model the evolution of the Hubble parameter is given by

$$H(t)=\frac{\sqrt{(C_1+3\kappa t)^2-24\alpha\kappa}+C_1+3\kappa t}{12\alpha}, \qquad (10.43)$$

where $C_1 > 0$ is an arbitrary integration constant. The scale factor of the Universe is given by

$$a(t) = C_2 \frac{\exp\left[\frac{(C_1+3\kappa t)\sqrt{(C_1+3\kappa t)^2-24\alpha\kappa+9\kappa t^2+6\kappa C_1 t}}{72\alpha\kappa}\right]}{\sqrt[3]{\sqrt{(C_1+3\kappa t)^{2}-24\alpha\kappa}+C_1+3\kappa t}}, \tag{10.44}$$

where $C_2 > 0$ is an arbitrary integration constant. In order to have a physical solution the integration constant C_1 must satisfy the constraint $C_1 \geq \sqrt{24\kappa\alpha}$.

The values of the integration constant can be determined from the condition $H(0) = H_0$, and $a(0) = a_0$, where H_0 and a_0 are the initial values of the Hubble parameter and of the scale factor of the Universe, respectively. This condition immediately provides for C_1 the following value $C_1 = (6\alpha H_0^2 + \kappa)/H_0$. For the integration constant C_2 we obtain

$$C_2 = a_0 \sqrt[3]{\sqrt{\frac{(\kappa - 6\alpha H_0^2)^2}{H_0^2}} + 6\alpha H_0 + \frac{\kappa}{H_0}} \exp\left[-\frac{\sqrt{\frac{\left(\kappa-6\alpha H_0^2\right)^2}{H_0^2}}\,(6\alpha H_0^2+\kappa)}{72\alpha H_0\kappa}\right].$$

In the small time limit, the scale factor is approximated by

$$a(t) \approx a_0 \left(1 + \frac{\kappa}{6H_0\alpha}t\right). \tag{10.45}$$

The deceleration parameter is obtained as

$$1 + q(t) = -\frac{36\alpha\kappa}{\sqrt{\bar{H}_0^2/H_0^2 - 24\alpha\kappa}\left[\bar{H}_0/H_0 + \sqrt{\bar{H}_0^2/H_0^2 - 24\alpha\kappa}\right]}, \tag{10.46}$$

where $\bar{H}_0 = 6\alpha H_0^2 + 3H_0\kappa t + \kappa$ was defined for notational convenience, and $q(t)$ can be represented in the form of a power series as

$$q(t) \approx -1 - \frac{18\alpha H_0^2}{\kappa - 6\alpha H_0^2} + \frac{6H_0\kappa^2}{(\kappa - 6\alpha H_0^2)^3}t. \tag{10.47}$$

For small values of time, if $24\alpha H_0^2 \ll \kappa$, $q \approx -1$, and the Universe starts its expansion from a de Sitter-like phase, entering, after a finite time interval, into a decelerating phase. On the other hand, if $\kappa > 6\alpha H_0^2$, $q < -1$, and the non-singular Universe experiences an initial super-accelerating phase.

The Case of Dust

Next, we consider the case of a low density cosmological matter, with $p = 0$. Moreover, we assume again that the condition $\alpha\rho \ll 1$ holds. Then the gravitational field equations, Eqs. (10.38) and (10.39), corresponding to a FLRW Universe, take the approximate form

$$3H^2 = \kappa\rho + \frac{3}{2}\alpha H\dot{\rho}, \tag{10.48}$$

$$2\dot{H} + 3H^2 = 2\alpha H\dot{\rho} + \frac{1}{2}\alpha\ddot{\rho}. \tag{10.49}$$

First we consider the matter dominated phase of the model, in which the non-accelerating expansion of the Universe can be described by a power law form of the scale factor, so that $a = t^m$, $m =$ constant, and $H = m/t$, respectively. The deceleration parameter is given by $q = 1/m - 1$. Therefore, Eq. (10.48) gives for the time evolution of the energy density the equation

$$\frac{3\alpha m}{2t}\dot{\rho} + \kappa\rho - 3\frac{m^2}{t^2} = 0, \tag{10.50}$$

with the general solution provided by

$$\rho(t) = \frac{1}{3\alpha}\, e^{-\frac{\kappa t^2}{3\alpha}}\left[3\rho_0\alpha e^{\frac{\kappa t_0^2}{3\alpha}} + \mathrm{Ei}\left(\frac{t^2\kappa}{3\alpha}\right) - \mathrm{Ei}\left(\frac{t_0^2\kappa}{3\alpha}\right)\right], \tag{10.51}$$

where $\mathrm{Ei}(z) = -\int_{-z}^{\infty} e^{-t}dt/t$ is the exponential integral function, and we have used the initial condition $\rho(t_0) = \rho_0$. By substituting the expressions of the energy density and of the Hubble parameter into Eq. (10.49), in the first order of approximation we obtain the following constraint on m, $(9m^2 - 10m + 1)/3t^2 + O(t^2) \approx 0$, which is (approximately) satisfied if m is given by the algebraic equation $9m^2 - 10m + 1 = 0$, having the solutions $m_1 = 1$, and $m_2 = 1/9$, respectively. The deceleration parameters corresponding to these solutions are $q_1 = 0$, and $q_2 = 8$, respectively. Since a value of the deceleration parameter of the order of $q = 8$ seems to be ruled out by the observations, the physical solution has a scale factor $a = t$, and $q = 0$. The cosmological solutions with zero value of the deceleration parameter are called marginally accelerating, and they describe the pre-accelerating phase of the cosmic expansion.

Now we look for a de Sitter-type solution of the field equations for the pressureless matter, Eqs. (10.48) and (10.49), by taking $H = H_0 =$ constant. Then it follows that, in order to have an accelerated expansion, the matter density must satisfy the equation

$$\ddot{\rho} - H_0\dot{\rho} + \frac{2\kappa}{\alpha}\rho = 0, \tag{10.52}$$

with the general solution given by

$$\rho(t) = e^{\frac{1}{2}H_0(t-t_0)}\left\{\frac{\sqrt{\alpha}\,(2\rho_{01} - H_0\rho_0)}{\sqrt{\alpha H_0^2 - 8\kappa}}\sinh\left[\frac{\sqrt{\alpha H_0^2 - 8\kappa}}{2\sqrt{\alpha}}(t - t_0)\right]\right.$$
$$\left. + \cosh\left[\frac{\sqrt{\alpha H_0^2 - 8\kappa}}{2\sqrt{\alpha}}(t - t_0)\right]\right\}, \tag{10.53}$$

where we have used the initial conditions $\rho(t_0) = \rho_0$, and $\dot{\rho}(t_0) = \rho_{01}$, respectively. Therefore, in the presence of a nontrivial curvature-matter coupling, once the evolution of the matter density is given by Eq. (10.53), the time evolution of the Universe is of the de Sitter type.

10.4 Further Applications

The $f(R,T)$ gravitational theory presented in this chapter has been given a great amount of recent attention. We will briefly outline a few cosmological applications. In [256], the cosmological reconstruction of $f(R,T)$ gravity describing matter dominated and accelerated phases was analyzed. Special attention was paid to a specific case of $f(R,T) = f_1(R) + f_2(T)$. Further examples were analyzed in [243], where specific solutions were presented. For instance, the use of an auxiliary scalar field was considered with two known examples for the scale factor corresponding to an expanding universe. In the first example, where ordinary matter is usually neglected for obtaining the unification of matter dominated and accelerated phases with $f(R)$ gravity, it was shown that this unification can be obtained in the presence of ordinary matter. In the second example, as in $f(R)$ gravity, the model of $f(R,T)$ gravity with a transition of the matter dominated phase to the acceleration phase was obtained.

The cosmology of $f(R,T)$ gravity is also extensively analyzed in the literature, and one may mention a few specific solutions. In [268], it was shown that the dust fluid reproduces ΛCDM, phantom–non-phantom era, and the phantom cosmology. Furthermore, different cosmological models were reconstructed, including the Chaplygin gas, and scalar field with some specific forms of $f(R,T)$. The numerical simulation for the Hubble parameter shows a good agreement with the baryon acoustic oscillations (BAO) observational data for low redshifts $z < 2$. In [258], through a numerical reconstruction, it was shown that specific $f(R,T)$ models are able to reproduce the same expansion history generated, in standard GR, by dark matter and holographic dark energy. It was further shown that these theories are able to reproduce the four known types of future finite-time singularities [257]. In [464], a nonequilibrium picture of thermodynamics was discussed at the apparent horizon of an FLRW Universe in $f(R,T)$ gravity, where the validity of the first and second law of thermodynamics in this scenario were checked. It was shown that the Friedmann equations can be expressed in the form of the first law of thermodynamics and that the second law of thermodynamics holds both in the phantom energy and non-phantom phases.

The energy conditions have also been extensively explored in $f(R,T)$ gravity, for instance, in an FLRW universe with perfect fluid [463], and in the context of exact power-law solutions [465], where certain constraints were found that have to be satisfied to ensure that power-law solutions may be stable and match the bounds prescribed by the energy conditions. In [15], it was also shown that the energy conditions were satisfied for specific models. Furthermore, an analysis of the perturbations and stabilities of de Sitter solutions and power-law solutions was performed, and it was shown that for those models in which the energy conditions are satisfied, de Sitter solutions and power-law solutions may be stable.

Solutions of the Gödel universe in the framework of $f(R,T)$-modified theories of gravity were also obtained [447]. Algorithms were derived for constructing five dimensional Kaluza–Klein cosmological space-times in the presence of a perfect fluid source in $f(R,T)$ gravity [428]. Spatially homogeneous and anisotropic Bianchi type-V cosmological models in a scalar–tensor theory of gravitation [374] were also explored. The Bianchi type-I model with perfect fluid as a matter content in $f(R,T)$ gravity was analyzed, where the physical and kinematical properties of the model were also discussed [467]. In [128], the specific case of the conservation of the energy-momentum tensor was considered, and cosmological solutions were obtained for an homogeneous and isotropic model of the universe.

In [14], the evolution of scalar cosmological perturbations in the metric formalism were analyzed. According to restrictions on the background evolution, a specific model within these theories was assumed in order to guarantee the standard continuity equation. Using a completely general procedure, the complete set of differential equations for the matter density perturbations was found. In the case of sub-Hubble modes, the density contrast evolution reduces to a second-order equation, and it was shown that for well-motivated $f(R,T)$ Lagrangians the quasistatic approximation provide different results from the ones derived in the framework of the concordance ΛCDM model, which severely constrains the viability of such theories. In [466], the Ricci and modified Ricci dark energy models were considered in the context of $f(R,T)$ gravity, and it was found that specific models can reproduce the expansion history of the Universe in accordance with the present observational data. The Dolgov–Kawasaki stability condition was also obtained for the specific reconstructed $f(R,T)$ functions.

In fact, astrophysical and cosmological applications of $f(R,T)$ gravity have been extensively analyzed in the literature. The aim of the present chapter is to essentially provide an introduction to the reader of the theory and specific applications and extensions. Rather than provide an extensive review of the existing literature, we refer the reader [237] for a review and further consideration.

11

Dark Matter as a Curvature-Matter Coupling Effect

11.1 Introduction

According to Newton's gravitation theory, the rotation of hydrogen clouds around galaxies should show a Keplerian decrease with distance r of the orbital rotational speed v_{tg} outside the luminous baryonic matter, $v_{tg}^2 \propto M(r)/r$, where $M(r)$ is the dynamical mass of the galaxy. However, observations show that the rotation curves are rather flat [65, 85, 414, 441]. The rotational velocities increase near the center of the galaxy, and then they remain nearly constant at a value of $v_{tg\infty} \sim 200 - 300$ km/s. Hence the observations give a general mass profile $M(r) \approx rv_{tg\infty}^2/G$ [65, 85, 414, 441]. Consequently, even at large distances, where very little luminous matter can be detected, the mass within a distance r from the center of the galaxy increases linearly with the radial distance r.

This unusual behavior of the galactic rotation curves is usually explained by postulating the existence of some dark (invisible) matter, distributed in a spherical halo around the galaxies. The rotation curves are obtained by measuring the frequency shifts z of the 21-cm radiation emission from the neutral hydrogen gas clouds. Usually, astronomers report the resulting z in terms of a velocity field v_{tg} [85, 414, 441]. In the standard cosmological model, dark matter is assumed to be a cold, pressureless medium. Many possible candidates for dark matter have been proposed, the most popular ones being the weakly interacting massive particles (WIMP) (for a review of the particle physics aspects of dark matter see [400]). The interaction cross section with normal baryonic matter is expected to be small, but nonzero, and we expect that dark matter particles can be detected directly. However, after more than 20 years of intense experimental and observational effort, presently no convincing *nongravitational* evidence for dark matter exists.

In the present chapter, we will briefly analyze, following [217], the possibility that the dynamic behavior of hydrogen clouds rotating around the galactic center can be explained due to the presence of a curvature-matter coupling. We will restrict our analysis to the case of modified gravity models with linear curvature-matter coupling. The extra-terms in the gravitational field equations modify the equations of motion of test particles, and induce a supplementary gravitational interaction, which can account for the observed behavior of the galactic rotation curves.

Note that the modified gravity approach has been explored, as an alternative to dark matter, in different contexts [74, 75, 106, 105], such as in $f(R)$ gravity and the recently proposed hybrid metric-Palatini gravity [104, 109, 110, 224]. In this chapter, we build on the analysis outlined in Section 6.5, related to dark matter in $f(R)$ gravity.

11.2 Stable Circular Orbits and Frequency Shifts in Modified Gravity

In the following section, we assume that gas clouds behave like massive test particles, moving in a static and spherically symmetric space-time outside the galactic baryonic mass distribution. In the galactic space-time we consider two observers O_E and O_∞, with four-velocities u_E^μ and u_∞^μ, respectively. Observer O_E is the light emitter (i.e., the gas clouds placed at a point P_E of the space-time), and O_∞ represents the detector at point P_∞, located far from the emitter, at "spatial infinity" [387].

Without any loss of generality, we can assume that the gas clouds move in the galactic plane $\theta = \pi/2$, so that $u_E^\mu = \left(\dot{t}, \dot{r}, 0, \dot{\phi}\right)_E$, where the dot stands for derivation with respect to the affine parameter s. On the other hand, we suppose that the detector is static (i.e., O_∞'s four-velocity is tangent to the static Killing field $\partial/\partial t$), and in the chosen coordinate system its four-velocity is $u_\infty^\mu = \left(\dot{t}, 0, 0, 0\right)_\infty$ [387].

The static spherically symmetric metric outside the galactic baryonic mass distribution is given by

$$ds^2 = -e^{\nu(r)}dt^2 + e^{\lambda(r)}dr^2 + r^2\left(d\theta^2 + \sin^2\theta d\phi^2\right), \tag{11.1}$$

where the metric coefficients ν and λ are functions of the radial coordinate r only. The motion of a test particle in the gravitational field in modified gravity with a linear curvature-matter coupling is described by the Lagrangian given by Eq. (8.15),

$$L = Q\left[e^{\nu(r)}\left(\frac{dt}{ds}\right)^2 - e^{\lambda(r)}\left(\frac{dr}{ds}\right)^2 - r^2\left(\frac{d\Omega}{ds}\right)^2\right], \tag{11.2}$$

where $d\Omega^2 = d\theta^2 + \sin^2\theta d\phi^2$, and the function Q is given by

$$Q = \left[1 + \zeta f_2(R)\right]\frac{dL_m(\rho)}{d\rho}, \tag{11.3}$$

where ζ is a constant.

For $\theta = \pi/2$, $d\Omega^2 = d\phi^2$. From the Lagrange equations it follows that we have two constants of motion, the energy E

$$E = Qe^{\nu(r)}\dot{t}, \tag{11.4}$$

and the angular momentum l, given by

$$l = Qr^2\dot{\phi}, \tag{11.5}$$

where a dot denotes the derivative with respect to the affine parameter s. The condition of the normalization of the four-velocity $U^\mu U_\mu = -1$ gives

$$-1 = -e^{\nu(r)}\dot{t}^2 + e^{\lambda(r)}\dot{r}^2 + r^2\dot{\phi}^2, \tag{11.6}$$

from which, with the use of the constants of motion, we obtain

$$E^2 = Q^2 e^{\nu+\lambda}\dot{r}^2 + e^{\nu}\left(\frac{l^2}{r^2} + Q^2\right). \tag{11.7}$$

Equation (11.7) shows that the radial motion of the particles in modified gravity with a linear curvature-matter coupling is the same as that of a particle in ordinary Newtonian mechanics, with velocity $\dot{r}$, position-dependent mass $m_{eff} = 2Q^2e^{\nu+\lambda}$, and energy E^2, respectively, moving in the effective potential given by

$$V_{eff}(r) = e^{\nu(r)}\left(\frac{l^2}{r^2} + Q^2\right). \tag{11.8}$$

The conditions for stable circular orbits $\partial V_{eff}/\partial r = 0$ and $\dot{r} = 0$ determine the energy and the angular momentum of the particle as

$$l^2 = \frac{1}{2}\frac{r^3 Q\left(\nu' Q + 2Q'\right)}{1 - r\nu'/2}, \tag{11.9}$$

and

$$E^2 = \frac{e^{\nu}Q\left(rQ' + Q\right)}{1 - r\nu'/2}, \tag{11.10}$$

respectively.

The line element (11.1) can be rewritten in terms of the spatial components of the velocity, normalized with the speed of light, measured by an inertial observer far from the source, as

$$ds^2 = dt^2\left(1 - v^2\right), \tag{11.11}$$

where

$$v^2 = e^{-\nu}\left[e^{\lambda}\left(\frac{dr}{dt}\right)^2 + r^2\left(\frac{d\Omega}{dt}\right)^2\right]. \tag{11.12}$$

For a stable circular orbit $dr/dt = 0$, the tangential velocity of the test particle is obtained as

$$v_{tg}^2 = e^{-\nu} r^2 \left(\frac{d\Omega}{dt}\right)^2. \tag{11.13}$$

In terms of the conserved quantities E and l the angular velocity is given, for $\theta = \pi/2$, by

$$v_{tg}^2 = \frac{e^{\nu}}{r^2}\frac{l^2}{E^2}. \tag{11.14}$$

With the use of Eqs. (11.9) and (11.10) we obtain

$$v_{tg}^2 = \frac{1}{2}\frac{r(\nu' Q + 2Q')}{rQ' + Q}. \tag{11.15}$$

Thus, the rotational velocity of the test body in modified gravity with a linear curvature-matter coupling is determined by the metric coefficient $\exp(\nu)$, and by the function Q and its derivative with respect to the radial coordinate r. In the standard general relativistic limit $\zeta = 0$, $Q = 1$, and we obtain $v_{tg}^2 = r\nu'/2$.

11.3 Curvature-Matter Coupling Effects on Light Shifts

The velocities of the hydrogen clouds rotating around galactic centers are determined from the red and blueshifts of the radiation emitted by the gas moving on circular orbits on both sides of the central region. The radiation travels on null geodesics with tangent k^μ. We may restrict, without any loss of generality, k^μ to lie on the equatorial plane $\theta = \pi/2$, and evaluate the frequency shift for a light signal emitted from O_E, which is in motion in a circular orbit, and detected by O_∞ at infinity. The frequency shift associated to the emission and detection of the light signal is given by

$$z = 1 - \frac{\omega_E}{\omega_\infty}, \tag{11.16}$$

where $\omega_I = -k_\mu u_I^\mu$, and the index I refers to emission ($I = E$) or detection ($I = \infty$) at the corresponding space-time point [387, 301]. Two frequency shifts, corresponding to the frequency shifts of a receding or approaching gas cloud, respectively, are associated with light propagation in the same and opposite direction of motion of the emitter, respectively. In terms of the tetrads

$$e_{(0)} = e^{-\nu/2}\partial/\partial t, \quad e_{(1)} = e^{-\lambda/2}\partial/\partial r, \quad e_{(2)} = r^{-1}\partial/\partial\theta, \quad e_{(3)} = (r\sin\theta)^{-1}\partial/\partial\phi, \tag{11.17}$$

the frequency shifts take the form [387]

$$z_\pm = 1 - e^{[\nu_\infty - \nu(r)]/2}(1 \mp v)\Gamma, \tag{11.18}$$

where $v = \left[\sum_{i=1}^{3}\left(u_{(i)}/u_{(0)}\right)^{2}\right]^{1/2}$, with $u_{(i)}$ the components of the particle's four velocity along the tetrad (i.e., the velocity measured by an Eulerian observer whose world line is tangent to the static Killing field). $\Gamma = \left(1-v^{2}\right)^{-1/2}$ is the usual Lorentz factor, and $\exp\left(\nu_{\infty}\right)$ is the value of $\exp\left[\nu\left((r)\right)\right]$ for $r \to \infty$. In the case of circular orbits in the $\theta = \pi/2$ plane, we obtain

$$z_{\pm} = 1 - e^{\left[\nu_{\infty}-\nu(r)\right]/2}\frac{1 \mp \sqrt{r\left(\nu' Q + 2Q'\right)/2\left(rQ'+Q\right)}}{\sqrt{1 - r\left(\nu' Q + 2Q'\right)/2\left(rQ'+Q\right)}}. \tag{11.19}$$

It is convenient to define two other quantities, $z_D = \left(z_{+} - z_{-}\right)/2$, and $z_A = \left(z_{+} + z_{-}\right)/2$, respectively [387]. In the modified gravity model with a linear curvature-matter coupling, the redshift factors are given by

$$z_D\left(r\right) = e^{\left[\nu_{\infty}-\nu(r)\right]/2}\frac{\sqrt{r\left(\nu' Q + 2Q'\right)/2\left(rQ'+Q\right)}}{\sqrt{1 - r\left(\nu' Q + 2Q'\right)/2\left(rQ'+Q\right)}}, \tag{11.20}$$

and

$$z_A\left(r\right) = 1 - \frac{e^{\left[\nu_{\infty}-\nu(r)\right]/2}}{\sqrt{1 - r\left(\nu' Q + 2Q'\right)/2\left(rQ'+Q\right)}}, \tag{11.21}$$

respectively, which can be easily related to the galactic observations [387]. The quantities z_A and z_D satisfy the relation

$$\left(z_A - 1\right)^{2} - z_D^{2} = \exp\left[2\left(\nu_{\infty} - \nu\left(r\right)\right)\right], \tag{11.22}$$

and thus in principle, by assuming that the metric tensor component $\exp\left[\nu\left((r)\right)\right]$ is known, Q and Q' can be obtained directly from astrophysical observations. Hence the observations of the red and blueshifts of the radiation emitted by hydrogen clouds rotating around the galactic center could provide a direct observational test of the galactic geometry and, implicitly, of the modified gravity models with linear coupling between curvature and matter.

11.4 Galactic Rotation Curves

The tangential velocities v_{tg} of gas clouds, considered as massive test particles moving around the centers of the galaxies, cannot be measured directly. Instead, they can be inferred from the redshift z_{∞} of the radiation observed at spatial infinity, for which

$$1 + z_{\infty} = \exp\left[\left(\nu_{\infty} - \nu\right)/2\right]\left(1 \pm v_{tg}\right)/\sqrt{1 - v_{tg}^{2}}. \tag{11.23}$$

Since the velocities of the gas clouds are nonrelativistic, with $v_{tg} \leq (4/3)\times 10^{-3}$, we observe $v_{tg} \approx z_{\infty}$ (as the first term of a geometric series), with the consequence that the lapse function $\exp\left(\nu\right)$ necessarily tends at infinity to unity, i.e., $e^{\nu} \approx e^{\nu_{\infty}}/\left(1 - v_{tg}^{2}\right) \approx e^{\nu_{\infty}} \to 1$. The observations show that, at distances large enough from the galactic center, $v_{tg} \approx$ constant.

Next we use this observational constraint to reconstruct the coupling term between curvature and matter in the "dark matter"-dominated region, far away from the baryonic matter distribution. By assuming that v_{tg} = constant, Eq. (11.15) can be written as

$$v_{tg}^2 \frac{1}{rQ}\frac{d}{dr}(rQ) = \frac{\nu'}{2} + \frac{Q'}{Q}, \tag{11.24}$$

which can be immediately integrated to give

$$Q(r) = \left(\frac{r}{r_0}\right)^{v_{tg}^2/\left(1-v_{tg}^2\right)} \exp\left[-\frac{\nu}{2\left(1-v_{tg}^2\right)}\right], \tag{11.25}$$

where r_0 is an arbitrary constant of integration. Since hydrogen clouds are a pressureless dust ($p = 0$) that can be characterized by their density ρ only, the Lagrangian of the matter (gas cloud) is given by $L_{mat}(\rho) = \rho$. Therefore, from Eqs. (11.3) and (11.25), we obtain

$$\begin{aligned} \zeta f_2(R) &= \left(\frac{r}{r_0}\right)^{v_{tg}^2/2\left(1-v_{tg}^2\right)} \exp\left[-\frac{\nu}{4\left(1-v_{tg}^2\right)}\right] - 1 \\ &\approx \frac{v_{tg}^2}{2\left(1-v_{tg}^2\right)} \ln\left(\frac{r}{r_0}\right) - \frac{\nu}{4\left(1-v_{tg}^2\right)} - \frac{v_{tg}^2}{8\left(1-2v_{tg}^2\right)}\nu \ln\left(\frac{r}{r_0}\right). \end{aligned} \tag{11.26}$$

For an arbitrary velocity profile $v_{tg} = v_{tg}(r)$, the general solution of Eq. (11.24) is given by

$$\sqrt{Q(r)} = 1 + \zeta f_2(R) = \sqrt{Q_0} \exp\left[\frac{1}{2}\int \frac{v_{tg}^2(r)/r - \nu'/2}{1 - v_{tg}^2(r)} dr\right], \tag{11.27}$$

where Q_0 is an arbitrary constant of integration.

For v_{tg}^2 we assume the simple empirical dark halo rotational velocity law [445]

$$v_{tg}^2 = \frac{v_0^2 x^2}{a^2 + x^2}, \tag{11.28}$$

where $x = r/r_{opt}$, and r_{opt} is the optical radius containing 83% of the galactic luminosity. The parameters a, the ratio of the halo core radius and r_{opt}, and the terminal velocity v_0 are all functions of the galactic luminosity L. For spiral galaxies $a = 1.5\left(L/L_*\right)^{1/5}$ and $v_0^2 = v_{opt}^2\left(1-\beta_*\right)\left(1+a^2\right)$, where $v_{opt} = v_{tg}\left(r_{opt}\right)$, and $\beta_* = 0.72 + 0.44\log_{10}\left(L/L_*\right)$, with $L_* = 10^{10.4}L_\odot$.

One can assume that the coupling between the neutral hydrogen clouds and the geometry is small, $\zeta f_2(R)L_{mat} \ll 1$, and consequently the galactic geometry is not significantly modified in the vacuum outside the baryonic mass distribution with mass M_B, corresponding to $L_{mat} \approx 0$. Moreover, we assume for simplicity, that outside the baryonic matter distribution the galactic metric is given by the Schwarzschild metric (which is also a solution of the vacuum field equations of

$f(R)$ gravity [425], and still gives the dominant contribution to the total metric, even if the curvature-matter coupling is not small), written as

$$e^{\nu} = e^{-\lambda} = 1 - \frac{2R_0}{x}, \tag{11.29}$$

where $R_0 = GM_B/r_{opt}$, from Eq. (11.27) we obtain

$$1 + \zeta f_2(R) = \exp\left[\alpha \times \operatorname{arctanh}\left(\frac{\sqrt{1-v_0^2}}{a}x\right)\right] \times \left(1 - \frac{2R_0}{x}\right)^{\beta} \frac{\left[\left(1-v_0^2\right)x^2 + a^2\right]^{\gamma}}{Q_0^{-1/2}x^{1/4}}, \tag{11.30}$$

where

$$\alpha = -\frac{aR_0v_0^2}{2\sqrt{1-v_0^2}\left[a^2 + 4\left(1-v_0^2\right)R_0^2\right]}, \tag{11.31}$$

$$\beta = \frac{a^2 - 4R_0^2}{4\left[a^2 + 4\left(1-v_0^2\right)R_0^2\right]}, \tag{11.32}$$

and

$$\gamma = -\frac{v_0^2\left[a^2 + 6\left(1-v_0^2\right)R_0^2\right]}{4\left(1-v_0^2\right)\left[a^2 + 4\left(1-v_0^2\right)R_0^2\right]}, \tag{11.33}$$

respectively. Thus the geometric part of the curvature-matter coupling can be completely reconstructed from the observational data on the galactic rotation curves.

As one can see from Eq. (11.15), in the limit of large r, when $\nu' \to 0$ (in the case of the Schwarzschild metric $\nu' \approx 2M_B/r^2$), the tangential velocity of test particles at infinity is given by

$$v_{tg\infty}^2 = \frac{r}{r + Q/Q'}, \tag{11.34}$$

which, due to the presence of the curvature-matter coupling, does not decay to zero at large distances from the galactic center, a behavior that is perfectly consistent with the observational data, and is usually attributed to the existence of dark matter.

By using the simple observational fact of the constancy of the galactic rotation curves, the curvature-matter coupling function can be completely reconstructed, without any supplementary assumption. If, for simplicity, we consider again that the metric in the vacuum outside the galaxy can be approximated by the Schwarzschild metric, with $\exp(\nu) = 1 - 2GM_B/r$, where M_B is the mass of the baryonic matter of the galaxy, then, in the limit of large r, we have $\nu \to 0$. Therefore from Eq. (11.26) we obtain

$$\zeta \lim_{r\to\infty} f_2(R) \approx \frac{v_{tg}^2}{2\left(1-v_{tg}^2\right)} \ln\frac{r}{r_0}. \tag{11.35}$$

If the galactic rotation velocity profiles and the galactic metric are known, the coupling function can be reconstructed exactly over the entire mass distribution of the galaxy.

One can formally associate an approximate "dark matter" mass profile $M_{DM}(r)$ to the tangential velocity profile, which is determined by the nonminimal curvature-matter coupling, and is given by

$$M_{DM}(r) \approx \frac{1}{2G}\frac{r^2\left(\nu' Q + 2Q'\right)}{rQ' + Q}. \tag{11.36}$$

The corresponding "dark matter" density profile $\rho_{DM}(r)$ can be obtained as

$$\rho_{DM}(r) = \frac{1}{4\pi r^2}\frac{dM}{dr} = \frac{1}{4\pi G}\times\left[\frac{\nu' Q + 2Q'}{r\left(rQ' + Q\right)} + \frac{\nu'' Q + \nu Q' + 2Q''}{2\left(rQ' + Q\right)} - \frac{\left(\nu' Q + 2Q'\right)\left(rQ'' + 2Q'\right)}{2\left(rQ' + Q\right)^2}\right]. \tag{11.37}$$

11.5 Constraining the Curvature-Matter Coupling with Galactic Stellar Distributions

Other observational constraints on M_{DM} and ρ_{DM} can be obtained from the study of the galactic stellar populations. We assume that each galaxy consists of a single, pressure-supported stellar population that is in dynamic equilibrium, and traces an underlying gravitational potential resulting from the nonminimal curvature-matter coupling. In spherical symmetry, the equivalent mass profile induced by the curvature-matter coupling (the mass profile of the "dark matter" halo) relates to the moments of the stellar distribution function via the Jeans equation [65]

$$\frac{d}{dr}\left[\rho_s\left\langle v_r^2\right\rangle\right] + \frac{2\rho_s(r)\,\beta}{r} = -\frac{G\rho_s M_{DM}(r)}{r^2}, \tag{11.38}$$

where $\rho_s(r)$, $\langle v_r^2\rangle$, and $\beta(r) = 1 - \langle v_\theta^2\rangle / \langle v_r^2\rangle$ describe the three-dimensional density, radial velocity dispersion, and orbital anisotropy of the stellar component, where $\langle v_\theta^2\rangle$ is the tangential velocity dispersion. By assuming that the anisotropy is a constant, the Jeans equation has the solution [341]

$$\rho_s\left\langle v_r^2\right\rangle = Gr^{-2\beta}\int_r^\infty s^{2(1-\beta)}\rho_s(s)\,M_{DM}(s)\,ds. \tag{11.39}$$

With the use of Eq. (11.36), we obtain for the stellar velocity dispersion the equation

$$\rho_s\left\langle v_r^2\right\rangle \approx \frac{1}{2}r^{-2\beta}\int_r^\infty s^{2(2-\beta)}\rho_s(s)\,\frac{\nu'(s)Q(s) + 2Q'(s)}{sQ'(s) + Q(s)}ds. \tag{11.40}$$

After projection along the line of sight, the "dark matter" mass profile can be related to two observable profiles, the projected stellar density $I(R)$, and to the stellar velocity dispersion $\sigma_p(R)$ according to the relation [65]

$$\sigma_P^2(R) = \frac{2}{I(R)}\int_R^\infty \left(1-\beta\frac{R^2}{r^2}\right)\frac{\rho_s\left\langle v_r^2\right\rangle r}{\sqrt{r^2-R^2}}dr. \tag{11.41}$$

Given a projected stellar density model $I(R)$, one recovers the three-dimensional stellar density from [65].

$$\rho_s(r) = -(1/\pi)\int_r^\infty \left(\frac{dI}{dR}\right)\left(R^2-r^2\right)^{-1/2}dR. \tag{11.42}$$

Therefore, once the stellar density profile $I(R)$, the stellar velocity dispersion $\langle v_r^2\rangle$, and the galactic metric are known, with the use of the integral equation Eq. (11.40) one can obtain the explicit form of the curvature-matter coupling function Q, and the equivalent mass profile induced by the nonminimal coupling between matter and curvature. The simplest analytic projected density profile is the Plummer profile [65], given by

$$I(R) = \mathcal{L}\left(\pi r_{half}^2\right)^{-1}\left(1+R^2/r_{half}^2\right)^{-2}, \tag{11.43}$$

where $\mathcal{L}$ is the total luminosity, and r_{half} is the projected half-light radius (the radius of the cylinder that encloses half of the total luminosity).

11.6 Stability of the Stable Circular Orbits

An important physical requirement for the circular orbits of the test particles moving around galaxies is that they must be stable. Let r_0 be a circular orbit and consider a perturbation of it of the form $r = r_0 + \delta$, where $\delta \ll r_0$ [301]. Taking expansions of $V_{eff}(r)$, $\exp(\nu+\lambda)$ and $Q^2(r)$ about $r = r_0$, it follows from Eq. (11.7) that

$$\ddot{\delta} + \frac{1}{2}Q^2(r_0)\,e^{\nu(r_0)+\lambda(r_0)}V_{eff}''(r_0)\,\delta = 0. \tag{11.44}$$

The condition for stability of the simple circular orbits requires $V_{eff}''(r_0) > 0$ [301]. This gives for the coupling function Q the constraint

$$\left.\frac{d^2Q}{dr^2}\right|_{r=r_0} < \left.\left[\nu''\left(\frac{l^2}{r^2}+Q^2\right)+\nu'\left(-\frac{2l^2}{r^2}+2QQ'\right)-\frac{6l^2}{r^4}\right]\right|_{r=r_0}, \tag{11.45}$$

a condition that must be satisfied at any point r_0 of the galactic space-time.

From the observational as well as from the theoretical point of view an important problem is to estimate an upper bound for the cutoff of the constancy of the tangential velocities. If in the large r limit the coupling function satisfies the condition $Q'/Q \to 0$, then $Q/Q' \to \infty$, and in this limit the tangential velocity decays to zero. If the exact functional form of Q is known, the value of r at which the rotational velocity becomes zero can be accurately estimated.

11.7 Conclusion

In this chapter, we have investigated the possibility that the behavior of the rotational velocities of test particles gravitating around galaxies can be explained in the framework of modified theories of gravity with a nonminimal curvature-matter coupling. Generally, the dynamics of test particles around galaxies, as well as the corresponding mass deficit, is explained by postulating the existence of dark matter. The extra-terms in the gravitational field equations with a curvature-matter coupling modify the equations of motion of test particles, and induce a supplementary gravitational interaction. Starting from the variational principle describing the particle motion in the presence of the non-minimal coupling, the expression of the tangential velocity of a test particle, moving in the vacuum on a stable circular orbit in a spherically symmetric geometry, was derived. The tangential velocity depends on the metric tensor components, as well as of the coupling function between matter and geometry. The Doppler velocity shifts were also obtained in terms of the coupling function. If the tangential velocity profile is known, the coupling term between matter and geometry can be obtained explicitly in an analytical form. Therefore, these results open the possibility of directly testing the modified gravity models with nonminimal coupling between matter and curvature by using direct astronomical and astrophysical observations at the galactic or extra-galactic scale.

12

Thermodynamical Interpretation of Curvature-Matter Coupling

12.1 Introduction

Both $f(R, L_m)$- and $f(R, T)$-modified gravity theories have the intriguing property that the four-divergence of the energy-momentum tensor does not vanish, as is the case in standard General Relativity, or in $f(R)$ gravity. This is in fact a general feature that immediately emerges once curvature-matter coupling are introduced in any gravitational theory. But then we must raise the important question of the physical interpretation of this effect, and of its compatibility with the general laws of physics. It is the main goal of the present chapter to try to give a physical interpretation of the nonconservation of the energy-momentum tensor, and to propose a thermodynamic interpretation of the gravitational theories that are based on curvature-matter coupling. We will formulate our physical interpretation of the $f(R, L_m)$- and $f(R, T)$-modified theories of gravity in the theoretical framework of the irreversible thermodynamics of open systems, as initially introduced and developed in [419, 420]. In our presentation of the thermodynamic interpretation of modified gravity theories with curvature-matter coupling, we follow the approach proposed in [220] and [241], respectively, where it was shown that the nonconservation of the matter energy-momentum tensor can be naturally interpreted from a physical point of view as describing irreversible matter production processes. The coupling of matter and geometry that leads to the nonconservation of the energy-momentum tensor also determines a transfer of energy from the gravitational field to matter, which results in the production of new particles in the given geometric framework.

The investigation of the possible role that irreversible matter creation processes may play in GR and cosmology was initiated in the studies [419] and [420]. The phenomenological description of a particle is based on the reinterpretation (at the level of classical physics) of the matter energy-momentum tensor T_ν^μ in Einstein's gravitational field equations. This has been achieved through the modification of the usual energy conservation laws for adiabatic thermodynamic systems, in which the irreversible matter production processes were also included. Hence, in this formulation of the laws of thermodynamics, matter production is caused by the irreversible energy flow from the gravitational field to the newly produced particles. On the other hand, in a gravitational system the second law of thermodynamics may allow that geometry generates matter, but the opposite transformation cannot take place, i.e., the annihilation of the particles cannot generate new geometric structures. As first shown in [420], the general relativistic initial cosmological singularity existing at the moment of the big bang is structurally unstable with respect to irreversible matter production. Including dissipative and irreversible thermodynamic processes into the energy-momentum tensor containing the matter part of the Einstein field equations leads to the possibility of obtaining cosmological models beginning with a vacuum state, and building up gradually their matter and entropy content from geometry. Moreover, in this approach gravitational entropy has a simple physical meaning and interpretation as corresponding to the entropy necessary to create particles. This interpretation also leads to an interesting extension of the third law of thermodynamics, since now the zero point of entropy corresponds to a space-time geometric structure from which matter is continuously created [419].

The phenomenological approach to particle production developed in [419, 420] was reconsidered in [97], where an explicit covariant formulation of the thermodynamics of open systems was introduced. A general covariant form for the entropy creation rates and the temperature evolution equation were also obtained and analyzed. For the analysis of the effects and implications of the thermodynamics of open systems on the cosmological evolution see [115, 126, 129, 200, 236, 246, 270, 314, 338, 359, 388, 415, 429].

It is the goal of this chapter to briefly review the application of the thermodynamics of open systems for the physical interpretation of modified theories of gravity with a curvature-matter coupling. We will begin our analysis by presenting the basic ideas of the irreversible thermodynamics and of the description of matter creation in a full covariant formulation. Then, we will proceed to the investigation of the thermodynamic aspects of $f(R, L_m)$ and $f(R, T)$ theories of gravity. For each theory we will obtain the expression of the particle balance equation and of the source terms. The behavior of the entropy will also be considered.

12.2 Thermodynamical Description of Matter Creation in General Relativity

12.2.1 Thermodynamics of Irreversible Processes

In the presence of specific physical conditions, cosmological matter production is allowed to take place from the quantum vacuum. Such an external factor is, for example, due to the expansion of the Universe. The cosmological quantum creation processes were first investigated a long time ago by Schrödinger [454], who realized that the cosmological evolution could lead to a mixing of positive and negative energy states. From a quantum mechanical point of view this effect would lead to the production or the annihilation of matter in an evolving Universe. On the other hand, similar processes arise in quantum field theory in curved space-times, where matter creation is closely related with the geometric curvature of the Universe. In quantum field theory the equations of motion can be formulated in the form of the harmonic oscillator equation, with the possible friction terms ignored. In this case, the effect of the gravitational field is absorbed in the effective mass of the particle. From a cosmological perspective it turns out that the quantum rate of particle production has a maximum value in the vicinity of the initial cosmological singularity. Moreover, the post-inflationary reheating of the Universe may be explained by gravitational matter production [220, 241].

Now we briefly review the basic physical concepts of the thermodynamics of irreversible processes. We begin our analysis by considering an arbitrary cosmological volume element $V = a^3(t)$ in the comoving frame, where $a(t)$ denotes the scale factor of the expanding Universe. The volume V contains a number of N particles. In the case of a closed thermodynamical system, the particle number N is constant, and, consequently, the conservation of the total internal energy E can be formulated as the first law of thermodynamics, which can be formulated mathematically as [420]

$$dE = dQ - pdV, \tag{12.1}$$

where dQ is the heat received/released by the system during a time interval dt, and p is the thermodynamic pressure, respectively. Next we introduce the basic thermodynamic parameters of the system, which are the energy density ρ, given by $\rho = E/V$, the particle number density n, defined as $n = N/V$, and the heat per unit particle dq, with $dq = dQ/N$. Hence, in term of these quantities Eq. (12.1) takes the form

$$d\left(\frac{\rho}{n}\right) = dq - pd\left(\frac{1}{n}\right). \tag{12.2}$$

Equation (12.2) is generally valid for all thermodynamic systems, including open systems, in which N is a function of time, so that $N = N(t)$.

We will now proceed to study the general relativistic covariant formulation of thermodynamics. In the general-relativistic fully covariant approach the basic macroscopic variables that characterize the thermodynamic properties of a relativistic system, which for simplicity we take as a fluid, are the energy-momentum tensor $T_{\mu\nu}$, the particle flux four-vector N^μ, and the entropy flux four-vector S^μ. The energy-momentum tensor must satisfy the basic conservation law $\nabla_\nu T^{\mu\nu} = 0$. In the case of an open thermodynamic system in which particle production takes place, generally the energy-momentum tensor can be written down in the form

$$T^{\mu\nu} = (\rho + p + p_c)\, u^\mu u^\nu + (p + p_c)\, g^{\mu\nu}, \tag{12.3}$$

where the new term p_c, designated as the creation pressure, takes into account particle and other dissipative thermodynamical effects.

In relativistic thermodynamics the particle flux four-vector is defined as $N^\mu = nU^\mu$, where by n we denote the particle number density, while U^μ is the four-velocity of the fluid. N^μ must satisfy the general balance equation

$$\nabla_\mu N^\mu = \frac{1}{\sqrt{-g}} \frac{\partial}{\partial x^\mu} \left(\sqrt{-g} n U^\mu\right) = \Psi, \tag{12.4}$$

where $-g$ is the determinant of the metric tensor, and the function Ψ, giving the variation of the particle numbers, is called a particle source for $\Psi > 0$, and a particle sink for $\Psi < 0$, respectively. In the standard approach to cosmology Ψ is usually taken to be zero, or assumed to have negligibly small values. Another important thermodynamic quantity is the entropy flux S^μ, given by $S^\mu = n\sigma U^\mu$ [97], where σ is called the specific entropy per particle. A strong restriction on the entropy flux is imposed by the second law of thermodynamics, which states that $\nabla_\mu S^\mu \geq 0$. The Gibbs equation, a basic relation in thermodynamics, is given for an open thermodynamic system with temperature $\mathcal{T}$ and in the presence of particle production by

$$n\mathcal{T} d\sigma = d\rho - \frac{\rho + p}{n} dn. \tag{12.5}$$

With the use of the Gibbs equation we can obtain the important entropy balance equation as [97]

$$\nabla_\mu S^\mu = -\frac{p_c \Theta}{\mathcal{T}} - \frac{\mu \Psi}{\mathcal{T}}, \tag{12.6}$$

where by $\Theta = \nabla_\mu u^\mu$ we have denoted the expansion of the fluid, and we have also introduced the chemical potential μ, which is given by Euler's relation

$$\mu = \frac{1}{n}(\rho + p) - \mathcal{T}\sigma. \tag{12.7}$$

In our approach, we will assume that the cosmological matter is produced from the gravitational field in such a way that it is in thermal equilibrium with the already existing matter. This means that the entropy creation is a result of the particle production only. Moreover, to introduce a specific thermodynamical description we will assume that the creation pressure p_c can be represented by the following phenomenological ansatz [97, 420]

$$p_c = -\alpha \frac{\Psi}{\Theta}, \tag{12.8}$$

where the function α is assumed to be positive for all values of its argument, $\alpha > 0$. By using this choice for p_c for the entropy balance we obtain the basic equation

$$\nabla_\mu S^\mu = \frac{\Psi}{\mathcal{T}}(\alpha - \mu) = \Psi\sigma + \left(\alpha - \frac{\rho + p}{n}\right)\frac{\Psi}{\mathcal{T}} = \Psi\sigma + n\dot{\sigma}, \tag{12.9}$$

where $\dot{\sigma} = u^\mu \nabla_\mu \sigma = d\sigma/ds$. If we use now Eq. (12.6), we find for the specific entropy production $\dot{\sigma}$ the equation [97]

$$\dot{\sigma} = \frac{\Psi}{n\mathcal{T}}\left(\alpha - \frac{\rho + p}{n}\right). \tag{12.10}$$

We require now that the specific entropy per particle should be constant, $\sigma =$ constant. This assumption helps us to constrain our formalism. Thus, with a constant σ, Eq. (12.10) determines the functional form of α as $\alpha = (\rho + p)/n$. This gives for the creation pressure the representation [97]

$$p_c = -\frac{\rho + p}{n\Theta}\Psi. \tag{12.11}$$

Moreover, under the assumption of the constancy of the specific entropy, for the Gibbs equation we find the form

$$\dot{\rho} = (\rho + p)\frac{\dot{n}}{n}. \tag{12.12}$$

In the thermodynamics of open systems the second law of thermodynamics must be written down by decomposing the total change of the entropy into an entropy flow term $d_e S$, and an entropy creation term $d_i S$. Hence for an open system the differential of total entropy S is given by [419, 420]

$$dS = d_e S + d_i S, \tag{12.13}$$

where $d_i S > 0$. To calculate the entropy flow and the entropy creation terms, we consider the total differential of the entropy, which is given by the second law of thermodynamics as [420],

$$\mathcal{T} d\left(sa^3\right) = d\left(\rho a^3\right) + p da^3 - \mu d\left(na^3\right), \tag{12.14}$$

with $s = S/a^3$, and μ is the chemical potential rewritten as

$$\mu n = h - \mathcal{T} s, \tag{12.15}$$

where we have introduced the enthalpy (per-unit volume) h of the system, defined as $h = (\rho + p)$.

In the case of closed systems and for adiabatic transformations, the second law of thermodynamics tells us that $dS = 0$, and $d_i S = 0$, respectively. But the presence of particle production will give a nonzero contribution to the entropy. Homogeneous systems still satisfy the condition $d_e S = 0$. However, particle production will give a significant contribution to the entropy creation. In this case the entropy time variation can be calculated as [420]

$$\mathcal{T}\frac{d_iS}{dt} = \mathcal{T}\frac{dS}{dt} = \frac{h}{n}\frac{d}{dt}\left(na^3\right) - \mu\frac{d}{dt}\left(na^3\right) = \mathcal{T}\frac{s}{n}\frac{d}{dt}\left(na^3\right) \geq 0. \quad (12.16)$$

In the case of an expanding Universe, from Eq. (12.16) we obtain for the time variation of the entropy in the form

$$\frac{dS}{dt} = \frac{S}{n}\left(\dot{n} + 3Hn\right) \geq 0, \quad (12.17)$$

where $H = \dot{a}/a$ is the Hubble function. The entropy creation rate must satisfy the second law of thermodynamics, $\nabla_\mu S^\mu \geq 0$, where the entropy flux vector is given by $S^\mu = n\sigma U^\mu$. Since the specific entropy obeys the Gibbs relation [97]

$$n\mathcal{T}d\sigma = d\rho - \frac{h}{n}dn, \quad (12.18)$$

by also taking into account the definition of the chemical potential μ of the system as $\mu = h/n - \mathcal{T}\sigma$, we obtain

$$\begin{aligned}\nabla_\mu S^\mu &= \left(\dot{n} + 3nH\right)\sigma + nU^\mu\nabla_\mu\sigma \\ &= \frac{1}{\mathcal{T}}\left(\dot{n} + 3Hn\right)\left(\frac{\rho + p}{n} - \mu\right),\end{aligned} \quad (12.19)$$

where we have made use of the relation

$$n\mathcal{T}\dot{\sigma} = \dot{\rho} - \frac{\rho + p}{n}\dot{n} = 0, \quad (12.20)$$

which immediately follows from Eq. (12.5). Note that this relation can be expressed as

$$\dot{\rho} + 3H(\rho + p) = (\rho + p)\Gamma, \quad (12.21)$$

where Γ is the particle production rate, and is given by

$$\Gamma = \frac{\dot{n}}{n} + 3H. \quad (12.22)$$

Generally, a perfect fluid is characterized thermodynamically by two basic variables, the particle number density n, and the temperature $\mathcal{T}$, respectively. In this case the energy density ρ and the pressure p can be expressed in terms of n and $\mathcal{T}$ by using the equilibrium equations of state of the fluid,

$$\rho = \rho(n, \mathcal{T}), \qquad p = p(n, \mathcal{T}). \quad (12.23)$$

Hence the energy conservation equation of the perfect fluid Eq. (12.21) becomes

$$\frac{\partial\rho}{\partial n}\dot{n} + \frac{\partial\rho}{\partial\mathcal{T}}\dot{\mathcal{T}} + 3(\rho + p)H = \Gamma n. \quad (12.24)$$

By using the general thermodynamic relation [97]

$$\frac{\partial \rho}{\partial n}=\frac{\rho+p}{n}-\frac{\mathcal{T}}{n}\frac{\partial p}{\partial \mathcal{T}}, \tag{12.25}$$

giving the variation of the density with respect to the particle number n, we find the equation giving the temperature evolution of the newly created particle as

$$\frac{\dot{\mathcal{T}}}{\mathcal{T}}=\frac{\dot{n}}{n}\frac{\partial p}{\partial \rho}=(\Gamma-3H)\frac{\partial p}{\partial \rho}. \tag{12.26}$$

12.2.2 Matter Creation in Homogeneous and Isotropic Cosmological Models

In order to discuss the applications of the thermodynamics of open systems to cosmology, in the following section we will adopt the cosmological principle, according to which the matter distribution in the Universe must be homogeneous and isotropic. Hence, the line element of the Universe must be given by the FLRW metric, defined as

$$d\tilde{s}^2=-dt^2+a^2(t)\left(dx^2+dy^2+dz^2\right), \tag{12.27}$$

where $a(t)$ is the scale factor. Moreover, for the following discussion we will use the comoving coordinate system, in which the components of the four-velocity of the cosmic matter are given by $U^{\mu}=(1,0,0,0)$.

The cosmological principal also imposes the constraint that all the thermodynamics, as well as the geometric parameters are functions of the cosmological time t only. Then the derivative of any arbitrary time dependent function $f(t)$ with respect to the line element s will coincide with the ordinary time derivative, so that $\dot{f}=U^{\mu}\nabla_{\mu}f=df/dt$. For the expansion of the cosmic fluid we find $\nabla_{\mu}U^{\mu}=\dot{V}/V$, where $V=a^3$ is the comoving volume. In the cosmological context Eq. (12.12) can be reformulated in a number of equivalent ways as

$$\dot{\rho}=\left(\frac{h}{n}\right)\dot{n}, \tag{12.28}$$

where $h=\rho+p$ is the enthalpy (per unit volume) of the fluid, or in an alternative form, as

$$p=\dot{\rho}-\rho\frac{\dot{n}}{n}. \tag{12.29}$$

To write down the Einstein field equations we need to know the expression of the macroscopic energy-momentum tensor $T_{\mu\nu}$, for which, taking into account that we are dealing with the cosmological case, we will adopt a perfect fluid form. Hence $T_{\mu\nu}$ can be completely characterized by an energy density ρ of the cosmic matter, and by its total pressure $\bar{p}$, which in the general case also includes the matter creation effects. In the comoving frame and in the presence of matter creation the components of the energy-momentum tensor are given by

$$T_0^0 = -\rho, \qquad T_1^1 = T_2^2 = T_3^3 = \bar{p}. \tag{12.30}$$

The Einstein field equations $G_\nu^\mu = R_\nu^\mu - (1/2)R\delta_\nu^\mu = 8\pi G T_\nu^\mu$ imply the conservation of the energy-momentum tensor, which follows directly from the Bianchi identities, which give $\nabla_\mu G_\nu^\mu = 0 = \nabla_\mu T_\nu^\mu$. From the conservation of the energy-momentum tensor we find the relation

$$d(\rho V) = -\bar{p}dV. \tag{12.31}$$

In the presence of irreversible adiabatic particle production processes the analysis of the cosmological evolution must be given by using the thermodynamics of open systems. This requires the inclusion of the supplementary particle/annihilation pressure p_c in the mathematical formalism of the cosmological theory. Then we can write Eq. (12.12) in a form similar to Eq. (12.31), or more specifically [420]

$$d(\rho V) = -\left(p + p_c\right) dV. \tag{12.32}$$

From Eq. (12.11) it immediately follows that, in a homogeneous and isotropic cosmology with matter creation, the creation pressure p_c is obtained as

$$p_c = -\left(\frac{h}{n}\right)\frac{d(nV)}{dV} = -\left(\frac{h}{n}\right)\frac{V}{\dot{V}}\left(\dot{n} + \frac{\dot{V}}{V}n\right). \tag{12.33}$$

The production of particles corresponds to a negative pressure $p_c < 0$, which must be included as a supplementary term in the total cosmological pressure $\bar{p}$ that appears in the matter energy-momentum tensor

$$\bar{p} = p + p_c. \tag{12.34}$$

Furthermore, note that the decay of matter can be similarly analyzed by including a positive decay pressure in the total pressure p.

Another important thermodynamic parameter, the entropy change dS can be decomposed in an open thermodynamic system into two components, an entropy flow term d_0S, and the entropy creation term d_iS, so that

$$dS = d_0S + d_iS, \tag{12.35}$$

where d_iS must satisfy the condition $d_iS \geq 0$. To calculate dS we begin by writing down the total differential of the entropy, which is given by

$$\mathcal{T}d(sV) = d(\rho V) + pdV - \mu d(nV), \tag{12.36}$$

where $s = S/V \geq 0$, and $\mu n = h - \mathcal{T}s$, with $\mu \geq 0$. The homogeneity of the Universe imposes the condition $d_0S = 0$, and thus only particle production can contribute to the entropy creation. By using Eqs. (12.35) and (12.36) we obtain first [420]

$$\mathcal{T}\frac{dS}{dt} = \mathcal{T}\frac{d_iS}{dt} = \mathcal{T}\frac{s}{n}\frac{d(nV)}{dt}. \tag{12.37}$$

In order to completely solve the problem we still need the relation between the particle number n and the comoving volume V. Such a relation should describe the time evolution of n as a result of particle production (decay) processes. We have already introduced this relation, which is given by Eq. (12.4), and which in the case of a homogeneous and isotropic cosmological model can be reformulated as

$$\frac{1}{V}\frac{d(nV)}{dt} = \Psi(t), \tag{12.38}$$

where $\Psi(t)$ is the particle production (or decay) rate. $\Psi(t) > 0$ indicates matter production, while $\Psi(t) < 0$ corresponds to particle decay [97, 420]. The creation pressure p_c (12.33) is also a function of the particle production (decay) rate $\Psi(t)$. The entropy creation can also be formulated in terms of the matter creation rate as

$$S(t) = S(t_0)\, e^{\int_{t_0}^{t} \frac{\Psi(t)}{n} dt}. \tag{12.39}$$

Hence, in the case of particle with $\Psi(t) > 0$ the entropy of the Universe monotonically increases in time.

12.3 Thermodynamic Interpretation of $f(R, L_m)$ Gravity

In the present section, we will investigate $f(R, L_m)$ gravity from the point of view of the irreversible thermodynamics of open systems, and of particle processes. In the $f(R, L_m)$-modified theory of gravity the matter energy-momentum tensor satisfies the nonconservation relation given by Eq. (9.6), which we rewrite here, for convenience

$$\begin{aligned}\nabla^\mu T_{\mu\nu} &= \nabla^\mu \ln\left[f_{L_m}(R, L_m)\right]\left(L_m g_{\mu\nu} - T_{\mu\nu}\right)\\ &= 2\nabla^\mu \ln\left[f_{L_m}(R, L_m)\right]\frac{\partial L_m}{\partial g^{\mu\nu}}.\end{aligned} \tag{12.40}$$

In a general approach, the matter Lagrangian L_m is an arbitrary function of the matter energy density ρ, of the pressure p, and possibly a function of other thermodynamic variables, such as, for example, the specific entropy s, or the baryon number n, so that $L_m = L_m(\rho, p, s, n, ...)$.

As a first step in our study of the thermodynamic interpretation of modified gravity theories with a curvature-matter coupling we derive the generalized energy conservation equation in $f(R, L_m)$ gravity. The conservation equation has a new feature, that is, as compared to the standard adiabatic conservation equation of GR or $f(R)$ gravity, it contains an extra term. As mentioned before, this new term can be interpreted thermodynamically as a particle production rate. Moreover, particle production also acts as a source for the entropy, with the nonminimal curvature-matter coupling also strongly influencing the behavior of the entropy flux, as well as the temperature evolution of the Universe regarded as an open system.

By assuming that the matter content of the Universe is represented by a perfect cosmological fluid with energy density ρ, thermodynamic pressure p, and four-velocity U^{ν}, normalized according to the condition $U_{\nu}U^{\nu}=-1$, it follows that the energy-momentum tensor $T_{\mu\nu}$ of the Universe can be represented as

$$T_{\mu\nu}=(\rho+p)\,U_{\mu}U_{\nu}+pg_{\mu\nu}. \tag{12.41}$$

We will also make use of the projection operator h_{λ}^{ν}, defined as

$$h_{\lambda}^{\nu}=\delta_{\lambda}^{\nu}+U_{\lambda}U^{\nu}, \tag{12.42}$$

with the basic property $U_{\nu}h_{\lambda}^{\nu}=0$.

12.3.1 The Energy Conservation Equation

With the help of the operator ∇ we take first the covariant divergence of Eq. (12.41) to obtain

$$\begin{aligned}\nabla^{\mu}T_{\mu\nu}=&\left(\nabla^{\mu}\rho+\nabla^{\mu}p\right)U_{\mu}U_{\nu}+(\rho+p)\,\nabla^{\mu}U_{\mu}U_{\nu}\\&+(\rho+p)\,U_{\mu}\nabla^{\mu}U_{\nu}-\nabla^{\mu}pg_{\mu\nu}.\end{aligned} \tag{12.43}$$

Therefore the nonconservation equation (12.40) of the matter in $f(R,L_m)$ gravity can be written as

$$\begin{aligned}&\left(\nabla^{\mu}\rho+\nabla^{\mu}p\right)U_{\mu}U_{\nu}+(\rho+p)\,\nabla^{\mu}U_{\mu}U_{\nu}+(\rho+p)\,U_{\mu}\nabla^{\mu}U_{\nu}+\nabla^{\mu}pg_{\mu\nu}\\&=\nabla^{\mu}\ln\left[f_{Lm}\left(R,L_m\right)\right]\times\left[\left(L_m+p\right)g_{\mu\nu}+(\rho+p)\,U_{\mu}U_{\nu}\right].\end{aligned} \tag{12.44}$$

By multiplying Eq. (12.44) with h_{λ}^{ν}, given by Eq. (12.42), we obtain the momentum balance equation for a perfect fluid as

$$\begin{aligned}U^{\mu}\nabla_{\mu}U^{\lambda}&=\frac{d^2x^{\lambda}}{ds^2}+\Gamma_{\mu\nu}^{\lambda}U^{\mu}U^{\nu}\\&=\frac{1}{\rho+p}\left\{h^{\nu\lambda}\nabla_{\nu}p+h_{\mu}^{\lambda}\nabla^{\mu}\ln\left[f_{Lm}\left(R,L_m\right)\right]\left(L_m+p\right)\right\}.\end{aligned} \tag{12.45}$$

We multiply now Eq. (12.44) with U^{ν}. With the use of the mathematical identity $U^{\nu}\nabla^{\mu}U_{\nu}=0$, which can be obtained easily from the normalization condition of the four-velocity, we obtain the energy balance equation in the $f\left(R,L_m\right)$ gravity as

$$\dot{\rho} + 3(\rho + p) H = \frac{d}{ds} \ln \left[f_{Lm}(R, L_m) \right] (L_m - \rho), \tag{12.46}$$

where we have denoted $H = (1/3)\nabla^\mu U_\mu$, and the overdot $\dot{} = U^\mu \nabla_\mu = d/d\tilde{s}$, respectively, where $d\tilde{s}$ is the line element corresponding to the general relativistic metric represented generally as $g_{\mu\nu}$, $d\tilde{s}^2 = g_{\mu\nu} dx^\mu dx^\nu$.

In the flat FLRW geometry that describes a homogeneous and isotropic Universe, we have $H = \dot{a}/a$, and $U^\mu \nabla_\mu = \dot{} = d/dt$, respectively. Therefore, Eq. (12.46) can be written in the equivalent form, which is appropriate for cosmological applications as

$$\frac{d}{dt}\left(\rho a^3\right) + p \frac{d}{dt} a^3 = a^3 \frac{d}{dt} \ln \left[f_{Lm}(R, L_m) \right] (L_m - \rho). \tag{12.47}$$

To obtain the general relativistic limit of the $f(R, L_m)$ theory we take $f(R, L_m) = R/2 + L_m$, which gives $f_{Lm}(R, L_m) = 1$. Hence Eq. (12.47) reduces to the adiabatic matter conservation law $d\left(\rho a^3\right) + p da^3 = 0$ of standard cosmology. On the other hand, due to our choice of the geometry and of the cosmological model, all the non-diagonal components of the energy-momentum tensor are identically equal to zero, so that $T_{\mu\nu} = 0$, $\mu \neq \nu$. Since the condition $T_{0i} \equiv 0$, $i = 1, 2, 3$ must always hold, the diagonal form of the energy-momentum tensor implies, in particular, the impossibility of heat transfer processes in the FLRW models of modified gravity with curvature-matter coupling.

12.3.2 The Matter and Entropy Creation Rates

For a thermodynamic system containing N particles in a volume $V = a^3$, the second law of thermodynamics is given in its general form by [420]

$$\frac{d}{dt}\left(\rho a^3\right) + p \frac{d}{dt} a^3 = \frac{dQ}{dt} + \frac{h}{n} \frac{d}{dt}\left(n a^3\right), \tag{12.48}$$

where, as we have already seen, $n = N/V$ is the particle number density, dQ denotes the heat received or released by the system during the small time interval dt, and $h = \rho + p$ is the enthalpy per unit volume. We will restrict our investigation of the cosmological matter and entropy creation to the case of the adiabatic transformations, which satisfy the condition $dQ = 0$. From a physical point of view this means that we neglect proper heat transfer processes in the cosmological fluid filling the Universe. For adiabatic transformations Eq. (12.48) gives the formulation of the second law of thermodynamics in which the variation of the number of particles of the system, described by the term $(h/n) d\left(n a^3\right)/dt$, is explicitly taken into account. Thus it follows that in the thermodynamic description of open systems with particle the "heat" (internal energy) produced in the system is due to the particle number variation. From a gravitational and cosmological point of view, this change of the internal energy is entirely due to the transfer of energy from gravitation to particles, a process that acts as a source of internal energy for the cosmological system. Hence, by taking into account that for adiabatic transformations $dQ/dt = 0$, Eq. (12.48) becomes

$$\dot{\rho} + 3(\rho + p)H = \frac{\rho + p}{n}(\dot{n} + 3Hn). \tag{12.49}$$

Therefore, if we adopt a thermodynamic perspective, we can interpret Eq. (12.47), which gives the energy conservation equation in $f(R, L_m)$ gravity, as describing particle production in a homogeneous and isotropic Universe. The time variation of the particle number density of the created particles is given by the equation

$$\dot{n} + 3nH = \frac{n}{\rho + p}\frac{d}{dt}\ln\left[f_{L_m}(R, L_m)\right](L_m - \rho) = \Gamma n, \tag{12.50}$$

where we have introduced the cosmological particle creation rate Γ, defined as

$$\Gamma = \frac{1}{\rho + p}\frac{d}{dt}\ln\left[f_{L_m}(R, L_m)\right](L_m - \rho). \tag{12.51}$$

Therefore, in $f(R, L_m)$ gravity, the energy conservation equation can be formulated as Eq. (12.21), which we rewrite

$$\dot{\rho} + 3(\rho + p)H = (\rho + p)\Gamma. \tag{12.52}$$

For the particular case of the adiabatic transformations with $dQ = 0$, Eq. (12.48), describing matter production in an open thermodynamic system, can be reformulated as an effective energy conservation equation, which can be written as

$$\frac{d}{dt}\left(\rho a^3\right) + (p + p_c)\frac{d}{dt}a^3 = 0, \tag{12.53}$$

or, equivalently, as

$$\dot{\rho} + 3(\rho + p + p_c)H = 0, \tag{12.54}$$

where the creation pressure p_c can be easily obtained, and is given by [420]

$$p_c = -\frac{\rho + p}{n}\frac{d\left(na^3\right)}{da^3} = -\frac{\rho + p}{3nH}(\dot{n} + 3nH) = -\frac{\rho + p}{3}\frac{\Gamma}{H}. \tag{12.55}$$

Hence, we have arrived to the important result that in $f(R, L_m)$ gravity the creation pressure p_c is entirely determined by the curvature-matter coupling and, once the specific form of the coupling is known, it can be obtained from the relation

$$p_c = -\frac{1}{3H}\frac{d}{dt}\ln\left[f_{L_m}(R, L_m)\right](L_m - \rho). \tag{12.56}$$

Taking into account Eq. (12.50), which gives the particle number balance, it is easy to show that the entropy production rate associated to the particle creation processes in $f(R, L_m)$ gravity can be obtained from the expression

$$\nabla_\mu S^\mu = \frac{1}{calT} \frac{n}{\rho + p} \frac{d}{dt} \ln\left[f_{L_m}(R, L_m)\right](L_m - \rho)\left(\frac{\rho + p}{n} - \mu\right). \quad (12.57)$$

With the use of Eq. (12.50), giving the particle number balance, for the entropy production rate we obtain the equation

$$\frac{1}{S}\frac{dS}{dt} = \frac{1}{\rho + p}\frac{d}{dt} \ln\left[f_{L_m}(R, L_m)\right](L_m - \rho) \geq 0. \quad (12.58)$$

Finally, with the help of Eq. (12.26), for the temperature evolution of the newly created particles we find the relation

$$\frac{\dot{\mathcal{T}}}{\mathcal{T}} = \frac{1}{\rho + p}\frac{\partial p}{\partial \rho}\left\{\frac{d}{dt} \ln\left[f_{L_m}(R, L_m)\right](L_m - \rho) - 3H(\rho + p)\right\}. \quad (12.59)$$

12.3.3 The case $L_m = p$

The exact form of the matter Lagrangian is an interesting and open theoretical problem in present-day physics, with important implications in GR and cosmology, as extensively discussed in Section 8.2. Several possible choices for the matter Lagrangian have been proposed. One of them is to assume that the matter Lagrangian is the pressure (with minus sign), $L_m = p$, a choice that was used in [452, 457, 93] to derive the standard general relativistic equations of motion of test fluids. In this case, as it was pointed out in [470], the extra-force, present in the right hand side of Eq. (12.45), and whose presence is one of the most important features of $f(R, L_m)$ gravity, identically vanishes. On the other hand, as discussed in detail in [53], other alternative choices for the matter Lagrangian, leading to the same equations of motion, such as $L_m = -\rho$ or $L_m = -na$, where a is the thermodynamic free energy, having the expression $a = \rho/n - Ts$, are also physically acceptable [93, 250]. Actually all these expressions proposed for the relativistic matter Lagrangian are just on shell representations of a more general Lagrangian density, which is obtained by substituting back the equations of motion in the corresponding action [93].

In the following discussion, we will study the thermodynamic and cosmological implications of the choice $L_m = p$ for the Lagrangian of the matter. For this choice the extra-force of the theory identically vanishes, which shows that in this implementation of $f(R, L_m)$ gravity the motion of the test fluids takes place on a geodesic path. On the other hand, even with this choice the right hand side of the energy balance equation given by Eq. (12.46) does not vanish. This shows that even in the absence of the extra-force, the matter production processes are still present in the theory. Hence with the choice $L_m = p$ the particle number balance, given by Eq. (12.50), takes the form

$$\dot{n} + 3nH = -n\frac{d}{dt}\ln\left[f_p\left(R, p\right)\right], \tag{12.60}$$

where by f_p we have denoted the derivative of the function $f\left(R, p\right)$ with respect to the parameter p. Then Eq. (12.60) can be immediately integrated to obtain the particle number relation as

$$na^3 = \frac{n_0 a_0^3}{f_p\left(R, p\right)}, \tag{12.61}$$

where a_0 and n_0 are arbitrary constants of integration. Hence it turns out that the variation of the total number of particles within a volume V is inversely proportional to the derivative of the function f (the total Lagrangian of the theory), with respect to the matter Lagrangian. For the energy conservation equation (12.52) we find

$$\dot{\rho} + 3(\rho + p)H = -(\rho + p)\frac{d}{dt}\ln\left[f_p\left(R, p\right)\right]. \tag{12.62}$$

With the use of Eq. (12.56) the creation pressure p_c can be calculated as

$$p_c = \frac{\rho + p}{3H}\frac{d}{dt}\ln\left[f_{L_m}\left(R, L_m\right)\right]. \tag{12.63}$$

For a radiation-like fluid obeying the equation of state $p = \rho/3$, for the variation of the energy density of the particle content of the Universe we find the expression

$$\rho a^4 = \frac{\rho_0 a_0^4}{\left[f_p\left(R, p\right)\right]^{4/3}}, \tag{12.64}$$

where ρ_0 and a_0 are arbitrary constants of integration. For the total entropy variation we find from Eq. (12.58) the relation

$$S = \frac{S_0}{f_p\left(R, p\right)}, \tag{12.65}$$

where S_0 is an arbitrary constant of integration to be determined from the initial conditions of the Universe. In a standard cosmological fluid in a homogeneous and isotropic Universe the density and pressure of the fluid are some given functions of the temperature $\mathcal{T}$ only, $\rho = \rho(\mathcal{T})$, $p = p(\mathcal{T})$. In this case the entropy of the fluid is given by [501]

$$S = \frac{(\rho + p)\,V}{\mathcal{T}}. \tag{12.66}$$

For a radiation-type fluid with equation of state $\rho = 3p$, it follows that the total gravitational Lagrangian density f of the theory satisfies the following partial differential equation

$$\frac{\partial}{\partial (p)} f(R,p) = -\frac{S_0 \mathcal{T}(a)}{4(p)a^3}, \tag{12.67}$$

which closely relates the gravitational, matter, and thermodynamical properties. By assuming a cosmological model with $f(R,p) = f_0(R(a)) g_0(p)$, that is, with the total Lagrangian density given by the product of two functions, one depending on the Ricci scalar only, and the second one on the matter Lagrangian only, Eq. (12.67) gives the relation

$$f_0(R(a)) = -\frac{S_0 \mathcal{T}(a)}{a^3}, \qquad \frac{dg_0(p)}{dp} = \frac{1}{4p}, \tag{12.68}$$

which, interestingly, fully determines the matter Lagrangian dependence of the total action as

$$g_0(L_m) = \frac{1}{4} \ln L_m. \tag{12.69}$$

12.4 Irreversible Thermodynamic Interpretation of $f(R,T)$ Gravity

In the previous section, we have considered the thermodynamic interpretation of $f(R, L_m)$ gravity. In the present section, we will move to consider the thermodynamic interpretation of the second gravitational theory involving a curvature-matter coupling, namely, the $f(R,T)$-modified theory of gravity [220]. As for the energy-momentum tensor, we assume that it is given by Eq. (12.41), that is, we model again matter as a perfect fluid characterized by two thermodynamic variables only. To obtain some concrete results we assume from the beginning that $L_m = p$, and in the present analysis we will not consider other possible choices.

Since for a perfect fluid the trace of the matter energy-momentum tensor can be obtained easily as $T = -\rho + 3p$, for the variation $\Theta_{\mu\nu}$ of the energy-momentum tensor $T_{\mu\nu}$ with respect to the metric tensor $g_{\mu\nu}$ we find the simple relation [243]

$$\Theta_{\mu\nu} = -2T_{\mu\nu} + pg_{\mu\nu} = -2(\rho+p)U_\mu U_\nu + pg_{\mu\nu}. \tag{12.70}$$

12.4.1 Energy and Particle Creation Rates

The divergence of the energy-momentum tensor in $f(R,T)$ gravity can be obtained immediately with the help of Eq. (12.70) as [243]

$$\nabla^\mu T_{\mu\nu} = -\frac{f_T(R,T)}{8\pi + f_T(R,T)} \left[(\rho+p) U_\nu U_\mu \nabla^\mu \ln f_T(R,T) + \nabla_\nu \frac{\rho - p}{2} \right]. \tag{12.71}$$

After multiplication with U^ν, Eq. (12.71) gives

$$U^\nu \nabla^\mu T_{\mu\nu} = -\frac{f_T(R,T)}{8\pi + f_T(R,T)}\left[(\rho + p)\, U_\mu \nabla^\mu \ln f_T(R,T) + U^\nu \nabla_\nu \frac{\rho - p}{2}\right]. \tag{12.72}$$

With the use of the matter energy-momentum tensor for the cosmological perfect fluid, from Eqs. (12.72) the energy conservation equation in $f(R,T)$ gravity can be found as

$$\dot{\rho} + 3(\rho + p)H = -\frac{f_T(R,T)}{8\pi + f_T(R,T)}\left[(\rho + p)\, U_\mu \nabla^\mu \ln f_T(R,T) + U^\nu \nabla_\nu \frac{\rho - p}{2}\right]. \tag{12.73}$$

We adopt again the point of view of the thermodynamic of open systems, and therefore we interpret Eq. (12.73) as describing adiabatic irreversible matter production in a cosmological setting, according to Eq. (12.4). Hence the particle balance equation in $f(R,T)$ gravity can be obtained as

$$\dot{n} + 3nH = \Gamma n, \tag{12.74}$$

where the particle creation rate Γ has been defined as

$$\Gamma = -\frac{f_T(R,T)}{8\pi + f_T(R,T)}\left[\frac{d}{dt}\ln f_T(R,T) + \frac{1}{2}\frac{\dot{\rho} - \dot{p}}{\rho + p}\right]. \tag{12.75}$$

Alternatively, the energy conservation equation can be reformulated as

$$\frac{d}{dt}\left(\rho a^3\right) + p\frac{d}{dt}a^3 = (\rho + p)\, a^3 \Gamma. \tag{12.76}$$

The creation pressure corresponding to the matter creation from the gravitational field, defined by Eq. (12.33), can be found in $f(R,T)$ gravity as

$$p_c = \frac{\rho + p}{3H}\frac{f_T(R,T)}{8\pi + f_T(R,T)}\left[\frac{d}{dt}\ln f_T(R,T) + \frac{1}{2}\frac{\dot{\rho} - \dot{p}}{\rho + p}\right]. \tag{12.77}$$

If the total Lagrangian density of $f(R,T)$ gravity is independent of the trace of the energy-momentum tensor T (a condition that gives the general relativistic limit), then $f_T(R,T) \equiv 0$, and consequently both Γ and p_c vanish. Hence in this limit we have reobtained the standard cosmological scenario without particle production, in which during its entire cosmological evolution the Universe strictly obeys the conservation law of the total energy.

The time variation of the entropy can be obtained from Eq. (12.17). Note that the entropy is entirely generated by the particle production processes, so that $S(t)$ is given by

$$S(t) = S_0 \exp\left[\int_0^t \Gamma\left(t'\right) dt'\right], \tag{12.78}$$

where S_0 is an arbitrary constant of integration. For the entropy production rate we find, with the use of Eq. (12.19), the expression

$$\nabla_\mu S^\mu = \frac{n}{\mathcal{T}}\Gamma\left(\frac{\rho+p}{n}-\mu\right) = -\frac{n}{\mathcal{T}}\left(\frac{\rho+p}{n}-\mu\right) \times \frac{f_T(R,T)}{8\pi+f_T(R,T)}\left[\frac{d}{dt}\ln f_T(R,T)+\frac{1}{2}\frac{\dot{\rho}-\dot{p}}{\rho+p}\right]. \tag{12.79}$$

The time variation of the temperature of the newly created particles can be obtained by assuming that their energy density and pressure are, according to Eqs. (12.23), functions of both the particle number n and of the temperature $\mathcal{T}$. Then the expression of $\dot{\mathcal{T}}$ follows immediately from Eq. (12.26), and in the open system thermodynamic interpretation of $f(R,T)$ gravity is given by

$$\frac{1}{c_s^2}\frac{\dot{\mathcal{T}}}{\mathcal{T}} = \Gamma - 3H = -\left\{\frac{f_T(R,T)}{8\pi+f_T(R,T)} \times \left[\frac{d}{dt}\ln f_T(R,T)+\frac{1}{2}\frac{\dot{\rho}-\dot{p}}{\rho+p}\right]+3H\right\}, \tag{12.80}$$

where $c_s^2 = \partial p/\partial \rho$ denotes the speed of sound in the cosmological fluid.

12.4.2 Cosmological Applications

As an application of the open system interpretation of the $f(R,T)$-modified gravity theory we will consider now the thermodynamic properties of a cosmological model discussed first in [243]. As a starting point we adopt for the function $f(R,T)$ the simple functional expression

$$f(R,T) = R + 2g(T), \tag{12.81}$$

where $g(T) = g(\rho - 3p)$ is an arbitrary function of the trace of the energy-momentum tensor of the cosmological fluid. The gravitational field equations corresponding to this form of $f(R,T)$ can be obtained as [243]

$$R_{\mu\nu} - \frac{1}{2}Rg_{\mu\nu} = 8\pi T_{\mu\nu} - 2g'(T)T_{\mu\nu} - 2g'(T)\Theta_{\mu\nu} + g(T)g_{\mu\nu}, \tag{12.82}$$

where a prime denotes the derivative of the given function with respect to its argument.

By adopting for the matter source the perfect fluid form, we obtain immediately $\Theta_{\mu\nu} = -2T_{\mu\nu} + pg_{\mu\nu}$, and thus the gravitational field equations become

$$R_{\mu\nu} - \frac{1}{2}Rg_{\mu\nu} = 8\pi T_{\mu\nu} + 2g'(T)T_{\mu\nu} + \left[2pg'(T) + g(T)\right]g_{\mu\nu}. \tag{12.83}$$

By assuming that the matter content of the Universe consists of pressureless dust with $p = 0$, the gravitational field equations of $f(R,T)$ gravity further simplify to

$$R_{\mu\nu} - \frac{1}{2}Rg_{\mu\nu} = 8\pi T_{\mu\nu} + 2g'(T)T_{\mu\nu} + g(T)g_{\mu\nu}. \tag{12.84}$$

The simplest cosmological model that can be constructed from Eqs. (12.84) corresponds to a pressureless dust dominated Universe, with $p = 0$. Then $T = \rho$. Furthermore, for the function $g(T)$ we assume the form $g(T) = \lambda T$, where λ is an arbitrary constant. Hence we have

$$f_T(R,T) = 2\lambda = \text{constant}. \tag{12.85}$$

The gravitational field equations of this simple cosmological model are given in the flat FLRW metric by

$$3H^2 = (8\pi + 3\lambda)\,\rho, \tag{12.86}$$

and

$$2\dot{H} + 3H^2 = -\lambda\rho, \tag{12.87}$$

respectively. Therefore this simple $f(R,T)$-modified gravity model is equivalent to a cosmological scenario with an effective cosmological constant given by $\Lambda_{\text{eff}} \propto H^2$ [243]. Also note that for the adopted simple form of $f(R,T)$ an effective time-dependent gravitational coupling is generated, given by

$$G_{\text{eff}} = G \pm 2g'(T). \tag{12.88}$$

Hence the modification of the gravitational Lagrangian by a term of the form $2g(T)$ changes the nature of the gravitational interaction in the presence of matter, by substituting the standard gravitational constant G of GR by a running, time-dependent gravitational coupling parameter G_{eff}.

After some simple mathematical transformations the cosmological field equations (12.86) and (12.87) can be reduced to a single equation for the Hubble function H,

$$2\dot{H} + 3\frac{8\pi + 4\lambda}{8\pi + 3\lambda}H^2 = 0, \tag{12.89}$$

having its general solution given by

$$H(t) = \frac{2\,(8\pi + 3\lambda)}{3\,(8\pi + 4\lambda)}\frac{1}{t} = \frac{H_0}{t}, \tag{12.90}$$

where we have denoted

$$H_0 = \frac{2\,(8\pi + 4\lambda)}{3\,(8\pi + 3\lambda)}. \tag{12.91}$$

The scale factor a of this model has a power law time dependence given by $a(t) = t^{H_0}$, while the energy density of matter in the Universe evolves as

$$\rho(t) = \frac{\rho_0}{t^2}, \tag{12.92}$$

where $\rho_0 = 3H_0^2/(8\pi + 3\lambda)$. The deceleration parameter q is defined as

$$q = \frac{d}{dt}\frac{1}{H} - 1, \tag{12.93}$$

and can be immediately calculated to find

$$q = \frac{1}{H_0} - 1 = \frac{8\pi + \lambda}{2(8\pi + 4\lambda)}. \tag{12.94}$$

If $q < 0$, the Universe experiences an accelerating era, while a positive deceleration parameter $q > 0$ indicates a decelerating phase.

The energy conservation equation can be obtained directly from Eq. (12.73), and can be formulated by

$$\dot{\rho} + 3\frac{8\pi + 2\lambda}{8\pi + 3\lambda}\rho H = 0. \tag{12.95}$$

We can also obtain immediately the particle creation rate Γ, which describes the energy transfer from the gravitational field to the newly created particles, and which is found as

$$\Gamma(t) = -\frac{\lambda}{8\pi + 2\lambda}\frac{\dot{\rho}}{\rho} = \frac{3\lambda}{8\pi + 3\lambda}H = \frac{\Gamma_0}{t}, \tag{12.96}$$

where $\Gamma_0 = (2\lambda/3)(8\pi + 4\lambda)/(8\pi + 3\lambda)^2$. The time variation of the particle number $n(t)$ in this open matter-gravitation system with energy transfer and particle production is described by the simple differential equation

$$\dot{n} = \frac{\Gamma_0 - 3H_0}{t}n, \tag{12.97}$$

with the solution given by

$$n(t) = n_0 t^{\Gamma_0 - 3H_0}, \tag{12.98}$$

where n_0 is an arbitrary constant of integration, which must be determined from the initial cosmological conditions. The creation pressure p_c can be calculated from Eq. (12.77), and it is found as

$$p_c(t) = \frac{\lambda}{3(8\pi + 2\lambda)}\frac{\dot{\rho}}{H} = -\frac{\lambda}{8\pi + 3\lambda}\rho = -\frac{p_{c0}}{t^2}, \tag{12.99}$$

where $p_{c0} = \lambda\rho_0/(8\pi + 3\lambda)$. The time variation of the entropy $S(t)$ is fully determined by the creation of matter from the gravitational field. Thus, we obtain

$$S(t) = S_0 t^{\Gamma_0}. \tag{12.100}$$

Hence we have developed a full thermodynamic interpretation of this simple $f(R,T)$ cosmological model in the framework of open irreversible systems. It turns out that in this model the matter production rate is related to the Hubble function via a simple proportionality relation. Therefore, similarly to the Hubble function, it is a linearly decreasing function of cosmological time.

On the other hand the number of particles produced during the early phases of the cosmological evolution decreases in time if the condition $\Gamma_0 < 3H_0$ is satisfied. The matter production processes determine the creation of a large amount of comoving entropy, which increases in time for $\Gamma_0 \neq 0$, and whose functional dependence is given by a power law function of the cosmological time t.

12.4.3 Bulk-Viscosity Description of Particle Production Processes

Another physical interpretation of matter production processes in cosmology was proposed initially in [519], and later further developed in [373] and [259]. In accordance with this thermodynamic interpretation, the bulk viscosity of the cosmological fluid could provide an effective phenomenological description of the production of matter from the time-dependent gravitational field of the expanding (or collapsing) Universe. Hence, from this alternative point of view, a nonvanishing particle creation rate is equivalent to the addition of an effective bulk viscous pressure to the energy-momentum tensor of the cosmological fluid. In the framework of a quantum mechanical approach, the effective viscous pressure can also be linked to the vacuum viscosity [259, 373, 519]. The theoretical interpretation of the equivalence between matter creation and some forms of viscosity is a consequence of the simple circumstance that in the energy balance equation of a general relativistic fluid any source term may be properly rewritten as an effective bulk viscosity term [333].

By assuming that bulk viscosity is the only dissipative process in the general relativistic cosmological fluid, its energy-momentum tensor is given by [333, 336]

$$T_{\mu\nu} = (\rho + p + \Pi) U_\mu U_\nu + (p + \Pi) g_{\mu\nu}, \tag{12.101}$$

where Π denotes the bulk viscous pressure. As usual, the particle flux four-vector N^μ is defined as $N^\mu = nU^\mu$, where U^μ is the four-velocity of the fluid. However, in the framework of the causal thermodynamic approach the entropy flux four-vector S^μ takes the generalized form [274]

$$S^\mu = sN^\mu - \frac{\tau \Pi^2}{2\xi T} U^\mu, \tag{12.102}$$

where τ denotes the relaxation time, and ξ is the coefficient of bulk viscosity, respectively. It is important to note that in describing the deviations from the thermodynamic equilibrium we have enclosed only second-order terms in Eq. (12.102). In the case of homogeneous and isotropic cosmological models, once the bulk viscous dissipative phenomena are included in the matter energy-momentum tensor, the energy conservation equation can be written as [333]

$$\dot{\rho} + 3(\rho + p + \Pi) H = 0. \tag{12.103}$$

From the simple comparison of Eq. (12.103), which for a bulk viscous cosmological fluid gives the energy conservation equation, for instance with Eq. (12.56), expressing the creation pressure in $f(R, L_m)$ gravity, it follows that these two equations are equivalent if one takes the creation pressure as equal to the bulk viscous pressure

$$p_c = \Pi = -\frac{1}{3H}\frac{d}{dt}\ln\left[f_{L_m}(R, L_m)\right](L_m - \rho). \tag{12.104}$$

Note that the creation pressure explicitly includes in the energy balance the production of matter from the gravitational field due to the curvature-matter coupling. Therefore matter production can indeed be described phenomenologically with the use of an alternative point of view based on the introduction of an effective bulk viscous pressure term in the energy-momentum tensor of the cosmological fluid. From a physical point of view it follows that the causal bulk viscous pressure Π can be interpreted as an effective creation pressure.

Hence, it would be of interest for the physical interpretation of modified gravity theories with a curvature-matter coupling to investigate particle production processes from the point of view of bulk viscous thermodynamic processes. Therefore, we will now consider that the change in the number of particles, due to particle production phenomena can be described with the help of the bulk viscous pressure that plays the role of an effective creation pressure. To further develop this idea we introduce a simple toy model in which the newly created matter obeys an equation of state of the form

$$\rho = \rho_0\left(\frac{n}{n_0}\right)^\gamma = kn^\gamma, \qquad p = (\gamma - 1)\rho, \tag{12.105}$$

which relates the matter density to the particle number density n, and the pressure to the energy density, respectively. In the above equations ρ_0, n_0 and γ are constants, and we have denoted $k = \rho_0/n_0^\gamma$, and $1 \leq \gamma \leq 2$, respectively. With the use of Eq. (12.105), it follows that Eq. (12.103) becomes a particle balance equation, which may be written as

$$\dot{n} + 3Hn = \Gamma n, \tag{12.106}$$

where

$$\Gamma = -\frac{\Pi}{\gamma H} \tag{12.107}$$

is the particle production rate, In the present approach the particle production rate Γ is proportional to the bulk viscous pressure. From the combination of the equation of state Eq. (12.105) with the Gibbs relation $Tds = d(\rho/n) + pd(1/n)$ it follows that $s = s_0 =$ constant, that is, in this model matter is created with constant entropy density.

However, it is important to note that there is a major difference between the description of the matter production in the framework of the irreversible thermodynamic of open systems, and the bulk viscosity description. This conceptual difference is closely related to the relation giving the entropy creation rate. While the entropy creation rate associated to particle production is given by [97]

$$\nabla_\mu S^\mu = -\frac{3Hp_c}{\mathcal{T}}\left(1+\frac{\mu\Gamma n}{3Hp_c}\right)\geq 0, \tag{12.108}$$

the entropy production rate in the presence of bulk viscous dissipative processes is obtained as [333]

$$\nabla_\mu S^\mu = -\frac{\Pi}{\mathcal{T}}\left[3H+\frac{\tau}{\xi}\dot{\Pi}+\frac{\tau}{2\xi}\Pi\left(3H+\frac{\dot{\tau}}{\tau}-\frac{\dot{\xi}}{\xi}-\frac{\dot{\mathcal{T}}}{\mathcal{T}}\right)\right]. \tag{12.109}$$

In the particle model based on the irreversible thermodynamics of open systems, the entropy creation rate is proportional to the creation pressure. On the other hand, in the thermodynamic interpretation of particle based on viscous dissipative processes, $\nabla_\mu S^\mu$ is a quadratic function of the creation pressure, $\nabla_\mu S^\mu \propto p_c^2/\xi\mathcal{T}$. Moreover, the entropy flux involves a new dynamical variable, the bulk viscosity coefficient, which does not have a physical analogue in the description of the particle processes in the framework of the open systems thermodynamic.

12.5 Discussion

In the present chapter, we have presented and discussed in a systematic way the possibility that, at least in principle, modified gravity theories with curvature-matter coupling are closely related to the phenomenological description of the particle processes that could take place in the expanding cosmological fluid filling the Universe. A common and intriguing characteristic of modified theories of gravity with curvature-matter coupling is the nonconservation of the energy-momentum tensor of the matter. By using the physical and mathematical formalism of the irreversible thermodynamics of open systems, we have proposed an interpretation of the extra-terms generated by the nonminimal matter-curvature coupling. In the present approach, the terms that break standard energy conservation are considered as describing particle, with the particle source being represented by the gravitational field. Both $f(R, L_m)$ and $f(R,T)$ gravity theories fit well in the framework of such a thermodynamic interpretation. In our analysis we have explicitly obtained the particle production rates, the creation pressure, the entropy, and the entropy generation rate as functions of the Lagrangian density of the models $f(X)$, where $X=(R,L_M)$, or $X=(R,T)$, and of their derivatives with respect to the function's arguments, respectively.

On the other hand, we would like to mention that the approach adopted in the present chapter cannot give any hints on the nature or type of the new matter created by the gravitational field. However, we may assume that such creation events should be consistent with the similar particle production processes that are an essential characteristic of quantum field theory in curved space-times.

It is a result of the quantum approach that generally, a static gravitational field does not create particles. But particle production can occur in a time-dependent gravitational field. In the standard quantum mechanical approach, tunneling of a Universe (described as a single particle) from a small scale factor a_s to a Universe having a large scale factor a_l would be characterized by a wave function that contains as $a \to \infty$ only outgoing waves [311]. In the classically forbidden region the decaying component of the wave function can be written as

$$\Psi(a) \propto \exp\left\{-\int \sqrt{2\left[V(a)-\rho\right]}da\right\}, \tag{12.110}$$

where $V(a)$ denotes the quantum potential, while ρ represents the energy of the scalar mode in the conformal frame.

Unlike in standard nonrelativistic quantum mechanics, when a massive scalar field with an action of the form

$$S_\Phi = \int \left(g^{\mu\nu}\nabla_\mu\Phi\nabla_\nu\Phi/2 - m_\Phi^2\Phi^2 + R\Phi^2/12\right)\sqrt{-g}d^4x, \tag{12.111}$$

is conformally coupled to gravity, it turns out that in the resulting Schrödinger equation a negative kinetic term for the scale factor will appear. Moreover, the potential $V(a)$ also enters in the equation, but with a negative sign. Therefore we obtain the basic result of quantum cosmology, and of the quantum particle creation processes, which tells us that due to the increase of the effective value of ρ, the excitation of the scalar field Φ makes the tunneling processes from vacuum much easier [311]. Hence, the scalar quanta Φ of the quantum field can be produced at a high rate. Moreover, based on the analogy with similar particle processes in quantum field theory, we propose that most of the matter generated due to the nonminimal curvature-matter coupling may be in the form of some low-mass scalar particles (bosons), which may correspond to the dark matter component of the Universe, which represents around 24% of the total matter content. In the analysis of the simple $f(R,T)$ cosmological model performed in the previous section, we have seen that the matter production rate has a maximum value at the beginning of the cosmological expansion. Therefore, under our assumption that cosmological models with a curvature-matter coupling can give an effective description of the quantum particle processes, we may also suggest that most (if not all) of the dark matter was created at the very early stages of the cosmological evolution. On the other hand, once the temperature of the Universe becomes smaller than the critical temperature, the bosons making up the self-interacting dark matter may condense to form a Bose–Einstein condensate.

The investigation of the irreversible particle processes in homogeneous and isotropic flat FLRW cosmological models in theories with curvature-matter coupling opens the possibility of testing their viability via astrophysical and cosmological observations. However, in order to confirm the validity of the thermodynamic interpretation, it is necessary to consider a much wider range of tests of the $f(R, L_m)$ and $f(R,T)$-type theories. In particular, an essential confirmation of these theories would be through the analysis of their predictions for the evolution of the linear density perturbations investigated, for example, in the Newtonian limit of small scales. Secondly, the fitting of the observations of the CMB anisotropies with the inclusion of the effects of particle would provide another strong constraint on the theory. However, the simple toy models addressed in the present chapter may also offer an alternative approach that could provide some insights in the physical interpretation of the theories with curvature-matter coupling, and for observationally testing alternative gravity theories.

13

Quantum Cosmology of $f(R, T)$ Gravity

13.1 Introduction

General Relativity is perhaps the most fundamental and beautiful, area of theoretical physics. It is expressed mathematically in terms of the Einstein field equations, and it established itself beyond doubt as a very productive gravitational theory at the scale of the solar system. With the help of GR we can understand and describe the gravitational dynamics and phenomena within the solar system with a high degree of precision. Intriguing observations and physical processes, like the perihelion precession of the planet Mercury, the bending of light while passing near the Sun, and the gravitational redshift of the light coming from massive stars, can be fully understood. The gravitational field equations can be derived from a variational principle through the variation of the Einstein–Hilbert action (we refer the reader to Chapter 3 for details). Fundamentally GR is a stunning geometric theory that substantiates the existence of a profound relation between matter fields, the geometry of the space-time, and the gravitational interaction.

On a larger astrophysical and cosmological scales of the Universe, we can numerically simulate, using GR, the formation and collisions of the galaxies. The results of these simulations can be checked by the large number of high precision astronomical data obtained due to the rapidly and significantly improving observational astronomical techniques and methods. The theoretical predictions of GR have been successful not only in the weak gravity approximation, but they have also been fully confirmed in the strong gravity regime by the observations of the binary pulsars system PSR 1913 + 16 that lead to the discovery of the gravitational wave emission in this system [262]. This discovery has opened a new window for testing GR, its modifications, and its extensions. A new era in observational astronomy began with the detection of the gravitational event GW150914 by LIGO, resulting from the inspiral and merger of a pair of black holes [2], an event which can be interpreted and understood within the physical and mathematical formalism of GR.

Thus, being tested and confirmed by a large number of astrophysical, astronomical, and cosmological observations, as well as terrestrial experiments in both weak and intense gravity regimes, it would seem that GR may give a complete description of the gravitational interaction at least at the classical level. However, a number of recent cosmological and astrophysical observations have raised the provocative prospect that GR may not be able to explain and mathematically model the gravitational mechanics and kinetics at cosmic scales much larger than the solar system [513].

Besides the major theoretical difficulties posed by the present day cosmological observations, a primal problem in physics is that of the *unification of quantum mechanics and gravitation.* The gravitational force dictates the motion of celestial objects at large astronomical scales, while quantum mechanics depicts the microscopic conduct of particles. The investigation of the Universe as a whole by using the methods and the theoretical framework of quantum mechanics is the topic of *quantum cosmology* [80, 81]. Quantum cosmology is based on the fundamental idea that *quantum mechanics must be able to describe all microscopic physical processes, including the Universe itself.* In the standard model of particle physics the unification of the weak force, the electromagnetic force, and the strong force is achieved in a very successful way, by using the mathematical and physical formalism of quantum field theory. The standard model of particle physics leaves the gravitational force as an interesting exception that seems to be impossible to unify with the other fundamental forces. This is a direct consequence of the fact that when trying to quantize gravitation, we must include in the quantization procedure not only ordinary matter, but also space and time, as physical objects. In GR space and time abide to some dynamical laws, and they have their own excitations, such as the gravitational waves that interact with each other. These physical aspects makes the quantization of the gravitational field and of the Universe understood as a quantum system extremely nontrivial. On the other hand the formation of cosmic structures is governed by the force of gravity, which is determined by the structure of the space-time interaction. Hence quantum cosmology, dealing with the description of the observed properties of the Universe, is closely correlated to quantum gravity, whose object of study is the quantum theory of space-time and of the gravitational force [369].

Quantum cosmology has a long history [167, 168, 169], during which it has tried to overcome several difficult conceptual problems. But it still remains today a speculative and controversial subject. Various competing attempts have been proposed to quantize the gravitational field, for instance canonical quantum gravity, loop quantum gravity, and string theory [80, 81]. However, the shortage of observations related to the quantum nature of gravity reduces significantly our capacity to resolve conceptual issues by comparing them with physical phenomenology. Since, according to the standard big bang model, the average radius of each particle, as well as the radius of the Universe itself, are infinitely small at their birth, while the curvature of the space-time is infinitely large, at the early stages of its evolution quantum gravitational effects control the full dynamics of the Universe. and therefore quantum effects cannot be ignored in the study of the very early Universe. One of the main hurdles in the implementation of the quantum cosmological program is the so-called *problem of time*. The latter arises from the imposition of the Hamiltonian constraint in the Arnowitz–Deser–Misner (ADM) formalism, with the use of the fundamental equation in canonical quantization of cosmology, namely, the Wheeler–DeWitt equation [167, 168, 169]. The investigation of the quantum cosmology of $f(R)$ gravity with the matter content of the Universe modeled as a Schutz's fluid was discussed in [485, 486], with new insights to the problem of time in quantum gravity.

The present chapter presents a brief introduction to the study of the quantum cosmology of $f(R,T)$ gravity, and of some of its physical and theoretical implications. As a first step in our study, and in order to familiarize the reader with the canonical quantization formalism for $f(R,T)$ gravity, we derive the general form of the gravitational Hamiltonian of the theory, of the corresponding quantum potential, and of the canonical momenta, respectively. Once these physical quantities are explicitly obtained, we introduce the full Wheeler–DeWitt equation of the $f(R,T)$ modified gravity theory, which describes the quantum properties of the very early Universe in the framework of this theory. The Wheeler–DeWitt equation can provide a satisfactory description of the Universe during the period when quantum effects had a dominant influence on its dynamic evolution. Then we propose and study in detail the quantum cosmological implications and properties of a particular implementation of the $f(R,T)$ theory, namely, the quantum cosmology of the model in which $f(R,T) = F^0(R) + \theta RT$, where $F^0(R)$ is an arbitrary function of the Ricci scalar, and θ is a function of the scale factor of the Universe only. For this model we obtain first the Hamiltonian form of the classical equations of motion, and then we write down the Wheeler–DeWitt equation, which gives the unfolding of the wave function of the very early Universe immediately after its birth. By appropriately interpreting from a physical point of view the Wheeler–DeWitt equation, we propose a time parameter for the $f(R,T)$ gravity quantum dynamical system, which allows us to introduce for the quantum-mechanical system under consideration a Schrödinger–Wheeler–DeWitt-type equation for the description of its specific attributes. In order to solve the Schrödinger–Wheeler–DeWitt-type equation of the quantum cosmology of $f(R,T)$ gravity, describing the properties of the early Universe, we come forth with a perturbative approach for the study of this quantum cosmological equation. The energy levels of the Universe are obtained perturbatively by using a standard twofold degenerate perturbation method. Finally, we study the problem of the quantum time in $f(R,T)$ gravity by introducing a second quantization approach of time.

13.2 The Wheeler–DeWitt Equation in $f(R,T)$ Gravity

To develop the quantum cosmology of $f(R,T)$ gravity we need to define first the classical parameters of the model. Thus, for the description of the classical background line element we adopt the FLRW metric, corresponding to a homogeneous and isotropic geometry. Moreover, we suppose that the matter content of the very early Universe can be modeled with the help of a perfect cosmological fluid, whose properties can be described by using two basic thermodynamic parameters only, taken to be the energy density, and the thermodynamic pressure, respectively. As a first step in our analysis we develop the Hamiltonian formulation of $f(R,T)$ gravity, which helps us to write down the Wheeler–DeWitt (WDW) equation, and gives the description of the evolution of the quantum Universe in its very early stages, in the presence of the curvature-matter coupling.

As we have already seen, the classical field equations of $f(R,T)$ gravity are given in an arbitrary geometry by Eq. (10.3), and for convenience, we reproduce here

$$\begin{aligned} f_R(R,T)R_{\mu\nu} - \frac{1}{2}f(R,T)g_{\mu\nu} + (g_{\mu\nu}\Box - \nabla_\mu\nabla_\nu)f_R(R,T) \\ = -8\pi T_{\mu\nu} - f_T(R,T)T_{\mu\nu} - f_T(R,T)\Theta_{\mu\nu}. \end{aligned} \tag{13.1}$$

Therefore, by taking the trace of the field equation Eq. (13.1), we immediately obtain

$$f_R(R,T)R - 2f(R,T) + 3\Box f_R(R,T) = \frac{1}{2}T + f_T(R,T)T - 4pf_T(R,T). \tag{13.2}$$

13.2.1 The Effective Cosmological Lagrangian and the Potential

We describe the geometry of the classical space-time, assumed to be homogeneous and isotropic, by the FLRW metric, given in spherical coordinates by

$$ds^2 = -N^2(t)dt^2 + a^2(t)\left[\frac{dr^2}{1-kr^2} + r^2(d\theta^2 + \sin^2\theta d\varphi^2)\right], \tag{13.3}$$

where $N(t)$ is called the lapse function, $a(t)$ is the cosmological scale factor, and the constant k, taking the values $k = 1, 0, -1$, represents the closed, flat, and open geometric models of the Universe, respectively. The components of the Ricci tensor for the FLRW metric (13.3) can be obtained by simple calculations as

$$R_{00} = -3\frac{\ddot{a}}{a} + 3\frac{\dot{N}\dot{a}}{Na}, \tag{13.4}$$

$$R_{ii} = \frac{g_{ii}}{N^2}\left[\frac{\ddot{a}}{a} + 2\left(\frac{\dot{a}}{a}\right)^2 + \frac{2kN^2}{a^2} - \frac{\dot{N}\dot{a}}{Na}\right], \qquad i = 1, 2, 3. \tag{13.5}$$

Hence, for the expression of the Ricci scalar we find

$$R = R^{\mu}_{\mu} = \frac{6}{N^2}\left[\frac{\ddot{a}}{a} + \left(\frac{\dot{a}}{a}\right)^2 + \frac{kN^2}{a^2} - \frac{\dot{N}\dot{a}}{Na}\right]. \tag{13.6}$$

Since we assume that the matter content of the Universe consists of a perfect fluid, in the comoving frame, in which the components of the four-velocity are $U_\mu = (N(t), 0, 0, 0)$ and $U^\mu = (-1/N(t), 0, 0, 0)$, respectively, the trace of the energy-momentum tensor is given by $T = -\rho + 3p$.

To further advance in our analysis, we derive now the expression of the effective Lagrangian for the $f(R,T)$ theory. Varying the Lagrangian with respect to its dynamical variables gives the equations of motion of our system.

By taking into account the explicit expressions of the components of the Ricci tensor $R_{\mu\nu}$, and with the use of the identity (13.2), the cosmological action for $f(R,T)$ gravity is given by

$$S_{grav} = \int dt \Bigg\{ Na^3 f(R,T) - \lambda \left[R - \frac{6}{N}\left(\frac{\ddot{a}}{a} + \left(\frac{\dot{a}}{a}\right)^2 + \frac{kN^2}{a^2} - \frac{\dot{N}\dot{a}}{Na}\right)\right]$$
$$-\mu\left[\frac{1}{2}T + f_T(R,T)T - f_R(R,T)R - 3\Box f_R(R,T) + 2f(R,T) - 4pf_T(R,T)\right]\Bigg\}. \tag{13.7}$$

In Eq. (13.7) we have introduced the two parameters λ and μ as *Lagrange multipliers*. The term containing the second Lagrange multiplier is chosen as the contracted field equation, since in this way it can be derived directly from the gravitational action, and no further assumptions are needed for it. Moreover, if we use other possible representations for the action, such as adopting for the term multiplying the second Lagrange multiplier μ the expression $T + \rho - 3p$, important information about the geometry of the modified gravity will be lost. Additionally, more conceptual problems will appear when the energy density of the matter ρ and its pressure p are related by the radiation equation of state $p = \rho/3$.

By varying Eq. (13.7) with respect to R and T, we obtain the expressions of the Lagrange multipliers λ and μ as given by

$$\frac{\mu}{Na^3} = \frac{f_T}{1/2 + 3f_T + f_{TT}T - f_{RT}R - 3\Box f_{RT} - 4pf_{TT}} \equiv \widetilde{\mu}, \tag{13.8}$$

$$\frac{\lambda}{Na^3} = f_R - \frac{\mu(f_{RT}T - f_{RR}R + f_R - 3\Box f_{RR} - 4pf_{RT})}{Na^3} \equiv \widetilde{\lambda}, \tag{13.9}$$

Therefore the gravitational part of the Lagrangian is obtained as

$$\mathcal{L}_{grav} = -\frac{6}{N}a\dot{a}^2\widetilde{\lambda} - \frac{6}{N}a^2\dot{a}\dot{\widetilde{\lambda}} + 6kNa\widetilde{\lambda} - Na^3V, \tag{13.10}$$

where the potential V is defined by the relation

$$V = -f(R,T) + \widetilde{\lambda} R + \widetilde{\mu}\left[\frac{1}{2}T + f_T(R,T)T - f_R(R,T)R \right.$$
$$\left. -3\Box f_R(R,T) + 2f(R,T) - 4pf_T(R,T)\right]. \quad (13.11)$$

In order to simplify the mathematical formalism of the problem we introduce the following notations

$$f_R = A, \qquad f_T = B, \qquad f_{RR} = C, \qquad f_{RT} = D,$$
$$f_{TT} = E, \qquad \Box f_{RR} = F, \qquad \Box f_{RT} = G, \qquad T - 4p = M.$$

By using the newly introduced variables $\widetilde{\lambda}$ can be expressed as

$$\widetilde{\lambda} = A - \frac{B(DM - CR + A - 3F)}{1/2 + 3B + EM - DR - 3G}. \quad (13.12)$$

We introduce now two new quantities $\mathcal{A}$ and $\mathcal{Z}$ defined as

$$\mathcal{A} = \frac{1}{2} + 3B + EM - DR - 3G, \qquad \mathcal{Z} = DM - CR + A - 3F. \quad (13.13)$$

Thus, for $\widetilde{\lambda}$ we obtain the simple representation

$$\widetilde{\lambda} = A - \frac{B\mathcal{Z}}{\mathcal{A}}. \quad (13.14)$$

By taking the derivative with respect to the cosmological time of the above expression we find

$$\dot{\widetilde{\lambda}} = \dot{A} - \frac{B\dot{\mathcal{Z}} + \dot{B}\mathcal{Z}}{\mathcal{A}} + \frac{B\mathcal{Z}\dot{\mathcal{A}}}{\mathcal{A}^2}. \quad (13.15)$$

13.2.2 The Cosmological Hamiltonian

By definition the canonical momentum P_q associated to the coordinate q is obtained as $P_q = \partial\mathcal{L}/\partial\dot{q}$. Hence, by taking into account the definition of the momenta associated to the canonical variables, the cosmological Hamiltonian of $f(R,T)$ gravity is found as

$$H_{grav} = \dot{a}P_a + \dot{A}P_A + \dot{B}P_B + \dot{C}P_C + \dot{D}P_D + \dot{E}P_E$$
$$+\dot{F}P_F + \dot{G}P_G + \dot{R}P_R + \dot{M}P_M + -\mathcal{L}_{grav} \quad (13.16)$$

The explicit forms of the canonical momenta associated to the cosmological action Eq. (13.7) of $f(R,T)$ gravity can be found, after some simple calculations, as being given by

$$P_a = -2\frac{6}{N}a\dot{a}\widetilde{\lambda} - \frac{6}{N}a^2\dot{\widetilde{\lambda}}, \qquad P_{\mathcal{A}} = -\frac{6}{N}a^2\dot{a}\left(1 - \frac{\mathcal{B}}{\mathcal{A}}\right), \tag{13.17}$$

$$P_{\mathcal{B}} = -\frac{6}{N}a^2\dot{a}\left(-\frac{\mathcal{Z}}{\mathcal{A}} + \frac{3\mathcal{B}\mathcal{Z}}{\mathcal{A}^2}\right), \qquad P_{\mathcal{C}} = -\frac{6}{N}a^2\dot{a}\left(\frac{\mathcal{B}R}{\mathcal{A}}\right), \tag{13.18}$$

$$P_{\mathcal{D}} = -\frac{6}{N}a^2\dot{a}\left[-\frac{\mathcal{B}(T-4p)}{\mathcal{A}} - \frac{\mathcal{B}\mathcal{Z}R}{\mathcal{A}^2}\right], \tag{13.19}$$

$$P_{\mathcal{E}} = -\frac{6}{N}a^2\dot{a}\left[\frac{\mathcal{B}\mathcal{Z}(T-4p)}{\mathcal{A}^2}\right], \qquad P_{\mathcal{F}} = -\frac{6}{N}a^2\dot{a}\left(\frac{3\mathcal{B}}{\mathcal{A}}\right), \tag{13.20}$$

$$P_{\mathcal{G}} = -\frac{6}{N}a^2\dot{a}\left(-\frac{3\mathcal{B}\mathcal{Z}}{\mathcal{A}^2}\right), \qquad P_R = -\frac{6}{N}a^2\dot{a}\left[\frac{\mathcal{B}\mathcal{C}}{\mathcal{A}} - \frac{\mathcal{B}\mathcal{Z}\mathcal{D}}{\mathcal{A}^2}\right], \tag{13.21}$$

$$P_T = -\frac{6}{N}a^2\dot{a}\left(-\frac{\mathcal{B}\mathcal{D}}{\mathcal{A}} + \frac{\mathcal{B}\mathcal{Z}\mathcal{E}}{\mathcal{A}^2}\right), \tag{13.22}$$

$$P_p = -\frac{6}{N}a^2\dot{a}\left(\frac{4\mathcal{D}\mathcal{B}}{\mathcal{A}} - \frac{4\mathcal{B}\mathcal{Z}\mathcal{E}}{\mathcal{A}^2}\right). \tag{13.23}$$

With the use of the canonical momenta, we can construct, via the Legendre transformation, the cosmological Hamiltonian of $f(R,T)$ gravity as

$$H_{grav} = \left(-\frac{6}{N}a\dot{a}^2\widetilde{\lambda} - 6kNa\widetilde{\lambda} + Na^3V\right) - \frac{6}{N}a^2\dot{a}\Bigg[\dot{\mathcal{A}} - \frac{\mathcal{B}\dot{\mathcal{A}}}{\mathcal{A}} - \frac{\dot{\mathcal{B}}\mathcal{Z}}{\mathcal{A}} + \frac{3\mathcal{B}\dot{\mathcal{B}}\mathcal{Z}}{\mathcal{A}^2}$$
$$+\frac{\mathcal{B}R\dot{\mathcal{C}}}{\mathcal{A}} - \frac{\mathcal{B}\dot{\mathcal{D}}M}{\mathcal{A}} - \frac{\mathcal{B}\mathcal{Z}R\dot{\mathcal{D}}}{\mathcal{A}^2} + \frac{\mathcal{B}\mathcal{Z}\dot{\mathcal{E}}M}{\mathcal{A}^2} + \frac{3\mathcal{B}\dot{\mathcal{F}}}{\mathcal{A}} - \frac{3\mathcal{B}\mathcal{Z}\dot{\mathcal{G}}}{\mathcal{A}^2} + \frac{\mathcal{B}\mathcal{C}\dot{R}}{\mathcal{A}}$$
$$-\frac{\mathcal{B}\mathcal{Z}\mathcal{D}\dot{R}}{\mathcal{A}^2} - \frac{\mathcal{B}\mathcal{D}\dot{M}}{\mathcal{A}} + \frac{\mathcal{B}\mathcal{Z}\mathcal{E}\dot{M}}{\mathcal{A}^2}\Bigg]. \tag{13.24}$$

To simplify the notation we will represent the Hamiltonian in the form

$$H_{grav} = (\cdots) - \frac{6}{N}a^2\dot{a}[\cdots]. \tag{13.25}$$

From the definition of the canonical momenta we can now easily derive the relation

$$P_aP_{\mathcal{A}} = \left(\frac{6}{N}\right)^2 a^3\dot{a}^2\widetilde{\lambda}\left(1 - \frac{\mathcal{B}}{\mathcal{A}}\right) + \left(\frac{6}{N}\right)^2 a^4\dot{a}[\cdots] + \left(\frac{6}{N}\right)^2 a^4\dot{a}\left(-\frac{\mathcal{B}}{\mathcal{A}}\right)\dot{\widetilde{\lambda}}. \tag{13.26}$$

Since

$$P_aP_{\mathcal{F}} = \left(\frac{6}{N}\right)^2 2a^3\dot{a}^2\widetilde{\lambda}\left(3\frac{\mathcal{B}}{\mathcal{A}}\right) + \left(\frac{6}{N}\right)^2 a^4\dot{a}\dot{\widetilde{\lambda}}\left(\frac{3\mathcal{B}}{\mathcal{A}}\right), \tag{13.27}$$

the combination of the above two equations yields

$$-\frac{6}{N}a^2\dot{a}[\cdots] = -\frac{N}{6a^2}\left(P_a P_A + \frac{1}{3}P_a P_F\right) + \frac{2\cdot 6}{N}a\dot{a}^2\widetilde{\lambda}. \tag{13.28}$$

Therefore, the gravitational Hamiltonian can be written as

$$H_{grav} = \frac{6}{N}a\dot{a}^2\left(A - \frac{B\mathcal{Z}}{\mathcal{A}}\right) - \frac{N}{6a^2}\left(P_a P_A + \frac{1}{3}P_a P_F\right) + Na^3V - 6kNa\widetilde{\lambda}. \tag{13.29}$$

Since we have $P_A = -\frac{6}{N}a^2\dot{a}\left(1 - \frac{B}{\mathcal{A}}\right)$, after taking the square of it, we obtain

$$\begin{aligned}\frac{6}{N}a\dot{a}^2 A &= \left[\frac{P_A^2}{\left(\frac{6}{N}\right)^2 a^4\dot{a}^2} - \frac{B^2}{\mathcal{A}^2} + \frac{2B}{\mathcal{A}}\right]\cdot\frac{6}{N}a\dot{a}^2 A \\ &= \frac{NP_A^2}{6a^3}A + \frac{6}{N}a\dot{a}^2 A\left(\frac{-B^2}{\mathcal{A}^2} + \frac{2B}{\mathcal{A}}\right).\end{aligned} \tag{13.30}$$

Thus, for the Hamiltonian we find the expression

$$\begin{aligned}H_{grav} = \frac{N}{6a^3}P_A^2 A - \frac{N}{6a^2}(P_a P_A + \frac{1}{3}P_a P_F) - \frac{6}{N}a\dot{a}^2\frac{AB^2}{\mathcal{A}^2} \\ -\frac{6}{N}a\dot{a}^2\frac{B(\mathcal{Z} - 2A)}{\mathcal{A}} + Na^3V - 6kNa\widetilde{\lambda}.\end{aligned} \tag{13.31}$$

Considering now that

$$P_C P_R = \left(-\frac{6}{N}a^2\dot{a}\right)^2\left(\frac{B^2RC}{\mathcal{A}^2} - \frac{B^2\mathcal{Z}RD}{\mathcal{A}^3}\right), \tag{13.32}$$

$$P_D P_M = \left(-\frac{6}{N}a^2\dot{a}\right)^2\left(-\frac{BM}{\mathcal{A}} - \frac{B\mathcal{Z}R}{\mathcal{A}^2}\right)\left(-\frac{BD}{\mathcal{A}} + \frac{B\mathcal{Z}E}{\mathcal{A}^2}\right), \tag{13.33}$$

$$\frac{1}{3}P_E P_F = (-\frac{6}{N}a^2\dot{a})^2\frac{B^2\mathcal{Z}M}{\mathcal{A}^3}, \tag{13.34}$$

$$P_G^2 = \left(-\frac{6}{N}a^2\dot{a}\right)^2\frac{9B^2\mathcal{Z}^2}{\mathcal{A}^4}, \tag{13.35}$$

$$P_C\left(P_A + \frac{P_F}{3}\right) = \left(-\frac{6}{N}a^2\dot{a}\right)^2\left(\frac{BR}{\mathcal{A}}\right), \tag{13.36}$$

we can also obtain the relation

$$\begin{aligned}-\frac{6}{N}a\dot{a}^2\frac{B(\mathcal{Z} - 2A)}{\mathcal{A}} &= -\frac{6}{N}a\dot{a}^2\frac{B\left(DM - CR + A - 3F - 2A\right)}{1/2 + 3B + EM - DR - 3G} \\ &= \frac{N}{6a^3}\left(P_A + \frac{P_F}{3}\right)\frac{P_F}{3}(A + 3F) - \frac{N}{6a^3}\frac{\mathcal{A}}{B}(P_C P_R + P_D P_M) \\ &\quad + \frac{6a\dot{a}^2}{N}\frac{\mathcal{A}}{B}\frac{B\mathcal{Z}E}{\mathcal{A}^2}\left(-\frac{BM}{\mathcal{A}} - \frac{B\mathcal{Z}R}{\mathcal{A}^2}\right) - \frac{6a\dot{a}^2}{N}\left(-\frac{2BCR}{\mathcal{A}}\right).\end{aligned} \tag{13.37}$$

Therefore, the gravitational part of the cosmological Hamiltonian of $f(R,T)$ gravity can be written as

$$\begin{aligned} H_{grav} &= \frac{N}{6a^3}\left(P_A + \frac{P_F}{3}\right)(AP_A + FP_F - aP_a) \\ &+ \frac{N}{6a^3}\frac{\mathcal{A}}{B}\left(P_C P_R + P_D P_M + \frac{P_E P_F}{3}E + \frac{P_G^2}{9}RE\right) \\ &+ \frac{N}{6a^3}2P_C\left(P_A + \frac{1}{3}P_F\right)C + Na^3V - 6kNa\left(A - \frac{B\mathcal{Z}}{\mathcal{A}}\right), \end{aligned} \tag{13.38}$$

where we have introduced the quantum potential V of the cosmological system defined as

$$V = -f + \widetilde{\lambda}R + \widetilde{\mu}\left(\frac{1}{2}T + f_T M - f_R R - 3\Box f_R + 2f\right). \tag{13.39}$$

For the matter part of the Hamiltonian we have [93]

$$H_{matt} = -\mathcal{L}_{matt} = -Na^3 p. \tag{13.40}$$

Thus, the total Hamiltonian of the Universe in $f(R,T)$ gravity is

$$H = H_{grav} + H_{matt}. \tag{13.41}$$

The gravitational Hamiltonian H we have constructed is a very general one, since it contains all the canonical momenta associated to all variables of $f(R,T)$ gravity. This general representation can lead to the full description of the complex dynamics and evolution of all the gravitational field variables, as well as their associated canonical momenta.

13.2.3 The Wheeler–DeWitt Equation

The basic equation of quantum cosmology, the WDW equation, can be immediately obtained from the Hamiltonian (13.41) of $f(R,T)$ gravity. Hence, by applying the standard quantization procedure we obtain the evolution equation of the quantum universe as

$$H\Psi = (H_{grav} + H_{matt})\Psi = N\mathcal{H}\Psi = 0. \tag{13.42}$$

The Hamiltonian operator $\mathcal{H}$ for $f(R,T)$ gravity takes the form

$$\begin{aligned} \mathcal{H} &= \frac{1}{6a^3}\left(P_A^2 A + P_A P_F F - P_A P_a a + \frac{A}{3}P_F P_A + \frac{F}{3}P_F^2 - \frac{a}{3}P_F P_a\right) \\ &- \frac{1}{6a^3}\frac{\mathcal{A}}{B}\left(P_C P_R + P_D P_M + \frac{P_E P_F}{3}E + \frac{P_G^2}{9}RE\right) \\ &+ \frac{2}{6a^3}P_C\left(P_A + \frac{P_F}{3}\right)C + a^3V - 6ka\left(A - \frac{B\mathcal{Z}}{\mathcal{A}}\right) - a^3 p. \end{aligned} \tag{13.43}$$

To quantize the model we need to carry out first parameter ordering. Several ways to perform it have been proposed in the literature [475]. In the following we adopt a method that maintains the Hamiltonian Hermitian [485]. Hence we can obtain the following relationships (we perform the quantization via the substitution $P_q \to -i(\partial/\partial q)$),

$$qP_q^2 = \frac{1}{2}\left(q^u P_q q^v P_q q^w + q^w P_q q^v P_q q^u\right) = -q\frac{\partial^2}{\partial q^2} + uw\frac{1}{q}, \tag{13.44}$$

where the parameters u, v, w, denoting the ambiguity in the ordering of the factors q and P_q, satisfy the condition $u + v + w = 1$, and the relation

$$qP_q = \frac{1}{2}\left(q^r P_q q^s + q^s P_q q^r\right) = -i\left(q\frac{\partial}{\partial q} + 1\right), \tag{13.45}$$

where the parameters r, s denote the ambiguity in the ordering of factors q and P_q, and satisfy the condition $r + s = 1$. Similarly one can show that

$$q^{-2}P_q = -i\left(-\frac{2}{q^3} + \frac{1}{q^2}\frac{\partial}{\partial q}\right). \tag{13.46}$$

Therefore, the quantized cosmological Hamiltonian in $f(R,T)$ gravity can be obtained as

$$\begin{aligned}\mathcal{H} = &-\frac{1}{6a^3}\left[\left(A\frac{\partial}{\partial A} + F\frac{\partial}{\partial F} + 2C\frac{\partial}{\partial C} - a\frac{\partial}{\partial a} + 5\right)\left(\frac{\partial}{\partial A} + \frac{1}{3}\frac{\partial}{\partial F}\right) - u_1 w_1\frac{1}{A}\right.\\ &\left. -u_2 w_2\frac{1}{3F}\right] + \frac{1}{6a^3}\left[\frac{A}{B}\left(\frac{\partial}{\partial C}\frac{\partial}{\partial R} + \frac{\partial}{\partial D}\frac{\partial}{\partial M} + \frac{E}{3}\frac{\partial}{\partial E}\frac{\partial}{\partial F} - \frac{RE}{9}\frac{\partial^2}{\partial G^2}\right)\right.\\ &\left.+\left(-\frac{D}{B}\frac{\partial}{\partial C} - \frac{R}{B}\frac{\partial}{\partial M} + \frac{E}{B}\frac{\partial}{\partial D} + \frac{A + EM}{3B}\frac{\partial}{\partial F} + \frac{RE}{3BG}u_3 w_3\right)\right]\\ &+a^3 V - 6ka\left(A - \frac{B\mathcal{Z}}{\mathcal{A}}\right) - a^3 p. \end{aligned}\tag{13.47}$$

Here u_1, w_1, u_2, w_2, and u_3, w_3 denote the quantum mechanical ambiguity in the ordering of the factors A, P_A, F, P_F, and G, P_G, respectively.

In the next section, we will apply the general formalism developed so far to investigate some particular quantum cosmological models in $f(R,T)$ gravity.

13.3 Specific Model: $f(R,T) = F^0(R) + \theta RT$

In the previous section, we derived the general form of the WDW equation in $f(R,T)$ gravity. As one can see from the form of this equation, an analytic general solution of the WDW equation for an arbitrary $f(R,T)$ would be extremely difficult to find. Instead of concentrating on the investigation of the general equation, in the present section we consider a particular case of the $f(R,T)$ theory, in which the gravitational action takes the simple form

$$f(R,T) = F^0(R) + \theta RT, \tag{13.48}$$

where $F^0(R)$ is an arbitrary function of the Ricci scalar only, and θ is an arbitrary function depending on the cosmological scale factor $a(t)$ of the classical universe. In this toy model, the term in the gravitational action involving the coupling of the curvature of space-time with the trace of the matter energy-momentum tensor could give a hint of the implications of such a coupling on the quantum cosmological evolution of the Universe.

13.3.1 The Hamiltonian and the Wheeler–DeWitt Equation

For the model described by the Eq. (13.48) we can easily obtain

$$\mathcal{A} = \frac{1}{2} + 2\theta R, \qquad B = \theta R, \qquad \frac{B}{\mathcal{A}} = \frac{\theta R}{1/2 + 2\theta R}. \tag{13.49}$$

Similarly, the other variables of physical and cosmological interest become

$$A = F^0_R + \theta T, \qquad C = F^0_{RR}, \qquad D = \theta, \qquad E = 0, \qquad G = \Box\theta. \tag{13.50}$$

In the early quantum Universe after the initial big bang, the space-time has a very high curvature, so that the condition $R \to \infty$ holds. In the limit $R \to \infty$, we have $B/\mathcal{A} = 1/2$. By using the definitions of $P_{\mathcal{A}}$ and P_F, we find $P_F = [3B/(\mathcal{A} - B)]\, P_{\mathcal{A}} \approx 3P_{\mathcal{A}}$. By also assuming that in the newly born quantum Universe the conditions $f_R \gg \Box f_{RR} \to A \gg F$ hold, we find for the gravitational Hamiltonian of the $f(R,T) = F^0(R) + \theta RT$ gravity model the expression

$$\begin{aligned}
H_{grav} &= \frac{N}{6a^3}P_{\mathcal{A}}^2\mathcal{A}\frac{\mathcal{A}}{\mathcal{A}-B} - \frac{N}{6a^2}P_aP_{\mathcal{A}}\frac{\mathcal{A}}{\mathcal{A}-B} - \frac{N}{6a^3}\frac{\mathcal{A}}{B}\left(P_CP_R + P_DP_M\right) \\
&\quad + \frac{2N}{6a^3}\frac{\mathcal{A}}{\mathcal{A}-B}P_{\mathcal{A}}P_CC + Na^3V - 6kNa\left(A - \frac{BZ}{\mathcal{A}}\right) \\
&= -2\frac{N}{6a^2}P_aP_{\mathcal{A}} + 2\frac{N}{6a^3}P_{\mathcal{A}}^2\mathcal{A} - 2\frac{N}{6a^3}\left(P_CP_R + P_DP_M\right) \\
&\quad + \frac{4N}{6a^3}P_{\mathcal{A}}P_CC + Na^3V - 6kNa\left(A - \frac{BZ}{\mathcal{A}}\right),
\end{aligned} \tag{13.51}$$

where the quantum potential V is defined as

$$V = \left(\frac{2R\theta}{1/2 + 2R\theta} - 1\right)\left(F^0 - F^0_RR\right) + \frac{R\theta}{1/2 + 2R\theta}\left(\frac{1}{2}T + CR^2 + 3FR - 3\Box f_R\right). \tag{13.52}$$

In the limit $R \to \infty$, and by assuming that $\theta \neq 0$, the quantum potential takes the form

$$V = \frac{1}{2}\left(\frac{1}{2}T + CR^2 + 3FR - 3\Box f_R\right). \tag{13.53}$$

Therefore in the approximation of the large curvature R the total Hamiltonian of the early Universe becomes

$$\begin{aligned} H = H_{grav} + H_{matt} = & \left(-\frac{2N}{6a^2}P_a P_A + \frac{2N}{6a^3}P_A^2 A - 3kNaA\right) \\ & + \left(-\frac{2N}{6a^3}P_C P_R + \frac{Na^3}{2}CR^2 - 3kNaCR\right) \\ & + \left(-\frac{2N}{6a^3}P_D P_M + \frac{Na^3}{4}M + 3kNaDM\right) \\ & + \frac{4N}{6a^3}P_A P_C C + \frac{3Na^3}{2}\left(FR - \Box f_R\right). \end{aligned} \tag{13.54}$$

It is a general property of the Lagrangian/Hamiltonian systems that if we have some terms that can be ignored in the total action, they can also be ignored in L and H, without leading to any physical differences in the dynamical evolution of the given system. Due to Gauss's theorem, in Eq. (13.54) we have the relations

$$\int dt\sqrt{-g}\frac{3Na^3}{2}\Box f_R = \int_M d^4x\sqrt{-g}\Box f_R = \oint_{\partial M} f_R^{;\mu}\sqrt{-g}d\sigma_\mu^3. \tag{13.55}$$

Then we immediately see that the variational derivative of this term cancels out from the equation of motion,

$$\frac{2}{\sqrt{-g}}\frac{\delta}{\delta g^{\mu\nu}}\int \sqrt{-g}\Box f_R d^4x = 0, \tag{13.56}$$

and the term can be removed from the Hamiltonian function. Thus, we can now find the WDW equation for the $f(R,T) = F^0(R) + \theta RT$ gravitational model, which in this specific case has the form

$$\begin{aligned} \mathcal{H}\Psi = & \left[\left(\frac{2}{6a^2}\frac{\partial^2}{\partial a\partial A} - \frac{4}{6a^3}C\frac{\partial}{\partial A}\frac{\partial}{\partial C} - \frac{8}{6a^3}\frac{\partial}{\partial A} - \frac{2}{6a^3}A\frac{\partial^2}{\partial A^2} + \frac{2}{6}u_1 w_1\frac{1}{Aa^3} - 3kaA\right)\right. \\ & + \left(\frac{2}{6a^3}\frac{\partial}{\partial C}\frac{\partial}{\partial R} + \frac{a^3}{2}CR^2 - 3kaCR\right) \\ & \left. + \left(\frac{2}{6a^3}\frac{\partial^2}{\partial D\partial M} + \frac{a^3}{4}M + 3kaDM\right) + \frac{3a^3}{2}FR\right]\Psi = 0. \end{aligned} \tag{13.57}$$

13.3.2 The Hamiltonian Form of the Field Equations

It is a well-known result of classical mechanics that the total time derivative of any function of the canonical variables can be calculated with the use of the Poisson bracket $\{,\}$ as

$$\frac{d}{dt}f = \frac{\partial f}{\partial t} + \{f, H\}. \tag{13.58}$$

If the physical variables of the system do not depend explicitly on the time t, the Poisson brackets simplifies to

$$\frac{d}{dt}f = \{f, H\}. \tag{13.59}$$

Therefore, since the Hamiltonian of $f(R,T)$ gravity has already been found, we can formulate the classical equations of motion of this modified gravity theory as

$$\dot{a} = \{a, H\} = -2\frac{N}{6a^3}P_A, \tag{13.60}$$

$$\begin{aligned}\dot{P}_a = \{P_a, H\} = N\Bigg\{ &\left(-\frac{2}{3a^3}P_A P_a + \frac{1}{a^4}P_A^2 A + 3kA\right) \\ &+\left(-\frac{1}{a^4}P_C P_R - \frac{3a^2}{2}CR^2 - 3kCR\right) + \left[-\frac{1}{a^4}P_D P_M \right. \\ &\left. -\frac{3a^2}{4}M + 3kDM\right] + \frac{12}{6a^4}P_A P_C C\Bigg\},\end{aligned} \tag{13.61}$$

$$\dot{A} = \{A, H\} = N\left[-\frac{1}{3a^2}P_a + \frac{2}{3a^3}AP_A + \frac{4}{6a^3}P_C C\right], \tag{13.62}$$

$$\dot{P}_A = \{P_A, H\} = N\left[-\frac{1}{3a^3}P_A^2 + 3ka\right], \tag{13.63}$$

$$\dot{C} = \{C, H\} = -\frac{N}{3a^3}\left[(P_R - 2P_A C\right], \tag{13.64}$$

$$\dot{P}_C = \{P_C, H\} = N\left[-\frac{a^3}{2}R^2 + 3kaR - \frac{4}{6a^3}P_A P_C\right], \tag{13.65}$$

$$\dot{R} = \{R, H\} = -\frac{N}{3a^3}P_C, \tag{13.66}$$

$$\dot{P}_R = \{P_R, H\} = N\left[-a^3 CR + 3kaC - \frac{3a^3}{2}F\right], \tag{13.67}$$

$$\dot{D} = \{D, H\} = -2\frac{N}{6a^3}P_M, \tag{13.68}$$

$$\dot{P}_D = -3kNaM, \tag{13.69}$$

$$\dot{M} = -2\frac{N}{6a^3}P_D, \tag{13.70}$$

$$\dot{P_M} = -N\left(\frac{a^3}{4} + 3kaD\right). \tag{13.71}$$

Now, we introduce a new time variable τ, which is defined with the help of the original time variable t by means of the relation

$$\tau = \int N(t)dt, \tag{13.72}$$

or, in an equivalent formulation, as $d\tau/dt = N(t)$. On the other hand, by using the definition of P_M, we have

$$P_M = \frac{3a^2\dot{a}}{N}D = 3a^3hD, \tag{13.73}$$

where

$$h \equiv \frac{1}{N}\frac{\dot{a}}{a}, \tag{13.74}$$

denotes the Hubble function associated to the cosmological model. Then in the new time variable τ the Hamilton equations of motion for the cosmological fluid, in the $f(R,T)$-gravity-dominated universe, become

$$D' = -h\theta = -\frac{a'}{a}D, \qquad P_D' = -3kaM, \tag{13.75}$$

$$M' = -2\frac{1}{6a^3}P_D, \qquad P_M' = \frac{a^3}{4} + 3kaD, \tag{13.76}$$

where a prime denotes the derivative with respect to the modified time variable τ. From Eq. (13.75) we obtain the coupling constant $\theta = D$ of our toy gravitational model as

$$D = \frac{\delta}{a(\tau)}, \qquad \delta = \text{constant}. \tag{13.77}$$

This result shows us that the nonminimal coupling between the gravitational field and the matter field decreases as the scale factor of the Universe increases. This may be the physical reason why the nonminimal coupling between gravity and matter becomes so weak in the limit of large cosmological times, and that presently $f(R,T)$ gravity behaves as the standard general relativistic gravity.

13.3.3 The problem of Time

Our attempts of understanding quantum cosmology in a way similar to standard quantum mechanics, or quantum field theory, are seriously hampered by the lack of the time evolution of the wave function of the Universe in the WDW equation. A possible way to transform the WDW equation into a Schrödinger-type equation can be formulated within the framework of $f(R,T)$ gravity, and such a transformation may be made possible by the presence of the curvature-matter coupling. To derive such a Schrödinger-type equation we note first that the product $P_D P_T$ can be written as

$$-\frac{2}{6a^3}P_D P_M = -hDP_D. \tag{13.78}$$

In the following, we introduce the fundamental hypothesis that this term can be expounded as P_τ, that is, *the canonical momentum for time.* This assumption can be justified in a convincing way, since after performing the standard quantization procedure $P_q = -i\frac{\partial}{\partial q}$, we obtain

$$P_\tau = -i\frac{d}{d\tau} = \frac{da(\tau)}{d\tau}\left(-i\frac{\partial}{\partial a}\right) = -\frac{a'}{a}\frac{\delta}{a}\left(-\frac{a^2}{\delta}\right)\left(-i\frac{\partial}{\partial a}\right) = -hDP_D. \tag{13.79}$$

This result is obviously true if we accept that in the quantum cosmological model the scale factor $a(t)$ is the only time dependent variable. The relation introduced in Eq. (13.79) provides us with the possibility to introduce the further transformation

$$-\frac{2}{6a^3}P_D P_M \to P_\tau. \tag{13.80}$$

The possibility of carrying out such a transformation indicates that the coupling between the gravitational field and the matter field may play an important role in the evolution of the very early Universe. We will accomplish now a further simplification of the gravitational action, by assuming it is of the form

$$f(R,T) = R + \theta RT, \tag{13.81}$$

and whose main properties we have already discussed. Then we have $C = F = P_R = 0$, while the cosmological Hamiltonian of the gravitational system is given by

$$\begin{aligned} H = H_{grav} + H_{matt} &= -\frac{2N}{6a^2}P_a P_A + \frac{2N}{6a^3}P_A^2 A - 3kNaA \\ &-\frac{N}{6a^3}P_D P_M + \frac{Na^3}{4}M + 3kNaM. \end{aligned} \tag{13.82}$$

This Hamiltonian is very similar in its structure to the Hamiltonian obtained in the framework of $f(R)$ gravity, and discussed in [486], except for the presence of a new term $-2\frac{N}{6a^3}P_D P_T + N(\frac{a^3}{4} + 3ka)M$. This term shows us the effect of the coupling between curvature and matter. Then the WDW equation $\mathcal{H}\Psi = 0$ for this $f(R,T)$ gravity model can be written down as

$$\mathcal{H}\Psi = \left[-\frac{2}{6a^2}P_a P_A + \frac{2}{6a^3}P_A^2 A - 3kaA - \frac{2}{6a^3}P_D P_M + \frac{a^3}{4}M + 3kaDM\right]\Psi = 0. \tag{13.83}$$

With the help of the transformation defined in Eq. (13.80), we immediately find

$$\mathcal{H}_{eff}\Psi = \left[-\frac{2}{6a^2}P_aP_A+\frac{2}{6a^3}P_A^2A-3kaA+\frac{a^3}{4}M+3kaDM\right]\Psi = -P_\tau\Psi. \quad (13.84)$$

To quantize the model, we substitute $P_q = -i\partial/\partial q$ in the above Hamiltonian. Thus, we obtain the Schrödinger–Wheeler–DeWitt (SWDW) equation, describing the quantum evolution of the very early Universe in the specific $f(R,T)$ gravity model (13.81) as

$$\mathcal{H}_{eff}\Psi = \left[\left(\frac{2}{6a^2}\frac{\partial^2}{\partial a\partial A} - \frac{4}{6a^3}\frac{\partial}{\partial A} - \frac{2}{6a^3}A\frac{\partial^2}{\partial A^2} + \frac{2}{6}u_1w_1\frac{1}{Aa^3} - 3kaA\right)\right.$$
$$\left. +\frac{a^3}{4}M + 3kaDM\right]\Psi = i\frac{\partial\Psi}{\partial\tau}. \quad (13.85)$$

This quantum evolution equation has the form of the standard quantum-mechanical Schrödinger equation,

$$\mathcal{H}_{eff}\Psi = i\frac{\partial\Psi}{\partial\tau}. \quad (13.86)$$

Therefore, we have obtained the important result that in the framework of $f(R,T)$ gravity, *we can generate a Schrödinger-type quantum mechanical evolution equation from the WDW equation*, which can find an answer to the challenge of time in quantum gravity. In the limiting case $d\tau/dt = N(t) = 1$, the SWDW equation takes the standard form of the Schrödinger equation with which we are acquainted, namely,

$$\mathcal{H}\Psi = i\frac{\partial\Psi}{\partial t}. \quad (13.87)$$

The Physical Meaning of the Effective Hamiltonian and of the SWDW Equation

Let us analyze now in more detail the physical meaning of the time τ, and of the effective Hamiltonian H_{eff} we have just introduced. This would also allow us to take a more profound sight into the quantum gravitational time problem. In the standard approach to cosmology described by the Wheeler–DeWitt equation $\mathcal{H}\psi = 0$ there seems to be no time evolution, or dynamics, of the quantum-dominated early Universe. Therefore it follows that the wave function of the Universe (or more specifically, the corresponding physical states) does not characterize quantum gravitational states of the Universe at a particular time, as it happens in standard quantum mechanics. Rather, the solution of the WDW equation describes states for all times or, in a more precise formulation, the wave function provides only information about the state of the very early quantum Universe that is invariant with respect to all space-time diffeomorphism transformations [30].

However, in the modified gravity model $f(R,T) = R+\theta RT$, the concept of time can be introduced locally by taking into account the coupling of the geometry and of the matter field. This is possible since the interaction between gravity (geometry) and matter is also local. The deep relation between thermodynamics and gravity tells us that the arrow of time may be generated via the second law of thermodynamics, since both processes are closely related to some forms of irreversible evolution. If we restrict ourselves to the consideration of a small volume of the Universe, and interpret it as an adiabatic system, with the WDW equation $\mathcal{H}\Psi = 0$ still holding in it, the coupling between curvature and matter will generate an arrow of time to measure the increase of the entropy of this local volume of the Universe, as determined by its matter content. The other components of the Hamiltonian function, included in the effective Hamiltonian $\mathcal{H}_{eff}$, determine the dynamics of the gravitational field, and of its normal matter components. Consequently, the effective Hamiltonian can be written in the general form

$$\mathcal{H}_{eff} \sim b_1 p_i p_j + b_2 x_i x_j \,, \tag{13.88}$$

which is similar to the Hamiltonian form we usually deal with in ordinary quantum mechanics. Therefore, we can propose that the standard quantum mechanical Schrödinger equation just represents an effective, locally valid version of the SWDW equation. In other words, we may propose the conjecture according to which the WDW equation in the presence of a curvature-matter coupling supplies the global quantum description for the Universe, while the standard Schrödinger equation, obtained in the limit of the vanishing curvature-matter intermixing, provides just the local quantum description of the microscopic regions of the present Universe.

13.4 A Perturbative Approach to the Cosmological SWDW Equation

In the following, we look for stationary solutions of the SWDW equation (13.85) giving the wave function of the Universe and, consequently, we separate the variables as

$$\Psi(a,A,D,M,\tau) = e^{-iE\tau}\psi(a,A,D,M), \tag{13.89}$$

where E = constant. Thus, from the description of the time evolution of the wave function of the Universe we obtain the following differential equation

$$\mathcal{H}\psi = \left[\left(\frac{2}{6a^2}\frac{\partial^2}{\partial a \partial A} - \frac{4}{6a^3}\frac{\partial}{\partial A} - \frac{2}{6a^3}A\frac{\partial^2}{\partial A^2} + \frac{2}{6}u_1 w_1 \frac{1}{Aa^3} - 3kaA\right)\right.$$
$$\left. + \frac{a^3}{4}M + 3kaDM - E\right]\psi = 0. \tag{13.90}$$

With the help of the new variables $x = aA^{1/2}$ and $y = A$, respectively, we can rewrite the differential equation for ψ as

$$\left[\frac{1}{4}x^2\frac{\partial^2}{\partial x^2} - \frac{1}{4}x\frac{\partial}{\partial x} - 2y\frac{\partial}{\partial y} - y^2\frac{\partial^2}{\partial y^2} + u_1w_1 - 9kx^4 \right.$$
$$\left. +3M\left(\frac{a^6}{4} + 3ka^4D\right)y - 3x^3y^{-\frac{1}{2}}E\right]\psi = 0. \tag{13.91}$$

In the following, the stiff equation of state $p = \rho$ will approximate that of the early universe. Hence we assume that the Universe was born in a state of very high density, in which we expect that the speed of sound c_s is of the same order of magnitude as the speed of light, $c_s = \sqrt{\partial\rho/\partial p} = 1$. By taking $p = \rho$, and because

$$\dot{R} = -\frac{N}{3a^3}P_C = 2\frac{\dot{a}}{a}\frac{1}{2}R = \frac{\dot{a}}{a}R, \tag{13.92}$$

we obtain the result that in this case the matter energy is conserved, $\rho' + 3(\rho + p)h = 0$. Hence we find

$$M = T - 4p = -\frac{w}{a^6}, \tag{13.93}$$

where $w > 0$ is necessarily a positive constant. By substituting the expression of M into Eq. (13.91), and by taking into account that $D = \delta/a(\tau)$, we obtain finally

$$\left[\frac{1}{4}x^2\frac{\partial^2}{\partial x^2} - \frac{1}{4}x\frac{\partial}{\partial x} - 2y\frac{\partial}{\partial y} - y^2\frac{\partial^2}{\partial y^2} + u_1w_1 - 9kx^4 - \frac{3w}{4}y \right.$$
$$\left. -9w\delta k\frac{y^{\frac{5}{2}}}{x^3} - 3x^3y^{-\frac{1}{2}}E\right]\psi = 0. \tag{13.94}$$

13.4.1 Time Evolution as a Nonconstant Energy Perturbation

To obtain the solution of the most recent SWDW equation (13.94), we begin by estimating the order of magnitude of the different terms appearing in the equation. We already know that $x = aA^{1/2} \sim 1/a^{5/2}$, $y = A \sim 1/a^7$, then we can immediately obtain

$$x^4 \sim 1/a^{10}, \qquad y^{\frac{5}{2}}/x^3 \sim 1/a^8, \qquad x^3/y^{\frac{1}{2}} \sim 1/a^4. \tag{13.95}$$

Since in the quantum gravity dominated era a is very small, the term $x^3/y^{1/2} \sim 1/a^4$ can be interpreted as a small perturbation. In order to obtain some analytic solutions of Eq. (13.94), we consider the case when $k = 0$, corresponding to a flat homogeneous and isotropic Universe. This choice can also help us to eliminate the large coupling term of the form $y^{5/2}/x^3 \sim 1/a^8$. Following the analysis of Hawking and Page [251], we also assume that the ordering parameter u_1w_1 can be neglected. In the present case this can be realized by rearranging it into the energy term or, since it is very small as compared to the variables related to $a(\tau)$, we can neglect it directly in the equation.

Hence, with the use of the approximations and assumptions earlier in the section, the Schrödinger-Wheeler-de Witt equation becomes

$$\frac{y^{\frac{1}{2}}}{3x^3}\left[\frac{1}{4}x^2\frac{\partial^2}{\partial x^2}-\frac{1}{4}x\frac{\partial}{\partial x}-2y\frac{\partial}{\partial y}-y^2\frac{\partial^2}{\partial y^2}-\frac{3w}{4}y\right]\Psi=i\frac{\partial}{\partial\tau}\Psi, \tag{13.96}$$

where $\Psi=\Psi(x,y,\tau)$. We look again for stationary solutions, and therefore we can represent the variables as $\Psi(x,y,\tau)=e^{-iE\tau}\psi(x,y)$. Thus, we obtain

$$\frac{y^{\frac{1}{2}}}{3x^3}\left[\frac{1}{4}x^2\frac{\partial^2}{\partial x^2}-\frac{1}{4}x\frac{\partial}{\partial x}-2y\frac{\partial}{\partial y}-y^2\frac{\partial^2}{\partial y^2}-\frac{3w}{4}y\right]\psi=E\psi. \tag{13.97}$$

Now we modify the above equation so that it takes an equivalent form. For this we multiply by $3x^3/y^{\frac{1}{2}}$ both sides of the equation, and obtain the equation

$$\left[\frac{1}{4}x^2\frac{\partial^2}{\partial x^2}-\frac{1}{4}x\frac{\partial}{\partial x}-2y\frac{\partial}{\partial y}-y^2\frac{\partial^2}{\partial y^2}-\frac{3w}{4}y-\frac{3Ex^3}{y^{\frac{1}{2}}}\right]\psi=0. \tag{13.98}$$

In the formulation of the SWDW equation as given by Eq. (13.97), we can interpret the time evolution of the wave function as resulting from the addition of a constant E in the equation, with E representing a small perturbation. But in the perturbation theory of standard quantum mechanics we have a result according to which a constant perturbation does not produce any change in the energy level and in the wave function of the quantum system. This means that a constant perturbation does not affect the physical observables. Now after rewriting the unperturbed Hamiltonian as follows,

$$\frac{y^{\frac{1}{2}}}{3x^3}\left[\frac{1}{4}x^2\frac{\partial^2}{\partial x^2}-\frac{1}{4}x\frac{\partial}{\partial x}-2y\frac{\partial}{\partial y}-y^2\frac{\partial^2}{\partial y^2}-\frac{3w}{4}y\right]\rightarrow$$
$$\left[\frac{1}{4}x^2\frac{\partial^2}{\partial x^2}-\frac{1}{4}x\frac{\partial}{\partial x}-2y\frac{\partial}{\partial y}-y^2\frac{\partial^2}{\partial y^2}-\frac{3w}{4}y\right],$$

we can write down a new version of the SWDW equation, in which the constant E does not appear anymore as induced by a time evolution effect. Instead, we interpret this term as providing a variable perturbation of the quantum system, with the perturbation term given by

$$V_{pert}=-\frac{3Ex^3}{y^{\frac{1}{2}}}. \tag{13.99}$$

Hence, the total Hamiltonian of the quantum Universe in this modified gravity model with a curvature-matter coupling is $H=H_0+V_{pert}$. Hence it follows that the effect of the time evolution on the wave function in a non-perturbative system, such as $f(R,T)$ gravity, is quantum mechanically equivalent to the splitting of the degenerate energy levels (loss of symmetries) in a perturbative, static quantum system. This result also shows that the time evolution of the Universe in the quantum cosmology of $f(R,T)$ gravity determines the splitting of the degenerate energy levels of the quantum system. Hence curvature-matter coupling also implies the existence of a deep relation between energy and time that appears at the level of the quantum evolution.

The unperturbed component H_0 of the new form of the quantum cosmological Hamiltonian satisfies the equation $H_0\psi = 0$. Hence we see that the eigenvalue of the energy of the unperturbed Hamiltonian is zero. In the Hamiltonian H_0 we can separate again the variables according to $\psi(x,y) = X(x)Y(y)$, and thus we find

$$\left[x^2\frac{\partial^2}{\partial x^2} - x\frac{\partial}{\partial x} + (1-v^2)\right]X(x) = 0, \tag{13.100}$$

$$\left[y^2\frac{\partial^2}{\partial y^2} + 2y\frac{\partial}{\partial y} + \frac{3w}{4}y - \frac{v^2-1}{4}\right]Y(y) = 0, \tag{13.101}$$

where we have denoted the separation constant as $(v^2-1)/4$. Thus for the expression of the wave function of the unperturbed component H_0 of the total Hamiltonian we obtain the expression

$$\Psi(x,y,\tau) = e^{-iE\tau}\left(A_1x^{1-v} + A_2x^{-1-v}\right)\frac{1}{\sqrt{y}}\left[B_1J_v(\sqrt{3wy}) + B_2J_{-v}(\sqrt{3wy})\right], \tag{13.102}$$

where A_1, A_2, B_1, B_2 are arbitrary constants of integration, and $J_v(x)$ denotes the Bessel function. As is known from the theory of Bessel functions, $J_v(x)$ and $J_{-v}(x)$ are linearly independent functions. Since in the early quantum Universe the variable x diverges as $x \to \infty$, in order to find analytic solutions that are non-singular on the whole plane, we take $A_1 = B_2 = 0$. Moreover, we also assume $v \geq 0$.

13.4.2 The Twofold Degenerate Case

To investigate the splitting of the energy levels of the quantum Universe, we consider only the simplest case, namely, the twofold degenerate case [304]. Then, assuming that the energy levels of the Universe are twofold degenerate, at the beginning of the cosmic time $\tau = 0$, the wave function can be written as

$$\Psi = c_1\psi_1 + c_2\psi_2, \tag{13.103}$$

where c_1, c_2 are constants satisfying the relation $|c_1|^2 + |c_2|^2 = 1$, and

$$\psi_1 = x^{-1-v_1}\frac{1}{\sqrt{y}}J_{v_1}\left(\sqrt{3wy}\right), \qquad \psi_2 = x^{-1-v_2}\frac{1}{\sqrt{y}}J_{v_2}\left(\sqrt{3wy}\right), \tag{13.104}$$

where v_1 and v_2 are two positive constants of integration. For the quantum system with the perturbation $V_{pert} = -3Ex^3/y^{1/2}$, we introduce the quantities V_{ij} defined as

$$V_{ij} = \int \psi_i^* V_{pert} \psi_j dx dy. \tag{13.105}$$

By using the expressions of the unperturbed wave functions, we find

$$V_{11} = \int \psi_1^* V_{pert} \psi_1 dx dy = \int (-3E) x^{1-2v_1} \frac{1}{y^{\frac{3}{2}}} [J_{v_1}(\sqrt{3wy})]^2 dx dy, \tag{13.106}$$

$$V_{22} = \int \psi_2^* V_{pert} \psi_2 dx dy = \int (-3E) x^{1-2v_2} \frac{1}{y^{\frac{3}{2}}} [J_{v_2}(\sqrt{3wy})]^2 dx dy, \tag{13.107}$$

$$\begin{aligned} V_{12} = V_{21}^* &= \int \psi_1^* V_{pert} \psi_2 dx dy \\ &= \int (-3E) x^{1-v_1-v_2} \frac{1}{y^{\frac{3}{2}}} \left[J_{v_1}(\sqrt{3wy}) J_{v_2}(\sqrt{3wy})\right] dx dy. \end{aligned} \tag{13.108}$$

It is important to note that since the scale factor and the curvature scalar are always positive, we cannot define wave functions of the Universe that are analytic on the whole space. Thus, the conditions of the orthogonality and the normalization of the wave function are not satisfied in the present quantum cosmological model. To simplify the mathematical notation we define the following functions

$$S_{11} = \int \psi_1^* \psi_1 = \int x^{-2-2v_1} \frac{1}{y} [J_{v_1}(\sqrt{3wy})]^2 dx dy, \tag{13.109}$$

$$S_{22} = \int \psi_2^* \psi_2 = \int x^{-2-2v_2} \frac{1}{y} [J_{v_2}(\sqrt{3wy})]^2 dx dy, \tag{13.110}$$

$$S_{12} = S_{21}^* = \int \psi_1^* \psi_2 = \int x^{-2-v_1-v_2} \frac{1}{y} J_{v_1}(\sqrt{3wy}) J_{v_2}(\sqrt{3wy}) dx dy. \tag{13.111}$$

In the mathematical theory of the special functions there is an integral identity, called the Schafheitlin integral [499], which is expressed as

$$\int_0^\infty \frac{J_\mu(at) J_\nu(bt)}{t^\lambda} dt = \frac{\Gamma(\lambda)\Gamma(\frac{\mu+\nu-\lambda+1}{2})(\frac{a}{2})^{\lambda-1}}{2\Gamma(\frac{\mu-\nu+\lambda+1}{2})\Gamma(\frac{\nu-\mu+\lambda+1}{2})\Gamma(\frac{\mu+\nu+\lambda+1}{2})}, \tag{13.112}$$

where a and b are positive constants, while the arbitrary number μ, ν, and λ satisfy the relations given by

$$\mathrm{Re}(\mu + \nu + 1) > \mathrm{Re}(\lambda) > 0. \tag{13.113}$$

These conditions are necessary to make the integral convergent. For $z \in \Re$ the Gamma function $\Gamma(z)$ satisfies the identities $\Gamma(z+1) = z\Gamma(z)$, $\Gamma(1) = 1$, $\Gamma(\frac{1}{2}) = \sqrt{\pi}$.

In the following, in order to obtain some explicit analytic results, we will restrict our investigations to the simplest case

$$v_1 = \frac{3}{2}, \qquad v_2 = \frac{5}{2}, \tag{13.114}$$

for the indices of the Bessel functions. With the use of these numerical values, and by using the properties of the Bessel functions, we obtain

$$\begin{aligned} V_{11} &= -3E \int \frac{1}{x^2} \frac{[J_{\frac{3}{2}}(\sqrt{3wy})]^2}{y^{\frac{3}{2}}} dxdy \\ &= \eta \int_l^\infty dx \int_0^\infty \frac{1}{x^2} \frac{[J_{\frac{3}{2}}(z)]^2}{z^2} dz = \eta \int_l^\infty \frac{dx}{x^2} \frac{1}{2\pi} = \frac{\eta}{2\pi l}, \end{aligned} \tag{13.115}$$

$$\begin{aligned} V_{22} &= -3E \int \frac{1}{x^4} \frac{[J_{\frac{5}{2}}(\sqrt{3wy})]^2}{y^{\frac{3}{2}}} dxdy \\ &= \eta \int_l^\infty dx \int_0^\infty \frac{1}{x^4} \frac{[J_{\frac{5}{2}}(z)]^2}{z^2} dz = \eta \int_l^\infty \frac{dx}{x^4} \frac{1}{6\pi} = \frac{\eta}{18\pi l^3}, \end{aligned} \tag{13.116}$$

$$\begin{aligned} V_{12} = V_{21} &= -3E \int \frac{1}{x^3} \frac{J_{\frac{3}{2}}(\sqrt{3wy}) J_{\frac{5}{2}}(\sqrt{3wy})}{y^{\frac{3}{2}}} dxdy \\ &= \eta \int_l^\infty dx \int_0^\infty \frac{1}{x^3} \frac{J_{\frac{3}{2}}(z) J_{\frac{5}{2}}(z)}{z^2} dz = \eta \int_l^\infty \frac{dx}{x^3} \frac{1}{15} = \frac{\eta}{30 l^2}, \end{aligned} \tag{13.117}$$

$$\begin{aligned} S_{11} &= \int \frac{1}{x^5} \frac{[J_{\frac{3}{2}}(\sqrt{3wy})]^2}{y} dxdy \\ &= \int_l^\infty dx \int_0^\infty \frac{2}{x^5} \frac{[J_{\frac{3}{2}}(z)]^2}{z} dz = \int_l^\infty \frac{2dx}{x^5} \frac{1}{3} = \frac{1}{6 l^4}, \end{aligned} \tag{13.118}$$

$$\begin{aligned} S_{22} &= \int \frac{1}{x^7} \frac{[J_{\frac{5}{2}}(\sqrt{3wy})]^2}{y} dxdy \\ &= \int_l^\infty dx \int_0^\infty \frac{2}{x^7} \frac{[J_{\frac{5}{2}}(z)]^2}{z} dz = \int_l^\infty \frac{2dx}{x^7} \frac{1}{5} = \frac{1}{15 l^6}, \end{aligned} \tag{13.119}$$

$$\begin{aligned} S_{12} = S_{21} &= \int \frac{1}{x^6} \frac{J_{\frac{3}{2}}(\sqrt{3wy}) J_{\frac{5}{2}}(\sqrt{3wy})}{y} dxdy \\ &= \int_l^\infty dx \int_0^\infty \frac{2}{x^6} \frac{J_{\frac{3}{2}}(z) J_{\frac{5}{2}}(z)}{z} dz = \int_l^\infty \frac{2dx}{x^6} \frac{1}{2\pi} = \frac{1}{5\pi l^5}, \end{aligned} \tag{13.120}$$

where we have denoted $z = \sqrt{3wy} \to dz = \frac{\sqrt{3w}}{2\sqrt{y}} dy$, and $\eta = -6E\sqrt{3w}$, respectively. In performing the calculations we have also assumed that x and y are independent variables. Both upper limits in the integrals are ∞, and they correspond to the limit $a(\tau) \to 0$. As for the lower limits of integration we have introduced some cutoffs, and assumed they are some very small positive numbers l and l_1. These lower limits correspond to the transition from the quantum regime to the classical regime. Thus they also indicate the limits of validity of this $f(R,T)$-type quantum description of the Universe. From the numerical estimations of the integrals involving the Bessel functions it follows that when $l_1 < 0.5$, the numerical values of the integral change significantly, and therefore to compute the integrals we take $l_1 = 0$.

13.4.3 The Energy Levels of the Quantum Universe

In order to find the energy levels in the specific quantum cosmological model of $f(R,T)$ gravity we are considering we will use the perturbation theory of the degenerate levels. The wave function can be written down as $\Psi = c_1\psi + c_2\psi_2$, and the eigenvalues and the eigenfunctions of the unperturbed Hamiltonian H_0, which we have already discussed, can be obtained as solutions of the equation

$$H_0\psi_i = E^{(0)}\psi_i, \qquad i = 1, 2. \tag{13.121}$$

Substituting the unperturbed Hamiltonian in the Schrödinger equation, $H\Psi = (H_0 + V)\Psi = E\Psi$, we find

$$(H_0 + V)(c_1\psi_1 + c_2\psi_2) = E(c_1\psi_1 + c_2\psi_2). \tag{13.122}$$

Multiplication of Eq. (13.122) by ψ_1^* and ψ_2^*, and integration over an arbitrary volume V yields the equations

$$c_1\left(E_1^{(0)}S_{11} + V_{11} - ES_{11}\right) + c_2\left(E_2^{(0)}S_{12} + V_{12} - ES_{12}\right) = 0, \tag{13.123}$$

$$c_1\left(E_1^{(0)}S_{21} + V_{21} - ES_{21}\right) + c_2\left(E_2^{(0)}S_{22} + V_{22} - ES_{22}\right) = 0. \tag{13.124}$$

For the analysis of the degenerate energy levels, we represent the energy as $E = E^{(0)} + E^{(1)}$, where $E^{(1)}$ is the perturbation of the energy. Moreover, we assume $c_n = c_n^{(0)}$, that is, for the coefficients c_n we take the zero-order (unperturbed) approximation. Then we have

$$c_1\left(V_{11} - E^{(1)}S_{11}\right) + c_2\left(V_{12} - E^{(1)}S_{12}\right) = 0, \tag{13.125}$$

$$c_1\left(V_{21} - E^{(1)}S_{21}\right) + c_2\left(V_{22} - E^{(1)}S_{22}\right) = 0. \tag{13.126}$$

After the perturbation of the energy level $E^{(1)}$ is obtained, from Eqs. (13.125) and (13.126) we can find the coefficients c_1 and c_2, respectively. The secular equation, obtained from the requirement of the existence of nontrivial solutions of the system, is

$$\begin{vmatrix} V_{11} - E^{(1)}S_{11} & V_{12} - E^{(1)}S_{12} \\ V_{21} - E^{(1)}S_{21} & V_{22} - E^{(1)}S_{22} \end{vmatrix} = 0. \tag{13.127}$$

Then the first-order perturbations of the energy are obtained as

$$E^{(1)} = \frac{-B \pm \sqrt{B^2 - 4AC}}{2A}, \tag{13.128}$$

where we have denoted

$$A = S_{11}S_{22} - S_{12}S_{21}, \tag{13.129}$$

$$B = S_{12}V_{21} + S_{21}V_{12} - S_{11}V_{22} - S_{22}V_{11}, \tag{13.130}$$

$$C = V_{11}V_{22} - V_{12}V_{21}. \tag{13.131}$$

For the previously considered case, with $v_1 = 3/2, v_2 = 5/2$, we find

$$A = \left(\frac{1}{90} - \frac{1}{25\pi^2}\right)\frac{1}{l^{10}}, \qquad B = -\frac{79}{2700\pi}\frac{\eta}{l^7}, \tag{13.132}$$

$$C = \left(\frac{1}{36\pi^2} - \frac{1}{900}\right)\frac{\eta^2}{l^4}. \tag{13.133}$$

The perturbations to the two energy levels are given by

$$\begin{aligned} E_{\pm}^{(1)} &= \frac{-B \pm \sqrt{B^2 - 4AC}}{2A} \\ &= \frac{\frac{79}{2700\pi} \pm \sqrt{\left(\frac{79}{2700\pi}\right)^2 - 4\left(\frac{1}{90} - \frac{1}{25\pi^2}\right)\left(\frac{1}{36\pi^2} - \frac{1}{900}\right)}}{2\left(\frac{1}{90} - \frac{1}{25\pi^2}\right)}\eta l^3 \\ &= \beta_{\pm}\eta l^3 \approx (0.660 \pm 0.440)\eta l^3, \end{aligned} \tag{13.134}$$

where $\beta_{\pm}$ are constants introduced for notational convenience. From this equation we can immediately see that the perturbations to the energy levels have the same sign, and that they are proportional to l^3. This implies that once we know the numerical value of l, we can estimate the lower limit of the size of the energy gap of the quantum system. From Eq. (13.125) the coefficients c_1 and c_2 of the wave function can be found as given by

$$\begin{aligned} c_1 &= \sqrt{\frac{\left(\frac{\eta}{30l^2} - E_{\pm}^{(1)}\frac{1}{5\pi l^5}\right)^2}{\left(\frac{\eta}{2\pi l} - E_{\pm}^{(1)}\frac{1}{6l^4}\right)^2 + \left(\frac{\eta}{30l^2} - E_{\pm}^{(1)}\frac{1}{5\pi l^5}\right)^2}} \\ &= \sqrt{\frac{\left(\frac{1}{30} - \frac{\beta_{(\pm)}}{5\pi}\right)^2}{\left(\frac{1}{2\pi} - \frac{\beta_{\pm}}{6}\right)^2 l^2 + \left(\frac{1}{30} - \frac{\beta_{\pm}}{5\pi}\right)^2}}, \end{aligned} \tag{13.135}$$

and

$$c_2 = \pm\sqrt{\frac{\left(\frac{1}{2\pi}-\frac{\beta_\pm}{6}\right)^2 l^2}{\left(\frac{1}{2\pi}-\frac{\beta_\pm}{6}\right)^2 l^2+\left(\frac{1}{30}-\frac{\beta_\pm}{5\pi}\right)^2}}, \tag{13.136}$$

respectively.

Equations (13.135) and (13.136) indicate the relation between the coefficients c_1 and c_2 and the length scale l. In the limit of large l, the Universe will be, with a higher probability, in the state ψ_2. When l is small, the probability for the Universe to be in the state ψ_1 will be higher. It is important to note at this moment that both coefficients c_1 and c_2 are not functions of η, which means that the total energy E will not influence the behavior of the wave function Ψ.

13.4.4 The Transition Probability in the Quantum Universe

In the previous section, we have modeled the early Universe at the initial time $\tau = 0$ as a quantum system with twofold degenerate energy levels, matching the wave functions ψ_1, and ψ_2, respectively. In the following, we will investigate the transition probabilities of the Universe, from the initial state ψ_1 at $\tau = 0$, to a final state ψ_2, reached at the time τ. We assume that the quantum transition is due to a slight perturbation of the initial state ψ_1.

In the unperturbed state (zero-order approximation), the wave functions can be represented as

$$\psi = c_1\psi_1 + c_2\psi_2, \qquad \psi' = c_1'\psi_1 + c_2'\psi_2, \tag{13.137}$$

where c_1, c_2 and c_1', c_2' are the two pairs of coefficients we have already determined in the previous section. The wave functions ψ and ψ' amount to two energy states $E_0+E_+^{(1)}$ and $E_0+E_-^{(1)}$, respectively, where in our case $E_0 = 0$, and $E_+^{(1)}$ and $E_-^{(1)}$ are the alterations of the energy resulted from the effects of the perturbation. From the equations (13.137) we can express the wave function ψ_1 of the initial state of the Universe as

$$\psi_1 = \frac{c_2'\psi - c_2\psi'}{c_1c_2' - c_1'c_2}. \tag{13.138}$$

For the time dependent wave functions, after reintroducing their time dependence, we find the expressions

$$\begin{aligned}\Psi_1 &= \frac{e^{-\frac{i}{\hbar}E_0\tau}}{c_1c_2' - c_1'c_2}\left(c_2'\psi e^{-\frac{i}{\hbar}E_+^{(1)}\tau} - c_2\psi' e^{-\frac{i}{\hbar}E_-^{(1)}\tau}\right)\\ &= \frac{1}{c_1c_2' - c_1'c_2}\left(c_2'\psi e^{-\frac{i}{\hbar}E_+^{(1)}\tau} - c_2\psi' e^{-\frac{i}{\hbar}E_-^{(1)}\tau}\right).\end{aligned} \tag{13.139}$$

As one can see from the above equations at $\tau = 0$ we have $\Psi_1 = \psi_1$. By equating ψ_1 and ψ_2 with ψ and ψ', respectively, it follows that Ψ_1 can be represented as the linear combination of the wave functions ψ_1, ψ_2. In this representation the combination coefficients are time dependent functions. The transition probability w_{21} can be obtained, after integrating over the volume V, from the absolute value of the coefficient multiplying ψ_2, which after squaring is given by

$$\begin{aligned} w_{21} &= \frac{1}{c_1c_2' - c_1'c_2}\left[c_1c_2'S_{12}e^{-i\frac{E_+^{(1)}}{\hbar}\tau} + c_1'c_2S_{12}e^{-i\frac{E_-^{(1)}}{\hbar}\tau}\right. \\ &\quad \left. + c_2c_2'S_{22}\left(e^{-i\frac{E_+^{(1)}}{\hbar}\tau} - e^{-i\frac{E_-^{(1)}}{\hbar}\tau}\right)\right] \\ &= \left|\frac{Q_1}{Q_3}\left(e^{-i\frac{E_+^{(1)}}{\hbar}\tau} - e^{-i\frac{E_-^{(1)}}{\hbar}\tau}\right) + Q_2e^{-i\frac{E_-^{(1)}}{\hbar}\tau}\right|^2 \\ &= 2\left(\frac{Q_1^2}{Q_3^2} - \frac{Q_1Q_2}{Q_3}\right)\left[1 - \cos\left(\frac{E_+^{(1)} - E_-^{(1)}}{\hbar}\tau\right)\right] + Q_2^2, \end{aligned} \tag{13.140}$$

where we have denoted

$$Q_1 = c_1c_2'S_{12} + c_2c_2'S_{22}, \qquad Q_2 = S_{12}, \qquad Q_3 = c_1c_2' - c_1'c_2. \tag{13.141}$$

Hence, by using our previous results, we can obtain the transition probability w_{21} in ordinary units as

$$w_{21} = \frac{0.0106435}{l^{10}} - \frac{0.00659062}{l^{10}}\cos\left(\frac{0.880802}{\hbar}\eta l^3\tau\right). \tag{13.142}$$

From Eq. (13.142) one can see that the transition probability has a periodic time behavior, being represented by a cosine function of the time τ. In the limit of very small times τ, w_{21} is proportional to τ^2. The condition that the transition probability must be smaller than one imposes the restriction $0.0106/l^{10} - 0.0066/l^{10} \leq 1$, a circumstance that gives for the upper limit of length scale l the numerical value

$$l \geq 0.666 \rightarrow a(\tau) \leq 1.177. \tag{13.143}$$

Therefore, we have obtained an upper limit of the size of the Universe for which our quantum model is relevant. As for the lower limit of y, it can be formulated as $0.320 < 0.5$, which gives us the possibility of taking the lower limits in all considered integrals as $l_1 = 0$.

13.5 The Second Quantization of Time

Time is not an operator in standard quantum mechanics. But the energy of a quantum system is always the eigenvalue of the Hamiltonian operator that, in the case of the simple harmonic oscillator, can be written in the form

$$\hat{H} = \hbar\omega\left(a^+ a + \frac{1}{2}\right), \tag{13.144}$$

where a and a^+ are the particle creation and annihilation operators, respectively. The existing deep connections between energy and time, as expressed in the energy-time uncertainty relations, may suggest a way to introduce a representation of quantum time by using creation and annihilation operators. Once these operators are determined, we can remove from our theoretical formalism of cosmology the concept of time singularity at the beginning/birth of the Universe. Moreover, we can accurately describe the geometric distance between each pair of the slices of the time variable. As a further consequence of this possibility, we can also derive the quantum frequencies (energies) of the 3-space evolution, and therefore analyze from a full quantum mechanical point of view the physical mechanisms and processes that took place at the birth of the Universe.

In the previous section, with the use of the mathematical formalism of $f(R,T)$ gravity, we have introduced in the WDW equation a 'time' variable, thus transforming it into a SWDW equation containing explicit time dynamics. The time τ defined in the context of the SWDW equation is based on the operator relation

$$P_\tau = -\frac{2}{6a^3} P_D P_M, \tag{13.145}$$

and therefore we have assumed in our previous analysis that the concept of time in $f(R,T)$ gravity is interdependent with the gravitational field variables D and M. In the following we would like to introduce the time creation and annihilation operators by using the term

$$3kaDM - \frac{2}{6a^3} P_D P_M, \tag{13.146}$$

in the WDW equation (13.85).

The procedure we are proposing can be described as follows. We assume that the standard analogy between classical and quantum mechanics also holds for $f(R,T)$ gravity. The existence of this analogy grants us the possibility to replace the classical Poisson brackets with quantum commutators, according to the rule $\{...\} \to [...]$. Thus, in the framework of $f(R,T)$ gravity we *postulate* the existence of the following quantum commutation relations:

$$\left[\hat{D}, \hat{P_D}\right] = \left[\hat{M}, \hat{P_M}\right] = i\hbar. \tag{13.147}$$

Considering that

$$\left(\hat{D} + i\hat{M}\right)\left(\hat{P_D} - i\hat{P_M}\right) = \hat{D}\hat{P_D} - i\hat{D}\hat{P_M} + i\hat{M}\hat{P_D} + \hat{M}\hat{P_M}, \tag{13.148}$$

$$\left(\hat{P_D} - i\hat{P_M}\right)\left(\hat{D} + i\hat{M}\right) = \hat{P_D}\hat{D} - i\hat{P_M}\hat{D} + i\hat{P_D}\hat{M} + \hat{P_M}\hat{M}, \tag{13.149}$$

we have

$$\left(\hat{D} + i\hat{M}\right)\left(\hat{P_D} - i\hat{P_M}\right) - \left(\hat{P_D} - i\hat{P_M}\right)\left(\hat{D} + i\hat{M}\right) = 2i\hbar. \tag{13.150}$$

We denote in the following

$$\hat{v} = \hat{D} + i\hat{M}, \qquad \hat{P}_v = \hat{P}_D - i\hat{P}_M. \tag{13.151}$$

Hence, with the use of the easily derivable mathematical identities $\left[\hat{P}_D, \hat{M}\right] = \left[\hat{P}_M, \hat{D}\right] = 0$, $\hat{D} = \hat{D}^*, \hat{M} = \hat{M}^*, \hat{P}_D = \hat{P}_D{}^*, \hat{P}_M = \hat{P}_M{}^*$, where $*$ denotes the complex conjugate, we can find the commutation relations

$$[\hat{v}, \hat{P}_v] = 2i\hbar, \tag{13.152}$$

$$\begin{aligned}\hat{P}_v{}^*\hat{P}_v{}^* - \hat{P}_v\hat{P}_v &= \left(\hat{P}_D + i\hat{P}_M\right)\left(\hat{P}_D + i\hat{P}_M\right) - \left(\hat{P}_D - i\hat{P}_M\right)\left(\hat{P}_D - i\hat{P}_M\right)\\ &= 4i\hat{P}_D\hat{P}_M,\end{aligned} \tag{13.153}$$

$$\begin{aligned}\hat{v}^*\hat{v}^* - \hat{v}\hat{v} &= \left(\hat{D} - i\hat{M}\right)\left(\hat{D} - i\hat{M}\right) - \left(\hat{D} + i\hat{M}\right) \times \left(\hat{D} + i\hat{M}\right)\\ &= -4i\hat{M}\hat{D},\end{aligned} \tag{13.154}$$

and

$$\begin{aligned}\hat{v}\hat{v} - \hat{v}^*\hat{v}^* + \hat{P}_v\hat{P}_v - \hat{P}_v{}^*\hat{P}_v{}^* &= \left(\hat{v} - i\hat{P}_v\right)\left(\hat{v} + i\hat{P}_v\right)\\ &\quad - \left(\hat{v}^* + i\hat{P}_v{}^*\right)\left(\hat{v}^* - i\hat{P}_v{}^*\right),\end{aligned} \tag{13.155}$$

respectively. Also

$$\left[\hat{v}^*, \hat{P}_v{}^*\right] = (2i\hbar)^*. \tag{13.156}$$

Let us now introduce the assumption about the existence of *two directions of time*, defined as

$$\hat{\tau}_1 = \hat{v} - i\hat{P}_v, \qquad \hat{\tau}_2 = \hat{v} + i\hat{P}_v. \tag{13.157}$$

Then we have

$$[\hat{\tau}_1, \hat{\tau}_2] = \left[\hat{v} - i\hat{P}_v, \hat{v} + i\hat{P}_v\right] = -4\hbar, [\hat{\tau}_1, \hat{\tau}_1] = [\hat{\tau}_2, \hat{\tau}_2] = 0. \tag{13.158}$$

We move now a step forward in our analysis by interpreting $\hat{\tau}_1$ as a creation operator, $\hat{\tau}^+$, and $\hat{\tau}_2$ as an annihilation operator $\hat{\tau}$, respectively. Thus we immediately find

$$\begin{aligned}\hat{v}\hat{v} - \hat{v^*}\hat{v^*} + \hat{P}_v\hat{P}_v - \hat{P}_v^*\hat{P}_v^* &= \left(\hat{v} - i\hat{P}_v\right)\left(\hat{v} + i\hat{P}_v\right) - \left(\hat{v}^* + i\hat{P}_v{}^*\right)\left(\hat{v^*} - i\hat{P}_v^*\right)\\ &= \hat{\tau}^+\hat{\tau} - (\hat{\tau}^+\hat{\tau})^* = \hat{N}_\tau - \hat{N}_\tau^*\\ &= 2i\mathrm{Im}\hat{N}_\tau = 2i\hat{N}_{obs}.\end{aligned} \tag{13.159}$$

Now we introduce the complex time "number" operator $\hat{N}_\tau$, which we define as

$$\hat{N}_\tau \equiv \hat{\tau}^+\hat{\tau}. \tag{13.160}$$

From this definition we obtain immediately the result according to which the time "number" operator, given by

$$\hat{N}_{obs} \equiv \mathrm{Im}\hat{N}_{\tau}, \tag{13.161}$$

is a quantum mechanical observable. Although the complex time flows in two opposite directions, the actual physics experiments can measure only one real time observable.

Now let us go back to the case of the WDW equation in $f(R,T)$ gravity, where

$$-\frac{2}{6a^3}P_D P_M + 3kaDM = \frac{2}{6a^3}\left(-\hat{P}_D\hat{P}_M + 9ka^4\hat{D}\hat{M}\right). \tag{13.162}$$

In the following, for the sake of concreteness, we consider again the specific choice of the gravitational action as given by $f(R,T) = R + \theta RT$. We also introduce a rescaling of $\hat{v}$ as

$$\hat{v} = \sqrt{3a^2k}\hat{v}. \tag{13.163}$$

Then it turns out that the time-related terms in the WDW equation of the considered simple $f(R,T)$ gravity model can be written as

$$\begin{aligned}-\frac{2}{6a^3}\hat{P}_D\hat{P}_M + 3ka\hat{D}\hat{M} &= \frac{-i}{12a^3}\left(\hat{v}\hat{v} - \hat{v^*}\hat{v^*} + \hat{P}_v\hat{P}_v - \hat{P}_v^*\hat{P}_v^*\right)\\ &= \frac{-i}{12a^3}\left[\hat{\tau}^+\hat{\tau} - \left(\hat{\tau}^+\hat{\tau}\right)^*\right] = \frac{-1}{6a^3}\mathrm{Im}\hat{N}_{obs}.\end{aligned} \tag{13.164}$$

This relation shows us that the coupling between the geometry and the matter field provides a method for estimating the quantum time number of the Universe. With the use of Eq. (13.164) the WDW equation can be reformulated as

$$\left[-2aP_a P_A + 2P_A^2 A - 18ka^4 A - \frac{3a^6}{2}M\right]\Psi = \hat{N}_{obs}\Psi. \tag{13.165}$$

Equation (13.165) allows the introduction of a wave function Ψ_N, describing N quanta of time, which has the basic property

$$\hat{N}_{obs}\Psi_N = N\Psi_N. \tag{13.166}$$

Hence, the WDW equation for the wave function Ψ_N becomes

$$\left[-2aP_a P_A + 2P_A^2 A - 18ka^4 A - \frac{3a^6}{2}M\right]\Psi_N = N\Psi_N. \tag{13.167}$$

Equation (13.167) provides a clear account of the dynamics of the very early quantum Universe in the Arnowitz–Deser–Misner (ADM) formalism. Its basic physical result shows that the evolution of the Universe is strongly influenced by the variation of the time quanta N. An important consequence of our formalism is that at different time moments, different Schrödinger-type equations describe the local evolution of the quantum Universe.

Obviously, we can also introduce the *observable time vacuum* (the beginning of the Universe), and define it as

$$\tau|0_\tau\rangle = \tau^*|0_{\tau^*}\rangle = 0, \qquad |0_{obs}\rangle = |0_\tau\rangle|0_{\tau^*}\rangle \,. \tag{13.168}$$

In the present section, we have proposed a mathematical procedure aimed at introducing the Fock space representation of the quantum time variable τ in $f(R,T)$ gravity. If we are able to define in a consistent and rigorous manner the creation/annihilation operators of the quantum time variable in the Fock space, then we can remove the major problem of the singularity of time (and of the other physical variables) at the beginning of the Universe. Moreover, this may also give us the possibility of having a profound understanding of the discrete nature of time. But from a mathematical point of view the introduction of the Fock space representation will also require the conversion of the wave function into the quantum occupation number picture. However, we must note that this quantum mechanical technique is effective only when a canonical position variable and its corresponding canonical momentum are not coupled to each other. One of the major difficulties in the canonical quantization approach of the gravitational field is that the Hilbert space cannot be properly defined. The procedure based on the second quantization of time may offer a potential technique to be used for solving the problems of quantum gravity in the Fock space, instead of using the Hilbert space.

13.6 Conclusions

Quantum cosmology provides us with a large number of theoretical challenges, whose solutions would give us fascinating insights into the fundamental nature of space-time and of the geometry it defines. We might understand much better the physical questions related to the birth and very early evolution of the Universe if we can include in their description the mathematical formalism of quantum mechanics and quantum field theory, including the study of the symmetries of the Universe, the possible existence of discrete quantum structures, or semi classical properties drawn out from a generally covariant quantum theory of strongly interacting fields.

In this chapter, we have explored in detail the quantum cosmology of $f(R,T)$ gravity, a modified gravity theory that significantly extends GR by assuming that the gravitational action is given by an arbitrary function f of the Ricci scalar R, and of the trace of the matter energy-momentum tensor T. To obtain some understanding of the quantum properties of the Universe we have considered a simple $f(R,T)$-type extension of standard GR, in which we have introduced a "deformation" of the form θRT of the standard Einstein–Hilbert gravitational Lagrangian, given by the Ricci scalar R. We have presented and discussed the physical properties of this quantum cosmological model in detail. The possibility of the definition of a quantum time, as well as of an associated canonical momentum operator, is the most interesting feature of this quantum cosmological model. These physical properties allow the reformulation of the WDW equation of this model as a Schrödinger-type equation, in which the quantum states of the Universe explicitly evolve in time. The mathematical as well as the physical properties of the Schrödinger-type equation have been studied in detail, by using a perturbative quantum mechanical approach, under the assumption that the small perturbation is proportional to the energy of the system. For simplicity we have investigated only the case of a twofold degenerate quantum system, but the generalization to arbitrary number degenerate quantum states is immediate. We have presented and discussed the probability of transition of the Universe between different quantum states. As an interesting theoretical possibility we have very briefly discussed the feasibility of the second quantization of time at the beginning of the Universe, which leads to the appealing prospect of the substitution of the canonical quantization method, formulated in a Hilbert space, with the Fock space depiction of the quantum events that took place in the very early Universe.

Hence the basic theoretical tools that we have introduced in the present chapter could be used for the investigation of the complex interaction of curvature and matter in the quantum Universe, as well as for the description of the origin and evolution of the very early Universe. This approach may also help to understand the properties of the gravitational interaction at the beginning of the Universe, when geometry interacting with matter may have given birth to time, entropy, and irreversibility.

14

Modified Gravity from Quantum Metric Fluctuations

14.1 Introduction

An interesting alternative to the common explanation of present-day observations that have shaken the well-entrenched foundations of modern theoretical physics may be provided by modified theories of gravity. As we have already seen, and discussed in detail in the previous chapters, gravitational theories with curvature-matter coupling have the compelling (and perhaps appealing) property of the nonconservation of the matter energy-momentum tensor. From a thermodynamic point of view, this property can be understood by using the formalism of open thermodynamic systems and irreversible processes [220, 241]. In this framework we can ascertain that in gravitational theories with a curvature-matter coupling the generalized conservation equations imply *irreversible particle production processes*. In this way the nonconservation of the energy-momentum tensor corresponds to an irreversible energy flux from the geometric background of the space-time (describing the gravitational field) to the newly produced particles. The second law of thermodynamics does not forbid the transformation of the space-time into matter, while the reverse process is prohibited. In Chapter 12, we obtained the corresponding entropy production rates, particle number creation rates, and the creation pressures for both $f(R, L_m)$- and $f(R,T)$-modified gravity theories [220] (see [241] for the specific case of a linear curvature-matter coupling). The evolution equations describing the temperature evolution of the newly produced matter were also presented. As a result of the nonconservation of the energy-momentum tensor, along the cosmological evolution of the Universe, a large amount of comoving entropy is also generated. Hence in these theories the significant increase of the entropy of the Universe is also a direct consequence of the curvature-matter coupling.

One of the noteworthy outcomes of quantum field theory in curved space-times is the prediction of the creation of material particles from the cosmological vacuum [199, 405, 406, 407]. Hence in the quantum field, theoretical approaches to gravity, matter production processes naturally arises, and they are supposed to play a fundamental role for our understanding of the quantum properties of gravity. A standard result of quantum field theory in curved space-times shows that quanta of the minimally coupled scalar fields must be created in an expanding Universe with geometry given by the FLRW metric [407]. Therefore, the existence of matter production phenomena in both modified gravity with curvature-matter coupling and quantum field theoretical approaches to gravity may advocate the presence of a deep affinity between these two seemingly very different physical theories that are trying to understand the profound properties of the gravitational force. Intriguingly enough, such a relation has been found in [514], where by applying a non-perturbative approach for the quantization of the metric, suggested in [180, 181, 179], it was pointed out that a distinct type of $f(R,T)$-modified gravity theory, with Lagrangian density given by

$$L=\left[\frac{(1-\alpha)}{2\kappa^2}R+\left(L_m-\alpha\frac{T}{2}\right)\right]\sqrt{-g}, \tag{14.1}$$

where α is a constant, naturally implies the quantum fluctuations of the metric. This result alludes that an equivalent *microscopic quantum* depiction of the particle production processes in $f(R,T)$ or $f\left(R,L_m\right)$ gravity may be possible. Such an interpretation could yield some light on the possible quantum physical mechanisms leading to particle generation via gravity and curvature-matter coupling. In fact, such quantum processes indeed exist and, at least qualitatively, they can be inferred from the theoretical structure of some quantum/semiclassical gravity models.

In this chapter, we present the theoretical and cosmological consequences of the modification of gravity that can be obtained by considering the quantum fluctuations of the gravitational metric [320]. The basic idea of this approach is the assumption that the quantum metric can be generally represented as the sum of a classical and of a fluctuating part, the latter being of a quantum (or stochastic) nature. If such a representation is indeed possible, then at the classical level the related Einstein quantum gravity induces modified gravity models with a nonminimal curvature-matter coupling. In our present investigation we assume that the expectation value of quantum corrections can be typically expressed as a function of an arbitrary second-order tensor $K_{\mu\nu}$. This tensor can be made up from the metric tensor, and from the thermodynamic quantities that represent the matter content of the Universe. Once the functional form of $K_{\mu\nu}$ is adopted, the first-order quantum corrected gravitational action can be obtained. We investigate two cases of cosmological models generated by taking into account the quantum fluctuations of the metric tensor. First, we suppose that the quantum correction tensor $K_{\mu\nu}$ can represented by the product of the metric tensor and of a scalar field, and of a simple scalar function, respectively. Secondly, we assume that $K_{\mu\nu}$ can be represented as being proportional to the energy-momentum tensor of the matter. It turns out that the first selection of $K_{\mu\nu}$ generates a particular version of $f(R,T)$ gravity, while the second adopted form of $K_{\mu\nu}$ leads to a concrete case of the $f\left(R,T,R_{\mu\nu}T^{\mu\nu},T_{\mu\nu}T^{\mu\nu}\right)$-modified theory of gravity.

14.2 Quantum Metric Fluctuations and Gravitational Action

In the following, we describe the fluctuating metric avenue to quantum gravity, and we will introduce its relation with the usual semiclassical formalism. Moreover, we will show the quantum mechanical sources of the modified gravity models with curvature-matter coupling. We also derive, for an arbitrary form of the quantum fluctuation tensor $K_{\mu\nu}$, the general system of gravitational field equations induced by the quantum fluctuations of the metric.

14.2.1 Modified Gravity as the Semiclassical Approximation of Quantum Gravity

In the standard quantum mechanical approach operators must represent all geometrical or physical quantities. Therefore, in a full non-perturbative quantum strategy for the quantization of gravity, the Einstein field equations must be formulated in an operator form, given generally by [180, 181, 179]

$$\hat{R}_{\mu\nu}-\frac{1}{2}\hat{R}\,\hat{g}_{\mu\nu}=\frac{8\pi G}{c^4}\hat{T}_{\mu\nu}. \tag{14.2}$$

To obtain meaningful physical information from the Einstein operator equations, it is necessary to average them over all possible products of the metric tensor operators $\hat{g}\left(x_1\right)...\hat{g}\left(x_n\right)$ [180, 181, 179]. Thus we arrive at the infinite system of equations

$$\begin{aligned}
< Q|\hat{g}(x_1)\hat{G}_{\mu\nu}|Q> &= < Q|\hat{g}(x_1)\hat{T}_{\mu\nu}|Q>,\\
< Q|\hat{g}(x_1)\hat{g}(x_2)\hat{G}_{\mu\nu}|Q> &= < Q|\hat{g}(x_1)\hat{g}(x_2)\hat{T}_{\mu\nu}|Q>,\\
\ldots &= \ldots,
\end{aligned}$$

that must be solved to determine the Green functions $\hat{G}_{\mu\nu}$. In the above equations, we have denoted $|Q>$ as a quantum state that might be different from the ordinary vacuum state of standard quantum field theory. Generally, the averaged-value equations for the Green functions of the gravitational field cannot be solved analytically, and hence we must resort to the use of some approximate methods [180, 181, 179]. Such a method, based on the decomposition of the metric tensor operator into the sum of an average metric $g_{\mu\nu}$, and a fluctuating part $\delta\hat{g}_{\mu\nu}$, according to

$$\hat{g}_{\mu\nu}=g_{\mu\nu}+\delta\hat{g}_{\mu\nu}, \tag{14.3}$$

was suggested in [180]. The fluctuating part of the metric may be assumed now to be of the form

$$<\delta\hat{g}_{\mu\nu}>=K_{\mu\nu}\neq 0, \tag{14.4}$$

where by $K_{\mu\nu}$ we have denoted a classical tensor quantity. Hence, by ignoring higher-order fluctuations, the gravitational Lagrangian density of the gravitational field will be modified into [180]

$$\begin{aligned}\mathcal{L} &= \mathcal{L}_g\left(\hat{g}_{\mu\nu}\right) + \sqrt{-g}\mathcal{L}_m\left(\hat{g}_{\mu\nu}\right) \\ &\approx \mathcal{L}_g + \frac{\delta \mathcal{L}_g}{\delta g^{\mu\nu}}\delta\hat{g}^{\mu\nu} + \sqrt{-g}\mathcal{L}_m + \frac{\delta(\sqrt{-g}\mathcal{L}_m)}{\delta g^{\mu\nu}}\delta\hat{g}^{\mu\nu} \\ &= -\frac{1}{2\kappa^2}\sqrt{-g}\left(R + G_{\mu\nu}\delta\hat{g}^{\mu\nu}\right) + \sqrt{-g}\left(\mathcal{L}_m + \frac{1}{2}T_{\mu\nu}\delta\hat{g}^{\mu\nu}\right), \end{aligned} \tag{14.5}$$

where $\kappa^2 = 8\pi G/c^4$. The above equation also contains the energy-momentum tensor of the gravitating matter, defined by Eq. (8.3), which we reproduce here

$$T_{\mu\nu} = \frac{2}{\sqrt{-g}}\frac{\delta\left(\sqrt{-g}\mathcal{L}_m\right)}{\delta g^{\mu\nu}}. \tag{14.6}$$

Even if the theoretical formalism presented above represents initially a full quantum approach of gravity, after the decomposition of the metric into two components is introduced, we have arrived at a semiclassical theory. In the following, we will follow this line of thought, and consider several functional forms of the quantum perturbation tensor $K^{\mu\nu}$. For each case we will obtain the quantum-corrected gravitational field equations and investigate the physical properties of the corresponding gravity theory. We will also consider in detail the cosmological implications of these classes of modified gravity theories.

After substituting into the gravitational action given by Eq. (14.5) the adopted representation of the quantum metric fluctuation as $\langle\delta\hat{g}^{\mu\nu}\rangle = K^{\mu\nu}$, the gravitational Lagrangian density $\mathcal{L}$ takes the form

$$\mathcal{L} = -\frac{1}{2\kappa^2}\sqrt{-g}\left(R + G_{\mu\nu}K^{\mu\nu}\right) + \sqrt{-g}\left(\mathcal{L}_m + \frac{1}{2}T_{\mu\nu}K^{\mu\nu}\right). \tag{14.7}$$

By varying the action with respect to the classical metric tensor $g_{\mu\nu}$ we find

$$\delta\mathcal{L} = \sqrt{-g}\delta g^{\mu\nu}\left(\frac{1}{2}T_{\mu\nu} - \frac{1}{2\kappa^2}G_{\mu\nu}\right) - \frac{1}{2\kappa^2}\delta\left(\sqrt{-g}G_{\mu\nu}K^{\mu\nu}\right) + \frac{1}{2}\delta\left(\sqrt{-g}T_{\mu\nu}K^{\mu\nu}\right). \tag{14.8}$$

Next, we compute the second term in the equation as

$$\begin{aligned}\delta\left(\sqrt{-g}G_{\mu\nu}K^{\mu\nu}\right) &= -\sqrt{-g}\delta g^{\mu\nu}\frac{1}{2}g_{\mu\nu}G_{\alpha\beta}K^{\alpha\beta} + \sqrt{-g}\left[\delta\left(R_{\mu\nu}K^{\mu\nu}\right) - \frac{1}{2}\delta(RK)\right] \\ &= -\sqrt{-g}\delta g^{\mu\nu}\frac{1}{2}g_{\mu\nu}G_{\alpha\beta}K^{\alpha\beta} + \sqrt{-g}\delta g^{\mu\nu}\frac{1}{2}\left(\Box K_{\mu\nu} + \nabla_\alpha\nabla_\beta K^{\alpha\beta}g_{\mu\nu} - \nabla_\alpha\nabla_{(\mu}K^\alpha_{\nu)}\right) \\ &+\sqrt{-g}\delta g^{\mu\nu}(\gamma^{\alpha\beta}_{\mu\nu}R_{\alpha\beta}) - \frac{1}{2}\sqrt{-g}\delta g^{\mu\nu}\Big(RK_{\mu\nu} + KR_{\mu\nu} + g_{\alpha\beta}\left(\gamma^{\alpha\beta}_{\mu\nu}R\right) \\ &+\nabla_\mu\nabla_\nu K + g_{\mu\nu}\Box K\Big), \end{aligned} \tag{14.9}$$

where we have defined $K = g_{\mu\nu}K^{\mu\nu}$ and $A_{\alpha\beta}\delta K^{\alpha\beta} = \delta g^{\mu\nu}(\gamma^{\alpha\beta}_{\mu\nu}A_{\alpha\beta})$. Here $A_{\alpha\beta} = R_{\alpha\beta}$, or $A_{\alpha\beta} = T_{\alpha\beta}$, and by $\gamma^{\alpha\beta}_{\mu\nu}$ we have denoted a tensor quantity, an operator, or a combination of them. The variation of the quantity $\sqrt{-g}T_{\mu\nu}K^{\mu\nu}$ can be calculated as

$$\delta\left(\sqrt{-g}T_{\mu\nu}K^{\mu\nu}\right) = -\sqrt{-g}\delta g^{\mu\nu}\frac{1}{2}g_{\mu\nu}T_{\alpha\beta}K^{\alpha\beta} + \sqrt{-g}\delta g^{\mu\nu}\Bigg[\gamma^{\alpha\beta}_{\mu\nu}T_{\alpha\beta} + K^{\alpha\beta}\left(2\frac{\delta^2 L_m}{\delta g^{\mu\nu}\delta g^{\alpha\beta}} - \frac{1}{2}g_{\alpha\beta}T_{\mu\nu} - \frac{1}{2}g_{\mu\nu}g_{\alpha\beta}L_m - L_m\frac{\delta g_{\alpha\beta}}{\delta g^{\mu\nu}}\right)\Bigg]. \tag{14.10}$$

Hence, for the variation with respect to the metric of the first-order quantum-corrected gravitational Lagrangian density, we finally obtain the result

$$\begin{aligned}\delta\mathcal{L} = {}& \sqrt{-g}\delta g^{\mu\nu}\left(\frac{1}{2}T_{\mu\nu} - \frac{1}{2\kappa^2}G_{\mu\nu}\right) - \frac{1}{2\kappa^2}\Bigg\{-\sqrt{-g}\delta g^{\mu\nu}\frac{1}{2}g_{\mu\nu}G_{\alpha\beta}K^{\alpha\beta}\\
&+\frac{1}{2}\sqrt{-g}\delta g^{\mu\nu}\left(\Box K_{\mu\nu} + \nabla_\alpha\nabla_\beta K^{\alpha\beta}g_{\mu\nu} - \nabla_\alpha\nabla_{(\mu}K^{\alpha}_{\nu)}\right) + \sqrt{-g}\delta g^{\mu\nu}(\gamma^{\alpha\beta}_{\mu\nu}R_{\alpha\beta})\\
&-\frac{1}{2}\sqrt{-g}\delta g^{\mu\nu}(RK_{\mu\nu} + KR_{\mu\nu} + \gamma^{\alpha\beta}_{\mu\nu}(Rg_{\alpha\beta}) + \nabla_\mu\nabla_\nu K + g_{\mu\nu}\Box K)\Bigg\}\\
&+\frac{1}{2}\Bigg\{-\sqrt{-g}\delta g^{\mu\nu}\frac{1}{2}g_{\mu\nu}T_{\alpha\beta}K^{\alpha\beta} + \sqrt{-g}\delta g^{\mu\nu}\Bigg[\gamma T_{\mu\nu} + K^{\alpha\beta}(2\frac{\delta^2 L_m}{\delta g^{\mu\nu}\delta g^{\alpha\beta}}\\
&-\frac{1}{2}g_{\alpha\beta}T_{\mu\nu} - \frac{1}{2}g_{\mu\nu}g_{\alpha\beta}L_m - L_m\frac{\delta g_{\alpha\beta}}{\delta g^{\mu\nu}})\Bigg]\Bigg\}.\end{aligned} \tag{14.11}$$

Therefore, the first-order-corrected quantum gravitational field equations that correspond to the quantum Lagrangian density (14.5) can be written, in a general form, as

$$\begin{aligned}G_{\mu\nu} = {}& \kappa^2 T_{\mu\nu} - \Bigg\{\frac{1}{2}g_{\mu\nu}G_{\alpha\beta}K^{\alpha\beta} + \frac{1}{2}\left(\Box K_{\mu\nu} + \nabla_\alpha\nabla_\beta K^{\alpha\beta}g_{\mu\nu} - \nabla_\alpha\nabla_{(\mu}K^{\alpha}_{\nu)}\right)\\
&+\gamma^{\alpha\beta}_{\mu\nu}R_{\alpha\beta} - \frac{1}{2}\left[RK_{\mu\nu} + KR_{\mu\nu} + \gamma^{\alpha\beta}_{\mu\nu}(Rg_{\alpha\beta}) + \nabla_\mu\nabla_\nu K + g_{\mu\nu}\Box K\right]\Bigg\}\\
&+\kappa^2\Bigg\{-\frac{1}{2}g_{\mu\nu}T_{\alpha\beta}K^{\alpha\beta} + \Bigg[\gamma^{\alpha\beta}_{\mu\nu}T_{\alpha\beta} + K^{\alpha\beta}\Bigg(2\frac{\delta^2\mathcal{L}_m}{\delta g^{\mu\nu}\delta g^{\alpha\beta}} - \frac{1}{2}g_{\alpha\beta}T_{\mu\nu}\\
&-\frac{1}{2}g_{\mu\nu}g_{\alpha\beta}\mathcal{L}_m - \mathcal{L}_m\frac{\delta g_{\alpha\beta}}{\delta g^{\mu\nu}}\Bigg)\Bigg]\Bigg\}.\end{aligned} \tag{14.12}$$

14.2.2 The Semiclassical Gravity Approach

It would be interesting to compare the theoretical formalism based on the quantum metric fluctuation as introduced in the previous section to the standard semiclassical approach to gravity. This approach also predicts the nonconservation of the matter energy-momentum tensor, and therefore it also leads to particle production via curvature-matter coupling [241]. Semiclassical gravitational models are usually obtained from the fundamental hypothesis that even in the high energy regime the gravitational field still remains *classical*, while in the same high energy scale the *classical matter (bosonic) fields* ϕ must be *quantized*. The coupling of the quantized matter fields to the classical gravitational fields is directly realized via the substitution of the quantum energy momentum tensor of the matter $\hat{T}_{\mu\nu}$ by its expectation value $\left\langle \hat{T}_{\mu\nu} \right\rangle$, which is defined with respect to some arbitrary quantum state Ψ. Hence, a set of effective semiclassical Einstein equations can be *postulated* as [117]

$$G_{\mu\nu} = R_{\mu\nu} - \frac{1}{2} g_{\mu\nu} R = \frac{8\pi G}{c^4} \langle \Psi | \hat{T}_{\mu\nu} | \Psi \rangle . \tag{14.13}$$

From Eq. (14.13) it immediately follows that the classical energy-momentum tensor $T_{\mu\nu}$ of the quantized gravitating system can be calculated from its quantum counterpart by using the relation

$$\langle \Psi | \hat{T}_{\mu\nu} | \Psi \rangle = T_{\mu\nu}. \tag{14.14}$$

On the other hand, the semiclassical Einstein equations (14.13) can be obtained from the variational principle [282]

$$\delta \left(S_g + S_\psi \right) = 0, \tag{14.15}$$

where $S_g = (1/16\pi G) \int R\sqrt{-g} d^4x$ represents the standard Einstein–Hilbert geometric classical general relativistic action for the gravitational field, while the quantum component S_Ψ of the total action is given by

$$S_\Psi = \int \left[\mathrm{Im} \left\langle \dot{\Psi} | \Psi \right\rangle - \left\langle \Psi | \hat{H} | \Psi \right\rangle + \alpha \left(\langle \Psi | \Psi \rangle - 1 \right) \right] dt. \tag{14.16}$$

In Eq. (14.16) we have denoted by $\hat{H}$ the Hamiltonian operator of the gravitating system, and α is a Lagrange multiplier that imposes the normalization condition of the wave function. The variation of Eq. (14.15) with respect to the Lagrange multiplier α leads to the normalization condition for the quantum wave function $\langle \Psi | \Psi \rangle = 1$, while the variation of the action with respect to the quantum wave function Ψ gives the Schödinger equation for the wave function Ψ,

$$i \left| \dot{\Psi}(t) \right\rangle = \hat{H}(t) \left| \Psi(t) \right\rangle - \alpha(t) \left| \Psi(t) \right\rangle . \tag{14.17}$$

The variation of the action with respect to the metric tensor results in the semiclassical Einstein equations (14.13), respectively. Hence, in this simple phenomenological model of semiclassical gravity, the Bianchi identities satisfied identically by the Einstein tensor $G_{\mu\nu}$ still impose the conservation of the effective energy-momentum tensor of the matter, $\nabla_\mu \langle \Psi | \hat{T}^{\mu\nu} | \Psi \rangle = 0$.

A more general system of semiclassical Einstein equations that have many similarities with the modified gravity models with curvature-matter coupling can be obtained by introducing a supplementary coupling between the classical curvature scalar R of the space-time, and the quantum fields, respectively. A theoretical model based on this approach was introduced in [282], where the quantum matter-geometry coupling term in the total semiclassical gravitational action was assumed to take the form

$$\int RF\left(\langle f(\phi)\rangle\right)_{\Psi}\sqrt{-g}d^4x, \tag{14.18}$$

where F and f are arbitrary functions of the field ϕ, and

$$\left(\langle f(\phi)\rangle\right)_{\Psi} = \langle\Psi(t)|\, f[\phi(x)]\, |\Psi(t)\rangle\,. \tag{14.19}$$

The introduction of such a curvature-matter interaction term changes the Hamiltonian $H(t)$ in the Schrödinger equation (14.17), which now becomes [282]

$$\hat{H}(t) \to \hat{H}_{\Psi} = \hat{H}(t) - \int NF'\left(\langle f(\phi)\rangle\right)_{\Psi} f(\phi)\sqrt{\varsigma}d^3\xi, \tag{14.20}$$

where N is denoted the lapse function, while ξ^i are intrinsic space-time coordinates defined on the given manifold. The coordinates ξ^i are chosen so that on the whole geometric space-time manifold the normal vector to a space-like surface is always time-like. The scalar function ς can be calculated as $\varsigma = \det \varsigma_{rs}$, where ς_{rs} is the induced metric on a space-like surface $\sigma(t)$, which globally slices the entire space-time manifold into specific space-like surfaces. Then, by also including the effect of the curvature-quantum matter coupling in the total gravitational action, it turns out that the system of the effective semiclassical Einstein equations becomes [282]

$$R_{\mu\nu} - \frac{1}{2}Rg_{\mu\nu} = 16\pi G\Big[\left\langle \hat{T}_{\mu\nu}\right\rangle_{\Psi} + G_{\mu\nu}F - \nabla_\mu\nabla_\nu F + g_{\mu\nu}\Box F\Big]. \tag{14.21}$$

As one can easily see, in the semiclassical gravitational theory introduced by Eq. (14.21), the energy-momentum tensor of the matter is not conserved, since from the generalized Einstein equations it follows that $\nabla_\mu \left\langle \hat{T}^{\mu\nu}\right\rangle_{\Psi} \neq 0$. Hence, Eq. (14.21) portrays an effective matter creation process, which is due to a quantum-mechanically generated energy flow from the space-time geometry to matter. This equation also provides an effective semiclassical characterization of the quantum mechanical processes taking place in a strong gravitational field, which are closely related to the energy and matter nonconservation phenomena.

Equations (14.12) and (14.21) have been obtained based on some particular suppositions on the quantum nature of gravity. While Eq. (14.12) is derived by considering a first-order approximation to the fluctuating metric in quantum gravity, Eq. (14.21) was introduced by using the postulate of the existence of a quantum coupling between curvature and matter. Moreover, we should point out that while the coupling in Eq. (14.21) is introduced with the use of a scalar function, the presence in Eq. (14.12) of the quantum effects is a result of the intrinsic fluctuations of the metric, with the fluctuations having the algebraic structure of a tensorial quantity. However, a common feature of both models is that the energy-momentum tensor is generally not conserved. Such a nonconservation indicates that in both models the possibility of the energy flow between geometry and matter does exist. Therefore, the main theoretical source, and the physical justification of the modified gravity theories with curvature-matter coupling, which "automatically" leads to particle creation processes in varying gravitational fields, may be found in the semiclassical approximation of quantum gravity in a Riemannian curved space-time geometry.

14.3 Modified Gravity from Quantum Perturbations Proportional to the Metric

In this section, we present the simplest modified gravity theory generated by the quantum fluctuations of the metric tensor, in which in the first-order approximation the quantum fluctuation tensor is given by $K^{\mu\nu} = \alpha(x)g^{\mu\nu}$. The cosmological implications of this theory will also be investigated.

14.3.1 The Gravitational Field Equations

As a first example of a modified gravity theory induced by the quantum metric fluctuations we investigate a simple model in which the expectation value of the quantum fluctuation tensor is proportional to the classical metric tensor [514, 180], so that, by definition

$$K^{\mu\nu} = \alpha(x)g^{\mu\nu}, \tag{14.22}$$

where the proportionality coefficient $\alpha(x)$ is an arbitrary function of the space-time coordinates $x = x^\mu = (x^0, x^1, x^2, x^3)$. The case $\alpha =$ constant was analyzed in [514] and [180], respectively. This choice of the tensor $K^{\mu\nu}$ implies that the additional part of the metric, which describes the quantum perturbation effects, is proportional to the classical metric.

In the following we will analyze two different cases. First, we assume that $\alpha(x)$ is a self-interacting *scalar field*, having a particular self-interaction potential. We will also investigate a second model, in which we assume that $\alpha(x)$ is a simple scalar function.

Scalar Field-Metric Coupling

Under the assumption that $\alpha(x)$ is a scalar field, we need to add an additional Lagrangian term L_α, given by

$$L_\alpha = \sqrt{-g}\left[\frac{1}{2}\nabla_\mu\alpha\nabla^\mu\alpha - V(\alpha)\right], \tag{14.23}$$

as the supplementary matter source into the general quantum perturbed Lagrangian (14.5), containing first-order quantum corrections. Therefore the Lagrangian of our model can be written as

$$\begin{aligned} L_{total} = & -\frac{1}{2\kappa^2}\sqrt{-g}\left[(1-\alpha)R\right] + \sqrt{-g}\left(\mathcal{L}_m + \frac{1}{2}\alpha T\right) \\ & + \sqrt{-g}\left[\frac{1}{2}\nabla_\mu\alpha\nabla^\mu\alpha - V(\alpha)\right]. \end{aligned} \tag{14.24}$$

By varying this Lagrangian with respect to α, the generalized Klein–Gordon equation for the scalar field can be obtained immediately from the corresponding Euler–Lagrange equation in the form

$$\Box\alpha - \frac{1}{2\kappa^2}R - \frac{1}{2}T + \frac{\partial V}{\partial\alpha} = 0. \tag{14.25}$$

The variation with respect to the metric tensor of Eq. (14.24) gives the Einstein field equations as

$$\begin{aligned} R_{\mu\nu} - \frac{1}{2}Rg_{\mu\nu} = & \frac{2\kappa^2}{1-\alpha}\left[\frac{1+\alpha}{2}T_{\mu\nu} - \frac{1}{4}\alpha T g_{\mu\nu} + \frac{1}{2}\alpha\theta_{\mu\nu}\right. \\ & \left. -\frac{1}{4}g_{\mu\nu}(\nabla_\beta\alpha\nabla^\beta\alpha - 2V) + \frac{1}{2}\nabla_\mu\alpha\nabla_\nu\alpha\right], \end{aligned} \tag{14.26}$$

where

$$\theta_{\mu\nu} = g^{\alpha\beta}\frac{\delta T_{\alpha\beta}}{\delta g^{\mu\nu}} = -g_{\mu\nu}\mathcal{L}_m - 2T_{\mu\nu}. \tag{14.27}$$

By contracting the above Einstein field equations we obtain the trace equation of the model as

$$R = -\kappa^2\left[T + \frac{\alpha\theta}{1-\alpha} - \frac{1}{1-\alpha}\left(\nabla_\mu\alpha\nabla^\mu\alpha - 4V\right)\right]. \tag{14.28}$$

Thus, with the use of the trace equation we can rewrite the basic gravitational field equations as

$$R_{\mu\nu} = \frac{\kappa^2}{2(1-\alpha)}\left[2(1+\alpha)T_{\mu\nu} - Tg_{\mu\nu} + 2\alpha\theta_{\mu\nu} - \alpha\theta g_{\mu\nu} - 2Vg_{\mu\nu} + 2\nabla_\mu\alpha\nabla_\nu\alpha\right]. \tag{14.29}$$

From Eq. (14.26) we can obtain the divergence of the matter energy-momentum tensor, which is given by

$$\begin{aligned} \nabla^\nu T_{\mu\nu} = & -\frac{1}{2(1+\alpha)}\left\{\alpha\left(2\nabla^\nu\theta_{\mu\nu} - g_{\mu\nu}\nabla^\nu T\right) + \frac{\nabla^\nu\alpha}{1-\alpha}\left[4T_{\mu\nu} - Tg_{\mu\nu} + 2\theta_{\mu\nu}\right.\right. \\ & \left.\left. + g_{\mu\nu}\left[2V + 2(1-\alpha)\Box\alpha - \nabla_\beta\alpha\nabla^\beta\alpha\right] + 2\nabla_\mu\alpha\nabla_\nu\alpha\right] + 2\nabla^\nu V g_{\mu\nu}\right\}. \end{aligned} \tag{14.30}$$

For the matter energy-momentum tensor we will adopt, in the following, the perfect fluid form

$$T_{\mu\nu} = (\rho + p)\, u_\mu u_\nu + p g_{\mu\nu}, \tag{14.31}$$

where, as usual, u^μ is the matter four velocity, satisfying the normalization condition $u_\mu u^\mu = -1$, ρ denotes the matter energy density, and p is the thermodynamic pressure, respectively. Moreover, to obtain the basic equations of our model for the matter Lagrangian, we adopt the representation $\mathcal{L}_m = p$. Then with these forms of the matter Lagrangian and of the energy-momentum tensor we can calculate immediately the tensor $\theta_{\mu\nu}$, which is given by

$$\theta_{\mu\nu} = g^{\alpha\beta}\frac{\delta T_{\alpha\beta}}{\delta g^{\mu\nu}} = -g_{\mu\nu}\mathcal{L}_m - 2T_{\mu\nu} = -p g_{\mu\nu} - 2\,(\rho + p)\, u_\mu u_\nu. \tag{14.32}$$

The scalar θ is given by

$$\theta = 2\,(\rho - p)\,. \tag{14.33}$$

By using Eq. (14.28), we can reformulate the Klein–Gordon equation, which now reads

$$\Box\alpha + \frac{1}{2(1-\alpha)}\left(\alpha\theta - \nabla_\mu\alpha\nabla^\mu\alpha + 4V\right) + \frac{\partial V}{\partial\alpha} = 0. \tag{14.34}$$

Cosmological Applications

To investigate the cosmological implications of the modified gravity theories induced by the quantum fluctuations, we adopt the FLRW metric, describing a homogeneous and isotropic universe, and which is given by

$$ds^2 = -c^2dt^2 + a^2(t)\left(\frac{1}{1-kr^2}dr^2 + r^2d\theta^2 + r^2 sin^2\theta d\phi^2\right), \tag{14.35}$$

where $a(t)$ denotes the scale factor of the Universe, and $k = 0, \pm1$ describes the geometric curvature of the space-time. For this metric the components of the Ricci tensor can be easily calculated, and are obtained as

$$R_{00} = -\frac{3\ddot{a}}{a}, \qquad R_{ij} = \frac{2k + 2\dot{a}^2 + a\ddot{a}}{a^2} g_{ij}, \qquad i, j = 1, 2, 3. \tag{14.36}$$

An important cosmological parameter, the Hubble function, is given by $H = \dot{a}/a$. As an indicator of the possible accelerated expansion of the Universe, we introduce another cosmological observable, the deceleration parameter q, which is defined as

$$q = \frac{d}{dt}\frac{1}{H} - 1. \tag{14.37}$$

An accelerating evolution is indicated by negative values of q, while a decelerating expansion is characterized by positive values of the deceleration parameter. The comparison between the observational results and the theoretical prediction of the model are very much simplified if instead of the time variable t we introduce the cosmological redshift z, defined as

$$1 + z = \frac{1}{a}. \tag{14.38}$$

In the definition of the redshift for the scale factor $a(z)$ we have adopted the normalization condition $a(t_0) = 1$, where t_0 denotes the present age of the Universe. Hence, we can express the cosmological time derivative operator as a function of the redshift as

$$\frac{d}{dt} = -H(z)(1+z)\frac{d}{dz}. \tag{14.39}$$

We will restrict in the following our investigations to the case of the flat FLRW universes, and hence we take $k = 0$ in the cosmological gravitational field equations. Therefore, the Friedmann and the Klein–Gordon equations, corresponding to our generalized gravity model in the presence of a fluctuating metric, with fluctuation tensor proportional to the metric tensor, take the form

$$3H^2 = \frac{\kappa^2}{1-\alpha}\left[\rho - \frac{1}{2}\alpha(3\rho - p) + \frac{1}{2}\dot{\alpha}^2 + V\right], \tag{14.40}$$

$$2\dot{H} + 3H^2 = \frac{-\kappa^2}{1-\alpha}\left[p + \frac{\alpha}{2}(\rho - 3p) + \frac{1}{2}\dot{\alpha}^2 - V\right], \tag{14.41}$$

$$\ddot{\alpha} + 3H\dot{\alpha} + \frac{1}{2(1-\alpha)}\left[2\alpha(p-\rho) - \dot{\alpha}^2 + 4V\right]\frac{\partial V}{\partial \alpha} = 0, \tag{14.42}$$

where

$$\Box\alpha = \nabla_\nu\nabla^\nu\alpha = \frac{1}{\sqrt{-g}}\frac{\partial(\sqrt{-g}\nabla^\nu\alpha)}{\partial x^\nu} = \left(\ddot{\alpha} + \dot{\alpha}\frac{3\dot{a}}{a}\right). \tag{14.43}$$

The Energy Conservation Equation

We multiply now Eq. (14.40) with a^3, and take the time derivative of both its sides, and obtain

$$\begin{aligned} 3H^2 + 6\frac{\ddot{a}}{a} = \frac{\kappa^2}{1-\alpha}\Bigg\{ & 3\left[\rho - \frac{\alpha}{2}(3\rho - p) + \frac{\dot{\alpha}^2}{2} + V\right] \\ & + \frac{a}{\dot{a}}\left[\dot{\rho} - \frac{\dot{\alpha}}{2}(3\rho - p) - \frac{\alpha}{2}(3\dot{\rho} - \dot{p}) + \dot{\alpha}\ddot{\alpha} + \frac{\partial V}{\partial \alpha}\dot{\alpha}\right] \\ & + \frac{a}{\dot{a}}\frac{\dot{\alpha}}{1-\alpha}\left[\rho - \frac{1}{2}\alpha(3\rho - p) + \frac{1}{2}\dot{\alpha}^2 + V\right]\Bigg\}. \end{aligned} \tag{14.44}$$

Using Eq. (14.41) we find

$$\left[(1-\alpha)(\rho+p)+\dot{\alpha}^2\right]\frac{d}{dt}a^3+\left(1-\frac{3}{2}\alpha\right)a^3\frac{d}{dt}\rho=-\frac{a^3}{2}\left\{\alpha\dot{p}-\dot{\alpha}(3\rho-p)\right.$$

$$\left.+2\dot{\alpha}\ddot{\alpha}+2\frac{\partial V}{\partial\alpha}\dot{\alpha}+\frac{2\dot{\alpha}}{1-\alpha}\left[\rho-\frac{1}{2}\alpha(3\rho-p)+\frac{1}{2}\dot{\alpha}^2+V\right]\right\}, \tag{14.45}$$

which can be reformulated in a mathematically equivalent form as

$$\frac{d}{dt}\left[(1-\alpha)\rho a^3\right]+(1-\alpha)p\frac{d}{dt}a^3=\frac{\alpha}{2}a^3(\dot{\rho}-\dot{p})-\frac{a^3}{2}\frac{\dot{\alpha}}{1-\alpha}\left(p-\rho+2V+\dot{\alpha}^2\right)$$

$$-a^3\frac{\partial V}{\partial\alpha}\dot{\alpha}-a^3\left(\dot{\alpha}\ddot{\alpha}+\dot{\alpha}^2\frac{3\dot{a}}{a}\right)-a^3\rho\dot{\alpha}. \tag{14.46}$$

From the field equations (14.30) we can also obtain directly the same conservation equation. By considering now the explicit form of the energy-momentum tensor of matter, as given by Eq. (14.31), and the expression of $\theta_{\mu\nu}$, written as

$$\theta_{\mu\nu}=-pg_{\mu\nu}-2(\rho+p)u_\mu u_\nu, \tag{14.47}$$

we find

$$\nabla^\nu T_{\mu\nu}=\dot{\rho}+3(\rho+p)\frac{\dot{a}}{a}, \tag{14.48}$$

and

$$\nabla^\nu\theta_{\mu\nu}=-2\dot{\rho}-\dot{p}-6(\rho+p)\frac{\dot{a}}{a}, \tag{14.49}$$

respectively. Hence Eq. (14.30) can be written as

$$-2(1-\alpha)(\rho+p)\frac{3\dot{a}}{a}-2\left(1-\frac{3}{2}\alpha\right)\dot{\rho}-\alpha\dot{p}=2\frac{\partial V}{\partial\alpha}\dot{\alpha}$$

$$+\frac{\dot{\alpha}}{1-\alpha}\left[p-\rho+2V+2(1-\alpha)\Box\alpha+\dot{\alpha}^2\right]. \tag{14.50}$$

With the use of Eq. (14.43) we can substitute $\Box\alpha$ in the above equation, and thus through this calculation we reobtain again the conservation equation Eq. (14.46).

The Dimensionless form of the Cosmological Evolution Equation

To simplify the mathematical formulation of the present cosmological model we rescale first the field α and its potential V according to $\alpha\to\kappa\alpha$ and $V\to\kappa^2V$, respectively. As a next step we introduce a set of dimensionless cosmological variables (τ,h,r,P,v), defined by the scaling transformations

$$\tau=H_0t,\qquad H=H_0h,\qquad \rho=\frac{3H_0^2}{8\pi G}r,\qquad p=\frac{3H_0^2}{8\pi G}P,\qquad v=\frac{1}{H_0^2}V, \tag{14.51}$$

where H_0 denotes the present-day value of the Hubble function, $H_0 = H(t_0)$. Then, in terms of the new variables, the cosmological evolution equation can be transformed into the dimensionless form

$$3h^2 = \frac{1}{1-\alpha}\left[r - \frac{1}{2}\alpha(3r-P) + \frac{1}{2}\left(\frac{d\alpha}{d\tau}\right)^2 + v\right], \tag{14.52}$$

$$2\frac{dh}{d\tau} + 3h^2 = \frac{1}{1-\alpha}\left\{-\left[P + \frac{\alpha}{2}(r-3P)\right] - \left[\frac{1}{2}\left(\frac{d\alpha}{d\tau}\right)^2 - v\right]\right\}, \tag{14.53}$$

$$\frac{d^2\alpha}{d\tau^2} + 3h\frac{d\alpha}{d\tau} + \frac{1}{2(1-\alpha)}\left[2\alpha(P-r) - \left(\frac{d\alpha}{d\tau}\right)^2 + 4V\right] + \frac{\partial v}{\partial \alpha} = 0. \tag{14.54}$$

To close the system of cosmological evolution equations (14.52)–(14.54) we must provide the equation of state of the cosmological matter $P = P(r)$, and the mathematical form of the self-interaction potential v of the scalar field α. In the following, we will limit our discussion to the case of the dust Universe, satisfying the equation of state $P = 0$. Then, after introducing the notation $u = d\alpha/d\tau$, from Eq. (14.52) we find for the dimensionless energy density r of the matter the expression

$$r = \frac{1}{1-3\alpha/2}\left[3(1-\alpha)h^2 - \left(\frac{1}{2}u^2 + v\right)\right]. \tag{14.55}$$

By introducing instead of the cosmological time variable the redshift z as the independent variable, we obtain the following system of equations describing the cosmological evolution in the modified gravity theory with the perturbation tensor proportional to the metric tensor,

$$\frac{d\alpha}{dz} = -\frac{1}{1+z}\frac{u}{h}, \tag{14.56}$$

$$\frac{dh}{dz} = \frac{1}{2(1+z)h}\left[\frac{u^2}{1-\alpha} - \frac{6(\alpha-1)h^2 + u^2 + 2v}{2(1-3\alpha/2)}\right], \tag{14.57}$$

$$\frac{du}{dz} = \frac{3}{1+z}u - \frac{4(\alpha-1)(2v-3\alpha h^2) + (2-5\alpha)u^2}{4(1+z)(1-\alpha)(1-3\alpha/2)h} + \frac{1}{(1+z)h}\frac{\partial v}{\partial \alpha}. \tag{14.58}$$

The deceleration parameter can be represented as

$$q = (1+z)\frac{1}{h}\frac{dh}{dz} - 1 = \frac{1}{h^2}\left[\frac{u^2}{1-\alpha} - \frac{6(\alpha-1)h^2 + u^2 + 2v}{2(1-3\alpha/2)}\right] - 1. \tag{14.59}$$

For the self-interaction potential v of the scalar field we choose a Higgs-type expression, so that it is given by

$$v(\alpha) = \frac{\mu^2}{2}\alpha^2 - \frac{\lambda}{4}\alpha^4, \tag{14.60}$$

where $\mu^2 > 0$ and $\lambda > 0$ are constants.

In the following discussion, two cases will be considered. Firstly, we suppose that the cosmological evolution of the Universe dominated by the quantum fluctuations of the metric takes place in the minimum of the Higgs potential, so that $\partial v/\partial \alpha = 0$. This fixes the value of α as $\alpha = \pm\sqrt{\mu^2/\lambda}$, and the value of the potential as $v(\alpha) = \mu^4/4\lambda =$ constant. Secondly, we will analyze the cosmological expansion of the Universe in the presence of a time variable scalar field self-interaction potential having the "full" Higgs form as given by Eq. (14.60). After the functional form of the potential has been adopted, the system of nonlinear differential equations (14.56)–(14.58) must be integrated together with an appropriate set of initial conditions $\alpha(0) = \alpha_0$, $h(0) = 1$, and $u(0) = u_0$, respectively.

Cosmological evolution of the Universe in the minimum of the Higgs potential: The results of the numerical integration of the system of cosmological evolution equations (14.56)–(14.58) are presented, for different numerical values of the constant self-interaction potential $v = v_0 =$ constant, in Figs. 14.1–14.4. The initial values of the scalar field α and of u used in the integration of the system of the cosmological evolution equation are $\alpha(0) = 0.001$ and $u(0) = 0.34$, respectively.

The time evolution of the Hubble function is represented in Fig. 14.1. As one can see from the figure, the Hubble function is a monotonically increasing function of the redshift z (as a function of the cosmological time it is a monotonically decreasing function). The increasing nature of the Hubble function indicates that the Universe is expanding. The redshift variation of h is basically independent on the adopted numerical values of v_0, and at $z \in [0, 0.10]$, h the Hubble function

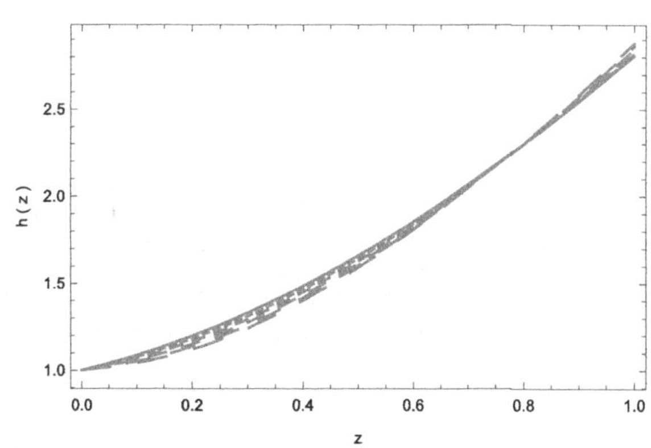

Figure 14.1 Variation with respect to the redshift of the dimensionless Hubble function for the Universe in the modified gravity model induced by the coupling between the metric and a Higgs-type scalar field, in the presence of a constant self-interaction potential $v = v_0$, for different values of v_0: $v_0 = 1.5$ (solid curve), $v_0 = 1.7$ (dotted curve), $v_0 = 1.9$ (short dashed curve), $v_0 = 2.1$ (dashed curve), $v_0 = 2.3$ (long dashed curve), and $v_0 = 2.5$ (ultra-long dashed curve), respectively.

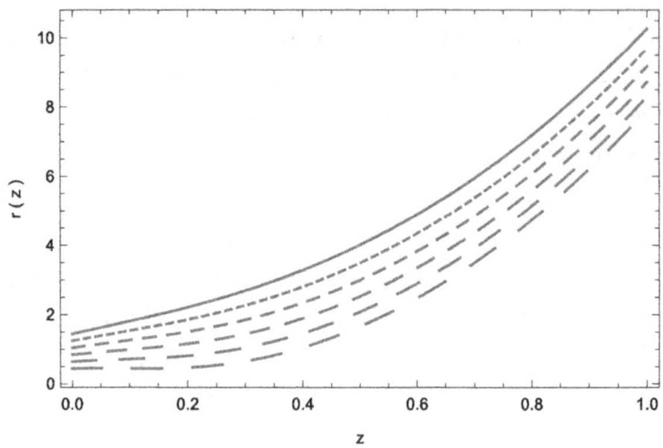

Figure 14.2 Variation with respect to the redshift of the dimensionless matter energy density for the Universe in the modified gravity model induced by the coupling between the metric and a Higgs-type scalar field, in the presence of a constant self-interaction potential $v = v_0$, for different values of v_0: $v_0 = 1.5$ (solid curve), $v_0 = 1.7$ (dotted curve), $v_0 = 1.9$ (short dashed curve), $v_0 = 2.1$ (dashed curve), $v_0 = 2.3$ (long dashed curve), and $v_0 = 2.5$ (ultra-long dashed curve), respectively.

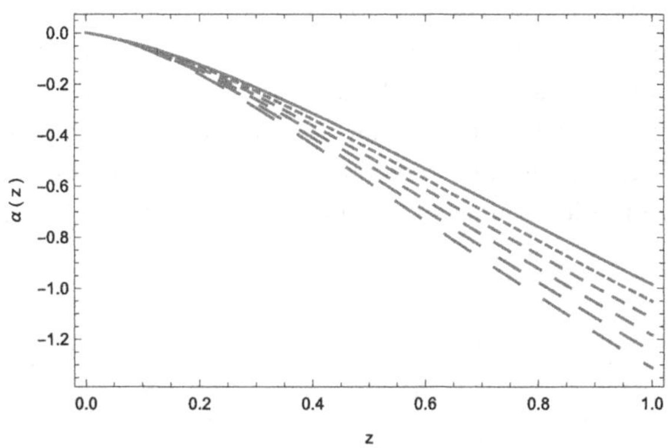

Figure 14.3 Variation with respect to the redshift of the scalar field α for the Universe in the modified gravity model induced by the coupling between the metric and a Higgs-type scalar field, in the presence of a constant self-interaction potential $v = v_0$, for different values of v_0: $v_0 = 1.5$ (solid curve), $v_0 = 1.7$ (dotted curve), $v_0 = 1.9$ (short dashed curve), $v_0 = 2.1$ (dashed curve), $v_0 = 2.3$ (long dashed curve), and $v_0 = 2.5$ (ultra-long dashed curve), respectively.

becomes approximately a constant, or a slowly varying function of the redshift. This indicates that the Universe has entered or is approaching a de Sitter-type phase, with exponential time increase of the scale factor. The time evolution of the energy density of the matter is presented in Fig. 14.2. Similarly to the Hubble function, the matter energy density r is a monotonically increasing function of the redshift. However, its variation indicates a significant dependence on the

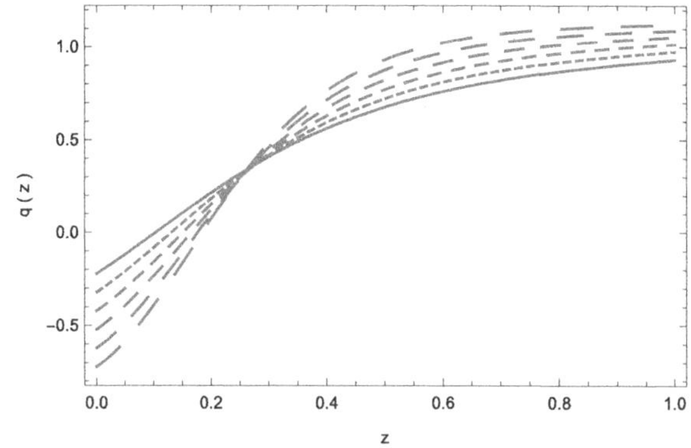

Figure 14.4 Variation with respect to the redshift of the deceleration parameter q for the Universe in the modified gravity model induced by the coupling between the metric and a Higgs-type scalar field, in the presence of a constant self-interaction potential $v = v_0$, for different values of v_0: $v_0 = 1.5$ (solid curve), $v_0 = 1.7$ (dotted curve), $v_0 = 1.9$ (short dashed curve), $v_0 = 2.1$ (dashed curve), $v_0 = 2.3$ (long dashed curve), and $v_0 = 2.5$ (ultra-long dashed curve), respectively.

numerical value of v_0. At the present time, and in this cosmological scenario, for the considered range of values of v_0, the density of the matter can either have values of the order of the critical cosmological density, corresponding to $r(0) \approx 1$, or alternatively, it may become negligibly small, or take values greater than the critical density. The scalar field α, whose behavior is shown in Fig. 14.3, monotonically decreases with the redshift, and for the adopted set of values of the cosmological parameters it becomes smaller than zero for all times. In the initial stages of the evolution of the Universe, corresponding to large redshifts values, the dynamics of the scalar field α indicates an important dependence on the numerical values of the constant potential v_0. But, once the redshift reaches the range of values $z \leq 0.2$, the variations in α, determined by the modifications of the numerical values of v_0, are negligibly small. For the redshift range $z \in [0, 0.05]$, the scalar field α becomes approximately a constant. On the other hand, the cosmological dynamics of the deceleration parameter q, portrayed in Fig. 14.4, indicates a complicated behavior. The Universe begins its evolution at a redshift $z = 1$ in a decelerating regime, with the decelerating parameter having values of the order $q \approx 1$. Then the Universe starts to accelerate, and it reaches the marginally accelerating phase, corresponding to $q = 0$, at a redshift having the numerical value of $z \approx 0.25$. This value is largely independent on the numerical values of v_0. Overall the redshift variation of the deceleration parameter is strongly dependent on the numerical values of the scalar field potential v_0. Depending on these numerical values of v_0, at the present time the evolution of the Universe can either be of de Sitter type, corresponding to $q = -1$, or it may have higher values of the deceleration parameter q, of the order of $q \in (-0.5, -0.25)$.

Cosmological evolution in the presence of the Higgs potential: The redshift evolution of the physical and cosmological parameters of a Universe containing a scalar field with a full Higgs potential given by Eq. (14.60), with the scalar field coupled to the fluctuating part of the quantum metric, is presented in Figs. 14.5–14.8. The gravitational field equations (14.56)–(14.58) have been integrated numerically in the redshift range $z \in [0, 2]$ with the use of the initial conditions $\alpha(0) = 0.11$ and $u(0) = 0.27$, respectively. The value of the coefficient

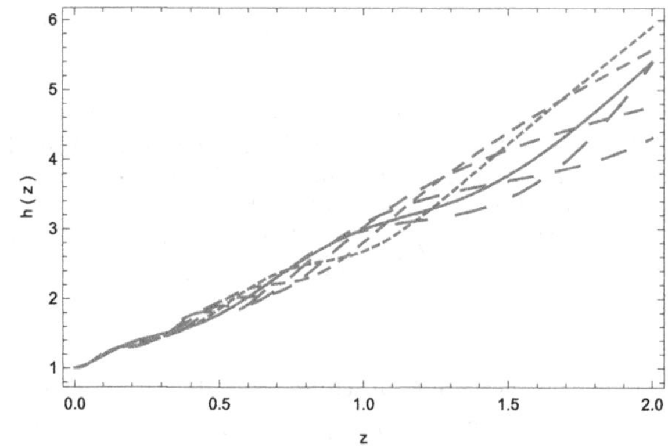

Figure 14.5 Variation with respect to the redshift z of the dimensionless Hubble function for the Universe filled with a Higgs-type scalar field coupled to the fluctuating quantum metric for $\lambda = 175$, and for different values of μ^2: $\mu^2 = 200$ (solid curve), $\mu^2 = 250$ (dotted curve), $\mu^2 = 300$ (short dashed curve), $\mu^2 = 350$ (dashed curve), $\mu^2 = 400$ (long dashed curve), and $\mu^2 = 450$ (ultra-long dashed curve), respectively.

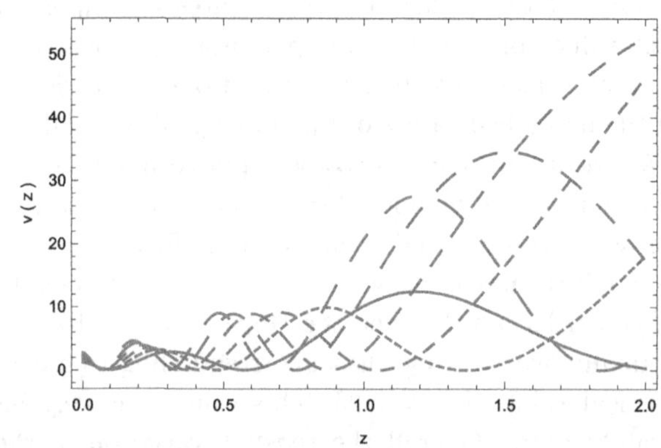

Figure 14.6 Variation with respect to the redshift z of the potential v in the modified gravity model induced by the coupling between the metric and a Higgs-type scalar field for $\lambda = 150$, and for different values of μ^2: $\mu^2 = 200$ (solid curve), $\mu^2 = 250$ (dotted curve), $\mu^2 = 300$ (short dashed curve), $\mu^2 = 350$ (dashed curve), $\mu^2 = 400$ (long dashed curve), and $\mu^2 = 450$ (ultra-long dashed curve), respectively.

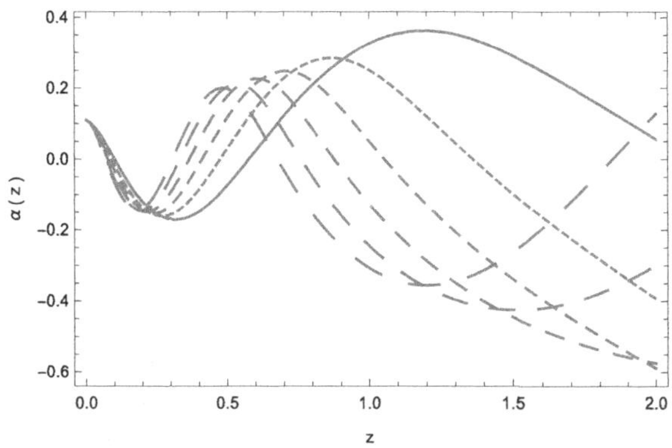

Figure 14.7 Variation with respect to the redshift z of the scalar field α in the modified gravity model induced by the coupling between the metric and a Higgs-type scalar field, for $\lambda = 150$, and for different values of μ^2: $\mu^2 = 200$ (solid curve), $\mu^2 = 250$ (dotted curve), $\mu^2 = 300$ (short dashed curve), $\mu^2 = 350$ (dashed curve), $\mu^2 = 400$ (long dashed curve), and $\mu^2 = 450$ (ultra-long dashed curve), respectively.

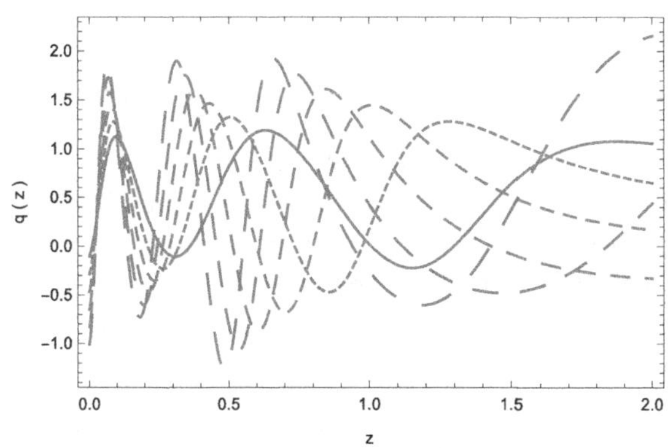

Figure 14.8 Variation with respect to the redshift z of the deceleration parameter q in the modified gravity model induced by the coupling between the metric and a Higgs-type scalar field, for a $\lambda = 150$, and for different values of μ^2:$\mu^2 = 200$ (solid curve), $\mu^2 = 250$ (dotted curve), $\mu^2 = 300$ (short dashed curve), $\mu^2 = 350$ (dashed curve), $\mu^2 = 400$ (long dashed curve), and $\mu^2 = 450$ (ultra-long dashed curve), respectively.

λ in the Higgs potential was fixed as $\lambda = 175$, and through numerical integration different cosmological scenarios have been obtained by varying the mass $\mu^2 > 0$ of the Higgs field.

The redshift evolution of the Hubble function, depicted in Fig. 14.5, shows that h is an increasing function of the redshift. This result indicates the presence of an expansionary dynamics of the Universe. On the other hand, it is important to note that in this model h has a complex behavior, showing a strong dependence at high redshifts on the numerical value of the mass μ^2 of the scalar field. In

Fig. 14.6 one can see the variation with z of the Higgs potential. The Higgs potential has a damped harmonic oscillator-type dynamics, with the amplitude of the oscillations decreasing once we are coming closer to the present age of the Universe, corresponding to $z = 0$. The redshift evolution of the scalar field α, portrayed in Fig. 14.7, shows the same damped oscillator-type pattern. The variation with respect to the redshift z of the deceleration parameter q, depicted in Fig. 14.8, indicates that an oscillatory-type behavior is also characteristic for q. The accelerating/decelerating nature of the cosmological evolution is strongly influenced by the numerical values of the mass of the scalar field μ^2. For example, depending on the value of μ^2, at $z = 2$ the Universe can be in either an accelerating phase, with $q < 0$, or in a decelerating state, with $q > 0$. The accelerating and decelerating epochs alternate in the high redshift range, but the Universe enters a very rapid accelerating phase at $z \approx 0.10$, which is (almost) independent of the values of μ^2. During this very short cosmological time interval the deceleration parameter significantly decreases from $q \approx 1.75$ to $q \approx -1$. Hence for this model the de Sitter solution is also an attractor, and independently of its initial state the cosmological evolution always ends in an exponentially expanding phase.

Scalar Function Coupling to the Metric

With the assumption that α is a simple function of the coordinates, without any specific physical/material field associated to it, we can obtain a second example of a modified gravity theory induced by a quantum perturbation tensor of the form (14.22). Then, from Eq. (14.5), by using the least action principle, we can easily obtain the gravitational field equations in the form

$$R_{\mu\nu} - \frac{1}{2}Rg_{\mu\nu} = \frac{2\kappa^2}{1-\alpha(x)}\left\{\frac{1}{2}\left[1+\alpha(x)\right]T_{\mu\nu} - \frac{1}{4}\alpha(x)Tg_{\mu\nu} + \frac{1}{2}\alpha(x)\theta_{\mu\nu}\right\}, \tag{14.61}$$

where we have denoted $\theta_{\mu\nu} = g^{\alpha\beta}\left(\delta T_{\alpha\beta}/\delta g^{\mu\nu}\right)$, and

$$\frac{\delta T}{\delta g^{\mu\nu}} = T_{\mu\nu} + \theta_{\mu\nu}, \tag{14.62}$$

respectively. As usual, $T = T_{\mu\nu}g^{\mu\nu}$ denotes the trace of the energy-momentum tensor, and we have also introduced the notation $\theta = \theta_{\mu\nu}g^{\mu\nu}$. By contracting Eq (14.61) we find

$$R = -\kappa^2\left[T + \frac{\alpha(x)}{1-\alpha(x)}\theta\right]. \tag{14.63}$$

The combination of Eqs. (14.61) and (14.63) provides the gravitational field equations in the form

$$R_{\mu\nu} = \frac{2\kappa^2}{1-\alpha(x)}\left[\frac{1}{2}(1+\alpha(x))T_{\mu\nu} - \frac{1}{4}Tg_{\mu\nu} + \frac{1}{2}\alpha(x)\theta_{\mu\nu} - \frac{1}{4}\alpha(x)\theta g_{\mu\nu}\right]. \tag{14.64}$$

We suppose now that the matter Lagrangian density $\mathcal{L}_m$ is a function of the metric tensor only, and it does not depend on the derivatives of $g_{\mu\nu}$. Thus for the energy-momentum tensor we find the expression

$$T_{\mu\nu} = 2\frac{\partial \mathcal{L}_m}{\partial g^{\mu\nu}} - g_{\mu\nu}\mathcal{L}_m, \tag{14.65}$$

while for the tensor $\theta_{\mu\nu}$ we obtain

$$\theta_{\mu\nu} = 2g^{\alpha\beta}\frac{\partial^2 \mathcal{L}_m}{\partial g^{\alpha\beta}\partial g^{\mu\nu}} - g_{\mu\nu}\mathcal{L}_m - 2T_{\mu\nu}, \tag{14.66}$$

respectively. Hence different forms of the Lagrangian densities of the matter may lead to different functional forms of $\theta_{\mu\nu}$, therefore leading to different classes of gravitational theories.

We take now the covariant divergence of Eq. (14.61). Using the Bianchi identities, $\nabla^\nu G_{\mu\nu} = 0$, we obtain

$$\begin{aligned}
0 = {} & \frac{\kappa^2\left(\nabla^\nu\alpha\right)}{2(1-\alpha)^2}\left[2(1+\alpha)T_{\mu\nu} - \alpha T g_{\mu\nu} + 2\alpha\theta_{\mu\nu}\right] \\
& + \frac{\kappa^2}{2(1-\alpha)}\Big[2\left(\nabla^\nu\alpha\right)T_{\mu\nu} + 2(1+\alpha)\nabla^\nu T_{\mu\nu} - \left(\nabla^\nu\alpha\right)Tg_{\mu\nu} \\
& -\alpha(\nabla^\nu T)g_{\mu\nu} + 2\left(\nabla^\nu\alpha\right)\theta_{\mu\nu} + 2\alpha(\nabla^\nu\theta_{\mu\nu})\Big] \\
= {} & \left(\nabla^\nu\alpha\right)\left[\frac{2(1+\alpha)T_{\mu\nu} - \alpha T g_{\mu\nu} + 2\alpha\theta_{\mu\nu}}{1-\alpha} + 2T_{\mu\nu} - Tg_{\mu\nu} + 2\theta_{\mu\nu}\right] \\
& + \alpha\left[2\nabla^\nu T_{\mu\nu} - g_{\mu\nu}\nabla^\nu T + 2\nabla^\nu\theta_{\mu\nu}\right] + 2\nabla^\nu T_{\mu\nu}.
\end{aligned} \tag{14.67}$$

Hence, it is straightforward to check now that the divergence of the energy-momentum tensor matter is given by

$$\nabla^\nu T_{\mu\nu} = -\frac{1}{2(1+\alpha)}\left[\left(\nabla^\nu\alpha\right)\left(\frac{4T_{\mu\nu} - Tg_{\mu\nu} + 2\theta_{\mu\nu}}{1-\alpha}\right) + \alpha\left(-g_{\mu\nu}\nabla^\nu T + 2\nabla^\nu\theta_{\mu\nu}\right)\right]. \tag{14.68}$$

Cosmological Implications of the Scalar Function-Metric Coupling

In the following analysis, we consider again that the Universe is filled with a perfect fluid characterized by two thermodynamic parameters only, the matter energy density ρ, and the thermodynamic pressure p. In the comoving frame, associated to the expansionary evolution of the Universe, the components of the matter four-velocity are $u^\mu = (1,0,0,0)$. Therefore the energy-momentum tensor of the matter is diagonal, and has the components $T^\mu_\nu = \mathrm{diag}\,(\rho, -p, -p, -p)$. Moreover, we will assume that the matter content of the Universe satisfies at all times a linear barotropic equation of state given by $p = \omega\rho$, with $\omega = \text{constant}$, and $0 \leq \omega \leq 1$. Then from the gravitational field equations (14.61) we can obtain the generalized Friedmann equations that describe the cosmological evolution of the Universe, and which are given by

$$H^2 = \frac{\kappa^2}{3}\left[\frac{2-(3-\omega)\alpha}{2(1-\alpha)}\right]\rho, \tag{14.69}$$

$$\frac{\ddot{a}}{a} = -\frac{\kappa^2}{6}\left[\frac{1+(3-4\alpha)\omega}{1-\alpha}\right]\rho, \tag{14.70}$$

or, equivalently,

$$\frac{\ddot{a}}{a} = -\frac{1+(3-4\alpha)\omega}{2-(3-\omega)\alpha}H^2. \tag{14.71}$$

Alternatively, Eq. (14.70) can be reformulated as

$$2\dot{H}+3H^2 = -\kappa^2\left[\frac{2\omega+\alpha(1-3\omega)}{2(1-\alpha)}\right]\rho. \tag{14.72}$$

Therefore the time variation of the Hubble function is described by the equation

$$\dot{H} = -\frac{\kappa^2}{2}(1+\omega)\,\rho = \frac{3(\alpha-1)(\omega+1)}{\alpha(\omega-3)+2}H^2. \tag{14.73}$$

The deceleration parameter q of the cosmological model can be obtained as

$$q = \frac{-\alpha(7\omega+3)+6\omega+4}{\alpha(\omega-3)+2}. \tag{14.74}$$

The accelerated expansion requires $\ddot{a} > 0$, which imposes on the function $\alpha(t)$ the condition

$$\frac{\alpha(t)}{1-\alpha(t)} > \frac{1+3\omega}{4\omega}\frac{1}{1-\alpha(t)}, \qquad \forall t \geq t_a. \tag{14.75}$$

Conservative Models: $\nabla^\nu \mathrm{T}_{\mu\nu} = 0$

Standard General Relativity has the important feature that the matter energy-momentum tensor is conserved, i.e., $\nabla^\nu T_{\mu\nu} = 0$. On the other hand, due to the curvature-matter coupling, as well as in the presence of quantum effects, in modified theories of gravity the classically defined energy-momentum tensor of matter is not always conserved. However, due to the inclusion of the function α in the theoretical formalism, in the modified gravity theories induced by the quantum fluctuations of the metric, with the expectation value of the fluctuation tensor $K_{\mu\nu}$ proportional to the metric tensor $g_{\mu\nu}$, we can still keep the conservation of the energy and momentum by imposing some appropriate constraints on the function α. In the following argument we suppose, as before, for the matter Lagrangian density the relation $L_m = p$. Then, because the second derivatives of the matter Lagrangian L_m with respect to the metric tensor identically vanish, for the tensor $\theta_{\mu\nu}$ we obtain the expression

$$\theta_{\mu\nu} = -g_{\mu\nu}L_m - 2T_{\mu\nu}. \tag{14.76}$$

Imposing the matter conservation condition $\nabla^\nu T_{\mu\nu} = 0$, we obtain

$$0 = (\nabla^\nu \alpha)\left[\frac{2(1+\alpha)T_{\mu\nu} - \alpha T g_{\mu\nu} + 2\alpha\theta_{\mu\nu}}{1-\alpha} + 2T_{\mu\nu} - Tg_{\mu\nu} + 2\theta_{\mu\nu}\right]$$
$$+\alpha\left(2\nabla^\nu\theta_{\mu\nu} - g_{\mu\nu}\nabla^\nu T\right). \tag{14.77}$$

Multiplying both sides of Eq (14.77) by the four-velocity u^μ we find

$$\dot{\alpha}\left(\frac{-\rho+p}{1-\alpha}\right) + \alpha\left(-5\dot{\rho}+\dot{p}\right) = 0, \tag{14.78}$$

giving

$$\frac{\dot{\alpha}}{\alpha(1-\alpha)} = \frac{-5\dot{\rho}+\dot{p}}{\rho - p}. \tag{14.79}$$

The use in the above equations of the simple linear barotropic equation of state given by $p = \omega\rho$, yields for $\omega \neq 1$ the differential equation relating α and the matter density,

$$\frac{\dot{\alpha}}{\alpha(1-\alpha)} = -\left(\frac{5-\omega}{1-\omega}\right)\frac{\dot{\rho}}{\rho}, \tag{14.80}$$

from which we can immediately find the density as a function of α, so that

$$\rho = \rho_0\left(\frac{\alpha}{1-\alpha}\right)^{-\frac{1-\omega}{5-\omega}}, \tag{14.81}$$

where ρ_0 is an arbitrary constant of integration. The conservation of the energy-momentum tensor of the matter gives for the time variation of the density the equation

$$-3\frac{\dot{a}}{a}(1+\omega) = \frac{\dot{\rho}}{\rho}, \tag{14.82}$$

from which the matter energy density can be obtained with the use of the standard relation

$$\rho = \rho_0' a^{-3(1+\omega)}, \tag{14.83}$$

where by ρ_0' we have denoted an arbitrary constant of integration. The scale factor dependence of the function α can be obtained from Eqs. (14.81) and (14.83) as

$$\frac{\alpha}{1-\alpha} = \alpha_0 a^{3n}, \tag{14.84}$$

where we have denoted $n = (1+\omega)(5-\omega)/(1-\omega)$, and $\alpha_0 = (\rho_0'/\rho_0)^{-(5-\omega)/(1-\omega)}$, respectively. Then the first Friedmann equation Eq. (14.69) can be written as

$$\frac{\dot{a}}{a} = \frac{\kappa}{\sqrt{6}}\sqrt{\rho_0 a^{-3(\omega+1)}\left(\alpha_0(\omega-1)a^{3n}+2\right)}, \tag{14.85}$$

which can be easily integrated to give

$$\frac{\kappa}{\sqrt{6}}(t-t_0) = \frac{\sqrt{2\alpha_0(\omega-1)a^{3n}+4}}{3(\omega+1)} \frac{{}_2F_1\left(\frac{1}{2}, \frac{\omega+1}{2n}; \frac{\omega+1}{2n}+1; -\frac{1}{2}a^{3n}\alpha_0(\omega-1)\right)}{\sqrt{\rho_0 a^{-3(\omega+1)}\left(\alpha_0(\omega-1)a^{3n}+2\right)}}, \tag{14.86}$$

where ${}_2F_1(a,b;c,z)$ denotes the hypergeometric function, defined as

$${}_2F_1(a,b;c,z) = \sum_{k=0}^{\infty}\left[(a)_k(b)_k/(c)_k\right]\left(z^k/k!\right), \tag{14.87}$$

while t_0 is an arbitrary constant of integration. Finally, for the deceleration parameter q we obtain the expression

$$q(a) = \frac{1}{2}\left\{n\left[\frac{6}{\alpha_0(\omega-1)a^{3n}+2}-3\right]+3\omega+1\right\}. \tag{14.88}$$

In the important limit of small values of the scale factor, when the condition $\alpha_0(\omega-1)a^{3n} \ll 2$ is satisfied, for the deceleration parameter we obtain the approximate expression $q \approx (1+3\omega)/2 > 0$. This result indicates a decelerating expansion during the very early stages of the cosmological evolution of the Universe. On the other hand, for $\alpha_0(\omega-1)a^{3n} \gg 2$, and for very large values of the scale factor a, the deceleration parameter can be approximated as $q \approx [3(\omega-n)+1]/2 = 4(\omega+2)/(1-\omega)$. Generally, for any physically realistic cosmological matter equation of state, the condition $\omega < 1$ must hold. Therefore from the analysis in this subsection, it follows that in both small and large time limits in the modified gravity theory induced by quantum metric fluctuations with the conserved matter energy-momentum tensor, the cosmological time evolution of the Universe is decelerating.

Nonconservative Cosmological Models with $\nabla^\nu T_{\mu\nu} \neq 0$

We consider now the class of the nonconservative models induced by the quantum metric fluctuations with a scalar function coupling to the metric. From Eq. (14.68), by taking into account that $T = -\rho + 3p$, we immediately find

$$-2(1+\alpha)\nabla^\nu T_{\mu\nu} = \nabla^\nu \alpha \frac{4T_{\mu\nu} - Tg_{\mu\nu} + 2\theta_{\mu\nu}}{1-\alpha} + \alpha\left(2\nabla^\nu\theta_{\mu\nu} - g_{\mu\nu}T\right). \tag{14.89}$$

Equivalently, the above equation can be written as

$$-2(1-\alpha)\nabla^\nu T_{\mu\nu} = \nabla^\nu \alpha \frac{g_{\mu\nu}(2L_m - T)}{1-\alpha} + \alpha g_{\mu\nu}\nabla^\nu(2L_m - T). \tag{14.90}$$

After multiplication of Eq. (14.90) with the four-velocity vector u^μ, with components defined in the comoving reference frame, we easily find

$$u^\mu \nabla^\nu T_{\mu\nu} = \frac{1}{2(1-\alpha)}\left[\dot{\alpha}\frac{\rho - p}{1-\alpha} + \alpha(\dot{\rho} - \dot{p})\right]. \tag{14.91}$$

With the help of the mathematical identity

$$u^\mu \nabla^\nu T_{\mu\nu} = \dot{\rho} + 3H(\rho + p) \tag{14.92}$$

we obtain immediately

$$\dot{\rho} + 3H(\rho + p) = \frac{1}{2(1-\alpha)}\left[\dot{\alpha}(\rho - p)\left(\frac{1}{1-\alpha}\right) + \alpha(\dot{\rho} - \dot{p})\right]. \tag{14.93}$$

We assume now that the matter content of the Universe satisfies a general linear barotropic equation of state, with the parameter of the equation of state ω a function of time, $\omega = \omega(t)$, so that $p = \omega(t)\rho$. Then Eq. (14.93) takes the form

$$\left(1 - \frac{3}{2}\alpha + \frac{1}{2}\omega\alpha\right)\dot{\rho} + \frac{1}{2}\alpha\rho\dot{\omega} = 3(\alpha - 1)(1+\omega)H\rho + \frac{\dot{\alpha}(\omega - 1)}{2(\alpha - 1)}\rho. \tag{14.94}$$

For $\omega =$ constant we find

$$\left(1 - \frac{3}{2}\alpha + \frac{1}{2}\omega\alpha\right)\dot{\rho} = 3(\alpha - 1)(1+\omega)H\rho + \frac{\dot{\alpha}}{2(\alpha - 1)}(\omega - 1)\rho. \tag{14.95}$$

We take now the derivative of Eq. (14.69) with respect to time. After replacing $\dot{H}$ by using Eq. (14.73), for the time derivative of the density of matter we obtain the evolution equation

$$\dot{\rho} = -\frac{6H^2\left[6(\omega+1)(\alpha-1)^2 H + (\omega - 1)\dot{\alpha}\right]}{\kappa^2\left[(\omega - 3)\alpha + 2\right]^2}. \tag{14.96}$$

Substitution of Eq. (14.96) into Eq. (14.95) provides the differential equation

$$\left[6(\omega+1)H(\alpha-1)^2 + (\omega-1)\dot{\alpha}\right]\left\{6H^2(\alpha-1) + \kappa^2\rho\left[(\omega-3)\alpha+2\right]\right\} = 0. \tag{14.97}$$

By taking into account the the first generalized Friedmann equation (14.69) of our cosmological model, it turns out that Eq. (14.97) is identically satisfied during the entire period of the cosmological evolution of the Universe. However, by imposing independently the condition that the first term in Eq. (14.97) also vanishes identically, a second solution of the gravitational field equations can be constructed. Here we present this solution, and investigate its cosmological properties.

The case $\alpha = 1 - \frac{1-\omega}{6(1+\omega)} \frac{1}{\ln(a/a_0)}$: By imposing the requirement that the first term in the left-hand side of Eq. (14.97) identically vanishes, for the time evolution of the scalar function $\alpha(t)$ we find the differential equation

$$\dot{\alpha} = \frac{6(1+\omega)(1-\alpha)^2}{1-\omega} H. \tag{14.98}$$

This equation can be immediately integrated to give

$$\alpha(t) = 1 - \frac{1-\omega}{6(1+\omega)} \frac{1}{\ln\left[a(t)/a_0\right]}, \tag{14.99}$$

where a_0 is an arbitrary constant of integration. By substituting this result for α into Eq. (14.71), for the description of the time evolution of the scale factor we obtain the following second-order differential equation

$$\frac{\ddot{a}}{\dot{a}} = \left[1 - \frac{3(\omega+1)}{6(\omega+1)\ln(Ca)+\omega-3}\right]\left(\frac{\dot{a}}{a}\right). \tag{14.100}$$

After a first integration we find

$$H = \frac{\dot{a}}{a} = \frac{\zeta}{\sqrt{6(\omega+1)\ln\left(a/a_0\right)+\omega-3}}, \tag{14.101}$$

where an arbitrary integration constant has been denoted by ζ. Hence the time evolution of the scale factor can be obtained as

$$a(t) = a_0 e^{-\frac{\omega-3}{6(1+\omega)}} e^{\frac{3^{1/3}\zeta^{2/3}}{2(1+\omega)^{1/3}}(t-t_0)^{2/3}}, \tag{14.102}$$

where t_0 is an arbitrary constant of integration. The time variation of the Hubble function is given by

$$H(t) = \frac{\zeta^{2/3}}{3^{2/3}\sqrt[3]{\omega+1}\sqrt[3]{t-t_0}}, \tag{14.103}$$

while for this nonconservative model the deceleration parameter q can be found as

$$q(t) = \frac{(1+\omega)^{1/3}}{3^{1/3}\zeta^{2/3}\left(t-t_0\right)^{2/3}} - 1. \tag{14.104}$$

In the limit of large cosmological times $t \to \infty$, for the deceleration parameter we obtain $q \to -1$, and therefore in this limit the Universe ends in an (approximately de Sitter) exponentially accelerating state. The time variation of the coupling function α can be found as

$$\alpha(t) = \frac{1-\omega}{-3^{4/3}\zeta^{2/3}(\omega+1)^{2/3}\left(t-t_0\right)^{2/3}+\omega-3} + 1. \tag{14.105}$$

It is easy to check now that in the limit of large times $\alpha(t)$ tends to a constant, and we have $\lim_{t\to\infty}\alpha(t)=1$.

14.4 Quantum Metric Fluctuations Proportional to the Energy-Momentum Tensor

In this second part of the book, we have investigated a number of extensions of GR, which are constructed starting from the premise that in certain astrophysical and cosmological processes matter and geometry may nonminimally interact with each each other [44, 211, 243]. Such classes of gravitational theories are also an outcome of the first-order approximation of quantum gravity theories with a fluctuating metric tensor. In the present section, we will introduce another such theory with geometry and matter coupling, which is obtained by assuming that the average of the metric quantum fluctuations tensor $K_{\mu\nu}$ is proportional to the energy-momentum tensor of the matter $T_{\mu\nu}$, so that $K^{\mu\nu}=\alpha T^{\mu\nu}$, where α is a constant. This modified gravitational theory also propounds that the presence of the classical matter may also significantly influence the nature and structure of the quantum perturbations of the geometric framework of the space-time.

14.4.1 The Gravitational Field Equations

After adopting the form $K^{\mu\nu}=\alpha T^{\mu\nu}$ for the classical form of the average of the quantum fluctuations of the metric tensor, for the first-order quantum corrected gravitational Lagrangian we find the representation

$$\begin{aligned}\mathcal{L}&=-\frac{1}{2k^2}\sqrt{-g}\left[R+\alpha\left(R_{\mu\nu}-\frac{1}{2}Rg_{\mu\nu}\right)T^{\mu\nu}\right]+\sqrt{-g}\left[L_m+\frac{1}{2}\alpha T_{\mu\nu}T^{\mu\nu}\right]\\&=-\frac{1}{2k^2}\sqrt{-g}\left[R\left(1-\frac{1}{2}\alpha T\right)+\alpha R_{\mu\nu}T^{\mu\nu}\right]+\sqrt{-g}\left[\mathcal{L}_m+\frac{1}{2}\alpha T_{\mu\nu}T^{\mu\nu}\right].\end{aligned}\tag{14.106}$$

In order to obtain the field equations of the $K_{\mu\nu}=\alpha T_{\mu\nu}$-modified gravity model, we need to vary the gravitational action (14.106) with respect to the metric tensor $g_{\mu\nu}$. The variation can be done easily, and thus we obtain

$$\begin{aligned}\delta\mathcal{L}=-\frac{1}{2k^2}&\left[\sqrt{-g}G_{\mu\nu}\left(1-\frac{1}{2}\alpha T\right)\delta g^{\mu\nu}-\frac{1}{2}\alpha\sqrt{-g}R\left(T_{\mu\nu}+\theta_{\mu\nu}\right)\delta g^{\mu\nu}\right.\\&\left.+\alpha\delta(\sqrt{-g}R_{\mu\nu}T^{\mu\nu})+\sqrt{-g}\left(1-\frac{1}{2}\alpha T\right)(g_{\mu\nu}\Box-\nabla_\mu\nabla_\nu)\delta g^{\mu\nu}\right]\\&+\left[\frac{1}{2}\sqrt{-g}T_{\mu\nu}\delta g^{\mu\nu}+\frac{1}{2}\alpha\delta(\sqrt{-g}T_{\mu\nu}T^{\mu\nu})\right]\end{aligned}$$

$$
\begin{aligned}
&= -\frac{1}{2k^2}\Bigg[\sqrt{-g}G_{\mu\nu}\left(1-\frac{1}{2}\alpha T\right)\delta g^{\mu\nu} - \frac{1}{2}\alpha\sqrt{-g}R(T_{\mu\nu}+\theta_{\mu\nu})\delta g^{\mu\nu} \\
&\quad -\frac{1}{2}\alpha g_{\mu\nu}\sqrt{-g}R_{\alpha\beta}T^{\alpha\beta}\delta g^{\mu\nu} + \sqrt{-g}\delta g^{\mu\nu}\left(g_{\mu\nu}\Box - \nabla_\mu\nabla_\nu\right)\left(1-\frac{1}{2}\alpha T\right)\Bigg] \\
&\quad -\left(\frac{1}{4}\alpha g_{\mu\nu}\sqrt{-g}T_{\alpha\beta}T^{\alpha\beta}\delta g^{\mu\nu} - \frac{1}{2}\sqrt{-g}T_{\mu\nu}\delta g^{\mu\nu}\right) \\
&\quad +\frac{1}{2}\alpha\sqrt{-g}\delta\left(T_{\mu\nu}T^{\mu\nu}\right) - \frac{\alpha}{2k^2}\sqrt{-g}\delta\left(R_{\mu\nu}T^{\mu\nu}\right).
\end{aligned} \tag{14.107}
$$

We calculate now the variation of the terms $\delta(T_{\mu\nu}T^{\mu\nu})$ and $\delta(R_{\mu\nu}T^{\mu\nu})$, respectively. The calculations can be performed by taking into account the following already introduced results, which for the reader's convenience are summarized again:

$$
\begin{aligned}
\delta\Gamma^\lambda_{\mu\nu} &= \frac{1}{2}g^{\lambda\sigma}\left(\nabla_\nu\delta g_{\mu\sigma} + \nabla_\mu\delta g_{\nu\sigma} - \nabla_\sigma\delta g_{\mu\nu}\right), \\
\nabla_\lambda\delta\Gamma^\lambda_{\mu\nu} &= \frac{1}{2}\left(\nabla^\sigma\nabla_\nu\delta g_{\mu\sigma} + \nabla^\sigma\nabla_\mu\delta g_{\nu\sigma} - \nabla^\sigma\nabla_\sigma\delta g_{\mu\nu}\right), \\
\delta\Gamma^\lambda_{\mu\lambda} &= \frac{1}{2}g^{\lambda\sigma}\left(\nabla_\lambda\delta g_{\mu\sigma} + \nabla_\mu\delta g_{\lambda\sigma} - \nabla_\sigma\delta g_{\mu\lambda}\right), \\
\nabla_\nu\delta\Gamma^\lambda_{\mu\lambda} &= \frac{1}{2}\left(\nabla_\nu\nabla^\sigma\delta g_{\mu\sigma} + g^{\lambda\sigma}\nabla_\nu\nabla_\mu\delta g_{\lambda\sigma} - \nabla_\nu\nabla^\lambda\delta g_{\mu\lambda}\right).
\end{aligned}
$$

Hence, we obtain

$$
\begin{aligned}
T^{\mu\nu}\delta R_{\mu\nu} &= \frac{1}{2}T^{\mu\nu}\left(\nabla^\sigma\nabla_\nu\delta g_{\mu\sigma} + \nabla^\sigma\nabla_\mu\delta g_{\nu\sigma} - \nabla^\sigma\nabla_\sigma\delta g_{\mu\nu} - g^{\lambda\sigma}\nabla_\nu\nabla_\mu\delta g_{\lambda\sigma}\right) \\
&= \frac{\delta g^{\mu\nu}}{2}\left(\Box T_{\mu\nu} + \nabla_\alpha\nabla_\beta T^{\alpha\beta}g_{\mu\nu} - \nabla_\alpha\nabla_{(\mu}T^\alpha_{\nu)}\right).
\end{aligned} \tag{14.108}
$$

The variation of the terms $R_{\mu\nu}\delta T^{\mu\nu}$ and $\delta R_{\mu\nu}T^{\mu\nu}$ can be found as

$$
\begin{aligned}
R_{\mu\nu}\delta T^{\mu\nu} &= R_{\mu\nu}\delta(g^{\mu\alpha}g^{\nu\beta}T_{\alpha\beta}) = R^{\alpha\beta}\delta T_{\alpha\beta} + R_{\alpha(\nu}T^\alpha_{\mu)}\delta g^{\mu\nu} = R_{\alpha(\nu}T^\alpha_{\mu)}\delta g^{\mu\nu} \\
&\quad +\delta g^{\mu\nu}\left[2R^{\alpha\beta}\frac{\partial^2\mathcal{L}_m}{\partial g^{\mu\nu}\partial g^{\alpha\beta}} - \frac{1}{2}R(g_{\mu\nu}\mathcal{L}_m + T_{\mu\nu}) - \mathcal{L}_m\frac{R^{\alpha\beta}\delta g_{\alpha\beta}}{\delta g^{\mu\nu}}\right] \\
&= \delta g^{\mu\nu}\left[R_{\alpha(\mu}T^\alpha_{\nu)} - \frac{1}{2}R(g_{\mu\nu}\mathcal{L}_m + T_{\mu\nu}) + \mathcal{L}_m R_{\mu\nu} + 2R^{\alpha\beta}\frac{\partial^2\mathcal{L}_m}{\partial g^{\mu\nu}\partial g^{\alpha\beta}}\right],
\end{aligned} \tag{14.109}
$$

$$
\begin{aligned}
\delta R_{\mu\nu}T^{\mu\nu} &= \delta g^{\mu\nu}\Bigg[R_{\alpha(\mu}T^\alpha_{\nu)} - \frac{1}{2}R\left(g_{\mu\nu}\mathcal{L}_m + T_{\mu\nu}\right) + \mathcal{L}_m R_{\mu\nu} + 2R^{\alpha\beta}\frac{\partial^2\mathcal{L}_m}{\partial g^{\mu\nu}\partial g^{\alpha\beta}} \\
&\quad +\frac{1}{2}\left(\Box T_{\mu\nu} + \nabla_\alpha\nabla_\beta T^{\alpha\beta}g_{\mu\nu} - \nabla_\alpha\nabla_{(\mu}T^\alpha_{\nu)}\right)\Bigg].
\end{aligned} \tag{14.110}
$$

Similarly,

$$T_{\mu\nu}\delta T^{\mu\nu} = \delta g^{\mu\nu}\left[T_{\alpha(\mu}T^{\alpha}_{\nu)} - \frac{1}{2}T\left(g_{\mu\nu}\mathcal{L}_m + T_{\mu\nu}\right) + \mathcal{L}_m T_{\mu\nu} + 2T^{\alpha\beta}\frac{\partial^2\mathcal{L}_m}{\partial g^{\mu\nu}\partial g^{\alpha\beta}}\right], \tag{14.111}$$

$$T^{\mu\nu}\delta T_{\mu\nu} = \delta g^{\mu\nu}\left[-\frac{1}{2}T(g_{\mu\nu}\mathcal{L}_m + T_{\mu\nu}) + L_m T_{\mu\nu} + 2T^{\alpha\beta}\frac{\partial^2\mathcal{L}_m}{\partial g^{\mu\nu}\partial g^{\alpha\beta}}\right], \tag{14.112}$$

$$\delta T_{\mu\nu}T^{\mu\nu} = \delta g^{\mu\nu}\left[T_{\alpha(\mu}T^{\alpha}_{\nu)} - T\left(g_{\mu\nu}L_m + T_{\mu\nu}\right) + 2\mathcal{L}_m T_{\mu\nu} + 4T^{\alpha\beta}\frac{\partial^2\mathcal{L}_m}{\partial g^{\mu\nu}\partial g^{\alpha\beta}}\right]. \tag{14.113}$$

Therefore the variation of the total action becomes

$$\begin{aligned}
\delta\mathcal{L} = -\frac{1}{2\kappa^2}\Bigg[&\sqrt{-g}G_{\mu\nu}\left(1 - \frac{1}{2}\alpha T\right)\delta g^{\mu\nu} \\
&-\frac{\alpha\sqrt{-g}}{2}\left[R(T_{\mu\nu} + \theta_{\mu\nu}) + g_{\mu\nu}R_{\alpha\beta}T^{\alpha\beta}\right]\delta g^{\mu\nu} \\
&+\sqrt{-g}\delta g^{\mu\nu}\left(g_{\mu\nu}\Box - \nabla_\mu\nabla_\nu\right)\left(1 - \frac{1}{2}\alpha T\right)\Bigg] \\
&-\delta g^{\mu\nu}\sqrt{-g}\left[\frac{1}{4}\alpha g_{\mu\nu}T_{\alpha\beta}T^{\alpha\beta} - \frac{1}{2}T_{\mu\nu}\right] \\
&+\frac{\alpha\sqrt{-g}}{2}\delta g^{\mu\nu}\Bigg[T_{\alpha(\mu}T^{\alpha}_{\nu)} - T\left(g_{\mu\nu}L_m + T_{\mu\nu}\right) \\
&+2\mathcal{L}_m T_{\mu\nu} + 4T^{\alpha\beta}\frac{\partial^2\mathcal{L}_m}{\partial g^{\mu\nu}\partial g^{\alpha\beta}}\Bigg] \\
&-\frac{\alpha\sqrt{-g}}{2\kappa^2}\delta g^{\mu\nu}\Bigg[R_{\alpha(\mu}T^{\alpha}_{\nu)} - \frac{1}{2}R\left(g_{\mu\nu}\mathcal{L}_m + T_{\mu\nu}\right) \\
&+\mathcal{L}_m R_{\mu\nu} + 2R^{\alpha\beta}\frac{\partial^2\mathcal{L}_m}{\partial g^{\mu\nu}\partial g^{\alpha\beta}} \\
&+\frac{1}{2}\left(\Box T_{\mu\nu} + \nabla_\alpha\nabla_\beta T^{\alpha\beta}g_{\mu\nu} - \nabla_\alpha\nabla_{(\mu}T^{\alpha}_{\nu)}\right)\Bigg].
\end{aligned} \tag{14.114}$$

From the condition $\delta\mathcal{L} = 0$, we obtain the basic result that the classical gravitational field equations that follow from the first-order quantum corrected gravitational action (14.106) have the form

$$\begin{aligned}
G_{\mu\nu}\left(1 - \frac{1}{2}\alpha T\right) = \Bigg[&\frac{1}{2}\alpha R\left(T_{\mu\nu} + \theta_{\mu\nu}\right) + \frac{1}{2}\alpha g_{\mu\nu}R_{\alpha\beta}T^{\alpha\beta} \\
&-\left(g_{\mu\nu}\Box - \nabla_\mu\nabla_\nu\right)\left(1 - \frac{1}{2}\alpha T\right)\Bigg] - \kappa^2\left(\frac{1}{2}\alpha g_{\mu\nu}T_{\alpha\beta}T^{\alpha\beta} - T_{\mu\nu}\right)
\end{aligned}$$

$$
\begin{aligned}
&+\kappa^2\alpha\left[T_{\alpha(\mu}T^{\alpha}_{\nu)} - T(g_{\mu\nu}\mathcal{L}_m + T_{\mu\nu}) + 2\mathcal{L}_m T_{\mu\nu}\right] \\
&-\alpha\Bigg[R_{\alpha(\mu}T^{\alpha}_{\nu)} - \frac{1}{2}R\left(g_{\mu\nu}\mathcal{L}_m + T_{\mu\nu}\right) + \mathcal{L}_m R_{\mu\nu} \\
&+\frac{1}{2}\left(\Box T_{\mu\nu} + \nabla_\alpha\nabla_\beta T^{\alpha\beta}g_{\mu\nu} - \nabla_\alpha\nabla_{(\mu}T^{\alpha}_{\nu)}\right)\Bigg].
\end{aligned}
\tag{14.115}
$$

In the field equations (14.115) we have ignored the terms containing the second derivatives of the matter Lagrangian $\mathcal{L}_m$ with respect to the metric tensor $g_{\mu\nu}$, since in most cases of astrophysical or cosmological interest they do not appear. The contraction of Eq. (14.115) gives

$$
R\left(1 - \frac{1}{2}\alpha T\right) - \alpha R\mathcal{L}_m + \alpha\left(\Box T - \nabla_\mu\nabla_\nu T^{\mu\nu}\right) + \kappa^2(T - 2\alpha T\mathcal{L}_m - \alpha T^2) = 0.
\tag{14.116}
$$

Equation (14.115) can then be reformulated as

$$
\begin{aligned}
&R_{\mu\nu}\left(1 - \frac{1}{2}\alpha T + \alpha\mathcal{L}_m\right) + \alpha R_{\alpha(\mu}T^{\alpha}_{\nu)} - \frac{\alpha}{2}g_{\mu\nu}R_{\alpha\beta}T^{\alpha\beta} = \\
&-\frac{1}{2}g_{\mu\nu}\frac{1 - \frac{1}{2}\alpha T}{1 - \frac{1}{2}\alpha T - \alpha\mathcal{L}_m}\left[\alpha\left(\Box T - \nabla_\alpha\nabla_\beta T^{\alpha\beta}\right) + 2\kappa^2\left(-\alpha T L_m - \frac{1}{2}\alpha T^2 + \frac{1}{2}T\right)\right] \\
&-\frac{1}{2}\alpha\left(\Box T_{\mu\nu} + g_{\mu\nu}\nabla_\alpha\nabla_\beta T^{\alpha\beta} - \nabla_\alpha\nabla_{(\mu}T^{\alpha}_{\nu)}\right) + \frac{1}{2}\alpha\left(g_{\mu\nu}\Box - \nabla_\mu\nabla\nu\right)T \\
&+2\kappa^2\Bigg[\frac{1}{2}\alpha\left(T_{\alpha(\mu}T^{\alpha}_{\nu)} - g_{\mu\nu}T\mathcal{L}_m - T_{\mu\nu}T + 2\mathcal{L}_m T_{\mu\nu}\right) \\
&-\frac{1}{4}\alpha g_{\mu\nu}T_{\alpha\beta}T^{\alpha\beta} + \frac{1}{2}T_{\mu\nu}\Bigg].
\end{aligned}
\tag{14.117}
$$

14.4.2 The Covariant Divergence of the Energy-Momentum Tensor

We take now the covariant divergence of the system of gravitational field equations (14.115), and obtain the following relation

$$
\begin{aligned}
&\left[\kappa^2\left(\alpha T - 2\alpha\mathcal{L}_m - 1\right) - \alpha R\right]\nabla^\nu T_{\mu\nu} \\
&= \frac{1}{2}\alpha\left[G_{\mu\nu}\nabla^\nu T + T_{\mu\nu}\nabla^\nu R + \nabla_\mu(R_{\alpha\beta}T^{\alpha\beta} - \kappa^2 T_{\alpha\beta}T^{\alpha\beta})\right] \\
&+\kappa^2\alpha\left[\nabla^\nu(T_{\alpha(\mu}T^{\alpha}_{\nu)}) - \nabla_\mu(T\mathcal{L}_m) - T_{\mu\nu}\nabla^\nu(T - 2\mathcal{L}_m)\right]
\end{aligned}
$$

$$
\begin{aligned}
&-\alpha\left[\nabla^\nu(R_{\alpha(\mu}T^\alpha_{\nu)})-\frac{1}{2}\nabla_\mu(R\mathcal{L}_m)-\frac{1}{2}T_{\mu\nu}\nabla^\nu R+\nabla^\nu(R_{\mu\nu}\mathcal{L}_m)\right.\\
&\left.+\frac{1}{2}\left(\nabla^\nu\Box T_{\mu\nu}+\nabla_\mu\nabla_\alpha\nabla_\beta T^{\alpha\beta}-\nabla^\nu\nabla_\alpha\nabla_{(\mu}T^\alpha_{\nu)}\right)\right].
\end{aligned}
\tag{14.118}
$$

Then in the modified gravity model induced by the quantum metric fluctuations proportional to the matter energy-momentum tensor we find for the divergence of the matter energy-momentum tensor the expression

$$
\begin{aligned}
\nabla^\nu T_{\mu\nu}=&\frac{1}{\kappa^2(\alpha T-2\alpha L_m-1)-\alpha R}\left\{\frac{1}{2}\alpha\left[T_{\mu\nu}\nabla^\nu R+\nabla_\mu(R_{\alpha\beta}T^{\alpha\beta}-\kappa^2 T_{\alpha\beta}T^{\alpha\beta})\right]\right.\\
&+\kappa^2\alpha\left[\nabla^\nu(T_{\alpha(\mu}T^\alpha_{\nu)})-\nabla_\mu(T\mathcal{L}_m)-T_{\mu\nu}\nabla^\nu(T-2\mathcal{L}_m)\right]\\
&-\alpha\left[\nabla^\nu(R_{\alpha(\mu}T^\alpha_{\nu)})-\frac{1}{2}\nabla_\mu(R\mathcal{L}_m)-\frac{1}{2}T_{\mu\nu}\nabla^\nu R+\nabla^\nu(R_{\mu\nu}\mathcal{L}_m)\right.\\
&\left.\left.+\frac{1}{2}(\nabla^\nu\Box T_{\mu\nu}+\nabla_\mu\nabla_\alpha\nabla_\beta T^{\alpha\beta}-\nabla^\nu\nabla_\alpha\nabla_{(\mu}T^\alpha_{\nu)})\right]\right\}\\
&+\frac{1}{2}\alpha\frac{2\nabla^\nu T}{2-\alpha T}\left\{\left[\frac{1}{2}\alpha R(T_{\mu\nu}+\theta_{\mu\nu})+\frac{1}{2}\alpha g_{\mu\nu}R_{\alpha\beta}T^{\alpha\beta}\right.\right.\\
&\left.-(g_{\mu\nu}\Box-\nabla_\mu\nabla_\nu)(1-\frac{1}{2}\alpha T)\right]-\kappa^2\left(\frac{1}{2}\alpha g_{\mu\nu}T_{\alpha\beta}T^{\alpha\beta}-T_{\mu\nu}\right)\\
&+\kappa^2\alpha\left[T_{\alpha(\mu}T^\alpha_{\nu)}-T(g_{\mu\nu}\mathcal{L}_m+T_{\mu\nu})+2\mathcal{L}_m T_{\mu\nu}\right]\\
&-\alpha\left[R_{\alpha(\mu}T^\alpha_{\nu)}-\frac{1}{2}R(g_{\mu\nu}\mathcal{L}_m+T_{\mu\nu})+\mathcal{L}_m R_{\mu\nu}\right.\\
&\left.\left.+\frac{1}{2}\left(\Box T_{\mu\nu}+\nabla_\alpha\nabla_\beta T^{\alpha\beta}g_{\mu\nu}-\nabla_\alpha\nabla_{(\mu}T^\alpha_{\nu)}\right)\right]\right\}.
\end{aligned}
\tag{14.119}
$$

Equation (14.119) shows that, in this class of modified gravity theories, generally the energy-momentum tensor $T_{\mu\nu}$ of the matter is not conserved. Once again, we note that from a physical point of view, the nonconservation of $T_{\mu\nu}$ can be related to particle creation effects that occur in the gravitational system and that are related to the quantum fluctuations of the space-time metric.

14.4.3 Cosmological Applications

In order to investigate the cosmological implications of the modified gravity theory induced by the quantum fluctuations of the metric proportional to the energy-momentum tensor of the matter, in the following we will consider a simple homogeneous and isotropic, spatially flat geometry, in which $k=0$, described by the FLRW metric. By taking into account the intermediate results

$$\Box T=(\ddot{\rho}-3\ddot{p})+(\dot{\rho}-3\dot{p})\frac{3\dot{a}}{a}, \tag{14.120}$$

$$\nabla^{\alpha}\nabla^{\beta}T_{\alpha\beta}=\frac{3\dot{a}}{a}\left[(2\dot{\rho}+\dot{p})+2(\rho+p)\frac{\dot{a}}{a}\right]+\ddot{p}+3(\rho+p)\frac{\ddot{a}}{a}, \tag{14.121}$$

$$\Box T_{\mu\nu}=\ddot{\rho}+\dot{\rho}\frac{3\dot{a}}{a}, \qquad \nabla_{\mu}\nabla_{\nu}T=\ddot{\rho}-3\ddot{p}, \tag{14.122}$$

$$\nabla_{\alpha}\nabla_{(\mu}T^{\alpha}_{\nu)}=2\ddot{\rho}+2\dot{\rho}\frac{3\dot{a}}{a}, \qquad \mu=\nu=0, \tag{14.123}$$

from the 00 component of Eq. (14.115), we first find

$$\begin{aligned}
&\frac{3\ddot{a}}{a}\left[1+\alpha(\rho+3p)\right]+\frac{\dot{a}^2}{a^2}3\alpha p=\frac{1}{2}\frac{1-\frac{1}{2}\alpha(\rho-3p)}{1-\frac{1}{2}\alpha(\rho-p)}\left[-\alpha\frac{3\dot{a}}{a}\left(\dot{\rho}+4\dot{p}+2(\rho+p)\frac{\dot{a}}{a}\right)\right.\\
&\left.-3\alpha\ddot{p}-3\alpha(\rho+p)\frac{\ddot{a}}{a}+\kappa^2(\rho-3p)(1-\alpha(\rho-p))\right]\\
&+\frac{\alpha}{2}\left\{3(\rho+p)\frac{\ddot{a}}{a}+\frac{3\dot{a}}{a}\left[4\dot{p}+2(\rho+p)\frac{\dot{a}}{a}\right]\right\}\\
&-\kappa^2\left[\rho-\alpha\left(\frac{1}{2}\rho^2+4\rho p+\frac{3}{2}p^2\right)\right].
\end{aligned} \tag{14.124}$$

Next for $\mu=\nu=1$ we find

$$\Box T_{\mu\nu}=a^2(\ddot{p}+\dot{p}\frac{3\dot{a}}{a}), \qquad \nabla_{\mu}\nabla_{\nu}T=0, \qquad \nabla_{\alpha}\nabla_{(\mu}T^{\alpha}_{\nu)}=0. \tag{14.125}$$

From the 11 component of Eq. (14.115) we obtain

$$\begin{aligned}
&\frac{\dot{a}^2}{a^2}\left[2-\alpha(\rho-5p)\right]+\frac{\ddot{a}}{a}\left[1-\alpha(2\rho-3p)\right]+2\alpha\dot{p}\frac{\dot{a}}{a}\\
&=\frac{1}{2}\frac{1-\frac{1}{2}\alpha(\rho-3p)}{1-\frac{1}{2}\alpha(\rho-p)}\left[-\alpha\frac{3\dot{a}}{a}\left[\dot{\rho}+4\dot{p}+2(\rho+p)\frac{\dot{a}}{a}\right]\right.\\
&\left.-3\alpha\ddot{p}-3\alpha(\rho+p)\frac{\ddot{a}}{a}+\kappa^2(\rho-3p)(1-\alpha\rho+\alpha p)\right]\\
&+\frac{1}{2}\alpha\left\{2\ddot{p}+3(\rho+p)\frac{\ddot{a}}{a}+\frac{3\dot{a}}{a}\left[\dot{\rho}+3\dot{p}+2(\rho+p)\frac{\dot{a}}{a}\right]\right\}\\
&+\kappa^2\left[\frac{1}{2}\alpha(\rho^2+3p^2)+p\right].
\end{aligned} \tag{14.126}$$

Hence, the generalized cosmological Friedmann equations for the modified gravity theory induced by the quantum fluctuations of the metric proportional to the matter energy-momentum tensor take the final form

$$3H^2 = -3\frac{\Pi \cdot \Gamma - \Sigma \cdot \Theta}{\Xi}, \tag{14.127}$$

$$2\dot{H} + 3H^2 = \frac{\Gamma \cdot (2\Delta - \Pi) + (\Sigma - l2) \cdot \Theta}{\Xi}, \tag{14.128}$$

where we have introduced the following notation

$$\lambda = \frac{1 - \frac{1}{2}\alpha(\rho - 3p)}{1 - \frac{1}{2}\alpha(\rho - p)}, \tag{14.129}$$

$$\begin{aligned}\Gamma = &\frac{3}{2}\alpha\lambda\left(\frac{\dot{a}}{a}\dot{\rho} + \ddot{p}\right) + 6\alpha\frac{\dot{a}}{a}(\lambda - 1)\dot{p} + \kappa^2\left[\rho - \alpha\left(\frac{3p^2}{2} + 4p\rho + \frac{\rho^2}{2}\right)\right]\\ &-\frac{1}{2}\lambda\kappa^2(\rho - 3p)\left(-\alpha\rho + \alpha p + 1\right),\end{aligned} \tag{14.130}$$

$$\Delta = 2 - \alpha(\rho - 5p) + 3\alpha(\lambda - 1)(p + \rho), \tag{14.131}$$

$$\begin{aligned}\Theta = &\frac{3}{2}\alpha(\lambda - 1)\frac{\dot{a}}{a}\dot{\rho} + 6\alpha\frac{\dot{a}}{a}\lambda\dot{p} - \frac{5\alpha\dot{p}}{2}\frac{\dot{a}}{a} - \kappa^2\left(\frac{\alpha\rho^2}{2} + \frac{3\alpha p^2}{2} + p\right) + \frac{3\alpha\lambda\ddot{p}}{2}\\ &-\alpha\ddot{p} - \frac{1}{2}\lambda\kappa^2(\rho - 3p)(-\alpha\rho + \alpha p + 1),\end{aligned} \tag{14.132}$$

$$\Lambda = 3\alpha\lambda(\rho + p) - 3\alpha\rho, \tag{14.133}$$

$$\begin{aligned}\Xi = 3\Bigg[&\frac{\alpha^2\rho^2}{2}(11 - 9\lambda) - 3\alpha\lambda(\rho + p) + \alpha(2\rho - 7p) - (5 + 7\lambda)\alpha^2 p^2\\ &+\frac{\alpha^2}{2}p\rho(15 - 23\lambda) - 2\Bigg],\end{aligned} \tag{14.134}$$

$$\Pi = \frac{3\alpha\lambda}{2}(\rho + p) + \frac{\alpha}{2}(3p - 7\rho) + 1, \tag{14.135}$$

$$\Sigma = 3\left[\frac{\alpha}{2}(\lambda - 1)(p + \rho) + \alpha(3p + \rho) + 1\right]. \tag{14.136}$$

The deceleration parameter q can be represented as

$$q = \frac{d}{dt}\frac{1}{H} - 1 = -\frac{\ddot{a}}{aH^2} = \frac{\Gamma \cdot \Delta - \Theta \cdot \Lambda}{\Pi \cdot \Gamma - \Sigma \cdot \Theta}. \tag{14.137}$$

For $\alpha = 0$, we obtain $\Gamma = \kappa^2(\rho c^2 + 3p)/2$, $\Delta = 2$, $\Theta = -\kappa^2(\rho c^2 + p)/2$, $\Lambda = 0$, $\Xi = -6$, $\Pi = 1$, and $\Sigma = 3$, respectively. Hence in the limit of the vanishing quantum fluctuation tensor we recover the Friedmann equations of standard GR.

14.4.4 Dust Cosmological Models with $p = 0$

We assume now that the Universe is filled with pressureless dust, having a negligibly small pressure, so that $p = 0$. Then it immediately follows that $\lambda = 1$. Hence, the cosmological field equations of this modified gravity theory can be written first as

$$\frac{\ddot{a}}{a} = -\frac{3\alpha H\dot{\rho} + \kappa^2\rho}{6(\alpha\rho + 1)}, \tag{14.138}$$

and

$$H^2 = -\frac{3\alpha H\dot{\rho}(1 - 2\alpha\rho) + \kappa^2\rho(4 + \alpha\rho)}{6(\alpha\rho - 2)(\alpha\rho + 1)}, \tag{14.139}$$

respectively. Therefore, the generalized Friedmann equations of the present model are obtained in the form

$$3H^2 = -\frac{3\alpha H\dot{\rho}(1 - 2\alpha\rho) + \kappa^2\rho(4 + \alpha\rho)}{2(\alpha\rho - 2)(\alpha\rho + 1)}, \tag{14.140}$$

$$2\dot{H} + 3H^2 = -\frac{\alpha(\kappa^2\rho^2 - 3H\dot{\rho})}{2(\alpha\rho - 2)(\alpha\rho + 1)}. \tag{14.141}$$

For the deceleration parameter we find the expression

$$q = \frac{(3\alpha H\dot{\rho} + k^2\rho)(\alpha\rho - 2)}{3\alpha H\dot{\rho}(1 - 2\alpha\rho) + \kappa^2\rho(4 + \alpha\rho)}. \tag{14.142}$$

After multiplication with a^3 of both sides of Eq. (129), by taking its derivative with respect to the time, and with the consideration of Eq. (130), for the time evolution of the matter energy density we find the equation

$$\begin{aligned} H\ddot{\rho} = -\dot{\rho}\frac{\ddot{a}}{a} + \frac{1}{-2\alpha + 3\alpha^2\rho + 3\alpha^3\rho^2 - 2\alpha^4\rho^3} \times \\ \left\{\dot{\rho}\left[\frac{\kappa^2}{3}(8 + 4\alpha\rho + 5\alpha^2\rho^2) + \alpha H^2(10 - 3\alpha\rho \right.\right. \\ \left. -9\alpha^2\rho^2 + 4\alpha^3\rho^3) - H\dot{\rho}\left(5\alpha^2 - 2\alpha^3\rho + 2\alpha^4\rho^2\right)\right] \\ \left. +\kappa^2 H(8\rho + \alpha\rho^2 - 4\alpha^2\rho^3)\right\}. \end{aligned} \tag{14.143}$$

de Sitter-type Evolution of the Dust Universe

We look first for de Sitter-type solutions of the field equations, by assuming that $H = H_0 =$ constant. Then the cosmological evolution equations (14.140) and (14.141) reduce to a single first-order differential equation for the time evolution of the matter density, given by

$$\dot{\rho}(t) = \frac{2\kappa^2\rho(t)}{3\alpha H_0[\alpha\rho(t) - 1]}. \tag{14.144}$$

The general solution of the above equation can be obtained as

$$\alpha\rho - \ln(\rho) = \frac{2\kappa^2}{3\alpha H_0}(t - t_0), \tag{14.145}$$

where t_0 is an arbitrary constant of integration. When the condition $\alpha\rho \gg \ln(\rho)$ is satisfied, the matter energy density increases linearly in time according to the relation $\rho(t) \propto t$, showing that the late-time de Sitter-type accelerated expansion of the Universe is driven by the (necessarily quantum) matter production processes.

Cosmological Evolution of the Dust Universe

In the case of a dust-filled Universe, the time evolution of the matter density follows from Eq. (14.139), and is governed by the differential equation

$$\dot{\rho} = \frac{6H^2(\alpha\rho - 2)(\alpha\rho + 1) + \kappa^2\rho(\alpha\rho + 4)}{3\alpha H(2\alpha\rho - 1)}. \tag{14.146}$$

By substituting this equation into Eq. (14.141) we find for the evolution of the Hubble function the equation

$$2\dot{H} = -3H^2 + \frac{3H^2 - \kappa^2\rho}{2\alpha\rho - 1}. \tag{14.147}$$

Now we introduce a set of dimensionless cosmological variables (τ, h, α_0, r), defined as

$$\tau = H_0 t, \qquad H = H_0 h, \qquad \alpha = \alpha_0 \frac{\kappa^2}{3H_0^2}, \qquad \rho = \frac{3H_0^2}{\kappa^2} r, \tag{14.148}$$

where H_0 is the present-day value of the Hubble function. We also change the independent cosmological time variable from τ to the redshift z, and after performing these rescalings Eqs. (14.146) and (14.147) take the form

$$\frac{dr}{dz} = -\frac{2h^2(\alpha_0 r - 2)(\alpha_0 r + 1) + r(\alpha_0 r + 4)}{\alpha_0(1+z)h^2(2\alpha_0 r - 1)}, \tag{14.149}$$

$$\frac{dh}{dz} = \frac{3}{2}\frac{h}{1+z} - \frac{3(h^2 - r)}{2(1+z)h(2\alpha_0 r - 1)}. \tag{14.150}$$

In the new variables the deceleration parameter q can be obtained as

$$q = \frac{1}{2} - \frac{3(h^2 - r)}{2h^2(2\alpha_0 r - 1)}. \tag{14.151}$$

The system of ordinary nonlinear differential equations (14.149) and (14.150) must be integrated numerically, after imposing the initial conditions $h(0) = h_0$ and $r(0) = r_0$, respectively. In this context, it is interesting to point out that for $h(0) = r(0) = 1$, the present-day deceleration parameter takes the positive value $q(0) = 1/2 > 0$. Thus, to obtain expanding and accelerating cosmological models we will adopt as an initial condition for the present-day energy density of

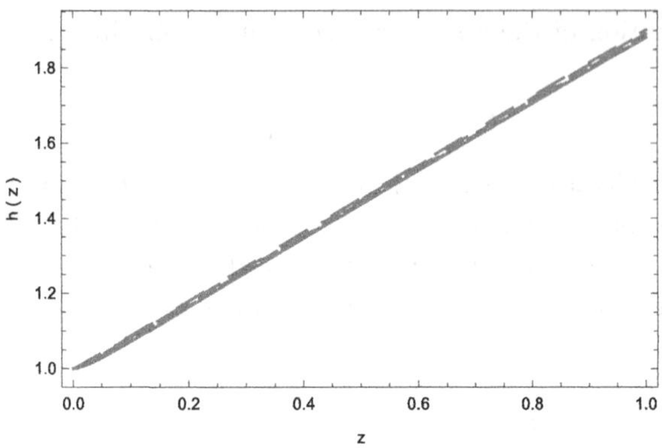

Figure 14.9 Variation with respect to the redshift z of the dimensionless Hubble function $h(z)$ for the modified gravity model with fluctuating quantum metric with $K_{\mu\nu} = \alpha T_{\mu\nu}$ for $h(0) = 1$, $r(0) = 0.70$, and for different values of α_0: $\alpha_0 = 0.93$ (solid curve), $\alpha_0 = 0.98$ (dotted curve), $\alpha_0 = 1.03$ (short dashed curve), $\alpha_0 = 1.08$ (dashed curve), $\alpha_0 = 1.13$ (long dashed curve), and $\alpha_0 = 1.18$ (ultra-long dashed curve), respectively.

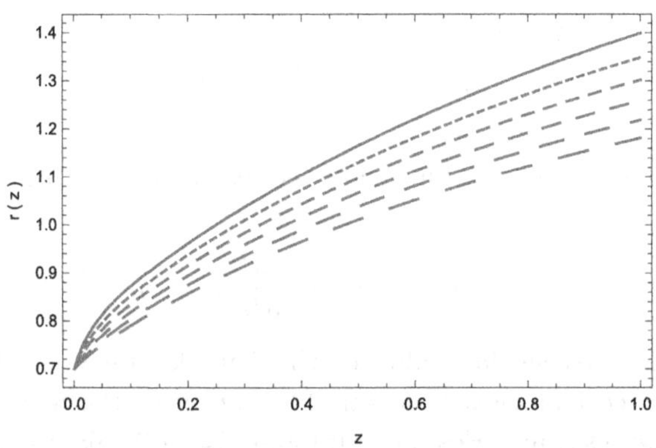

Figure 14.10 Variation with respect to the redshift z of the dimensionless matter energy density $r(z)$ for the modified gravity model with fluctuating quantum metric with $K_{\mu\nu} = \alpha T_{\mu\nu}$ for $h(0) = 1$, $r(0) = 0.70$, and for different values of α_0:$\alpha_0 = 0.93$ (solid curve), $\alpha_0 = 0.98$ (dotted curve), $\alpha_0 = 1.03$ (short dashed curve), $\alpha_0 = 1.08$ (dashed curve), $\alpha_0 = 1.13$ (long dashed curve), and $\alpha_0 = 1.18$ (ultra-long dashed curve), respectively.

the matter the numerical value $r(0) = 0.70$, while for the dimensionless Hubble function h we impose the initial condition $h(0) = 1$. In Figs. 14.9–14.11 we have represented the variations with respect to the redshift z of the dimensionless Hubble function h, of the dimensionless matter energy density r, and of the deceleration parameter q, respectively. The cosmological evolution equations have been numerically integrated in the redshift range $z \in [0, 1]$.

In the present cosmological model, the Hubble function, portrayed in Fig. 14.9, is a monotonically increasing function of the redshift z. Hence in the

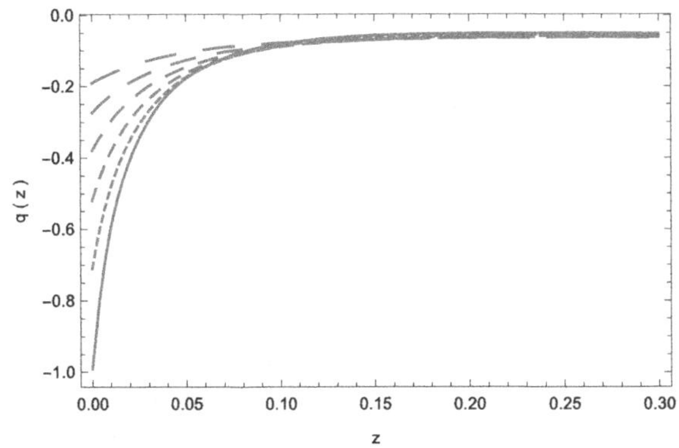

Figure 14.11 Variation with respect to the redshift z of the deceleration parameter $q(z)$ for the modified gravity model with fluctuating quantum metric with $K_{\mu\nu} = \alpha T_{\mu\nu}$ for $h(0) = 1$, $r(0) = 0.70$, and for different values of α_0: $\alpha_0 = 0.93$ (solid curve), $\alpha_0 = 0.98$ (dotted curve), $\alpha_0 = 1.03$ (short dashed curve), $\alpha_0 = 1.08$ (dashed curve), $\alpha_0 = 1.13$ (long dashed curve), and $\alpha_0 = 1.18$ (ultra-long dashed curve), respectively.

considered model the Universe experiences an expansionary evolution. In the adopted redshift range the numerical values of h display a very weak dependence on the numerical values of the independent model parameter α_0. In Fig. 14.10 we present the redshift evolution of the energy density r of the matter. As one can see from the figure, r is increasing monotonically as a function of the redshift z, which means that during the cosmological evolution r decreases in time. Note that the numerical values of r strongly depend on the model parameter α_0 at all redshifts. The deceleration parameter q is depicted in Fig. 14.11. The evolution of q is also strongly dependent on the coupling constant α_0. The Universe is already in a marginally decelerating state for redshifts in the range $z \in (0.14, 1]$, for which $q \approx 0$. In this range of z the evolution is essentially independent on the numerical values of the parameter α_0. The Universe enters in an accelerating phase for redshift values $z < 0.10$. By varying the numerical values of α_0, a large number of accelerating cosmological models can be built. With increasing α_0, the numerical values of the deceleration parameter q rapidly decrease, and for $\alpha_0 \approx 1$, it turns out that $q(0) \approx -1$. On the other hand, for $\alpha_0 \approx 1.4$, the deceleration parameter q becomes a constant, with $q \approx 0$ for all values of z in the range $z \in [0, 1]$.

14.5 Conclusion

By decomposing the quantum metric into two components, a classical and a quantum fluctuating one, and after the substitution of the fluctuating part by its (classical) average value $K_{\mu\nu}$, we have obtained a large class of modified gravity theories. This class of theories has the important property of the nonconservation of the matter energy-momentum tensor. This property can be understood in physical terms as related to the processes of matter production, which can be attributed to the presence of specific quantum effects in the curved space-time of GR and other modified gravity theories.

The physical and cosmological properties of two types of such theories, obtained from the quantum treatment of gravity, have been presented in detail. More exactly, we have investigated gravitational theories in which the quantum perturbation tensor can be taken as proportional to the metric tensor, and to the matter energy-momentum tensor, respectively. The second choice of the quantum perturbation tensor $K_{\mu\nu}$ gives rise to a generalization of $f\left(R, T, R_{\mu\nu}T^{\mu\nu}\right)$ gravity, in which a new term of the form $T_{\mu\nu}T^{\mu\nu}$ appears in the quantum corrected action of the $f\left(R, T, R_{\mu\nu}T^{\mu\nu}\right)$ theory. Therefore, a particular case of a general $f\left(R, T, R_{\mu\nu}T^{\mu\nu}, T_{\mu\nu}T^{\mu\nu}\right)$-modified theory of gravity can be obtained by regarding the effects of the quantum fluctuations of the metric as being proportional to the energy-momentum tensor of ordinary matter. The cosmological evolution equations for this theory have been investigated numerically, with the numerical analysis showing that for a dust-matter-filled Universe a large diversity of cosmological behaviors turns up at low redshifts. The cosmological dynamics essentially depend on the numerical values of the coupling constant α, relating the fluctuation tensor to the matter energy-momentum tensor. In this class of theories the deceleration parameter varies on a large redshift range between an (approximately zero) constant value, $q \approx 0$, and the value $q \approx -1$, corresponding to a de Sitter phase, reached at $z = 0$.

In concluding, in the present chapter, we have striven to show that a deeper understanding of the physical and theoretical basis of the modified gravity theories with a curvature-matter coupling can be obtained from the study of the quantum gravitational theories with first-order quantum corrections induced by a stochastically fluctuating metric. We have also examined some of the cosmological implications of these theories, and have elaborated some basic mathematical and theoretical instruments that can be put to work for the further investigation of the quantum effects in gravity.

15
Discussion and Conclusion

In this second part of the book, we have reviewed and extensively analyzed the cosmological and astrophysical applications of generalized curvature-matter coupling in $f(R)$-type gravity. Specific models were explored and presented in detail. The gravitational field equations in the metric and Palatini formalisms, in the presence of a nonminimal coupling between curvature and matter were presented, as well as the equations of motion for test particles, which follow from the covariant divergence of the energy-momentum tensor. Generally, the motion is non-geodesic, and takes place in the presence of an extra-force orthogonal to the four-velocity. The Newtonian limit of the equation of motion was also described, and a procedure for obtaining the matter energy-momentum tensor in the framework of these gravity models was presented. On the other hand, it was shown that the gravitational field equations are equivalent to the effective Einstein equations of $f(R)$ gravity in empty space-time, but differ from them, as well as from standard General Relativity, in the presence of matter. Therefore the predictions of these gravity theories could lead to some major differences, as compared to the predictions of standard GR, or to its extensions that ignore the role of matter, in several problems of current interest, such as cosmology, dark matter, dark energy, gravitational collapse, or the generation of gravitational waves.

We have also reviewed, based on the already existing literature, the implications of the curvature-matter coupling models on the galactic dynamics, and we have shown that the behavior of the neutral hydrogen gas clouds outside the galaxies, and their flat rotation curves, can be explained in terms of a nonminimal coupling between matter and curvature. We have also shown that in the curvature-matter coupling "dark matter" model all the relevant physical quantities, including the "dark mass" associated to the coupling, and which plays the role of dark matter, its corresponding density profile, as well as the curvature-matter coupling function, can be expressed in terms of observable parameters – the tangential velocity, the baryonic (luminous) mass, and the Doppler frequency shifts of test particles moving around the galaxy. Therefore, this opens the possibility of directly testing the modified gravity models with nonminimal coupling between matter and geometry by using direct astronomical and astrophysical observations at the galactic or extra-galactic scale. Since the observations on the galactic rotation curves are obtained from the Doppler frequency shifts, we have generalized the expression of the frequency shifts by including the effect of the curvature-matter coupling. Thus, at least in principle, the coupling function can be obtained directly from astronomical observations. We have also presented two other possibilities for testing these classes of gravity theories, at the level of the solar system, by using the perihelion precession of Mercury, and at astrophysical scales, from possible observations of the gravitational tidal effects.

The physical and mathematical formalisms of cosmological matter production have been extensively discussed in the literature within the theory of standard GR, from both phenomenological and theoretical points of view. However, it is important to note that there is an essential difference between the thermodynamics of open systems as formulated in the framework of standard GR, and in theories with a curvature-matter coupling. In GR the basic thermodynamic parameters, such as the particle creation rates, and the creation pressure, must be introduced in the theory either based on some plausible physical models, or they must be obtained from some microscopic description of the particle creation. On the other hand, as we have already seen, in both $f(R,T)$- and $f\left(R, L_m\right)$-modified gravity theories the nonminimal coupling between geometry and matter fully determines the matter production rate Γ, the creation pressure p_c, as well as the entropy production.

As it is easy to observe from Eqs. (12.51) and (12.56), in $f\left(R, L_m\right)$-modified gravity theory, Γ and p_c are fully determined by the partial derivatives of the Lagrangian density $f\left(R, L_m\right)$ of the theory with respect to the matter Lagrangian L_m, and by the matter Lagrangian itself. For the $f(R,T)$ gravity theory, the matter production rate and the creation pressure, which can be calculated from Eqs. (12.75) and (12.77), respectively, are also fully determined by the partial derivative of the Lagrangian density of the theory $f(R,T)$ with respect to the trace of the matter energy-momentum tensor T. Therefore, it follows that in these classes of theories with curvature-matter coupling all the physical quantities related to the introduction of the irreversible thermodynamic description are fully determined by the gravitational part of the action – note that in these theories the gravitational part also contains the matter component. If we go to the general relativistic limit by taking $f\left(R, L_m\right) = R/2 + L_m$, and $f(R,T)$ independent of T, all the particle creation rates Γ and the creation pressures p_c identically vanish. This result confirms that the coupling between matter and geometry may play an essential role in the phenomenological description of the cosmological particle production processes induced by the time variation of the gravitational field in the expanding Universe.

We have also explored in detail the quantum cosmology of $f(R,T)$ gravity. In our approach we have considered that the classical evolution of the Universe takes place in a homogeneous and isotropic background, described by the flat FLRW geometry, which was used in a systematic way as the classical counterpart in the investigation of the quantum properties of the very early Universe, immediately after its birth. To perform the quantization of the model, we introduced in the total action describing the Universe two Lagrange multipliers λ and μ. With the help of the first Lagrange multipliers we have imposed the definition of the Ricci scalar, a purely geometric quantity, while we have used the second Lagrange multiplier to impose the trace constraint of the $f(R,T)$ gravity theory. It is important to point out that this second constraint has a wider meaning than the simple introduction in the theory of the trace of the energy-momentum tensor T, since it gives the possibility for the search of the profound relation between geometry and matter at a higher level than the one allowed by the simple thermodynamic definition of T. From the gravitational action describing the cosmological evolution, we have constructed the corresponding Hamiltonian function, which, after its canonical quantization, leads us immediately to the formulation of the WDW equation in $f(R,T)$ gravity, which describes the evolution of the wave function of the quantum Universe in the presence of a curvature-matter coupling. Thus, it is possible that the theoretical tools introduced here could be used for the investigation of the complex interaction of curvature and matter in the quantum Universe, as well as for the description of the origin and evolution of the very early Universe. This approach may also help to understand the properties of the gravitational interaction at the beginning of the Universe, when geometry interacting with matter may have given birth to time, entropy, and irreversibility.

In fact, the greatest theoretical challenge present-day physics faces is the solution of the problem of quantum gravity. Unfortunately no exact solutions for this problem are known. Employing some approximate methods for the study of the quantum effects in gravity seems to be the best approach to follow in the search for the quantization of GR. A promising path for understanding the role that quantum mechanics may play in gravity consists in the inclusion of some fluctuating tensor terms in the gravitational metric, which are essentially stochastic in their physical structure. It is interesting that such a conceptualization of the problem leads us toward classical gravity theories with a compelling curvature-matter coupling, as well as the nonconservation of the energy-momentum tensors of the matter. Consequently, all such theories involve matter production processes. Hence a better comprehension of the physical and theoretical basis of the modified gravity theories with a curvature-matter coupling can be obtained even from the study of the quantum gravitational theories with first order quantum corrections induced by a stochastically fluctuating metric. In this work, we have also examined some of the cosmological implications of these theories, and we have elaborated some basic mathematical and theoretical instruments that can be put to work for the further investigation of the quantum effects in gravity.

As future avenues of research, one should aim to characterize as much as possible the phenomenology predicted by these theories with curvature-matter coupling in order to find constraints arising from observations. The study of these phenomena may also provide some specific signatures and effects, which could distinguish and discriminate between the various theories of modified gravity. We also propose to use a background metric to analyze the dynamic system for specific curvature-matter coupling models, and use the data of SNIa, BAO, CMB shift parameter to obtain restrictions for the respective models, and explore in detail the analysis of structure formation.

Part III

Hybrid Metric-Palatini Gravity

16

Introduction

One hundred years ago, Albert Einstein completed the mathematical formulation of his revolutionary view of the gravitational interactions in terms of curved space-time. The spirit, elegance, and experimental successes of the original theory has captivated the international scientific community and the theory has been accepted as the *standard model* for gravity [264]. At its centennial, pushed by new observational evidences, the theory is at a dramatic crossroad. Its continuation as the reference gravitational framework will imply that the Universe is mainly composed by exotic sources of matter and energy whose existence is purely inferred from their gravitational effects at the largest astrophysical and cosmological scales [43, 253, 412, 413, 436, 437, 438]. However, if such sources are not detected in any direct way, then we might be facing a failure of one of the most original theories of the twentieth century. The important implications of the two opposed alternatives, i.e., the search for unknown dark side constituents or the revision of gravitational theory demands a careful scrutiny of the different possible scenarios. In this book, we consider the second case, namely, the situation in which the gravitational dynamics may depart from that predicted by Einstein's theory of General Relativity (GR) at ultraviolet and infrared scales. We mainly focus on the latter regime.

Given the success of GR at relatively short scales (such as the solar system, stellar models, or compact binary systems), the idea that modified dynamics could arise at larger scales has been investigated in much detail over the last years. Theories in which the gravitational action consists of more general combinations of curvature invariants than the pure Einstein–Hilbert term have been investigated with special emphasis [34, 99, 103, 120, 140, 157, 324, 382, 384]. From these investigations it was soon noticed that the usual metric formulation of alternative theories of gravity is generically different from its Palatini (or metric-affine) counterpart (see [396] for a recent review on the Palatini approach). Whereas the metric approach typically leads to higher-order derivative equations, in the Palatini formulation the resulting field equations are always second-order.

The appealing character of the second-order equations of the Palatini formalism, however, is accompanied by certain algebraic relations between the matter fields and the affine connection, which is now determined by a set of equations coupled to the matter fields and the metric. The case of $f(R)$ theories is particularly useful to illustrate the differences between these two approaches. In the metric formulation, the object $\phi \equiv \mathrm{d}f/\mathrm{d}R$ behaves as a dynamical scalar field, which satisfies a second-order equation with self-interactions that depend on the form of the Lagrangian $f(R)$. In order to have an impact at large astrophysical and cosmological scales, the scalar field ϕ should have a very low mass, implying a long interaction range. It is well known, however, that light scalars do have an impact at shorter scales, where their presence is strongly constrained by laboratory and solar system observations unless some kind of screening mechanism is invoked [89, 90, 91, 275, 291]. In the Palatini case, a scalar–tensor representation is also possible, but with the scalar field satisfying an algebraic rather than a differential equation. One then finds that the scalar ϕ turns out to be an algebraic function of the trace of the stress–energy tensor of the matter, $\phi = \phi(T)$, which may lead, in models of late-time cosmic speed-up, to undesired gradient instabilities at various contexts, as has been shown by studies of cosmological perturbations [285, 289] and atomic physics [394, 395].

In this part we will review the *hybrid* variation of these theories, in which the (purely metric) Einstein–Hilbert action is supplemented with (metric-affine) correction terms constructed à la Palatini [109, 224]. Given that metric and Palatini $f(R)$ theories allow the construction of simple extensions of GR with interesting properties and, at the same time, suffer from different types of drawbacks, we initiated a program to establish bridges between these two seemingly disparate approaches hoping to find ways to cure or improve their individual deficiencies. For that purpose, in a number of works we considered a hybrid combination of metric and Palatini elements to construct the gravity Lagrangian and found that viable models sharing properties of both formalisms are possible. An interesting aspect of these theories is the possibility to generate long-range forces without entering into conflict with local tests of gravity and without invoking any kind of screening mechanism (which would however require that at the present time the cosmological evolution reduces to GR). The possibility of expressing these hybrid $f(R)$ metric-Palatini theories using a scalar–tensor representation simplifies the analysis of the field equations and the construction of solutions. In some sense, considering a theory like $R + f(\mathcal{R})$ means that we retain all the positive results of GR, represented by the Einstein–Hilbert part of the action R, while the further "gravitational budget" is endowed in the metric-affine $f(\mathcal{R})$ component, where $\mathcal{R}$ is the Palatini curvature scalar constructed in terms of an independent connection. In fact, it is well known that metric-affine and purely metric formalisms coincide in GR, i.e., considering the action R. On the contrary, the two formalisms lead to different results considering more generic functions $f(\mathcal{R})$ [396].

A related approach to studying $f(R)$ theories that interpolate between the metric and Palatini families is that of the so-called C-theories proposed in [18, 290]. There the space-time connection is associated to the metric $\hat{g}_{\mu\nu} = C(\mathcal{R})g_{\mu\nu}$ that is conformally related to the space-time metric $g_{\mu\nu}$, but the relation may depend upon the a scalar curvature $\mathcal{R}$. This framework contains the metric, $C(\mathcal{R}) = 1$, and the Palatini, $C(\mathcal{R}) = f'(R)$, formalisms as special limits, and one also finds that even when $f(\mathcal{R}) = \mathcal{R}$ physically distinct theories are possible. For further studies on variations of variational principles see, e.g., [40, 41, 205, 271, 272, 287, 446].

Other extensions of the $f(R)$ framework modify the coupling of matter to gravity by defining an action which depends linearly [284] or nonlinearly upon the matter Lagrangian [13, 44, 52, 53, 56, 223, 233, 398], or its trace [211, 243, 390]. These theories were extensively presented in the second part of this book. The new couplings generally induce non-geodesic motion mediated by an extra-force orthogonal to the four-velocity [235], which may have nontrivial effects already in flat Minkowski space. Instabilities due to new nonlinear interactions within the matter sector are thus common in these theories [27, 478]. We note, from this perspective, that in the hybrid metric-Palatini approach, considered in this work, it can be expected that such instabilities in the matter sector are absent, because the usual conservation laws are satisfied. In the following chapters, we review the formulation and the main applications of hybrid gravity models in cosmological and astrophysical scenarios.

17
The General Formalism

Let us start our considerations by giving the basic features of hybrid metric-Palatini gravity. In particular, we discuss the action and the field equations both in the theory and its equivalent scalar–tensor representations in both the Jordan and the Einstein frames. Next, the well-formulation and well-posedness of the Cauchy problem is considered. Finally, arbitrary hybrid gravity theories constructed from the metric and an independent connection are explored, and it is shown that the special hybrid models avoid otherwise generic pathologies.

17.1 Action and Gravitational Field Equations

The action of hybrid metric-Palatini gravity is specified as [104, 224]

$$S = \frac{1}{2\kappa^2}\int \mathrm{d}^4x\sqrt{-g}\,[R + f(\mathcal{R})] + \int \mathrm{d}^4\sqrt{-g}\,\mathcal{L}_m(g^{\mu\nu},\psi)\ , \tag{17.1}$$

where $\kappa^2 \equiv 8\pi G$, $\mathcal{L}_m$ is the standard minimally coupled matter Lagrangian, ψ collectively denotes the matter fields, R is the metric Einstein–Hilbert term, and $\mathcal{R} \equiv g^{\mu\nu}\mathcal{R}_{\mu\nu}$ is the Palatini curvature, defined in terms of an independent connection $\hat{\Gamma}^\alpha_{\mu\nu}$ as

$$\mathcal{R} \equiv g^{\mu\nu}\mathcal{R}_{\mu\nu} \equiv g^{\mu\nu}\left(\hat{\Gamma}^\alpha_{\mu\nu,\alpha} - \hat{\Gamma}^\alpha_{\mu\alpha,\nu} + \hat{\Gamma}^\alpha_{\alpha\lambda}\hat{\Gamma}^\lambda_{\mu\nu} - \hat{\Gamma}^\alpha_{\mu\lambda}\hat{\Gamma}^\lambda_{\alpha\nu}\right)\ , \tag{17.2}$$

that generates the Ricci curvature tensor $\mathcal{R}_{\mu\nu}$ as

$$\mathcal{R}_{\mu\nu} \equiv \hat{\Gamma}^\alpha_{\mu\nu,\alpha} - \hat{\Gamma}^\alpha_{\mu\alpha,\nu} + \hat{\Gamma}^\alpha_{\alpha\lambda}\hat{\Gamma}^\lambda_{\mu\nu} - \hat{\Gamma}^\alpha_{\mu\lambda}\hat{\Gamma}^\lambda_{\alpha\nu}\ . \tag{17.3}$$

Varying the action (17.1) with respect to the metric, one obtains the following gravitational field equation

$$G_{\mu\nu} + F(\mathcal{R})\mathcal{R}_{\mu\nu} - \frac{1}{2}f(\mathcal{R})g_{\mu\nu} = \kappa^2 T_{\mu\nu}\ , \tag{17.4}$$

where the matter stress-energy tensor is defined as usual,

$$T_{\mu\nu} \equiv -\frac{2}{\sqrt{-g}} \frac{\delta(\sqrt{-g}\mathcal{L}_m)}{\delta(g^{\mu\nu})}. \tag{17.5}$$

Varying the action with respect to the independent connection $\hat{\Gamma}^{\alpha}_{\mu\nu}$, one obtains the following equation of motion

$$\hat{\nabla}_\alpha \left(\sqrt{-g}F(R)g_{\mu\nu}\right) = 0, \tag{17.6}$$

which implies that $\hat{\Gamma}^{\alpha}_{\mu\nu}$ is the Levi–Civita connection of a metric $h_{\mu\nu} = F(R)g_{\mu\nu}$. Thus, $h_{\mu\nu}$ is conformally related to the physical metric $g_{\mu\nu}$, with the conformal factor given by $F(\mathcal{R}) \equiv df(\mathcal{R})/d\mathcal{R}$. This implies that the Palatini Ricci tensor $\mathcal{R}_{\mu\nu}$ is related to the metric one by the following relation

$$\mathcal{R}_{\mu\nu} = R_{\mu\nu} + \frac{3}{2}\frac{1}{F^2(\mathcal{R})}F(\mathcal{R})_{,\mu}F(\mathcal{R})_{,\nu} - \frac{1}{F(\mathcal{R})}\nabla_\mu F(\mathcal{R})_{,\nu} - \frac{1}{2}\frac{1}{F(\mathcal{R})}g_{\mu\nu}\Box F(\mathcal{R})\,. \tag{17.7}$$

The Palatini curvature, $\mathcal{R}$, can be obtained from the trace of the field equation (17.4), which yields

$$F(\mathcal{R})\mathcal{R} - 2f(\mathcal{R}) = \kappa^2 T + R \equiv X\,. \tag{17.8}$$

Note that we can express $\mathcal{R}$ algebraically in terms of X if the form of $f(\mathcal{R})$ allows analytic solutions. The variable X measures how much the theory deviates from the GR trace equation, $R = -\kappa^2 T$. These two observations shed light upon the structure of the theory and the central part that the "trace deviation" X plays in it. It is for this reason the $R + f(\mathcal{R})$ hybrid metric-Palatini theories have also been called simply the "$f(X)$ theories," and we use these two equivalent designations interchangeably throughout his work.

Indeed, we can express the field equation (17.4) solely in terms of the metric and X as

$$G_{\mu\nu} = \frac{1}{2}f(X)g_{\mu\nu} - F(X)R_{\mu\nu} + F'(X)\nabla_\mu X_{,\nu} + \left[F''(X) - \frac{3}{2}\frac{(F'(X))^2}{F(X)}\right]X_{,\mu}X_{,\nu} + \frac{1}{2}\left[F'(X)\Box X + F''(X)\,(\partial X)^2\right]g_{\mu\nu} + \kappa^2 T_{\mu\nu}\,. \tag{17.9}$$

Note that $(\partial X)^2 = X_{,\mu}X^{,\mu}$. The trace of the field equations is now

$$F'(X)\Box X + \left[F''(X) - \frac{1}{2}\frac{(F'(X))^2}{F(X)}\right](\partial X)^2 + \frac{1}{3}\left[X + 2f(X) - F(X)R\right] = 0\,,$$

while the relation between the metric scalar curvature R and the Palatini scalar curvature $\mathcal{R}$ is

$$\mathcal{R}(X) = R + \frac{3}{2}\left[\left(\frac{F'(X)}{F(X)}\right)^2 - 2\frac{\Box F(X)}{F(X)}\right], \tag{17.10}$$

which can be obtained by contracting Eq. (17.7).

17.2 Scalar–Tensor Representation

Analogously to the pure metric and Palatini cases [392, 393], the action (17.1) for hybrid metric-Palatini gravity can be turned into that of a scalar–tensor theory by introducing an auxiliary field A such that

$$S = \frac{1}{2\kappa^2}\int d^4x\sqrt{-g}\,[\Omega_A R + f(A) + f_A(\mathcal{R} - A)] + S_m\,, \tag{17.11}$$

where $f_A \equiv df/dA$, S_m is the matter action, and we have included a coupling constant Ω_A for generality. Note that $\Omega_A = 1$ for the original hybrid metric-Palatini theory [224].

The field A is dynamically equivalent to the Palatini scalar curvature $\mathcal{R}$ if $f''(\mathcal{R}) \neq 0$. Defining[1]

$$\phi \equiv f_A, \qquad V(\phi) = Af_A - f(A), \tag{17.15}$$

the action (17.11) becomes

$$S = \frac{1}{2\kappa^2}\int d^4x\sqrt{-g}\,[\Omega_A R + \phi\mathcal{R} - V(\phi)] + S_m\,. \tag{17.16}$$

It is easy to see that the action (17.16) is equivalent to our original starting point Eq. (17.1). Variation of the above action with respect to the metric, the scalar ϕ and the connection leads to the field equations

$$\Omega_A R_{\mu\nu} + \phi\mathcal{R}_{\mu\nu} - \frac{1}{2}\left(\Omega_A R + \phi\mathcal{R} - V\right)g_{\mu\nu} = \kappa^2 T_{\mu\nu}\,, \tag{17.17}$$

$$\mathcal{R} - V_\phi = 0\,, \tag{17.18}$$

$$\hat{\nabla}_\alpha\left(\sqrt{-g}\phi g^{\mu\nu}\right) = 0\,, \tag{17.19}$$

respectively, where $V_\phi = V'(\phi)$.

[1] It is useful to note that Eq. (17.15) is a Clairaut differential equation [523], that is,

$$Af_A - f(A) = V\left(f_A\right). \tag{17.12}$$

It admits a general linear solution

$$f\left(A\right) = h\,A - V\left(h\right), \tag{17.13}$$

for arbitrary $V\left(\phi\right)$ and a singular solution followed from the equation

$$\frac{\partial V\left(f_A\right)}{\partial f_A} - A = 0\,. \tag{17.14}$$

The solution of Eq. (17.19) implies that the independent connection is the Levi–Civita connection of a metric $h_{\mu\nu} = \phi g_{\mu\nu}$. This means that the relation Eq. (17.7) between the tensors $\mathcal{R}_{\mu\nu}$ and $R_{\mu\nu}$ can now be rewritten as

$$\mathcal{R}_{\mu\nu} = R_{\mu\nu} + \frac{3}{2\phi^2}\partial_\mu\phi\partial_\nu\phi - \frac{1}{\phi}\left(\nabla_\mu\nabla_\nu\phi + \frac{1}{2}g_{\mu\nu}\Box\phi\right) , \tag{17.20}$$

which can be used in the action (17.16) to get rid of the independent connection and to obtain the scalar–tensor representation of hybrid metric-Palatini gravity [286]. Thus, one finally arrives at the following action

$$S = \frac{1}{2\kappa^2}\int d^4x\sqrt{-g}\left[(\Omega_A + \phi)R + \frac{3}{2\phi}\partial_\mu\phi\partial^\mu\phi - V(\phi)\right] + S_m \, . \tag{17.21}$$

It is interesting to point out that, by the substitution $\phi \to -(\kappa\phi)^2/6$, the action (17.21) reduces to the well-known case of a conformally coupled scalar field with a self-interaction potential. Namely, this redefinition makes the kinetic term in the action (17.21) the standard one, and the action itself becomes that of a massive scalar-field conformally coupled to the Einstein gravity. Of course, it is not the Brans–Dicke gravity where the scalar field is massless, but now we have a nonzero $V(\phi)$ as given in Eq. (17.16).

In the limit $\Omega_A \to 0$, the theory given by Eq. (17.21) presents the Palatini-$f(\mathcal{R})$ gravity, and in the limit $\Omega_A \to \infty$ the metric $f(R)$ gravity [286]. Apart from these singular cases, the more generic theories with a finite Ω_A thus lie in the "hybrid" regime, which from this perspective provides a unique interpolation between the two a priori completely distinct classes of gravitational theories. In fact, we have arrived at Brans–Dicke-type theories, specified by the nontrivial coupling function

$$\omega_{BD} = \frac{3\phi}{2\phi - 2\Omega_A} , \tag{17.22}$$

that generalizes the $\omega_{BD} = 0$ and $\omega_{BD} = -3/2$ cases, which correspond to the scalar–tensor representations of the metric $f(R)$ and the Palatini-$f(\mathcal{R})$ gravities [103], respectively.

Using Eqs. (17.18) and (17.20) in (17.17), the metric field equation can be written as

$$\begin{aligned}(\Omega_A + \phi)R_{\mu\nu} = \kappa^2\left(T_{\mu\nu} - \frac{1}{2}g_{\mu\nu}T\right) + \frac{1}{2}g_{\mu\nu}\left(V + \Box\phi\right) \\ +\nabla_\mu\nabla_\nu\phi - \frac{3}{2\phi}\partial_\mu\phi\partial_\nu\phi \, ,\end{aligned} \tag{17.23}$$

or equivalently as

$$\begin{aligned}(\Omega_A + \phi)G_{\mu\nu} = \kappa^2 T_{\mu\nu} + \nabla_\mu\nabla_\nu\phi - \Box\phi g_{\mu\nu} - \frac{3}{2\phi}\nabla_\mu\phi\nabla_\nu\phi \\ +\frac{3}{4\phi}\nabla_\lambda\phi\nabla^\lambda\phi g_{\mu\nu} - \frac{1}{2}V g_{\mu\nu},\end{aligned} \tag{17.24}$$

from which it is seen that the space-time curvature is generated by both the matter and the scalar field. Note that the metric field equation (17.24) can also be written as an effective Einstein field equation, i.e., $G_{\mu\nu} = \kappa^2 T^{\text{eff}}_{\mu\nu}$, where the effective energy-momentum tensor is given by

$$T^{\text{eff}}_{\mu\nu} = \frac{1}{1+\phi}\Bigg\{T_{\mu\nu} - \frac{1}{\kappa^2}\Bigg[\frac{1}{2}g_{\mu\nu}\left(V + 2\Box\,\phi\right) + \nabla_\mu\nabla_\nu\phi \\ -\frac{3}{2\phi}\partial_\mu\phi\,\partial_\nu\phi + \frac{3}{4\phi}g_{\mu\nu}(\partial\phi)^2\Bigg]\Bigg\}. \tag{17.25}$$

The scalar field equation can be manipulated in two different ways that illustrate further how the hybrid models combine physical features of the $\omega_{BD} = 0$ and $\omega_{BD} = -3/2$ scalar–tensor models. First, tracing Eq. (17.17) with $g^{\mu\nu}$, we find $-\Omega_A R - \phi\mathcal{R} + 2V = \kappa^2 T$, and, using Eq. (17.18), it takes the following form:

$$2V - \phi V_\phi = \kappa^2 T + \Omega_A R\,. \tag{17.26}$$

Similarly as in the Palatini ($\omega_{BD} = -3/2$) case, this equation tells us that the field ϕ can be expressed as an algebraic function of the scalar $X \equiv \kappa^2 T + \Omega_A R$, i.e., $\phi = \phi(X)$. In the pure Palatini case, however, ϕ is just a function of T. The right-hand side of Eq. (17.23), therefore, besides containing new matter terms associated with the trace T and its derivatives, also contains the curvature R and its derivatives. Thus, this theory can be seen as a higher-derivative theory in both matter and metric fields. However, such an interpretation can be avoided if R is replaced in Eq. (17.26) with the relation

$$R = \mathcal{R} + \frac{3}{\phi}\Box\phi - \frac{3}{2\phi^2}\partial_\mu\phi\partial^\mu\phi\,, \tag{17.27}$$

together with $\mathcal{R} = V_\phi$. One then finds that the scalar field is governed by the second-order evolution equation that becomes, when $\Omega_A = 1$,

$$-\Box\phi + \frac{1}{2\phi}\partial_\mu\phi\partial^\mu\phi + \frac{\phi[2V - (1+\phi)V_\phi]}{3} = \frac{\phi\kappa^2}{3}T\,, \tag{17.28}$$

which is an effective Klein–Gordon equation. This last expression shows that, unlike in the Palatini ($\omega_{BD} = -3/2$) case, the scalar field is dynamical. The theory is therefore not affected by the microscopic instabilities that arise in Palatini models with infrared corrections [396].

Einstein frame: Finally, we can make a conformal transformation into the Einstein frame of these theories. The conformal rescaling we need to achieve this is given by

$$\hat{g}_{\mu\nu} \equiv (\phi + \Omega_A)\,g_{\mu\nu}\,, \tag{17.29}$$

and the Einstein frame Lagrangian then becomes

$$\hat{\mathcal{L}} = \hat{R} + \frac{3\Omega_A}{2\phi} \frac{\hat{g}^{\alpha\beta}\phi_{,\alpha}\phi_{,\beta}}{(\phi+\Omega_A)^2} - \frac{V(\phi)}{(\phi+\Omega_A)^2} . \tag{17.30}$$

This can be further put into its canonical form by introducing the rescaled field ψ as

$$\phi = \Omega_A \tan^2\left(\frac{\psi}{2\sqrt{3}}\right) . \tag{17.31}$$

The vacuum theory then becomes a canonical scalar theory with a very specific potential (stemming of course from the original function $f(\mathcal{R})$) in the Einstein frame.

17.3 The Weak Field Limit

In order to consider the weak field limit of the theory, it is of paramount importance to determine the post-Newtonian parameters of the theory as they determine the compatibility of the theory with the local precision gravity tests. For post-Newtonian analysis of the metric and Palatini $f(R)$ theories, see [392, 393], and for a unified analysis [288]. Here, in particular, we are interested in the parameter γ that is basically the fractional difference of the Newtonian potentials (see later sections) in the limit where the cosmological expansion can be neglected.

To this end, we need to consider the perturbations of Eqs. (17.23) and (17.28) in a Minkowskian background. The usual procedure is to assume $\phi = \phi_0 + \varphi(x)$, where ϕ_0 is the asymptotic value of the field far away from the local system (and should be given by the cosmological background solution), and to take a quasi-Minkowskian coordinate system in which $g_{\mu\nu} \approx \eta_{\mu\nu} + h_{\mu\nu}$, with $|h_{\mu\nu}| \ll 1$. Then it is easy to see that the quadratic terms $\partial_\mu\phi\partial_\nu\phi$ and $\partial_\mu\phi\partial^\mu\phi$ in Eqs. (17.23) and (17.28) do not contribute to the linear order. The potential terms in Eq. (17.28) can be linearized as follows:

$$\begin{aligned} \frac{\phi}{3}[2V - (1+\phi)V_\phi] \approx & \frac{\phi_0}{3}[2V_0 - (1+\phi_0)V_{\phi 0}] \\ & -\frac{\varphi}{3}\left[\phi_0(1+\phi_0)V_{\phi\phi}|_0 + V_{\phi 0} - 2V_0\right] . \end{aligned} \tag{17.32}$$

The zeroth-order term in this equation is due to the background, and can be absorbed into a coordinate redefinition. The coefficient of the first-order term can be interpreted as a mass squared. The linearized scalar field equation is thus given by

$$(\vec{\nabla}^2 - m_\varphi^2)\varphi = -\frac{\phi_0\kappa^2}{3}\rho , \tag{17.33}$$

where, as usual, in this order of approximation we have neglected the time derivatives and the pressure terms.

The linearization of the metric field equations is a bit more complicated because we need to establish suitable gauge conditions. Since the background is Minkowskian, the perturbed Ricci tensor is given by

$$\delta R_{\mu\nu} \equiv \frac{1}{2}\left(\partial_\mu \partial_\lambda \tilde{h}^\lambda_\nu + \partial_\nu \partial_\lambda \tilde{h}^\lambda_\mu\right) - \frac{1}{2}\vec{\nabla}^2 h_{\mu\nu}\,, \tag{17.34}$$

where $\tilde{h}^\lambda_\nu \equiv h^\lambda_\nu - (1/2)\delta^\lambda_\nu h^\alpha_\alpha$. The term $\nabla_\mu \nabla_\nu \phi$ on the right-hand side of Eq. (17.23) can be combined with the terms $\partial_\mu \partial_\lambda \tilde{h}^\lambda_\nu$ to give the following gauge conditions

$$\partial_\lambda \tilde{h}^\lambda_\mu - \frac{1}{1+\phi_0}\partial_\mu \varphi = 0\,. \tag{17.35}$$

With this choice, the linearized equations for the metric become

$$-\frac{1}{2}\vec{\nabla}^2 h_{\mu\nu} = \frac{1}{1+\phi_0}\left(T_{\mu\nu} - \frac{1}{2}T\eta_{\mu\nu}\right) + \frac{V_0 + \vec{\nabla}^2\varphi}{2(1+\phi_0)}\eta_{\mu\nu}\,. \tag{17.36}$$

For consistency, to this order $T_{00} = \rho$, $T_{ij} = 0$, $T = -\rho$, which leads to

$$\vec{\nabla}^2\left(h^{(2)}_{00} - \frac{\varphi^{(2)}}{1+\phi_0}\right) = -\frac{1}{1+\phi_0}\left(\rho - V_0\right), \tag{17.37}$$

$$\vec{\nabla}^2\left(h^{(2)}_{ij} + \frac{\varphi^{(2)}}{1+\phi_0}\delta_{ij}\right) = -\frac{\delta_{ij}}{1+\phi_0}\left(\rho + V_0\right)\,. \tag{17.38}$$

Before solving Eqs. (17.33), (17.37), and (17.38), it is worth noting that while the connection equation (17.18) is invariant under constant rescalings of the field $\phi \to c\phi$, the other field equations do not share this invariance. This is manifest in the combinations $(1+\phi_0)$ in the above perturbation equations.

Using the generic solution for:

$$(\vec{\nabla}^2 - m^2)f = -\rho, \tag{17.39}$$

given by

$$f = \frac{1}{4\pi}\int d^3\vec{x}'\frac{\rho(t,\vec{x}')}{|\vec{x}-\vec{x}'|}e^{-m|\vec{x}-\vec{x}'|}, \tag{17.40}$$

we find that

$$\varphi^{(2)}(t,\vec{x}) = \frac{\kappa^2\phi_0}{4\pi}\int d^3\vec{x}'\frac{\rho(t,\vec{x}')}{3|\vec{x}-\vec{x}'|}e^{-m_\varphi|\vec{x}-\vec{x}'|}, \tag{17.41}$$

$$h^{(2)}_{00}(t,\vec{x}) = \frac{\kappa^2}{4\pi(1+\phi_0)}\int d^3\vec{x}'\frac{\rho(t,\vec{x}')}{|\vec{x}-\vec{x}'|}\left(1 + \frac{\phi_0}{3}e^{-m_\varphi|\vec{x}-\vec{x}'|}\right)$$
$$+\frac{V_0}{(1+\phi_0)}\frac{|\vec{x}-\vec{x}_2|^2}{6}, \tag{17.42}$$

$$h_{ij}^{(2)}(t,\vec{x}) = \left[\frac{\kappa^2}{4\pi(1+\phi_0)}\int d^3\vec{x}'\frac{\rho(t,\vec{x}')}{|\vec{x}-\vec{x}'|}\left(1-\frac{\phi_0}{3}e^{-m_\varphi|\vec{x}-\vec{x}'|}\right)\right.$$
$$\left.-\frac{V_0}{(1+\phi_0)}\frac{|\vec{x}-\vec{x}_2|^2}{6}\right]\delta_{ij}. \tag{17.43}$$

In spherical symmetry and far from the sources, the above equations become

$$\varphi(r) = \frac{\kappa^2\phi_0}{12\pi}\frac{M}{r}e^{-m_\varphi r}, \tag{17.44}$$

$$h_{00}^{(2)}(r) = \frac{2G_{\rm eff}M}{r} + \frac{V_0}{(1+\phi_0)}\frac{r^2}{6}, \tag{17.45}$$

$$h_{ij}^{(2)}(r) = \left(\frac{2\gamma G_{\rm eff}M}{r} - \frac{V_0}{(1+\phi_0)}\frac{r^2}{6}\right)\delta_{ij}, \tag{17.46}$$

where we have denoted

$$G_{\rm eff} \equiv \frac{\kappa^2}{8\pi(1+\phi_0)}\left(1+\frac{\phi_0}{3}e^{-m_\varphi r}\right), \tag{17.47}$$

$$\gamma \equiv \frac{[1+\phi_0\exp(-m_\varphi r)/3]}{[1-\phi_0\exp(-m_\varphi r)/3]}, \tag{17.48}$$

$$m_\varphi^2 \equiv \left.\frac{2V - V_\phi - \phi(1+\phi)V_{\phi\phi}}{3}\right|_{\phi=\phi_0}. \tag{17.49}$$

These results represent the standard post-Newtonian metric up to second order for this class of theories.

We emphasize a striking feature of the hybrid metric-Palatini gravity. Note that in $f(R)$ gravity, we have

$$G_{\rm eff} \equiv \frac{G}{\phi_0}\left(1+\frac{1}{3}e^{-m_f r}\right), \qquad \gamma \equiv \left(1-\frac{e^{-m_f r}}{3}\right)\Big/\left(1+\frac{e^{-m_f r}}{3}\right),$$

which requires a large mass $m_f^2 \equiv (\phi V_{\phi\phi} - V_\phi)/3$ to make the Yukawa-type corrections negligible in local experiments. Thus, to obtain $\gamma \approx 1$, there is only one possibility [392, 393], namely, $m_\varphi r \gg 1$ from millimeters to astronomical scales, i.e., the range of the scalar interaction, $1/m_\varphi$, should be smaller than a few millimeters. In the current case, however, there are two possibilities to obtain $\gamma \approx 1$. The first one is the same as in $f(R)$ theories and involves a very massive scalar field. The second possibility implies a small value ϕ_0. If $\phi_0 \ll 1$, then the Yukawa-type corrections are very small regardless of the magnitude of m_φ. This could allow for the existence of a long-range scalar field able to modify the cosmological and galactic dynamics, but leaving unaffected the solar system. Subtle modifications could in the most optimistic case be detected as anomalies in the local gravitational field [265].

17.4 The Cauchy Problem

The dynamical equivalence with scalar–tensor theories shown in the previous section is useful to discuss the well-posedness of the Cauchy problem for hybrid gravity in vacuo and coupled to standard matter sources. For previous studies of the Cauchy problem in different formulations of $f(R)$ theories, see [112, 111, 113, 114, 397].

We begin by proving the well-posedness of the Cauchy problem in vacuo, making use of the equivalent formulations (17.24) and (17.28). As we shall argue, the same conclusions hold in the presence of standard matter sources satisfying the usual conservation laws $\nabla^\mu T_{\mu\nu} = 0$.

Borrowing definitions and notations from [135], the key point of our discussion is the introduction of suitable generalized harmonic coordinates, defined by the conditions

$$F^\mu_\phi := F^\mu - H^\mu = 0 \qquad \text{with} \qquad F^\mu := g^{\alpha\beta}\Gamma^\mu_{\alpha\beta}, \quad H^\mu := \frac{1}{(1+\phi)}\nabla^\mu\phi\,. \tag{17.50}$$

As we shall see, the gauge Eq. (17.50) allows us to develop a second-order analysis very similar to the one used in GR [135]. We notice that the generalized harmonic gauge Eq. (17.50) is a particular case of the one introduced in [442] to prove the well-posedness of the Cauchy problem for a certain class of scalar–tensor theories of gravity.

Let us start with rewriting Eq. (17.24) in the form [110]

$$R_{\mu\nu} = \frac{1}{(1+\phi)}\left[\Sigma_{\mu\nu} - \frac{1}{2}\Sigma g_{\mu\nu}\right], \tag{17.51}$$

where

$$\Sigma_{\mu\nu} := \nabla_\mu\nabla_\nu\phi - \Box\phi g_{\mu\nu} - \frac{3}{2\phi}\nabla_\mu\phi\nabla_\nu\phi + \frac{3}{4\phi}\nabla_\lambda\phi\nabla^\lambda\phi - \frac{1}{2}V g_{\mu\nu}\,, \tag{17.52}$$

plays the role of an effective energy-momentum tensor. We recall that the Ricci tensor can be expressed as [135]

$$R_{\mu\nu} = R^\phi_{\mu\nu} + \frac{1}{2}\left[g_{\mu\sigma}\partial_\nu\left(F^\sigma_\phi + H^\sigma\right) + g_{\nu\sigma}\partial_\mu\left(F^\sigma_\phi + H^\sigma\right)\right], \tag{17.53}$$

with

$$R^\phi_{\mu\nu} := -\frac{1}{2}g^{\alpha\beta}\partial^2_{\alpha\beta}g_{\mu\nu} + A_{\mu\nu}(g, \partial g)\,, \tag{17.54}$$

where only first-order derivatives appear in the functions $A_{\mu\nu}$. Due to the assumed gauge condition $F^\mu_\phi = 0$ and the explicit expression of H^μ, from Eqs. (17.53) and (17.54), we get the following representation

$$R_{\mu\nu} = -\frac{1}{2}g^{\alpha\beta}\partial^2_{\alpha\beta}g_{\mu\nu} + \frac{1}{(1+\phi)}\partial^2_{\mu\nu}\phi + B_{\mu\nu}(g, \phi, \partial g, \partial\phi)\,, \tag{17.55}$$

where the functions $B_{\mu\nu}$ depend on the metric g, the scalar field ϕ and their first-order derivatives. At the same time, using Eq. (17.28) to replace all terms depending on the divergence $g^{\alpha\beta}\nabla_\alpha\nabla_\beta\phi$, the right-hand side of Eq. (17.51) can be expressed as

$$\frac{1}{(1+\phi)}\left[\Sigma_{\mu\nu}-\frac{1}{2}\Sigma g_{\mu\nu}\right]=\frac{1}{(1+\phi)}\partial^2_{\mu\nu}\phi+C_{\mu\nu}(g,\phi,\partial g,\partial\phi)\,, \tag{17.56}$$

where, again, the functions $C_{\mu\nu}$ depend only on first-order derivatives. A direct comparison of Eqs. (17.55) with (17.56) shows that, in the considered gauge, Eq. (17.51) assumes the form

$$g^{\alpha\beta}\partial^2_{\alpha\beta}g_{\mu\nu}=D_{\mu\nu}(g,\phi,\partial g,\partial\phi)\,. \tag{17.57}$$

The conclusion follows that Eqs. (17.28) together with (17.57), form a quasi-diagonal, quasi-linear second-order system of partial differential equations, for which well-known theorems by Leray [135, 310, 493] hold. Given initial data on a space-like surface, the associated Cauchy problem is then well posed in suitable Sobolev spaces [135]. Of course, the initial data have to satisfy the gauge conditions $F^i_\phi=0$ as well as the Hamiltonian and momentum constraints

$$G^{0\mu}=\frac{1}{(1+\phi)}\Sigma^{0\mu}, \qquad \mu=0,\ldots,3\,, \tag{17.58}$$

on the initial space-like surface. In connection with this, we notice that, from Eq. (17.28), we can derive the expression of the second partial derivative $\partial^2_0\phi$ and replace it on the right hand side of Eq. (17.58), and thus obtaining constraints involving no higher than first-order partial derivatives with respect to the time variable x^0. To conclude, we have to prove that the gauge conditions $F^\mu_\phi=0$ are preserved in a neighborhood of the initial space-like surface. To this end, we first verify that the divergence of the gravitational field equation (17.24) vanishes, namely

$$\nabla^\mu\left[(1+\phi)G_{\mu\nu}-\Sigma_{\mu\nu}\right]=0\,. \tag{17.59}$$

Taking into account the identities

$$\nabla^\mu G_{\mu\nu}=0, \qquad (\nabla^\mu\phi)\,R_{\mu\nu}=(\nabla^\mu\nabla_\mu\nabla_\nu-\nabla_\nu\Box)\,\phi\,, \tag{17.60}$$

automatically satisfied by the Einstein and Ricci tensors, we have

$$\nabla^\mu\left[(1+\phi)G_{\mu\nu}-\Sigma_{\mu\nu}\right]=-\frac{1}{2}R\nabla_\nu\phi$$
$$+\nabla^\mu\left(\frac{3}{2\phi}\nabla_\mu\phi\nabla_\nu\phi-\frac{3}{4\phi}\nabla_\lambda\phi\nabla^\lambda\phi g_{\mu\nu}+\frac{1}{2}V(\phi)g_{\mu\nu}\right)\,. \tag{17.61}$$

On the other hand, inserting the content of Eq. (17.26) (in this case, with $T = 0$) into the trace of the field Eq. (17.24), we end up with the identity

$$R = \frac{dV}{d\phi} + \frac{3}{\phi}\Box\phi - \frac{3}{2\phi^2}\nabla_\lambda\phi\nabla^\lambda\phi\,. \tag{17.62}$$

The identities Eq. (17.59) follow then from a direct comparison of Eqs. (17.61) with (17.62).

Now, if $g_{\mu\nu}$ and ϕ solve the reduced field equation (17.57) and the scalar field Equation (17.28), then we have

$$(1+\phi)G^{\mu\nu} - \Sigma^{\mu\nu} = -\frac{(1+\phi)}{2}\left(g^{\mu\sigma}\partial_\sigma F^\nu_\phi + g^{\nu\sigma}\partial_\sigma F^\mu_\phi - g^{\mu\nu}\partial_\sigma F^\sigma_\phi\right)\,. \tag{17.63}$$

The identities (17.59) imply then that the functions F^μ_ϕ satisfy necessarily a linear homogeneous system of wave equations of the form

$$g^{pq}\partial^2_{pq}F^i_\varphi + E^{iq}_p\partial_q F^p_\varphi = 0\,, \tag{17.64}$$

where E^{iq}_p are known functions on the space-time. Since the constraints given Eq. (17.58) amount to the condition $\partial_0 F^i_\varphi = 0$ on the initial space-like surface, a well-known uniqueness theorem for differential systems such as Eq. (17.64) assures that $F^i_\varphi = 0$ in the region where solutions of Eqs. (17.28) and (17.57) exist (see also [135]).

The illustrated analysis also applies in the case of couplings to standard matter sources such as electromagnetic or Yang–Mills fields, (charged) perfect fluid, (charged) dust, Klein–Gordon scalar fields [114], so showing the well-posedness of the Cauchy problem for hybrid metric-Palatini gravity in the presence of standard matter fields. Indeed, when matter sources are present, Eqs. (17.28) and (17.57) have to be coupled with the matter field equations. Applying the same arguments developed for GR [135, 134, 195], it is easily seen that, in the generalized harmonic gauge (17.50), the matter field equations together with Eqs. (17.28) and (17.57) form a Leray hyperbolic and a causal differential system admitting a well-posed Cauchy problem [310]. In addition to the well-known results by Bruhat, the crucial point is again that the field equations of matter field imply the standard conservation laws $\nabla^\mu T_{\mu\nu} = 0$ [284]. In summary, the hybrid metric-Palatini gravity satisfies the well formulation and well-posedness of the Cauchy problem for standard forms of matter and then, in this sense, it is a viable theory.

17.5 General Hybrid Metric-Palatini Theories

17.5.1 Generalized Hybrid $f(R, \mathcal{R})$ gravity

Consider the action of the generalized hybrid metric-Palatini modified theory of gravity [477], given by

$$S = \frac{1}{2\kappa^2}\int_\Omega \sqrt{-g}f(R,\mathcal{R})\,d^4x + S_m(g_{\mu\nu},\psi), \tag{17.65}$$

where f is a function of metric scalar curvature R and the Palatini scalar curvature $\mathcal{R}$, and S_m is the matter action, in which matter is minimally coupled to the metric $g_{\mu\nu}$. The Palatini scalar curvature, $\mathcal{R}$, is defined in terms of an independent connection given by Eq. (17.2).

Now, varying the action (17.65) with respect to the metric $g_{\mu\nu}$ and the independent connection, yields the following field equations

$$\frac{\partial f}{\partial R}R_{\mu\nu} + \frac{\partial f}{\partial \mathcal{R}}\mathcal{R}_{\mu\nu} - \frac{1}{2}g_{\mu\nu}f\left(R,\mathcal{R}\right) - \left(\nabla_\mu\nabla_\nu - g_{\mu\nu}\Box\right)\frac{\partial f}{\partial R} = \kappa^2 T_{\mu\nu}, \qquad (17.66)$$

and

$$\hat{\nabla}_\alpha\left(\sqrt{-g}\,\frac{\partial f}{\partial \mathcal{R}}g^{\mu\nu}\right) = 0, \qquad (17.67)$$

respectively. Equation (17.67) implies that the independent connection is the Levi–Civita connection of a new metric tensor $h_{\mu\nu}$, which is conformally related to $g_{\mu\nu}$ by the relation $h_{\mu\nu} = g_{\mu\nu}\partial f/\partial\mathcal{R}$. Thus, the independent connection is the Levi–Civita connection related to the new auxiliary metric $h_{\mu\nu}$.

The action (17.65) can be written in a scalar–tensor representation, by considering an action with two auxiliary fields, α and β, respectively, in the following form

$$S = \frac{1}{2\kappa^2}\int_\Omega \sqrt{-g}\,\left[f\left(\alpha,\beta\right) + \frac{\partial f}{\partial\alpha}\left(R-\alpha\right) + \frac{\partial f}{\partial\beta}\left(\mathcal{R}-\beta\right)\right]d^4x + S_m. \qquad (17.68)$$

Substituting $\alpha = R$ and $\beta = \mathcal{R}$, one recovers action (17.65). We now define two scalar fields as

$$\varphi = \frac{\partial f}{\partial\alpha}, \qquad \psi = -\frac{\partial f}{\partial\beta}, \qquad (17.69)$$

where the negative sign in the second definition is imposed in order to guarantee a positive kinetic energy for the scalar field. Furthermore, we define the potential V as

$$V\left(\varphi,\psi\right) = -f\left(\alpha,\beta\right) + \varphi\alpha - \psi\beta\,, \qquad (17.70)$$

so that the action (17.68) is equivalent to

$$S = \frac{1}{2\kappa^2}\int_\Omega \sqrt{-g}\,\left[\varphi R - \psi\mathcal{R} - V\left(\varphi,\psi\right)\right]d^4x + S_m\,. \qquad (17.71)$$

Using $h_{\mu\nu} = -\psi\, g_{\mu\nu}$, we have the following relation

$$\mathcal{R} = R + \frac{3}{\psi^2}\partial^\mu\psi\partial_\mu\psi - \frac{3}{\psi}\Box\psi\,, \qquad (17.72)$$

so that we can replace $\mathcal{R}$ in the action (17.71), to obtain

$$S = \frac{1}{2\kappa^2}\int_\Omega \sqrt{-g}\Big[\left(\varphi-\psi\right)R - \frac{3}{2\psi}\partial^\mu\psi\partial_\mu\psi - V\left(\varphi,\psi\right)\Big]d^4x + S_m. \qquad (17.73)$$

In order to obtain the field equations, we vary the action (17.73) with respect to the metric $g_{\mu\nu}$ and the scalar fields φ and ψ, and eventually organizing terms, to obtain the following equations of motion

$$\begin{aligned}(\varphi-\psi)\,G_{\mu\nu} &= \kappa^2 T_{\mu\nu} + (\nabla_\mu\nabla_\nu - g_{\mu\nu}\Box)(\varphi-\psi) \\ &+\frac{3}{2\psi}\partial_\mu\psi\partial_\nu\psi + \left(\frac{1}{2}V + \frac{3}{4\psi}\partial^\alpha\psi\partial_\alpha\psi\right) g_{\mu\nu}\,, \end{aligned} \tag{17.74}$$

$$\Box\varphi + \frac{1}{3}\left(2V - \psi V_\psi - \varphi V_\varphi\right) = \frac{\kappa^2 T}{3}\,, \tag{17.75}$$

$$\Box\psi - \frac{1}{2\psi}\partial^\mu\psi\partial_\mu\psi - \frac{\psi}{3}\left(V_\varphi + V_\psi\right) = 0\,, \tag{17.76}$$

respectively, where V_φ and V_ψ are defined by

$$V_\varphi \equiv \frac{\partial V}{\partial\varphi}\,, \qquad V_\psi \equiv \frac{\partial V}{\partial\psi}\,, \tag{17.77}$$

respectively. Equation (17.75) shows an important difference between the two scalar fields, namely, φ is coupled to matter whereas ψ, given through Eq. (17.76), is not. This fact will have important consequences when searching for cosmological solutions in which matter is present. We refer the reader to [440] for cosmological solutions, where exact solutions were constructed in a FLRW Universe by either making an ansatz on the scale factor or on the effective potential. Among other relevant results, it was shown that it is possible to obtain exponentially expanding solutions for flat universes even when the cosmology is not purely vacuum. Furthermore, the classes of actions were derived for the original theory which generates these solutions.

17.5.2 More General Theories

The "hybrid" theory space is a priori large. In addition to the metric and its Levi–Civita connection, one also has an additional independent connection as a building block to construct curvature invariants from [108]. Thus one can consider various new terms such as

$$\mathcal{R}^{\mu\nu}\mathcal{R}_{\mu\nu}\,, \quad R^{\mu\nu}\mathcal{R}_{\mu\nu}\,, \quad \mathcal{R}^{\mu\nu\alpha\beta}\mathcal{R}_{\mu\nu\alpha\beta}\,, \qquad R^{\mu\nu\alpha\beta}\mathcal{R}_{\mu\nu\alpha\beta}\,, \quad \mathcal{R}R\,, \quad \text{etc} \tag{17.78}$$

Though an exhaustive analysis of such hybrid theories has not been performed, there is some evidence that the hybrid metric-Palatini class of theories we are focusing our attention upon here is a unique class of viable higher-order hybrid gravity theories. In the more restricted framework of purely metric theories, it is well known that the $f(R)$ class of theories is exceptional by avoiding the otherwise generic Ostrogradski instabilities by allowing a separation of the additional degrees of freedom into a harmless scalar degree of freedom [511]: as we have already seen, such a separation is possible also for the hybrid metric-Palatini theories. Furthermore, it turns out that this feature is a similar exception in the larger space of metric-affine theories, since a generic theory there is inhabited by ghosts, superluminalities or other unphysical degrees of freedom.

As a representative class of more general theories, actions of the form

$$S = \frac{1}{2\kappa^2} \int \mathrm{d}^4x\sqrt{-g}\, f(R, \mathcal{R}, \hat{Q}_H), \qquad \hat{Q}_H = R^{\mu\nu}\mathcal{R}_{\mu\nu} \tag{17.79}$$

were studied in [293], where the precise field content of this action in the weak-field limit were determined. Variation of Eq. (17.79) with respect to the metric produces the following field equations:

$$\begin{aligned} f_{,R}R_{\mu\nu} - \frac{1}{2}g_{\mu\nu}f + g_{\mu\nu}\Box f_{,R} - \nabla_\mu\nabla_\nu f_{,R} + f_{,\mathcal{R}}\mathcal{R}_{\mu\nu} + 2f_{,\hat{Q}}R^\lambda_\mu\mathcal{R}_{\nu\lambda} + \frac{1}{2}\Box\left(f_{,\hat{Q}}\mathcal{R}_{\mu\nu}\right) \\ + \frac{1}{2}g_{\mu\nu}\nabla_\alpha\nabla_\beta\left(f_{,\hat{Q}}\mathcal{R}^{\alpha\beta}\right) - \nabla_\lambda\nabla_{(\nu}\left(f_{,\hat{Q}}\mathcal{R}^\lambda_{\mu)}\right) = \kappa^2 T_{\mu\nu}\,, \end{aligned} \tag{17.80}$$

where $f_{,R}$, $f_{,\mathcal{R}}$ and $f_{,\hat{Q}}$ are the derivatives of f with respect to R, $\mathcal{R}$ and $\hat{Q}_H$ respectively. The solution to the equation of motion for the connection on the other hand dictates that it is the Levi–Civita connection of the metric $\hat{g}_{\mu\nu}$ given by

$$\hat{g}^{\mu\nu} = \frac{\sqrt{-r}}{\sqrt{-g}}r^{\mu\nu}, \quad \text{where} \quad r^{\mu\nu} = f_{,\mathcal{R}}g^{\mu\nu} + f_{,\hat{Q}}R^{\mu\nu}\,. \tag{17.81}$$

Using this, one can eliminate the auxiliary metric $\hat{g}_{\mu\nu}$ in terms of the physical metric $g_{\mu\nu}$.

Considering perturbations $h_{\mu\nu} = g_{\mu\nu} - \eta_{\mu\nu}$ around Minkowski space $g_{\mu\nu} = \eta_{\mu\nu}$, and inverting the linearized field equations for the physical metric, then gives us the propagators for the graviton and the additional degrees of freedom that may be present in $h_{\mu\nu}$. The propagator $\Pi^{\alpha\beta\gamma\delta}$ is defined by

$$\Pi^{-1\gamma\delta}_{\alpha\beta}h_{\gamma\delta} = \kappa^2\tau_{\alpha\beta}\,, \tag{17.82}$$

where $\tau_{\alpha\beta}$ represents the linearized stress energy source. In the formalism of the spin-projector operators employed in Ref. [68] and more pedagogically reviewed in Ref. [69], the result can be given in Fourier space (where basically $\Box \to -k^2$) in terms of two functions a and c as

$$k^2\Pi_{\alpha\beta\gamma\delta} = \frac{\mathcal{P}^2_{\alpha\beta\gamma\delta}}{a(-k^2)} - \frac{\mathcal{P}^0_{\alpha\beta\gamma\delta}}{a(-k^2) - 3c(-k^2)}\,, \tag{17.83}$$

where $\mathcal{P}^2_{\alpha\beta\gamma\delta}$ picks up the spin-2 and $\mathcal{P}^0_{\alpha\beta\gamma\delta}$ the scalar modes of the fluctuations (see [293, 68, 69] for details). The functions a and c can be determined immediately given a theory of the form Eq. (17.79), and depend upon the combinations

$$A = \frac{6f^{(0)}_{\mathcal{R}\mathcal{R}} + f^{(0)}_{,\hat{Q}}}{2f^{(0)}_{,\mathcal{R}}}\,, \quad \text{and} \quad B = \frac{f^{(0)}_{,\hat{Q}}}{f^{(0)}_{\mathcal{R}}}\,, \tag{17.84}$$

in the following way:

$$a(\Box) = f^{(0)}_{,R} + f^{(0)}_{,\mathcal{R}} - f^{(0)}_{,\hat{Q}} \frac{B}{4} \Box^2 , \tag{17.85}$$

$$c(\Box) = f^{(0)}_{,R} + f^{(0)}_{,\mathcal{R}} - 2\left(f^{(0)}_{,RR} + 4 f^{(0)}_{,R\mathcal{R}} + f^{(0)}_{,\hat{Q}}\right)\Box + \left[f^{(0)}_{,R\mathcal{R}} (6A + B) + f^{(0)}_{,\hat{Q}} \left(2A + \frac{B}{4}\right)\right]\Box^2 . \tag{17.86}$$

Let us then enumerate some special cases. To simplify things we assume $f^{(0)}_{\mathcal{R}\mathcal{R}} = 0$.

Metric $f(R)$ models: In the pure metric $f(R)$ case, $f^{(0)}_{,R\mathcal{R}} = A = 0$ and we have

$$\Pi^{\alpha\beta\gamma\delta}_{f(R)} = \Pi^{\alpha\beta\gamma\delta}_{GR} + \frac{1}{2\left(k^2 + (3 f^{(0)}_{,RR})^{-1}\right)} \mathcal{P}^{0\alpha\beta\gamma\delta} . \tag{17.87}$$

Thus we have an extra scalar degree of freedom, as we expect since the $f(R)$ models are known to be equivalent to Brans–Dicke theories with a vanishing parameter $\omega_{BD} = 0$. The mass of the "scalaron" is $m^2 = (3 f^{(0)}_{,RR})^{-1}$, and as long as $f''(R) > 0$ the theory is stable, otherwise a tachyonic mass spoils the stability around Minkowski space.

Palatini $f(\mathcal{R})$ models: As already discussed, the Palatini-type $f(\mathcal{R})$ models are equivalent to Brans–Dicke theories with the parameter $\omega_{BD} = -3/2$. This particular value corresponds to vanishing kinetic term of the field, which thus is nondynamical. Therefore we expect that no additional scalar degree of freedom should appear. For a proper normalization we may assume that $f^{(0)}_{,\mathcal{R}} = 1$, and we have now of course that $f^{(0)}_{,RR} = f^{(0)}_{,R\mathcal{R}} = f^{(0)}_{,\hat{Q}} = 0$. Hence,

$$\Pi^{\alpha\beta\gamma\delta}_{f(\mathcal{R})} = \Pi^{\alpha\beta\gamma\delta}_{GR} , \tag{17.88}$$

confirming our expectation.

Hybrid metric-Palatini models: It was already remarked in [286] that in Ricci-flat space-times the hybrid metric-Palatini theories share the properties of Palatini-$f(\mathcal{R})$ theories, which in vacuum reduce to GR with a possible cosmological constant. Therefore it is not a surprise that we find no new propagating degrees of freedom in Minkowski vacuum,

$$\Pi^{\alpha\beta\gamma\delta}_{f(X)} = \Pi^{\alpha\beta\gamma\delta}_{GR} . \tag{17.89}$$

Interestingly though, this class of theories is not equivalent to either of the previous two cases, since when one considers curved space-times, a new scalar degree of freedom appears. In this sense, the hybrid metric-Palatini gravity is a quite minimalistic scalar–tensor extension of GR, as the scalar propagates only in the presence of background curvature.

The hybrid $f(R, \mathcal{R})$ models: The generalized hybrid Ricci scalar theories were introduced in [192, 477] and found to have qualitatively different properties compared to the more restricted class of $f(X)$ models described earlier. In particular, the $f(R, \mathcal{R})$ were shown to be equivalent to a class of biscalar–tensor theories. These theories have an extra $\mathcal{P}^{0\alpha\beta\gamma\delta}$ spin-0 propagator with a double pole, corresponding to two propagating scalar degrees of freedom. We can easily deduce the masses of these scalar fields, which are given by

$$m_{\pm}^2 = \frac{f^{(0)}_{,\mathcal{R}}}{18\left(f^{(0)}_{,R\mathcal{R}}\right)^2}\left(f^{(0)}_{,RR} + 4f^{(0)}_{,\mathcal{R}\mathcal{R}} \pm S\right), \tag{17.90}$$

where, for convenience, the quantity S is defined by

$$S \equiv \sqrt{\left(f^{(0)}_{,RR} + 4f^{(0)}_{,\mathcal{R}\mathcal{R}}\right)^2 - 12\frac{\left(f^{(0)}_{,R\mathcal{R}}\right)^2}{f^{(0)}_{,\mathcal{R}}}}\,. \tag{17.91}$$

We note that the scalar particle with mass squared m_-^2 corresponds to the scalaron appearing in Equation (17.87) in the limit of pure $f(R)$ gravity, but in general now has a shifted mass. The other scalar is a new particle that occurs due to the nontrivial dependence upon $\mathcal{R}$, and unlike in the case of hybrid metric-Palatini gravity, it propagates also in Ricci-flat spaces. The condition that neither of the scalars has a tachyonic instability, is given by

$$f^{(0)}_{,\mathcal{R}} > 0\,, \quad \text{and} \quad f^{(0)}_{,RR} + 4f^{(0)}_{,\mathcal{R}\mathcal{R}} - S > 0\,. \tag{17.92}$$

The residues at the two poles corresponding to these masses are

$$r_{\pm} = \frac{S \pm \left(f^{(0)}_{,RR} + 4f^{(0)}_{,\mathcal{R}\mathcal{R}}\right)}{4S}\,. \tag{17.93}$$

In order for neither of these scalars to be a ghost, we should have both $r_+ > 0$ and $r_- > 0$. The second condition would require that

$$f^{(0)}_{,RR} + 4f^{(0)}_{,\mathcal{R}\mathcal{R}} - S < 0\,, \tag{17.94}$$

in contradiction with Eq. (17.92). It seems then that we cannot avoid both tachyons and ghosts in this theory.

The hybrid Ricci-squared $f(\mathcal{R}, \hat{Q}_H)$ theories: Let us finally consider the $\hat{Q}_H$-invariant. For simplicity, we restrict to models here without a nonlinear dependence on the metric Ricci scalar. Basically the graviton propagator acquires its structure from the function $a(\Box)$ in Eq. (17.85), and now only the higher-derivative term $\hat{Q}_H$ modifies it. We can arrange the result for the propagator in the form

$$\Pi^{\alpha\beta\gamma\delta}_{f(\mathcal{R},\hat{Q})} = \frac{3f^{(0)}_{,\hat{Q}}\left(1+\frac{3}{4}f^{(0)}_{,\hat{Q}}k^2\right)}{2\left(1-\frac{1}{4}\left(f^{(0)}_{,\hat{Q}}\right)^2 k^4\right)\left(1+3f^{(0)}_{,\hat{Q}}k^2+2\left(f^{(0)}_{,\hat{Q}}\right)^2 k^4\right)}\mathcal{P}^{0\alpha\beta\gamma\delta} + \frac{\Pi^{\alpha\beta\gamma\delta}_{GR}}{\left(1-\frac{1}{4}\left(f^{(0)}_{,\hat{Q}}\right)^2 k^4\right)}. \tag{17.95}$$

The sixth-order theory we have at hand has a modulated graviton propagator that adds two extra poles. In addition, there appears a scalar propagator that has five poles. This is in a quite drastic contrast with respect to the metric Q-theory, which contains only one additional spin-2 particle and features fourth-order field equations. We need not analyze in detail the properties of the new degrees of freedom here, since it is obvious the theory is seriously haunted by ghosts and thus not physical. It is easy to convince oneself that this occurs very generically once one builds the action from any hybrid curvature invariant – with the exception of $\mathcal{R}$ in the specific case of separable functional dependence $R+f(\mathcal{R})$.

These considerations corroborate our claim that the hybrid metric-Palatini theories are of special theoretical interest. In the rest of the chapters, we shall discuss their phenomenology.

18

Cosmological Applications

In order to study the cosmology of the metric-Palatini theories, we employ the scalar–tensor formulation given by the action (17.21), which we rewrite in the following form

$$S = \frac{1}{2\kappa^2}\int d^4x\sqrt{-g}\left[(\Omega_A + \phi)R + \frac{3}{2\phi}(\partial\phi)^2 - 2\kappa^2 V(\phi)\right] + S_m\,, \tag{18.1}$$

where

$$\kappa^2 V(\phi) = \frac{1}{2}\left[r(\phi)\phi - f(r(\phi))\right], \quad r(\phi) \equiv f'^{-1}(\phi)\,. \tag{18.2}$$

The cosmology of the equivalent theories has also been investigated using the $f(X)$ formulation in terms of the purely metric quantities, and we refer the reader to the recent phase space analysis for the most complete global analysis of the cosmological dynamics of these theories [98]. Here we will first write down the cosmological equations in the formulation (18.1) and then have a brief look at the phase space of exact solutions for these equations. Then we will analyze the formation of cosmological large-scale structure in these models. We follow Ref. [108] closely in this chapter.

18.1 Background Expansion

The flat FLRW metric is defined as

$$ds^2 = -dt^2 + a^2(t)\left(dx^2 + dy^2 + dz^2\right)\,, \tag{18.3}$$

where the rate of time-evolution of the scale factor $a(t)$ is conveniently parameterized by the Hubble rate $H(t) = (da(t)/dt)/a(t)$. In the following we will not write explicitly the time dependence of the cosmological background quantities, and we denote the time derivative by an overdot. This eases the notation and we can write for example $H = \dot{a}/a$. In the following we will mainly be interested in accelerating dark energy-like dynamics. For a study of Einstein static spaces, see [78].

18.1.1 The Friedmann Equations

The Friedmann equations that govern the evolution of H can be written in terms of the effective energy density and pressure, respectively, given by

$$3H^2 = \kappa^2 \rho_{\rm eff} \,, \tag{18.4}$$

$$\dot{H} = -\frac{\kappa^2}{2}\left(\rho_{\rm eff} + p_{\rm eff}\right) \,, \tag{18.5}$$

where for the theory defined by the action (18.1) we obtain the following effective source terms

$$(\Omega_A + \phi)\,\kappa^2 \rho_{\rm eff} = -\frac{3}{4\phi}\dot{\phi}^2 + \kappa^2 V(\phi) - 3H\dot{\phi} + \kappa^2 \rho_m \,, \tag{18.6}$$

$$(\Omega_A + \phi)\,\kappa^2 p_{\rm eff} = -\frac{3}{4\phi}\dot{\phi}^2 - \kappa^2 V(\phi) + \ddot{\phi} + 2H\dot{\phi} + \kappa^2 p_m \,, \tag{18.7}$$

respectively. The conservation equations for the matter component and the scalar field are

$$\dot{\rho}_m + 3H(\rho_m + p_m) = 0 \,, \tag{18.8}$$

$$\ddot{\phi} + 3H\dot{\phi} - \frac{\dot{\phi}^2}{2\phi} + \frac{1}{3}\phi R - \frac{2}{3}\kappa^2 \phi V'(\phi) = 0 \,. \tag{18.9}$$

Recalling that $R = 6(2H^2 + \dot{H})$ and using Eqs. (18.4) and (18.5), we can rewrite the Klein–Gordon equation as

$$\ddot{\phi} + 3H\dot{\phi} - \frac{\dot{\phi}^2}{2\phi} + U'(\phi) + \frac{\kappa^2 \phi}{3\Omega_A}\left(\rho_m - 3p_m\right) = 0 \,, \tag{18.10}$$

where, for notational simplicity, $U'(\phi)$ is defined by

$$U'(\phi) \equiv \frac{2\kappa^2 \phi}{3\Omega_A}\left[2V(\phi) - (\Omega_A + \phi)\,V'(\phi)\right] \,. \tag{18.11}$$

As a consistency check one can verify that the Klein–Gordon equation, together with the matter conservation, allows to derive Eq. (18.5) from (18.4). By combining Eqs. (18.9) and (18.10), we find that

$$2V(\phi) - V'(\phi)\phi = \frac{1}{2}\left(\Omega_A R + \kappa^2 T_m\right) \equiv \frac{1}{2}X \,. \tag{18.12}$$

The solution for $\phi = \phi(X = 0)$ gives us the natural initial condition for the field in the early Universe. The asymptotic value of the field in the far future may then be deduced by studying the minima of the function $U(\phi)$ defined by Eq. (18.11).

18.1.2 Dynamical System Analysis

Cosmological dynamics can be addressed by taking into account a suitable dynamical system. Let us introduce the dimensionless variables

$$\Omega_m \equiv \frac{\kappa^2 \rho_m}{3H^2}\,, \quad x \equiv \phi\,, \quad y = x_{,N}\,, \quad z = \frac{\kappa^2 V}{3H^2}\,, \tag{18.13}$$

where $N = \log a$ is the e-folding time. The Friedmann equation (18.4) can then be rewritten as

$$\Omega_A + x + y - z + \frac{y^2}{4x} = \Omega_m\,. \tag{18.14}$$

Due to this constraint, the number of independent degrees of freedom is three instead of four. We choose to span our phase space by the triplet $\{x, y, z\}$. The autonomous system of equations for them reads as

$$x_{,N} = y\,, \tag{18.15}$$

$$\begin{aligned} y_{,N} = \frac{2x+y}{8\Omega_A x}\Big\{ & (3w_m - 1)\, y^2 + 4x\,[(3w_m - 1)\, y - 3\,(1 + w_m)\, z] \\ & -4x^2\,(1 - 3w_m - 2u(x)z) + 4\Omega_A\big[3x\,(w_m - 1)\, y + y^2 \\ & -x^2\,(2 - 6w_m - 4u(x)z)\big]\Big\}\,, \end{aligned} \tag{18.16}$$

$$\begin{aligned} z_{,N} = \frac{z}{4\Omega_A x}\Big\{ & (3w_m - 1)\, y^2 + 4x\,[(3w_m - 1)\, y - 3\,(1 + w_m)\, z] \\ & +4\Omega_A x\,(3 + 3w_m + u(x)y) + 4x^2\,(3w_m - 1 + 2u(x)z)\Big\}\,. \end{aligned} \tag{18.17}$$

We have defined $u(x) \equiv V'(\phi)/V(\phi)$, which becomes a constant if the potential is exponential. The relevant fixed points appear in this system. In particular, we have the matter-dominated fixed point where $x = y = z = 0$ and $w_{\text{eff}} = w_m$, and the de Sitter fixed point (in addition, there exists the fixed point $x = -\Omega_A$ corresponding to some kind of singular evolution) that is described by $w_{\text{eff}} = -1$ and

$$x_* = (2 - \Omega_A u_*)/u*\,, \quad y_* = 0\,, \quad z_* = 2/u_*\,. \tag{18.18}$$

We denote the asymptotic values corresponding to this fixed point by a subscript star. In particular, the asymptotic value of the field x_* is solved from the first equation in Eq. (18.18) once the form of the potential is given. As expected, this value corresponds to a minimum of the effective potential (18.11), $U'(x_*) = 0$. To construct a viable model, the potential should be such that we meet the two requirements:

- The matter-dominated fixed point should be a saddle point, the de Sitter fixed point an attractor. Then we naturally obtain a transition to acceleration following standard cosmological evolution.

- At the present epoch the field value should be sufficiently close to zero. Then we avoid conflict with the solar system tests of gravity (see Section 17.3).

Note that the simplest metric $f(R)$ theories that provide acceleration fail in both predicting a viable structure formation era and the solar system as we observe it. The Palatini-$f(\mathcal{R})$ models on the other hand can be ruled out as a dark energy alternative by considering their structure formation or implications to microphysics, if such a theory is regarded consistent in the first place. As shown here and explored further in this chapter, the hybrid metric-Palatini gravity models exist that are free of these problems.

To summarize: the field goes from ϕ_i to ϕ_*, where the former is given by $2V(\phi_i) = V'(\phi_i)\phi_i$ and the latter by $2V(\phi_*) = (\Omega_A + \phi_*)V'(\phi_*)$. We just need a suitable function $V(\phi)$, i.e., $f(\mathcal{R})$, in such a way that the slope will be downwards and ϕ_* near the origin.

We refer the reader to [98] for a more complete and detailed phase space analysis of the cosmological background dynamics.

18.1.3 On Cosmological Solutions

As a simple example, let us consider in more detail the specific case of the de Sitter solution in vacuum, when $\Omega_A = 1$. Then the modified Friedmann equations take the form

$$3H^2 = \frac{1}{1+\phi}\left[\kappa^2\rho + \frac{V}{2} - 3\dot{\phi}\left(H + \frac{\dot{\phi}}{4\phi}\right)\right], \tag{18.19}$$

$$2\dot{H} = \frac{1}{1+\phi}\left[-\kappa^2(\rho_m + p_m) + H\dot{\phi} + \frac{3}{2}\frac{\dot{\phi}^2}{\phi} - \ddot{\phi}\right], \tag{18.20}$$

and the scalar field equation (18.10) becomes

$$\ddot{\phi} + 3H\dot{\phi} - \frac{\dot{\phi}^2}{2\phi} + \frac{2\kappa^2\phi}{3}[2V - (1+\phi)V_\phi] = -\frac{\phi\kappa^2}{3}(\rho_m - 3p_m). \tag{18.21}$$

To further specify the setup, consider a model that arises by demanding that matter and curvature satisfy the same relation as in GR. Taking

$$V(\phi) = V_0 + V_1\phi^2, \tag{18.22}$$

the trace equation automatically implies $R = -\kappa^2 T + 2V_0$ [104, 224]. As $T \to 0$ with the cosmic expansion, this model naturally evolves into a de Sitter phase, which requires $V_0 \sim \Lambda$ for consistency with observations. If V_1 is positive, the de Sitter regime represents the minimum of the potential. The effective mass for local experiments, $m_\varphi^2 = 2(V_0 - 2V_1\phi)/3$, is then positive and small as long as $\phi < V_0/V_1$. For sufficiently large V_1 one can make the field amplitude small enough to be in agreement with solar system tests. It is interesting that the exact de Sitter solution is compatible with dynamics of the scalar field in this model.

The accelerating dynamics that drives the hybrid metric-Palatini gravitational theory toward its general relativistic limits today have indeed been realized in several specific models [104, 224, 316, 317]. Our preliminary phase space analysis confirmed the existence of de Sitter attractor solutions, and the recent study of cosmology in terms of dynamical system analysis extends this result to more general models [98]. Analytic solutions were presented also in [104] as well as in [84], there using a Nöether symmetry technique. A designer approach was devised in [316] to reconstruct precisely the standard ΛCDM expansion history by a nontrivial hybrid metric-Palatini model, and finally, two families of models were constrained by confronting their predictions with a combination of cosmic microwave background, supernovae Ia, and baryonic acoustic oscillations background data [317].

18.2 Cosmological Perturbations

To understand the implications of these models to the cosmological structure formation, we will derive the perturbation equations and analyze them in some specific cases of interest. This paves the way for a detailed comparison of the predictions with the cosmological data on large-scale structure and the cosmic microwave background. For generality, we will keep the parameter Ω_A in the formulas in this section.

18.2.1 Field Equations and Conservation Laws

We work in the Newtonian gauge [332], which can be parameterized by the two gravitational potentials Φ and Ψ,

$$\mathrm{d}s^2 = -\left(1+2\Psi\right)\mathrm{d}t^2 + a^2(t)\left(1+2\Phi\right)\mathrm{d}\vec{x}^2 \,. \tag{18.23}$$

As a matter source we consider a perfect fluid, with the background equation of state w and with density perturbation $\delta = \delta\rho_m/\rho_m$, pressure perturbation $\delta p_m = c_s^2 \delta\rho_m$ and velocity perturbation v.

The 0-0 part of the field equations is

$$\frac{k^2}{a^2}\Phi + 3\left(H - \frac{\dot{\phi}}{2\left(\Omega_A+\phi\right)}\right)\dot{\Phi} - 3\left(H^2 + \frac{H\dot{\phi}}{\Omega_A+\phi} - \frac{\dot{\phi}^2}{4\phi\left(\Omega_A+\phi\right)}\right)\Psi$$
$$= \frac{1}{2\Omega_A+\phi}\left[\kappa^2\delta\rho_m + \left(\frac{3}{4\phi^2}\dot{\phi}^2 + V'(\phi) - 3H^2 - \frac{k^2}{a^2}\right)\varphi - 3\left(H + \frac{\dot{\phi}}{2\phi}\right)\dot{\varphi}\right], \tag{18.24}$$

where we have denoted $\varphi = \delta\phi$. The Raychaudhuri equation for the perturbations reads

$$\left[6\left(H^2+2\dot{H}\right)-2\frac{k^2}{a^2}+\frac{6}{\Omega_A+\phi}\left(\ddot{\phi}-\frac{\dot{\phi}^2}{\phi^2}+H\dot{\phi}\right)\right]\Psi$$

$$+3\left(2H-\frac{\dot{\phi}}{\Omega_A+\phi}\right)\left(\dot{\Phi}-\dot{\Psi}\right)-6\ddot{\Phi}=\frac{1}{\Omega_A+\phi}\left[\kappa^2\left(\delta\rho_m+3\delta p_m\right)\right.$$

$$\left.+\left(6H^2+6\dot{H}+3\frac{\ddot{\phi}}{\phi^2}-2V'(\phi)+\frac{k^2}{a^2}\right)\varphi+3\left(H-\frac{2\dot{\phi}}{\phi}\right)\dot{\varphi}+3\ddot{\varphi}\right]. \quad (18.25)$$

The 0-i equation is

$$-\left(H+\frac{\dot{\phi}}{2\left(\Omega_A+\phi\right)}\right)\Phi+\dot{\Phi}$$

$$=\frac{1}{2\left(\Omega_A+\phi\right)}\left[\kappa^2\left(\rho_m+p_m\right)av_m+\left(H+\frac{3\dot{\phi}}{2\phi}\right)\varphi+\dot{\varphi}\right]. \quad (18.26)$$

Note that the set of perturbed field equations is completed by the off-diagonal spatial piece:

$$\Psi+\Phi=-\frac{\varphi}{\Omega_A+\phi}. \quad (18.27)$$

Assuming a perfect fluid, the continuity and Euler equations for the matter component are

$$\dot{\delta}+3H\left(c_s^2-w\right)\delta=-\left(1+w\right)\left(3\dot{\Phi}-\frac{k^2}{a}v\right), \quad (18.28)$$

$$\dot{v}+\left(1-3c_a^2\right)Hv=\frac{1}{a}\left(\Psi+\frac{c_s^2}{1+w}\delta\right), \quad (18.29)$$

respectively. The linear part of the Klein–Gordon equation is then compatible with the above system. For completeness, it is given by

$$\ddot{\varphi}+\left(3H+\frac{1}{\phi}\right)\dot{\varphi}+\left(\frac{k^2}{a^2}+\frac{\dot{\phi}^2}{2\phi^2}-\frac{2}{3}V''(\phi)\right)\varphi=\left(2\ddot{\phi}+6H\dot{\phi}-\frac{3}{2\phi}\dot{\phi}^2\right)\Psi$$

$$+\dot{\phi}\left(\dot{\Psi}-3\dot{\Phi}\right)-\frac{\phi}{3}\delta R. \quad (18.30)$$

This completes the presentation of the field equations and the conservation laws. For the equations in the synchronous gauge, see [316].

18.2.2 Matter-dominated Cosmology

Let us consider the formation of structure in the matter-dominated universe, where $w=c_s^2=0$. In this subsection, we shall consider scales deep inside the Hubble radius. This so called quasi-static approximation is well known in the literature and indeed we will arrive at a similar result known to apply for scalar–tensor theories since early studies [79] that have been more recently generalized to a wide variety of coupled dark sector models [292]. The approximation neglects a fluctuating degree of freedom that is expected to be insignificant at small enough scales. For analyses of the applicability of the approximation with different assumptions on the cosmological models, see [8, 158, 292, 321, 450].

In the quasi-static subhorizon limit the spatial gradients are more important than the time derivatives, and, consequently, the matter density perturbations are much stronger than the gravitational potentials. Combining the continuity and the Euler equation in this approximation, one obtains

$$\ddot{\delta} = -2H\dot{\delta} - \frac{k^2}{a^2}\Psi\,. \tag{18.31}$$

We need then to solve the gravitational potential. Let us define $\Pi = a^2\rho_m\delta/k^2$ and write the field equations and the Klein–Gordon equation at this limit in a very simple way as

$$(\Omega_A + \phi)\,\Phi = \Pi - \varphi\,, \tag{18.32}$$
$$(\Omega_A + \phi)\,(\Psi + \Phi) = -\varphi\,, \tag{18.33}$$
$$-2\,(\Omega_A + \phi)\,\Psi = \Pi + \varphi\,, \tag{18.34}$$
$$3\varphi = -2\phi\,(\Psi + 2\Phi)\,. \tag{18.35}$$

We immediately see that one of the equations is (as expected) redundant, and that the Ψ is (as usual) proportional to Π, where now the proportionality is given as a function of the field ϕ. Our result is

$$\ddot{\delta} + 2H\dot{\delta} = 4\pi G_{\text{eff}}\rho_m\delta\,, \tag{18.36}$$

with

$$G_{\text{eff}} \equiv \frac{\Omega_A - \frac{1}{3}\phi}{\Omega_A\,(\Omega_A + \phi)}G\,. \tag{18.37}$$

This shows that instabilities can be avoided in the evolution of the matter inhomogeneities, in contrast to the Palatini-$f(\mathcal{R})$ models and some matter-coupled scalar field models (recall that the hybrid theory can be mapped into such in the Einstein frame). Equation (18.36) provides a very simple approximation to track the growth of structure accurately within the linear regime during matter-dominated cosmology. Confrontations of specific model predictions with the present large-scale structure data and forecasts for the constraints from future experiments, in particular the Euclid mission [17], is interesting work to be done.

18.2.3 Vacuum Fluctuations

The propagation of our scalar degree of freedom in vacuum is also a crucial consistency check on the theory. Let us set $\rho_m = 0$, and consider the curvature perturbation in the uniform-field gauge ζ. In terms of the Newtonian gauge perturbations, this is

$$\zeta = \Phi - \frac{H}{\dot{\phi}}\varphi . \tag{18.38}$$

After somewhat more tedious algebra than in the previous case, we obtain the exact (linear) evolution equation

$$\ddot{\zeta} + \left[3H - 2\frac{\ddot{\phi} + 2\dot{H}\left(\Omega_A + \phi\right) - \frac{\dot{\phi}^2}{\Omega_A + \phi}}{\dot{\phi} + 2H\left(\Omega_A + \phi\right)} + \frac{\phi\left(\Omega_A + \phi\right)}{\dot{\phi}^2} \right.$$
$$\left. \times \left(\frac{2\ddot{\phi}\dot{\phi}}{\phi\left(\Omega_A + \phi\right)} + \frac{\dot{\phi}^3\left(\Omega_A + \phi\right)^2 \phi}{1 - \phi^3\left(\Omega_A + \phi\right)^3} \right) \right] \dot{\zeta} = -\frac{k^2}{a^2}\zeta . \tag{18.39}$$

The friction term depends on the perturbation variable we consider, but the perturbations at small scales still propagate with the speed of light, as in canonical scalar field theory. This excludes also gradient and tachyon instabilities in the graviscalar sector. Now, Eq. (18.39) can be used to study generation of fluctuations in $f(X)$-inflation. Construction of specific models and their observational tests are left for forthcoming studies. The Einstein-frame formulation might present a convenient starting point for such studies, as it, given the function $f(\mathcal{R})$, presents directly the relevant inflationary potential in terms of the canonic field.

18.2.4 Summary

As mentioned earlier, much work along these lines still needs to done. However, one may mention previous works, for instance, on the evolution of linear perturbations in viable hybrid metric-Palatini theories. Indeed, the most promising observational tools for distinguishing $f(R)$ theories from the ΛCDM model are those intrinsically related to the growth of structure, such as weak lensing. In [316], the full set of linearized evolution equations were derived for the perturbed potentials in the Newtonian and synchronous gauges. The evolution of the lensing potential was analyzed for a model with a background evolution indistinguishable from ΛCDM, by introducing a designer approach that allows us to retrieve a family of $f(\mathcal{R})$ functions for which the effective equation of state is exactly $w_{eff} = -1$. For this particular model, it was found that the main deviations from standard GR and the cosmological constant model arise in the distant past, with an oscillatory signature in the ratio between the Newtonian potentials.

Furthermore, in [317] the background evolution of specific models were explored, and it was shown explicitly that one recovers standard GR with an effective cosmological constant at late times, as we also derived in this chapter. This occurs in [317] because the Palatini curvature scalar evolves toward and asymptotically settles at the minimum of its effective potential during cosmological evolution. A combination of cosmic microwave background, supernovae, and baryonic accoustic oscillations background data were then used to constrain the models' free parameters, and it was shown that it is possible to constrain the maximum deviation from the gravitational constant G one can have at early times to be around 1%.

The cosmological evolution of a specific model, given by $f(\mathcal{R}) \propto \mathcal{R}^2$, was also extensively analyzed in [305], where the constraints on the free parameters were obtained by using different sources of cosmological data. More specifically, the viability of the model was analyzed by combining the conditions imposed by the supernovae Ia and baryonic acoustic oscillations data, where the free parameters of the model were fitted through an MCMC analysis, and the results were compared with the local constraints. It was found that the value of the scalar field today determines whether the cosmological constraints and the local ones are both satisfied simultaneously, since the mass of the scalar field remains very small within the 1σ region of the free parameters. Hence, the value of the scalar field today becomes much better constrained, reducing its viable range noticeably around zero, where GR with a cosmological constant is recovered. In this sense, the model satisfactorily passes the local and cosmological restrictions. Future analysis of the perturbations and the growth of large-scale structure may provide additional information and constraints on this type of model.

Recently, the efficiency of screening mechanisms in the hybrid metric-Palatini gravity was also investigated [488]. The value of the field was computed around spherical bodies embedded in a background of constant density, and a thin shell condition was found for the field depending on the background scalar field value. In order to quantify how the thin shell effect was relevant, the behavior in the neighborhood of different astrophysical objects was analyzed, and it was found that the condition is very well satisfied. Furthermore, bounds were established on the model using data from solar system experiments such as the spectral deviation measured by the Cassini mission and the stability of the Earth–Moon system, which gives the best constraint to date on $f(R)$ theories. These bounds contribute to fix the range of viable hybrid gravity models.

19

Astrophysical Applications

In this chapter we develop methods to study dark matter phenomenology in the hybrid models; for studies of dark matter generated by metric $f(R)$ modification, see, e.g., [22, 127], and Section 6.5 for more details. The hybrid metric-Palatini gravitational theory allows to address several issues related to dark matter dynamics ranging from galaxies to galaxy clusters. These self-gravitating structures can be probed by studying the motion of test particles (stars for galaxies and galaxies themselves for galaxy clusters) moving in a gravitational potential. The behavior of rotational and dispersion velocities of such test particles can be explained within the framework of the gravitational potential derived from the theory. For example, the tangential velocity can be explicitly obtained as a function of the scalar field of the equivalent scalar–tensor description. The model predictions can be compared with samples of rotation curves of spiral galaxies and galaxy clusters, respectively. The possibility of constraining the form of the scalar field and the parameters of the model by using the stellar velocity dispersions is also analyzed. Furthermore, the Doppler velocity shifts are also obtained in terms of the scalar field. Finally, suitable generalizations of the virial theorem and the relativistic Boltzmann equation allow to construct a self-consistent theory for galaxy clusters. In conclusion, all the physical and geometrical quantities and the numerical parameters in the hybrid metric-Palatini model can be expressed in terms of observable/measurable quantities, such as the tangential velocity, the baryonic mass of the galaxy, the Doppler frequency shifts, the dispersion velocity, the geometrical quantities characterizing the clusters of galaxies, respectively. These results open the possibility of testing the hybrid metric-Palatini gravitational theories at the galactic or extragalactic scale by using direct astronomical and astrophysical observations.

19.1 Galactic Phenomenology: Stable Circular Orbits of Test Particles

The most direct method for studying the gravitational field inside a spiral galaxy is provided by the galactic rotation curves. They are obtained by measuring the frequency shifts z of the 21-cm radiation emission from the neutral hydrogen gas clouds. The 21-cm radiation also originates from stars. The 21-cm background from the epoch of reionization is also a promising cosmological probe: line-of-sight velocity fluctuations distort redshift, so brightness fluctuations in Fourier space depend upon angle, which linear theory shows can separate cosmological from astrophysical information (for a review see [421]). Instead of using z the resulting redshift is presented by astronomers in terms of a velocity field v_{tg} [414, 441].

In the following discussion, we will assume that the gas clouds behave like test particles, moving in the static and spherically symmetric geometry around the galaxy. Without a significant loss of generality, we assume that the gas clouds move in the galactic plane $\theta = \pi/2$, so that their four-velocity is given by $U^\mu = \left(\dot{t}, \dot{r}, 0, \dot{\phi}\right)$, where the overdot stands for derivation with respect to the affine parameter s. In this section we find it illustrative to restore the units of c.

The static spherically symmetric metric outside the galactic baryonic mass distribution is given by the following line element

$$ds^2 = -e^{\nu(r)}c^2 dt^2 + e^{\lambda(r)} dr^2 + r^2 \left(d\theta^2 + \sin^2\theta d\phi^2\right), \tag{19.1}$$

where the metric coefficients $\nu(r)$ and $\lambda(r)$ are functions of the radial coordinate r only. The motion of a test particle in the gravitational field with the metric given by Eq. (19.1), is described by the Lagrangian [217, 387]

$$L = \left[e^{\nu(r)}\left(\frac{cdt}{ds}\right)^2 - e^{\lambda(r)}\left(\frac{dr}{ds}\right)^2 - r^2\left(\frac{d\Omega}{ds}\right)^2\right], \tag{19.2}$$

where $d\Omega^2 = d\theta^2 + \sin^2\theta d\phi^2$, which simplifies to $d\Omega^2 = d\phi^2$ along the galactic plane $\theta = \pi/2$. From the Lagrange equations it follows that we have two constants of motion, namely, the energy E per unit mass, and the angular momentum l per unit mass, given by $E = e^{\nu(r)}c^3\dot{t}$ and $l = cr^2\dot{\phi}$, respectively. The normalization condition for the four-velocity $U^\mu U_\mu = -1$ gives $1 = e^{\nu(r)}c^2\dot{t}^2 - e^{\lambda(r)}\dot{r}^2 - r^2\dot{\phi}^2$, from which, with the use of the constants of motion, we obtain the energy of the particle as

$$\frac{E^2}{c^2} = e^{\nu+\lambda}\dot{r}^2 + e^{\nu}\left(\frac{l^2}{c^2r^2} + 1\right). \tag{19.3}$$

From Eq. (19.3) it follows that the radial motion of the test particles is analogous to that of particles in Newtonian mechanics, having a velocity $\dot{r}$, a position-dependent effective mass $m_{\text{eff}} = 2e^{\nu+\lambda}$, and an energy E. In addition to this, the test particles move in an effective potential provided by the following relationship

$$V_{\text{eff}}(r) = e^{\nu(r)}\left(\frac{l^2}{c^2r^2}+1\right). \tag{19.4}$$

The conditions for circular orbits, namely, $\partial V_{\text{eff}}/\partial r = 0$ and $\dot{r} = 0$ lead to

$$l^2 = \frac{c^2}{2}\frac{r^3\nu'}{1-r\nu'/2}, \qquad \frac{E^2}{c^4} = \frac{e^{\nu}}{1-r\nu'/2}, \tag{19.5}$$

respectively. Note that the spatial three-dimensional velocity is given by

$$v^2(r) = e^{-\nu}\left[e^{\lambda}\left(\frac{\mathrm{d}r}{\mathrm{d}t}\right)^2 + r^2\left(\frac{\mathrm{d}\Omega}{\mathrm{d}t}\right)^2\right]. \tag{19.6}$$

For a stable circular orbit $dr/dt = 0$, the tangential velocity of the test particle can be expressed as

$$v_{tg}^2(r) = e^{-\nu}r^2\left(\frac{\mathrm{d}\Omega}{\mathrm{d}t}\right)^2 = e^{-\nu}r^2\left(\frac{\mathrm{d}\Omega}{\mathrm{d}s}\right)^2\left(\frac{\mathrm{d}s}{\mathrm{d}t}^2\right). \tag{19.7}$$

In terms of the conserved quantities, and along the galactic plane $\theta = \pi/2$, the angular velocity is given by

$$\frac{v_{tg}^2(r)}{c^2} = c^2\frac{e^{\nu}}{r^2}\frac{l^2}{E^2}, \tag{19.8}$$

and taking into account Eq. (19.5), we finally obtain the following relationship [217, 387]

$$\frac{v_{tg}^2(r)}{c^2} = \frac{r\nu'}{2}. \tag{19.9}$$

Therefore, once the tangential velocity of test particles is known, the metric function $\nu(r)$ outside the galaxy can be obtained as

$$\nu(r) = 2\int\frac{v_{tg}^2(r)}{c^2}\frac{\mathrm{d}r}{r}. \tag{19.10}$$

The tangential velocity v_{tg}/c of gas clouds moving like test particles around the center of a galaxy is not directly measurable, but can be inferred from the redshift z_{∞} observed at spatial infinity [217, 387], for which

$$1+z_{\infty} = \exp\left[\frac{1}{2}(\nu_{\infty}-\nu)\right]\frac{(1\pm v_{tg}/c)}{\sqrt{1-v_{tg}^2/c^2}}. \tag{19.11}$$

Due to the nonrelativistic velocities of the gas clouds, with $v_{tg}/c \leq (4/3)\times 10^{-3}$, we observe that $v_{tg}/c \approx z_{\infty}$, as the first part of a geometric series. The observations show that at distances large enough from the galactic center the tangential velocities assume a constant value, i.e., $v_{tg}/c \approx$ constant [414, 441]. In the latter regions of the constant tangential velocities, Eq. (19.10) can be readily integrated to provide the following metric tensor component

$$e^{\nu}=\left(\frac{r}{R_{\nu}}\right)^{2v_{tg}^{2}/c^{2}}\approx 1+2\frac{v_{tg}^{2}}{c^{2}}\ln\left(\frac{r}{R_{\nu}}\right), \tag{19.12}$$

where R_{ν} is an arbitrary constant of integration. If we match the metric given by Eq. (19.12) with the Schwarzschild metric on the surface of the galactic baryonic matter distribution, having a radius R_B, $e^{\nu}\big|_{r=R_B}=1-2GM_B/c^2R_B$, we obtain the following relationship

$$R_{\nu}=\frac{R_B}{\left(1-2GM_B/c^2R_B\right)^{c^2/2v_{tg}^2}}. \tag{19.13}$$

An important physical requirement for the circular orbits of the test particle around galaxies is that they must be stable. Let r_0 be the radius of a circular orbit and consider a perturbation of it of the form $r=r_0+\delta$, where $\delta\ll r_0$ [301]. Taking expansions of $V_{\rm eff}(r)$ and $e^{\nu+\lambda}$ about $r=r_0$, it follows from Eq. (19.3) that

$$\ddot{\delta}+\frac{1}{2}e^{\nu(r_0)+\lambda(r_0)}V_{\rm eff}''(r_0)\,\delta=0. \tag{19.14}$$

The condition for stability of the simple circular orbits requires $V_{\rm eff}''(r_0)>0$ [301]. Hence, with the use of the condition $V_{\rm eff}'(r_0)=0$, we obtain the condition of the stability of the orbits as $\left[3\nu'+r\nu''>r\nu'^2/2\right]\big|_{r=r_0}$. By taking into account Eq. (19.9), it immediately follows that stable circular orbits always exist for massive test particles.

19.1.1 Galactic Geometry and Tangential Velocity Curves

The rotation curves only determine one, namely $\nu(r)$, of the two unknown metric functions, $\nu(r)$ and $\lambda(r)$, which are required to describe the gravitational field of the galaxy [105]. Hence, in order to determine $\lambda(r)$ we proceed to solve the gravitational field equations for the hybrid metric-Palatini gravitational theory outside the baryonic matter distribution. This allows us to take all the components of the ordinary matter energy-momentum tensor as being zero. Furthermore, our task becomes easier when we restrict to a perturbative weak-field treatment.

The weak field limit of the gravitational theories at the solar system level is usually obtained by using isotropic coordinates. However, it is useful to apply Schwarzschild coordinates in studying exact solutions and in the context of galactic dynamics, and this is what we do here. We assume that the gravitational field inside the halo is weak, so that $\nu(r)\sim\lambda(r)\sim(v_{tg}/c)^2$, which allows us to linearize the gravitational field equations retaining only terms linear in $(v_{tg}/c)^2$, and again consider the scalar field as $\phi=\phi_0+\varphi$, where $\varphi\ll 1$ is a small perturbation around the background value $\phi_0 > 0$. The Klein–Gordon equation was already solved in Section 17.3 at this limit and the result was that the interaction range is given by $r_\varphi = 1/m_\varphi$, where m_φ is given in Eq. (17.49). In a spherically symmetric configuration, the general solution then has the form

$$\varphi(r) = \Psi_0 \frac{e^{-r/r_\varphi}}{r}\,, \tag{19.15}$$

where Ψ_0 is an integration constant. Comparing this expression with the results obtained in [224] for the weak-field limit (taking into account the transformation from isotropic to Schwarzschild coordinates), we find that

$$\Psi_0 = -\frac{2GM_B\,\phi_0}{3c^2} e^{R_B/r_\varphi} < 0\,, \tag{19.16}$$

where M_B and R_B are the mass and the radius of the galactic baryonic distribution, respectively.

Within this linear approximation the energy-momentum tensor of the scalar field is given by

$$T^{(\phi)}_{\mu\nu} = \frac{1}{\kappa^2}\left[\nabla_\mu \nabla_\nu \varphi + (\alpha\varphi + \beta)\, g_{\mu\nu}\right]\,, \tag{19.17}$$

where α and β are defined by

$$\alpha = \frac{1}{r_\varphi^2} - \frac{1}{2}V'(\phi_0)\,, \qquad \beta = -\frac{1}{2}V(\phi_0)\,. \tag{19.18}$$

Therefore, the linearized gravitational field equations take the form

$$\frac{1}{r^2}\frac{d}{dr}(r\lambda) = \alpha\varphi + \beta\,, \tag{19.19}$$

$$-\frac{\nu'}{r} + \frac{\lambda}{r^2} = \varphi'' + \alpha\varphi + \beta\,, \tag{19.20}$$

$$-\frac{1}{2}\left(\nu'' + \frac{\nu' - \lambda'}{r}\right) = \alpha\varphi + \beta\,. \tag{19.21}$$

Equation (19.19) can be immediately integrated to provide

$$\begin{aligned}\lambda(r) &= \frac{\lambda_0}{r} + \frac{1}{r}\int^r (\alpha\varphi + \beta)\,\tilde{r}^2 d\tilde{r} \\ &= \frac{\lambda_0}{r} + \frac{\beta}{3}r^2 - \frac{\alpha r_\varphi^2 \Psi_0 e^{-r/r_\varphi}}{r}\left(1 + \frac{r}{r_\varphi}\right)\,,\end{aligned} \tag{19.22}$$

where λ_0 is an integration constant. Comparing again with the results obtained in [224] for the weak-field limit, we find that $\lambda_0 = 2GM/c^2$. The tangential velocity of the test particles in stable circular orbits moving in the galactic halo can be derived immediately from Eq. (19.20), and is given by

$$\frac{v_{tg}^2}{c^2} = \frac{r\nu'}{2} = \frac{\lambda}{2} - r^2\frac{\varphi''}{2} - \frac{\alpha}{2}r^2\varphi - \frac{\beta}{2}r^2, \tag{19.23}$$

which in terms of the solutions found above becomes

$$\frac{v_{tg}^2}{c^2} = \frac{V_0}{6}r^2 + \frac{GM_B}{c^2 r} - \frac{\Psi_0 e^{-r/r_\varphi}}{2r}\left[\left(1+\frac{r}{r_\varphi}\right)(2+\alpha r_\varphi^2) + \frac{r^2}{r_\varphi^2}(1+\alpha r_\varphi^2)\right], \tag{19.24}$$

where $V_0 = -\beta = V(\phi_0)/2$. The term proportional to r^2 corresponds to the cosmological background, namely the de Sitter geometry, and we assume that it has a negligible contribution on the tangential velocity of the test particles at the galactic level.

On the surface of the baryonic matter distribution the tangential velocity must satisfy the boundary condition

$$\frac{v_{tg}^2(R_B)}{c^2} \approx \frac{GM_B}{c^2 R_B}, \tag{19.25}$$

which, with the use of Eq. (19.16), gives the following constraint on the parameters of the model

$$\left(1+\frac{R_B}{r_\varphi}\right)(2+\alpha r_\varphi^2) + \frac{R_B^2}{r_\varphi^2}(1+\alpha r_\varphi^2) \approx 0. \tag{19.26}$$

In order to satisfy the above condition, it would require that $-2 < \alpha r_\varphi^2 < -1$, or equivalently,

$$V'(\phi_0) > 0, \qquad 2 < \frac{1}{2}V'(\phi_0)\, r_\varphi^2 < 3. \tag{19.27}$$

In the regions near the galactic baryonic matter distribution, where $R_B \leq r \ll r_\varphi$, we have $e^{-r/r_\varphi} \approx 1$, to a very good approximation. Hence in this region the tangential velocity can be approximated as

$$\frac{v_{tg}^2}{c^2} \approx \frac{2GM_B - c^2\Psi_0\left(\alpha r_\varphi^2+2\right)}{2c^2 r} - \Psi_0\frac{\alpha r_\varphi^2+2}{2r_\varphi} - \frac{\Psi_0}{2r_\varphi^2}\left(1+\alpha r_\varphi^2\right) r, \tag{19.28}$$

for $R_B \leq r \ll r_\varphi$.

If the parameters of the model satisfy the condition

$$2GM_B - c^2\Psi_0\left(\alpha r_\varphi^2+2\right) \approx 0, \tag{19.29}$$

the term proportional to $1/r$ becomes negligible, while for small values of Ψ_0, and $\left|\alpha r_{\varphi 2}\right| \approx 1$, the term proportional to r can also be neglected. Therefore for the tangential velocity of test particles rotating in the galactic halo we obtain

$$\frac{v_{tg}^2}{c^2} \approx -\Psi_0\frac{\alpha r_\varphi^2+2}{2r_\varphi} \approx -\frac{\Psi_0 \alpha r_\varphi}{2}, \qquad R_B \leq r \ll r_\varphi. \tag{19.30}$$

Since according to our assumptions, $r_\varphi \gg 1$, the coefficient α can be approximated as $\alpha \approx -V'(\phi_0)/2$, which provides for the rotation curve, in the constant velocity region, the following expression

$$\frac{v_{tg}^2}{c^2} \approx \frac{\Psi_0 V'(\phi_0)\, r_\varphi}{4}, \qquad R_B \leq r \ll r_\varphi. \tag{19.31}$$

Since $\Psi_0 < 0$, the scalar field potential must satisfy the condition $V'(\phi_0) < 0$. In the first order of approximation, with $\exp(-r/r_\varphi) \approx 1 - r/r_\varphi$, for the tangential velocity we obtain the expression

$$\frac{v_{tg}^2}{c^2} \approx \frac{2GM_B - c^2\Psi_0\left(\alpha r_\varphi^2 + 2\right)}{2c^2 r} + \frac{\Psi_0}{2r_\varphi^2} r + \frac{\Psi_0\left(\alpha r_\varphi^2 + 1\right)}{2r_\varphi^2} r^2. \tag{19.32}$$

Alternatively, in general we can write the tangential velocity as follows,

$$\frac{v_{tg}^2}{c^2} = \frac{V_0}{6} r^2 + \frac{GM_B}{c^2 r} \left\{1 + \frac{2\phi_0}{3} e^{\frac{GM_B/c^2 - r}{r_\varphi}} \left[\left(1 + \frac{r}{r_\varphi}\right)(2 + \alpha r_\varphi^2) + \frac{r^2}{r_\varphi^2}(1 + \alpha r_\varphi^2)\right]\right\}.$$

As compared to our previous results, in this representation we have $e^{(GM_B/c^2 - r)/r_\varphi}$ instead of $e^{(R_B - r)/r_\varphi}$. Since we are working in a regime in which $R_B \ll r_\varphi$, the choice of the constants R_B or M_B does not seem very relevant, since it just amounts to a rescaling of ϕ_0. From now on we will also assume that $e^{GM_B/c^2 r_\varphi} \approx 1$.

From the above equation we want to find the constraints on the model parameters that arise from the expected behavior at different scales. For that purpose, it is convenient to write the equation, equivalently, as follows:

$$\frac{v_{tg}^2}{c^2} = \frac{GM_B}{c^2 r}\left[1 + \frac{2\phi_0}{3}(2 + \alpha r_\varphi^2) e^{-\frac{r}{r_\varphi}}\right] + \frac{GM_B}{c^2 r_\varphi}\left(2 + \alpha r_\varphi^2\right) e^{-\frac{r}{r_\varphi}} + \frac{GM_B}{c^2 r_\varphi}(1 + \alpha r_\varphi^2)\left(\frac{r}{r_\varphi}\right) e^{-\frac{r}{r_\varphi}}. \tag{19.33}$$

At intermediate scales, the asymptotic tangential velocity tends to a constant. If we expand the exponential as $e^{-r/r_\varphi} \approx 1 - r/r_\varphi$, then we obtain the following three constraints on the free parameters of the model,

$$(a)\ 1 + \frac{2\phi_0}{3}\left(2 + \alpha r_\varphi^2\right) \approx 0, \tag{19.34}$$

$$(b)\ \left(2 + \alpha r_\varphi^2\right)\left(1 - \frac{2\phi_0}{3}\right) \approx C = \text{constant}, \tag{19.35}$$

$$(c)\ \frac{GM_B}{c^2 r_\varphi}\left(\frac{r}{r_\varphi}\right) \ll |C|. \tag{19.36}$$

With increasing r, and by assuming that the condition $r \ll r_\varphi$ still holds, the rotation curves will decay, at very large distances from the galactic center, to the zero value.

19.1.2 On Astrophysical Tests at the Galactic Level

In [105], some observational astrophysical tests of hybrid metric-Palatini gravity at the galactic level were discussed. More specifically, a comparison of the theoretical predictions of the model with a sample of rotation curves of low surface brightness galaxies was analyzed. Indeed, the predictions of the theoretical model with the observational results show that the contribution of the scalar field energy density to the tangential velocity of the test particles can explain the existence of a constant rotational velocity region around the baryonic matter, without requiring the presence of the dark matter. The possibility of observationally determining the functional form of the scalar field φ by using the velocity dispersion of stars in galaxies, and the red- and blueshifts of gas clouds moving in the galactic halo, could also be worth considering. It can be shown that one can constrain the explicit functional form of the scalar field, and the free parameters of the model, in order to adequately fit the observational data. We refer the reader to [105] for more details.

19.2 Galactic Clusters: The Generalized Virial Theorem

A large number of astronomical and astrophysical observations confirm that galaxies form complex hierarchical structures, in which galaxies concentrate in large objects, called clusters of galaxies, bounded by the gravitational interaction. The total mass of the galaxy cluster ranges from $10^{13} M_\odot$ for the so-called groups (formed by a few hundred of galaxies) up to a few $10^{15} M_\odot$ for very large clusters, containing thousands of galaxies. From a morphological point of view galactic clusters are usually formed by a main component, which is regular and centrally peaked [24, 432]. For fundamental physics research, the main importance of the galactic clusters consists of the fact that they are considered to be "dark matter"-dominated astrophysical objects. Moreover, their formation and evolution is almost entirely controlled by the gravitational force, a property that allows the testing of different dark matter models. On the other hand, the mass distribution inside the clusters is fully determined by the initial conditions of the mass distribution that originate in the early universe [33, 312, 456], thus allowing the use of galaxy cluster properties to also test cosmological models. In this section, we discuss the possibility of observationally testing hybrid metric-Palatini gravity by using observational evidence from galaxy clusters [106].

19.2.1 Galaxy Cluster as a System of Identical and Collisionless Point Particles

As a first step, which is necessary to obtain our main result, and consists in the generalization of the relativistic virial theorem for galaxy clusters in hybrid metric-Palatini gravity, we write down the field equations and the Boltzmann equation for a static and spherically symmetric distribution of matter. We adopt a simplified physical model, in which the galactic cluster consists of a self-gravitating system of identical, collisionless point particles (the galaxies) in random dynamical motion. The metric will thus be described by Eq. (19.1), and the fluid by a distribution function f_B that obeys the general relativistic Boltzmann equation [222].

The energy-momentum tensor of the matter in the cluster is thus determined by the distribution function f_B, and its components are given by the equation [64, 335]

$$T_{\mu\nu} = \int f_B \, m \, u_\mu u_\nu \, du, \tag{19.37}$$

where m is the mass of the particle (galaxy), and $u_\mu = (u_t, u_r, u_\theta, u_\varphi)$ is the corresponding galactic four-velocity, with u_t denoting the temporal component. Finally, by $du = du_r du_\theta du_\varphi / u_t$, we denote the invariant volume element in the velocity space. Alternatively, the energy-momentum tensor $T_{\mu\nu}$ describing the matter distribution in a cluster of galaxies can be represented in terms of an effective energy density ρ_{eff}, and of two effective anisotropic thermodynamic pressures, the radial $p_{\text{eff}}^{(r)}$ and the tangential $p_{\text{eff}}^{(\perp)}$ pressure, respectively. These thermodynamic parameters of the cluster are obtained by averaging over the matter and velocity distributions, and are given by

$$\rho_{\text{eff}} = \rho \langle u_t^2 \rangle, \qquad p_{\text{eff}}^{(r)} = \rho \langle u_r^2 \rangle, \qquad p_{\text{eff}}^{(\perp)} = \rho \langle u_\theta^2 \rangle = \rho \langle u_\varphi^2 \rangle, \tag{19.38}$$

where ρ is the mass density of the ordinary baryonic matter, and $\langle u_i^2 \rangle$ $(i = t, r, \theta, \varphi)$ denotes the average value of u_i^2 $(i = t, r, \theta, \varphi)$, representing the square of the components of the four-velocities of the galaxies in the cluster [267]. The full field equations for the metric (19.1) with the source, given by Eq. (19.38), were listed in [106], and take the following form

$$\begin{aligned} -e^{-\lambda}\left(\frac{1}{r^2} - \frac{\lambda'}{r}\right) + \frac{1}{r^2} = {} & 8\pi \frac{G}{1+\phi} \rho \langle u_t^2 \rangle - \frac{1}{2(1+\phi)} V(\phi) \\ & + \frac{1}{1+\phi} \left(\nabla_t \nabla^t - \Box\right) \phi - \frac{3}{2} \frac{1}{\phi(1+\phi)} \nabla_t \phi \nabla^t \phi \\ & + \frac{3}{4} \frac{1}{\phi(1+\phi)} \nabla_\lambda \phi \nabla^\lambda \phi, \end{aligned} \tag{19.39}$$

$$\begin{aligned} e^{-\lambda}\left(\frac{\nu'}{r} + \frac{1}{r^2}\right) - \frac{1}{r^2} = {} & 8\pi \frac{G}{1+\phi} \rho \langle u_r^2 \rangle + \frac{1}{2(1+\phi)} V(\phi) \\ & - \frac{1}{1+\phi} \left(\nabla_r \nabla^r - \Box\right) \phi + \frac{3}{2} \frac{1}{\phi(1+\phi)} \nabla_r \phi \nabla^r \phi \\ & - \frac{3}{4} \frac{1}{\phi(1+\phi)} \nabla_\lambda \phi \nabla^\lambda \phi, \end{aligned} \tag{19.40}$$

$$\begin{aligned} \frac{1}{2} e^{-\lambda}\left(\nu'' + \frac{\nu'^2}{2} + \frac{\nu' - \lambda'}{r} - \frac{\nu'\lambda'}{2}\right) = {} & 8\pi \frac{G}{1+\phi} \rho \langle u_\theta^2 \rangle + \frac{1}{2(1+\phi)} V(\phi) \\ & - \frac{1}{1+\phi} \left(\nabla_\theta \nabla^\theta - \Box\right) \phi + \frac{3}{2} \frac{1}{\phi(1+\phi)} \nabla_\theta \phi \nabla^\theta \phi \\ & - \frac{3}{4} \frac{1}{\phi(1+\phi)} \nabla_\lambda \phi \end{aligned} \tag{19.41}$$

$$
\begin{aligned}
&= 8\pi \frac{G}{1+\phi} \rho \left\langle u_\varphi^2 \right\rangle - \frac{1}{2(1+\phi)} V(\phi) \\
&- \frac{1}{1+\phi} \left(\nabla_\varphi \nabla^\varphi - \Box \right) \phi + \frac{3}{2} \frac{1}{\phi(1+\phi)} \nabla_\varphi \phi \nabla^\varphi \phi \\
&- \frac{3}{4} \frac{1}{\phi(1+\phi)} \nabla_\lambda \phi \nabla^\lambda \phi, \qquad (19.42)
\end{aligned}
$$

where no summation upon the pair of indices (t, r, θ, φ) is assumed.

An important relationship, which is very useful in applications, is obtained by adding the gravitational field equations Eqs. (19.39)–(19.42). Hence we obtain the following equation

$$
\begin{aligned}
e^{-\lambda} \left(\frac{\nu''}{2} + \frac{\nu'^2}{4} + \frac{\nu'}{r} - \frac{\nu'\lambda'}{4} \right) &= 4\pi \frac{G}{1+\phi} \rho \left\langle u^2 \right\rangle + \frac{1}{1+\phi} V(\phi) \\
&+ \frac{1}{1+\phi} \left(2\nabla_t \nabla^t + \Box \right) \phi - \frac{3}{\phi(1+\phi)} \nabla_t \phi \nabla^t \phi, \qquad (19.43)
\end{aligned}
$$

where we have denoted $\langle u^2 \rangle = \langle u_t^2 \rangle + \langle u_r^2 \rangle + \langle u_\theta^2 \rangle + \langle u_\varphi^2 \rangle$.

Since in the following discussion we are interested only in astrophysical applications at the extra-galactic cluster scale, we will adopt a Newtonian-type approximation, which consists in assuming that the deviations from standard GR (corresponding to the background value $\phi = 0$) are small for the systems we are considering. This approximation implies that $\phi \ll 1$. Thus, Eq. (19.43) can be approximated as

$$
e^{-\lambda} \left(\frac{\nu''}{2} + \frac{\nu'^2}{4} + \frac{\nu'}{r} - \frac{\nu'\lambda'}{4} \right) \simeq 4\pi G \rho \left\langle u^2 \right\rangle + 4\pi G \rho_\phi^{(eff)}, \qquad (19.44)
$$

where

$$
4\pi G \rho_\phi^{(eff)} \simeq V(\phi) + \left(2\nabla_t \nabla^t + \Box \right) \phi - \frac{3}{\phi} \nabla_t \phi \nabla^t \phi, \qquad (19.45)
$$

corresponds to an effective, geometric-type "energy" of the scalar field in the hybrid metric-Palatini gravitational theory.

19.2.2 The Relativistic Boltzmann Equation

Next we proceed to the second step in the derivation of the virial theorem for galaxy clusters, and determine the general relativistic Boltzmann equation that governs the evolution of the galactic distribution function f_B.

A basic result in statistical physics is the transport equation for the distribution function for a system of particles in a curved arbitrary Riemannian spacetime. This transport equation is given by the Boltzmann equation without a collision term, and which can be formulated as [64, 335]

$$
\left(p^\alpha \frac{\partial}{\partial x^\alpha} - p^\alpha p^\beta \Gamma^i_{\alpha\beta} \frac{\partial}{\partial p^i} \right) f_B = 0, \qquad (19.46)
$$

where p^α is the four-momentum of the galaxy (particle), and $\Gamma^i_{\alpha\beta}$ are the Christoffel symbols associated to the metric. An important consequence of the collisionless Boltzmann equation is that the local phase space density, as measured by an observer in a frame comoving with a galaxy, is conserved.

An important simplification of the mathematical formulation of the Boltzmann equation can be achieved by introducing an appropriately chosen orthonormal frame, or tetrad $\mathrm{e}_\mu^{\mathrm{a}}(\mathrm{x})$, $a = 0, 1, 2, 3$. The tetrad fields vary smoothly over some coordinates neighborhood U, and they satisfy the general condition $g^{\mu\nu}\mathrm{e}_\mu^{\mathrm{a}}\mathrm{e}_\nu^{\mathrm{b}} = \eta^{\mathrm{ab}}$ for all $x \in U$, where η^{ab} denotes the Minkowski metric tensor [64, 267, 335]. A basic property of the tetrad fields is that any tangent vector p^μ defined at an arbitrary point x can be represented as $p^\mu = p^a \mathrm{e}_{\mathrm{a}}^\mu$, a relation that defines the tetrad components p^a.

For the spherically symmetric line element given by Eq. (19.1), the frame of orthonormal vectors can be chosen in an appropriate way as [64, 267, 335]:

$$\mathrm{e}_\mu^0 = \mathrm{e}^{\nu/2}\delta_\mu^0, \qquad \mathrm{e}_\mu^1 = \mathrm{e}^{\lambda/2}\delta_\mu^1, \qquad \mathrm{e}_\mu^2 = \mathrm{r}\delta_\mu^2, \qquad \mathrm{e}_\mu^3 = \mathrm{r}\sin\theta\delta_\mu^3. \tag{19.47}$$

The tetrad components of the four velocity are $u^a = u^\mu \mathrm{e}_\mu^{\mathrm{a}}$. In the tetrad components corresponding to our present choices, the relativistic Boltzmann equation (19.46) is given by

$$u^a \mathrm{e}_{\mathrm{a}}^\mu \frac{\partial \mathrm{f_B}}{\partial \mathrm{x}^\mu} + \gamma_{\mathrm{bc}}^{\mathrm{a}} \mathrm{u}^{\mathrm{b}}\mathrm{u}^{\mathrm{c}} \frac{\partial \mathrm{f_B}}{\partial \mathrm{u}^{\mathrm{a}}} = 0, \tag{19.48}$$

where the distribution function $f_B = f_B(x^\mu, u^a)$ is a function of x^μ and u^a, respectively, and $\gamma_{bc}^a = \mathrm{e}_{\mu;\nu}^{\mathrm{a}}\mathrm{e}_{\mathrm{b}}^\mu \mathrm{e}_{\mathrm{c}}^\nu$ are the Ricci rotation coefficients [64, 267, 335]. Due to the spherical symmetry of the problem the distribution function depends only on the radial coordinate r, and hence Eq. (19.48) becomes [267]

$$u_1 \frac{\partial f_B}{\partial r} - \left(\frac{1}{2}u_0^2 \frac{\partial \nu}{\partial r} - \frac{u_2^2 + u_3^2}{r}\right)\frac{\partial f_B}{\partial u_1} - \frac{1}{r}u_1\left(u_2 \frac{\partial f_B}{\partial u_2} + u_3 \frac{\partial f_B}{\partial u_3}\right)$$
$$-\frac{1}{r}\mathrm{e}^{\lambda/2} u_3 \cot\theta \left(u_2 \frac{\partial f_B}{\partial u_3} - u_3 \frac{\partial f_B}{\partial u_2}\right) = 0. \tag{19.49}$$

This formalism was presented in Section 6.5.2, but we partially rewrite it here for self-consistency and self-completeness. Due to the spherical symmetry of our astrophysical system, the coefficient of $\cot\theta$ in Eq. (19.49) must be zero. From a mathematical point of view this implies that the distribution function f_B is only a function of r, u_1 and $u_2^2 + u_3^2$. As a next steps in our analysis we multiply Eq. (19.49) by $m u_r du$, and we integrate over the velocity space. Then, by taking into account that the distribution function f_B vanishes sufficiently rapidly as the velocities tend to $\pm\infty$, we find the equation

$$r\frac{\partial}{\partial r}\left[\rho \left\langle u_1^2 \right\rangle\right] + \frac{1}{2}\rho \left[\left\langle u_0^2 \right\rangle + \left\langle u_1^2 \right\rangle\right] r \frac{\partial \nu}{\partial r} - \rho \left[\left\langle u_2^2 \right\rangle + \left\langle u_3^2 \right\rangle - 2\left\langle u_1^2 \right\rangle\right] = 0. \tag{19.50}$$

Now we multiply Eq. (19.50) by $4\pi r^2$, and by integrating over the entire cluster volume, corresponding to a radius R, gives [267]

$$\int_0^R 4\pi\rho\left[\langle u_1^2\rangle + \langle u_2^2\rangle + \langle u_3^2\rangle\right] r^2 \mathrm{d}r - \frac{1}{2}\int_0^R 4\pi r^3\rho\left[\langle u_0^2\rangle + \langle u_1^2\rangle\right]\frac{\partial\nu}{\partial r}\mathrm{d}r = 0. \quad (19.51)$$

19.2.3 Geometrical Quantities Characterizing Galactic Clusters

In order to obtain analytical estimations of the main geometrical and physical quantities characterizing the galactic clusters, we introduce some approximations to the motion of both test particles in stable circular orbits around galaxies, as well as to the motion of galaxies in galactic clusters. As a first approximation we assume that ν and λ are slowly varying functions of the radial coordinate r. Then in Equation (19.44) we can neglect all the quadratic terms as being negligibly small as compared to the first-order terms. Secondly, astronomical observations show that the motion of the galaxies in clusters is nonrelativistic. Hence the galactic velocities are much smaller than the velocity of the light, that is, $\langle u_1^2\rangle \approx \langle u_2^2\rangle \approx \langle u_3^2\rangle \ll \langle u_0^2\rangle \approx 1$. Thus, Eqs. (19.44) and (19.51) can be written as

$$\frac{1}{2r^2}\frac{\partial}{\partial r}\left(r^2\frac{\partial\nu}{\partial r}\right) = 4\pi G\rho + 4\pi G\rho_\phi^{(eff)}, \quad (19.52)$$

and

$$2K - \frac{1}{2}\int_0^R 4\pi r^3\rho\frac{\partial\nu}{\partial r}\mathrm{d}r = 0, \quad (19.53)$$

respectively, where

$$K = \int_0^R 2\pi\rho\left[\langle u_1^2\rangle + \langle u_2^2\rangle + \langle u_3^2\rangle\right] r^2\mathrm{d}r, \quad (19.54)$$

is the total kinetic energy of the galaxies in the cluster. We define the total baryonic mass M_B of the galactic cluster as

$$M_B = \int_0^R \mathrm{d}M(r) = \int_0^R 4\pi\rho r^2\mathrm{d}r. \quad (19.55)$$

We assume that the main contribution to the baryonic mass M_B is due to the presence of the intra-cluster gas and of the stars. On the other hand we also include in M_B the mass contribution of other particles, such as massive neutrinos, which may also give a significant contribution to M_B.

By multiplying Eq. (19.52) by r^2, and integrating from 0 to r, we obtain

$$GM_B(r) = \frac{1}{2}r^2\frac{\partial\nu}{\partial r} - GM_\phi^{(eff)}(r), \quad (19.56)$$

where we have introduced the notation

$$M_{\phi}^{(eff)}(r) = 4\pi \int_0^r \rho_{\phi}^{(eff)}(r')r'^2 \mathrm{d}r'. \tag{19.57}$$

It is interesting to note at this point that in hybrid metric-Palatini gravity, the quantity $M_{\phi}^{(eff)}$ has essentially a geometric origin. Hence it is natural to denote it the *geometric mass* of the galactic cluster. In the following exploration we introduce the gravitational potential energies of the cluster by means of the definitions

$$\Omega_B = -\int_0^R \frac{GM_B(r)}{r} \mathrm{d}M_B(r)\,, \tag{19.58}$$

$$\Omega_{\phi}^{(eff)} = \int_0^R \frac{GM_{\phi}^{(eff)}(r)}{r} \mathrm{d}M_B(r)\,, \tag{19.59}$$

where R is the cluster radius. By multiplying Eq. (19.56) with $\mathrm{d}M_B(r)$, and by integrating from 0 to the cluster radius R, we obtain the important relation

$$\Omega_B = \Omega_{\phi}^{(eff)} - \frac{1}{2}\int_0^R 4\pi r^3 \rho \frac{\partial \nu}{\partial r} \mathrm{d}r\,. \tag{19.60}$$

19.2.4 The Generalized Virial Theorem

As a last step in our analysis, with the help of Eq. (19.53), we obtain the generalization of the virial theorem in hybrid metric-Palatini gravity, which can be formulated in the familiar form

$$2K + \Omega = 0\,. \tag{19.61}$$

In the above equation the total gravitational potential energy of the system, Ω, defined as

$$\Omega = \Omega_B - \Omega_{\phi}^{(eff)}\,, \tag{19.62}$$

contains a supplementary term $\Omega_{\phi}^{(eff)}$, which has a purely geometric origin.

It is useful to represent the generalized virial theorem, given by Eq. (19.61), in a more transparent physical form, which can be obtained by introducing the radii R_V and R_{ϕ}, defined by

$$R_V = M_B^2 \Big/ \int_0^R \frac{M_B(r)}{r} \mathrm{d}M_B(r), \tag{19.63}$$

and

$$R_{\phi}^{(eff)} = \left[M_{\phi}^{(eff)}\right]^2 \Big/ \int_0^R \frac{M_{\phi}^{(eff)}(r)}{r} \mathrm{d}M_B(r), \tag{19.64}$$

respectively. The quantity R_ϕ, having a geometric origin similar to the geometric mass considered above, may be called the *geometric radius* of the cluster of galaxies in hybrid metric-Palatini gravity. Hence, finally we obtain the baryonic potential energy Ω_B and the effective scalar field potential energy $\Omega_\phi^{(eff)}$ as given by

$$\Omega_B = -\frac{GM_B^2}{R_V}, \qquad \Omega_\phi^{(eff)} = \frac{G\left[M_\phi^{(eff)}\right]^2}{R_\phi^{(eff)}}, \tag{19.65}$$

respectively. Another important observational quantity, the virial mass M_V of the cluster of galaxies is defined as follows

$$2K = \frac{GM_BM_V}{R_V}. \tag{19.66}$$

Eventually, the fundamental relation between the virial and the baryonic mass of the galaxy cluster can be obtained after substitution of Eq. (19.66) into the virial theorem as

$$\frac{M_V}{M_B} = 1 + \frac{\left[M_\phi^{(eff)}\right]^2 R_V}{M_B^2 R_\phi^{(eff)}}. \tag{19.67}$$

If $M_V/M_B > 3$, a condition that is satisfied by the astrophysical parameters of most of the observed galactic clusters, then from Eq. (19.67) we obtain the virial mass of galactic clusters in hybrid metric-Palatini gravity, which can be approximated as

$$M_V \approx \frac{\left[M_\phi^{(eff)}\right]^2}{M_B} \frac{R_V}{R_\phi^{(eff)}}. \tag{19.68}$$

The virial mass M_V is determined observationally from the study of the velocity dispersion σ_r^2 of the stars and of the galaxies in the galactic clusters. An important consequence of the virial theorem in hybrid metric-Palatini gravity is that in a cluster with mass M_{tot} most of the mass is in the form of the geometric mass $M_\phi^{(eff)}$. Hence we can use the approximation $M_\phi^{(eff)} \approx M_{tot}$ in order to study the cluster dynamics. A fundamental question related to the possibility of the observational testing of hybrid metric-Palatini gravity is to find out what astrophysical processes can detect the presence of the geometric mass. Such an observational possibility may be provided by gravitational lensing. Through the study of the lensing properties of the galactic clusters one can obtain direct evidence of the existence of the geometric mass of its distribution properties, as well as of the gravitational effects associated to the presence of the scalar field. It is interesting to note that gravitational lensing can give us theoretical information even at cosmic scales extending far beyond the virial radius of the matter distribution of the galactic cluster.

19.3 Final Remarks

In concluding, in the framework of hybrid $f(X)$ gravity theory we have established the existence of a strict proportionality between the virial mass of the cluster and its baryonic mass, a relation that can also be tested observationally. One of the important, and observationally testable, predictions of the hybrid metric-Palatini gravitational "dark matter" model is that the geometric masses associated to the clusters, as well as its gravitational effects, extend beyond the virial radii of the clusters. Observationally, the virial mass M_V is obtained from the study of the velocity dispersions of the stars in the cluster. Due to the observational uncertainties, this method cannot give a reliable estimation of the numerical value of the total mass $M_B + M_\phi^{(eff)}$ in the cluster. However, a much more powerful method for the determination of the total mass distribution in clusters is the gravitational lensing of light, which may provide direct evidence for the gravitational effects at large distances from the cluster, and for the existence of the geometric mass.

The presence of hybrid metric-Palatini gravity effects at large distances from the cluster, and especially the large extension of the geometric mass, may lead to significantly different lensing observational signatures, as compared to the standard relativistic/dark matter model case. The bending angle in the hybrid metric-Palatini gravity models could be larger than the one predicted by the standard dark matter models. Therefore, the detailed observational study of the gravitational lensing could discriminate between the different theoretical models introduced to explain the motion of galaxies ("particles") in the clusters of galaxies, and the standard dark matter models. We refer the reader to [106] for more details.

Finally, it is worth pointing out that hybrid gravity can be precisely tested also at smaller scales like those around the galactic center. As reported in [83], the observed peculiar orbit of S2 star, moving around the center of our galaxy, is theoretically reconstructed if one adopts the gravitational potential coming from hybrid gravity. This result opens new perspectives in achieving precision tests for the theory.

20
Compact Stellar Objects

20.1 Introduction

Spherical symmetry plays an important role in the investigation of the astrophysical properties of stellar-type objects. Spherically symmetric systems are also of fundamental importance in General Relativity, since under this assumption one can obtain a large number of solutions of Einstein's gravitational field equations, which have successfully been used to describe the structure and interior properties of the general relativistic high-density objects. Since the birth of GR, the interests of the scientific community has been continuously attracted by the search for exact solutions of the gravitational field equations that could describe static neutral, isotropic or anisotropic, or charged stellar-type astrophysical configurations. A large number of exact analytical solutions of the Einstein field equations that can characterize the interior structure of the static compact fluid astrophysical objects were obtained in the past 100 years (for extensive reviews on the presently known interior static solutions of the Einstein field equations, see [156, 299]).

The theoretical investigation of stellar structure, combined with the astrophysical observations, can also provide essential restrictions on the predictions of modified theories of gravity. Due to a significant increase in the precision of the astronomical and astrophysical observations, and of the technological advances in the field of science, a large number of precise numerical values of neutron star masses are available presently [255, 401]. These high-precision astronomical observations have disclosed an essentially extremely complex distribution of the masses of the observed neutron stars, with the major implication that the twentieth century's paradigm, according to which there is single, $1.4M_{\odot}$ mass scale, is not backed anymore by the observational data. Actually, in the numerical observational data a bimodal, or an even more complex distribution mass distribution can be seen [255]. High precision pulsar timing observations, performed for a large number of stars, have confirmed that at least some neutron stars have masses of the order of $2M_{\odot}$ [19, 160]. On the other hand, robust observational bounds on the maximum and minimum values of the masses of the neutron stars existing in nature are still unknown. Additional information provided by the numerical values of the maximum masses and radii of neutron or other types of compact stars, the observations of the surface gravitational redshift of compact objects can also provide significant restrictions on extended theories of gravity that go beyond GR. The physical properties and structure of different classes of neutron, quark, and other exotic stars in various modified-gravity theories have attracted a lot of attention, and they have been extensively investigated in [20, 28, 92, 139, 138, 240, 277, 283, 354, 367, 434, 459].

In the present chapter, we investigate the properties of the relativistic high-density compact astrophysical objects in the framework of hybrid metric-Palatini gravity, in its scalar–tensor formulation. Our presentation is based on the results obtained in [154]. As a first step in our study, after adopting a spherically symmetric geometry, and a perfect fluid matter source, we derive the mass continuity equation and the Tolman–Oppenheimer–Volkoff (TOV) equation that, together with the generalized Klein–Gordon equation satisfied by the scalar field, describe the macroscopic properties of the compact astrophysical object. The structure equations describing stellar properties in hybrid metric-Palatini gravity are then solved numerically for several equations of state of the dense nuclear matter, which we specify in advance.

As concrete examples of high-density stellar-type objects, we study first stars characterized by the causal stiff fluid (Zeldovich) equation of state, which has the fundamental property that the speed of sound in the dense matter equals the speed of light. Then we consider the radiation-type equation of state, describing a photon gas, for which the trace of the energy-momentum tensor is zero. Next, we investigate stars characterized by the quark matter equation of state and, finally, we examine stars described by the Bose–Einstein condensate equation of state, corresponding to a polytropic equation of state with polytropic index $n = 1$. Furthermore, the astrophysical properties of two classes of specific hybrid metric-Palatini stellar models, described by two fixed forms of the scalar field, are also investigated in detail.

For all these astrophysical and nuclear physics models the global physical parameters (radius and mass) of the corresponding compact objects, as well as the scalar field, are obtained in both the standard general relativistic limit, and in the framework of hybrid metric-Palatini gravity. This methodology permits an in depth comparison of the two theoretical gravitational theories in the description of stellar structure and properties. As a general conclusion of this study we find that the hybrid metric-Palatini gravitational theory admits the presence of more massive stars, as compared to standard GR. Moreover, another interesting result of this investigation is that in the case of a constant scalar field, corresponding to the minimum of a Higgs-type potential, the equation of state of the high-density nuclear matter takes the form of the bag model equation of state, describing quark matter.

20.2 The Hydrostatic Equilibrium Equations for Spherically Symmetric Stars

In our investigation of stellar structure we assume that the line element that describes the interior of a compact object can be represented, in curvature coordinates, by the relation

$$ds^2 = -e^{\nu(r)}c^2dt^2 + e^{\lambda(r)}dr^2 + r^2\left(d\theta^2 + \sin^2\theta d\varphi^2\right), \tag{20.1}$$

which represents a static and spherically symmetric geometry. The metric functions $\nu(r)$ and $\lambda(r)$ are functions of the radial coordinate only, and they represent the redshift and the mass functions, respectively. For the radial coordinate r we assume that its range is $0 \leq r < \infty$. Asymptotically flat geometries, in which $\nu(r) \to 0$ and $\lambda(r) \to 0$ as $r \to \infty$ are also possible to construct. For the matter energy-momentum tensor $T_{\mu\nu}$ inside the star we adopt the perfect fluid form, so that in the comoving frame with four-velocity $u^\mu = \left(e^{-\nu/2}, 0, 0, 0\right)$, it has the nonzero components $T^\mu_\nu = \mathrm{diag}\left(-\rho c^2, p_r, p_t, p_t\right)$, where ρ is the energy density, and p_r and p_t are the radial and tangential pressures, respectively.

With the use of the metric given by Eq. (20.1), the effective Einstein field equation (17.25) yields the following gravitational field equations

$$\begin{aligned}\kappa^2\rho(r)c^2 = {} & \frac{1}{r^2}\left[1 - e^{-\lambda}\left(1 - r\lambda'\right)\right](1+\phi) - e^{-\lambda}\left[\phi'' - \frac{3\phi'^2}{4\phi}\right] \\ & + \frac{\phi'}{2r}e^{-\lambda}\left(r\lambda' - 4\right) - \frac{V}{2},\end{aligned} \tag{20.2}$$

$$\kappa^2 p_r(r) = \left[\frac{1}{r^2}\left(e^{-\lambda} - 1\right) + \frac{\nu'}{r}e^{-\lambda}\right](1+\phi) + \phi'\left(\frac{\nu'}{2} + \frac{2}{r} + \frac{3\phi'}{4\phi}\right)e^{-\lambda} + \frac{V}{2} \tag{20.3}$$

$$\begin{aligned}\kappa^2 p_t(r) = {} & \left[\left(\frac{\nu''}{2} + \left(\frac{\nu'}{2}\right)^2 + \frac{\nu'}{2r}\right)e^{-\lambda} - \frac{1}{2}\frac{\lambda' e^{-\lambda}}{r}\left(1 + r\frac{\nu'}{2}\right)\right](1+\phi) \\ & + \left[\phi'' + \frac{\phi'\nu'}{2} + \frac{3\phi'^2}{4\phi}\right]e^{-\lambda} + \frac{\phi'}{r}e^{-\lambda}\left(1 - \frac{r\lambda'}{2}\right) + \frac{V}{2},\end{aligned} \tag{20.4}$$

where by a prime we have denoted the derivative with respect to radial coordinate r. The effective Klein–Gordon equation (17.28) can be written down as

$$-\left[\phi'' + \frac{\phi'\nu'}{2} - \frac{\phi'^2}{2\phi} + \frac{2\phi'}{r}\right]e^{-\lambda} + \frac{\phi'\lambda'}{2}e^{-\lambda} + \frac{\phi}{3}\left[2V - (1+\phi)V_\phi\right] = \frac{\phi\kappa^2}{3}T. \tag{20.5}$$

The components of the energy-momentum tensor, and the metric tensor component ν, are related by the relation,

$$\nu' = -\frac{2p_r'}{\rho c^2 + p_r} + \frac{2\left(p_t - p_r\right)}{r}, \tag{20.6}$$

which follows from the conservation of the matter energy-momentum tensor.

At this moment it is important to point out that Eqs. (20.2)–(20.5) give four independent equations, for seven unknown quantities, i.e., $\rho(r)$, $p_r(r)$, $p_t(r)$, $\nu(r)$, $\lambda(r)$, $\phi(r)$, and $V(r)$. Thus, the system of equations describing the interior of compact stellar objects in hybrid metric-Palatini gravity is underdetermined. In the following subsection we will reduce the number of unknown functions by assuming some specific relations for the field and thermodynamic variables. Moreover, we will restrict our investigations only to the case of the isotropic pressure distribution, by assuming $p_r = p_t = p$.

20.2.1 The Mass Continuity and the TOV Equations

In order to simplify the mathematical formalism, as a first step in our investigation we divide Eq. (20.2) by $1 + \phi$, and we define the effective gravitational "constant" as $G_{\text{eff}} = G_0/(1 + \phi)$, where G_0 is the standard Newtonian and general relativistic gravitational constant. Hence, based on this definition, we can also introduce the effective gravitational coupling, which we denote as $\kappa^2_{\text{eff}} = 8\pi G_{\text{eff}}/c^4$. With the use of the mathematical identity

$$\phi'' - \frac{3\phi'^2}{4\phi} = \phi^{3/4}\frac{d}{dr}\frac{\phi'}{\phi^{3/4}} = 4\phi^{3/4}\frac{d^2}{dr^2}\phi^{1/4}, \tag{20.7}$$

and by denoting $\phi = e^{\Phi} - 1$ and $G_{\text{eff}} = G_0 e^{-\Phi}$, so that $\phi'/(1 + \phi) = \Phi'$, one arrives at the following relations

$$\begin{gathered}\frac{1}{1+\phi}\left(\phi'' - \frac{3\phi'^2}{4\phi}\right) = \frac{1}{4}\left(1 + \frac{3}{1 - e^{\Phi}}\right)\Phi'^2 + \Phi'' = f(\Phi), \\ V(\phi) = (1+\phi)\,U(\Phi) = e^{\Phi}U\left(\Phi\right).\end{gathered} \tag{20.8}$$

Equation (20.2) can be reformulated as

$$\frac{d}{dr}re^{-\lambda} = -\frac{rf(\Phi) + 3\Phi'/2}{1 + \Phi' r/2}re^{-\lambda} + \frac{1 - \kappa^2_{\text{eff}}\rho c^2 r^2 - Ur^2/2}{1 + \Phi' r/2}. \tag{20.9}$$

After introducing the representation of the metric tensor coefficient $e^{-\lambda}$ as

$$e^{-\lambda} = 1 - \frac{2G_0 m_{\text{eff}}(r)}{c^2 r}, \tag{20.10}$$

we obtain the result that the effective mass $m_{\text{eff}}(r)$, introduced above, satisfies the differential equation

$$\frac{dm_{\text{eff}}}{dr} = -\frac{rf(\Phi) + 3\Phi'/2}{1 + \Phi' r/2}m_{\text{eff}} + \frac{4\pi r^2}{\kappa^2 c^2\left[1 + \Phi' r/2\right]}\left[2\frac{\Phi'}{r} + \frac{U}{2} + f(\Phi) + \kappa^2_{\text{eff}}\rho c^2\right], \tag{20.11}$$

with the general solution given by

$$m_{\text{eff}}(r) = \frac{4\pi}{\kappa^2 c^2} \exp\left[-\int_0^r \frac{r' f(\Phi(r')) + 3\Phi'(r')/2}{1 + \Phi'(r')r'/2} dr'\right]$$
$$\times \int_0^r \exp\left[\int_0^{r'} \frac{r'' f(\Phi(r'')) + 3\Phi'(r'')/2}{1 + \Phi'(r'')r''/2} dr''\right] \frac{r'^2}{[1 + \Phi'(r')r'/2]}$$
$$\times \left[\frac{2\Phi'(r')}{r'} + \frac{U(\Phi(r'))}{2} + f(\Phi(r')) + \kappa_{\text{eff}}^2 \rho(r') c^2\right] dr', \tag{20.12}$$

where we have used the transformation $c^2/2G_0 = 4\pi/\kappa^2 c^2$. Equivalently, Eq. (20.11) can be written in the form

$$\frac{dm_{\text{eff}}}{dr} = 4\pi \rho_{\text{eff}} r^2, \tag{20.13}$$

where we have introduced the effective mass density of the star, defined as

$$\rho_{\text{eff}} = -\frac{r f(\Phi) + 3\Phi'/2}{4\pi r^2 [1 + \Phi' r/2]} m_{\text{eff}} + \frac{1}{\kappa^2 c^2 [1 + \Phi' r/2]} \left[2\frac{\Phi'}{r} + \frac{U(\Phi)}{2} + f(\Phi) + \kappa_{\text{eff}}^2 \rho c^2\right]. \tag{20.14}$$

We solve now Eq. (20.3) for ν', thus obtaining

$$\nu' = \frac{(\kappa^2 p e^{-\Phi} - U/2) r^2 - (1 - 2G_0 m_{\text{eff}}/c^2 r) \{1 + r [2 + r h(\Phi)] \Phi'\} + 1}{r (1 - 2G_0 m_{\text{eff}}/c^2 r)(1 + \Phi' r/2)}, \tag{20.15}$$

where we have defined

$$h(\Phi) = \frac{3 e^{\Phi} \Phi'}{4 (e^{\Phi} - 1)}. \tag{20.16}$$

By using the previous results, and by taking into account the energy-momentum conservation equation (20.6), the generalized TOV equation, describing the hydrostatic equilibrium properties of high-density astrophysical objects in hybrid metric-Palatini gravity, can be obtained as

$$\frac{dp}{dr} = -\frac{(\rho c^2 + p) \{(\kappa^2 p e^{-\Phi} - U/2) r^2 - (1 - 2G_0 m_{\text{eff}}/c^2 r) \{1 + r [2 + r h(\Phi)] \Phi'\} + 1\}}{r (1 - 2G_0 m_{\text{eff}}/c^2 r)(2 + \Phi' r)}. \tag{20.17}$$

Finally, the Klein–Gordon equation (20.5) takes, in the new scalar field variable, the form

$$-\Phi'' + \frac{1}{2}\frac{2 - e^{\Phi}}{e^{\Phi} - 1} \Phi'^2 - \Phi' \left[-\frac{p'}{\rho c^2 + p} + \frac{2}{r} - \frac{G_0}{c^2} \frac{4\pi \rho_{\text{eff}} r^3 - m_{\text{eff}}}{r^2 (1 - 2G_0 m_{\text{eff}}/c^2 r)}\right]$$
$$+ \frac{e^{\Phi} - 1}{3 (1 - 2G_0 m_{\text{eff}}/c^2 r)} \left[U(\Phi) - \frac{dU(\Phi)}{d\Phi} - \kappa_{\text{eff}}^2 T\right] = 0, \tag{20.18}$$

where the following relations,

$$\lambda' = \frac{2G_0}{c^2}\frac{1}{e^{-\lambda}r^2}\left(r\frac{dm_{\text{eff}}}{dr} - m_{\text{eff}}\right), \tag{20.19}$$

and

$$\frac{dV}{d\phi} = \frac{d}{d\phi}\left[U\left(1+\phi\right)\right] = U\left(\Phi\right) + \frac{dU\left(\Phi\right)}{d\Phi}, \tag{20.20}$$

respectively, have been used.

After specifying an equation of state for the matter inside the star, $p = p(\rho)$, the system of equations (20.5), (20.11), and (20.17) must be solved together with the boundary conditions $m_{\text{eff}}(0) = 0$, $\rho(0) = \rho_c$, $\Phi(0) = \Phi_0$, $\Phi'(0) = \Phi_0'(0)$, and $p(R) = 0$, respectively, where ρ_c is the central density, and R is the radius of the star, respectively. However, when numerically integrating the gravitational field equations, due to the singular nature of the center of the star, corresponding to $r = 0$, one must impose the boundary conditions at a small but nonzero radius $r = r_0$ [148], so that $m_{\text{eff}}(r_0) = 0$, $\rho(r_0) = \rho_c$, etc. On the other hand, the initial values of the radial derivatives of Φ at the center, $\Phi'(r_0)$, must be determined so that they are equivalent with a regular Taylor expansion of Φ at the origin $r = 0$, which can be given, for example, as [148]

$$\Phi(r) = \Phi\left(0\right) + \frac{1}{6}r^2\Delta\Phi(0) + \mathrm{O}\left(r^4\right), \tag{20.21}$$

where $\Delta\Phi(0) = \Phi(r) - \Phi(0)$. This series expansion determines the derivative of the scalar field as

$$\lim_{r\to r_0}\Phi'(r) \approx \frac{1}{3}r_0\Delta\Phi(0). \tag{20.22}$$

Near the center of the star we can represent the effective mass as $m_{\text{eff}}(r) \sim 4\pi r^3\rho_c/3$. By taking into account the following limits: $\lim_{r\to r_0}\Phi'^2(r) = 0$, $\lim_{r\to r_0}p'(r) = 0$, $\lim_{r\to r_0}m_{\text{eff}}/r = 0$, $\lim_{r\to r_0}m_{\text{eff}}/r^2 = 0$, as well as the relation $\Phi''(r) = \Delta\Phi(0)$, from Eq. (20.18) we obtain

$$\Delta\Phi(0) = \frac{e^{\Phi_0}-1}{3}\left[U\left(\Phi_0\right) - \left.\frac{dU(\Phi)}{d\Phi}\right|_{\Phi=\Phi_0} + \frac{8\pi G}{c^4}e^{-\Phi_0}\left(\rho_c c^2 - 3p_c\right)\right], \tag{20.23}$$

giving for the central value of the derivative of the scalar field the expression

$$\lim_{r\to r_0}\Phi'(r) = \frac{e^{\Phi_0}-1}{9}r_0\left[U\left(\Phi_0\right) - \left.\frac{dU(\Phi)}{d\Phi}\right|_{\Phi=\Phi_0} + \frac{8\pi G}{c^4}e^{-\Phi_0}\left(\rho_c c^2 - 3p_c\right)\right]. \tag{20.24}$$

20.2.2 Dimensionless Form of the Mass Continuity, TOV and Klein–Gordon Equations

In order to simplify the mathematical formalism in the following we will introduce a set of five dimensionless variables $(\eta, \theta, M_{\text{eff}}, P, u)$, defined by means of the scaling transformations

$$r = a\eta, \qquad \rho = \rho_c \theta, \qquad m_{\text{eff}} = M_0 M_{\text{eff}}, \qquad p = \rho_c c^2 P, \qquad u = a^2 U, \quad (20.25)$$

where

$$a = \frac{c}{\sqrt{8\pi G_0 \rho_c}}, \qquad M_0 = \frac{ac^2}{G_0} = \frac{c^3}{\sqrt{8\pi G_0^3 \rho_c}}. \quad (20.26)$$

In the new variables Eqs. (20.5), (20.11), and (20.17) take the following dimensionless form

$$\frac{dM_{\text{eff}}}{d\eta} = -\frac{\eta f(\Phi(\eta)) + (3/2)(d\Phi/d\eta)}{1 + \eta(d\Phi/d\eta)/2} M_{\text{eff}} + \frac{\eta^2}{2[1 + \eta(d\Phi/d\eta)/2]} \left[\frac{2}{\eta}\frac{d\Phi}{d\eta} + \frac{u}{2} + f(\Phi(\eta)) + \theta e^{-\Phi}\right], \quad (20.27)$$

$$\frac{dP}{d\eta} = -\frac{(\theta + P)\{(Pe^{-\Phi} - u/2)\eta^2 - (1 - 2M_{\text{eff}}/\eta)\{1 + \eta[2 + \eta h(\Phi(\eta))](d\Phi/d\eta)\} + 1\}}{\eta(1 - 2M_{\text{eff}}/\eta)[2 + \eta(d\Phi/d\eta)]}, \quad (20.28)$$

$$\frac{d^2\Phi}{d\eta^2} + \frac{1 - e^{\Phi}/2}{e^{\Phi} - 1}\left(\frac{d\Phi}{d\eta}\right)^2 - \frac{d\Phi}{d\eta}\left[-\frac{dP/d\eta}{\theta + P} + \frac{2}{\eta} - \frac{\rho_{\text{eff}}(\eta)\eta^3/2 - M_{\text{eff}}}{\eta^2(1 - 2M_{\text{eff}}/\eta)}\right] + \frac{(e^{\Phi} - 1)}{3(1 - 2M_{\text{eff}}/\eta)}\left\{\left[u(\Phi) - \frac{du(\Phi)}{d\Phi}\right] - e^{-\Phi}(-\theta + 3P)\right\} = 0, \quad (20.29)$$

respectively, where

$$\rho_{\text{eff}}(\eta) = -2\frac{\eta f(\Phi(\eta)) + 3(d\Phi/d\eta)/2}{\eta^2[1 + \eta(d\Phi/d\eta)/2]} M_{\text{eff}} + \frac{1}{1 + \eta(d\Phi/d\eta)/2}\left[\frac{2}{\eta}\frac{d\Phi}{d\eta} + \frac{u(\Phi)}{2} + f(\Phi(\eta)) + \theta e^{-\Phi}\right]. \quad (20.30)$$

Once the equation of state of the matter $P = P(\theta)$ has been adopted, the system of ordinary differential equations (20.27)–(20.29) must be integrated numerically with the boundary conditions $M_{\text{eff}}(0) = 0$, $\theta(0) = 1$, $\Phi(0) = \Phi_0$, $(d\Phi/d\eta)|_{\eta=0} = \Phi_0'$, respectively. The numerical values of the derivative of the scalar field at the center of the star $r = 0$ can be found in a dimensionless form as

$$\lim_{\eta \to \eta_0} \Phi'(\eta) = \frac{e^{\Phi_0} - 1}{9} \eta_0 \left[u(\Phi_0) - \left. \frac{du(\Phi)}{d\Phi} \right|_{\Phi = \Phi_0} + e^{-\Phi_0} (1 - 3P_c) \right], \tag{20.31}$$

where P_c is the value of the dimensionless pressure P at the center of the star.

In the next section we will consider specific numerical solutions, corresponding to a given equation of state of the high-density stellar matter, which could describe the structure of the compact astrophysical objects in hybrid metric-Palatini gravity.

20.3 Structure of High-Density Compact Objects in Hybrid Metric-Palatini Gravity

In the present section, we will begin our investigations of the properties of high-density stars in the hybrid metric-Palatini gravitational theory without introducing any constraints on the functional form of the scalar field Φ. In the next section we will consider the stellar structure equations under the hypothesis that the scalar field Φ admits an explicit mathematical representation, which is not dynamically determined by the field equations. In this latter case, from the field equations one can obtain, after imposing the functional form of ϕ, either the explicit form of the equation of state of the dense stellar matter, or the dynamical behavior of the scalar field potential, related to the a priori introduced representation of the scalar field.

As for the equation of state of the dense nuclear matter inside the star, we will study four important cases, corresponding to the stiff fluid equation of state, in which the pressure is given by $P = \theta$, to the radiation fluid equation of state, with $P = \theta/3$, to the quark matter equation of state $P = (\theta - 4b)/3$, where B is a constant, and to the Bose–Einstein condensate superfluid neutron matter equation of state $P \propto \theta^2$, respectively.

In the following, for all adopted equations of state we assume that the scalar field potential $U(\Phi)$ is of the Higgs type

$$U(\Phi) = -\frac{\mu^2}{2}\Phi^2 + \frac{\xi}{4}\Phi^4, \tag{20.32}$$

where μ^2 and ξ are constants. Similarly to the standard particle physics case, we also assume that the constant $\mu^2 < 0$ in the potential is related to the mass m_Φ^2 of the scalar particle of hybrid metric-Palatini gravity by the relation $m_\Phi^2 = 2\xi v^2 = -2\mu^2$, where $v^2 = -\mu^2/\xi$ gives the minimum value of the Higgs potential of the scalar field. The Higgs self-coupling constant $\lambda \approx 1/8$ for the case of the strong interactions [1], which is a value derived from the accelerator experiments that have determined the mass of the Higgs boson. The dimensionless form $u(\Phi)$ of the Higgs potential is given by

$$u(\Phi) = -\frac{\mu_0^2}{2}\Phi^2 + \frac{\xi_0}{4}\Phi^4, \tag{20.33}$$

where $\mu_0^2 = a^2\mu^2$, and $\xi_0 = a^2\xi$, respectively, with a^2 given by Eq. (20.26). At this point it is important to specify that in the dimensionless representation of the Higgs-type potential, the coefficients μ_0^2 and ξ_0 are some given functions of the central density of the star.

In all considered cases, we will contrast our results with the general relativistic static stellar models with spherical symmetry, described by the standard structure equations [462]

$$\frac{dm}{dr} = 4\pi\rho r^2, \tag{20.34}$$

$$\frac{dp(r)}{dr} = -\frac{(G_0/c^2)\left[\rho(r)c^2 + p(r)\right]\left[(4\pi/c^2)\, p(r)r^3 + m(r)\right]}{r^2\left(1 - 2G_0 m(r)/c^2 r\right)}. \tag{20.35}$$

In the scaled dimensionless variables defined in Eqs. (20.25), the general relativistic structure equations have the simple dimensionless representation

$$\frac{dM_{GR}(\eta)}{d\eta} = \eta^2\theta_{GR}(\eta), \tag{20.36}$$

and

$$\frac{dP_{GR}(\eta)}{d\eta} = -\frac{\left[\theta_{GR}(\eta) + P_{GR}(\eta)\right]\left[P_{GR}(\eta)\eta^3 + M_{GR}(\eta)\right]}{\eta^2\left[1 - 2M_{GR}(\eta)/\eta\right]}, \tag{20.37}$$

respectively. These equations must be integrated, after imposing an equation of state $P = P(\theta)$, with the use of the boundary conditions $\theta_{GR}(0) = 1$ and $\theta_{GR}(\eta_S) = 0$, respectively.

20.3.1 Explicit Form of $f(\mathcal{R})$

Before proceeding further with the numerical analysis of the spherically symmetric stellar models in hybrid metric-Palatini gravity and GR, respectively, it would be interesting to find, by taking into account the potential given by Eq. (20.32), a specific form of the function $f(\mathcal{R})$. To do this, after using Eq. (17.18), we first arrive at the equation

$$\mathcal{R} = \frac{dV(\phi)}{d\phi} = -\mu^2 \ln(1+\phi)$$
$$\left\{1 - \ln(1+\phi)\left[\frac{\xi}{\mu^2}\ln(1+\phi)\left(\frac{1}{4}\ln(1+\phi)+1\right) - \frac{1}{2}\right]\right\}, \tag{20.38}$$

Generally, with the use of the above result, one cannot solve Eq. (17.15) to obtain the explicit exact form of the function $f(\mathcal{R})$. However, an approximate solution can be found by performing a series expansion of Eq. (20.38) around the point $\phi = 0$. Such a series expansion is consistent with the analysis that led to Eqs. (17.48)–(17.49). Thus we obtain

$$\mathcal{R} = -\mu^2\phi + \left(\frac{\mu^2}{6} + \xi\right)\phi^3 - \frac{5}{24}\left(\mu^2 + 6\xi\right)\phi^4 + O\left(\phi^5\right). \tag{20.39}$$

For the potential $V(\phi)$ given by Eq. (20.8) we similarly obtain

$$V(\phi) = -\frac{\mu^2\phi^2}{2} + \frac{1}{4}\left(\frac{\mu^2}{6} + \xi\right)\phi^4 + O\left(\phi^5\right). \tag{20.40}$$

In the first order of approximation, the potential is given by $V(\phi) \approx -\mu^2\phi^2/2$, and therefore the Clairaut-type differential equation (17.15) takes the form

$$-\frac{\mu^2}{2}\left[f'(A)\right]^2 = Af'(A) - f(A), \tag{20.41}$$

where $A = \mathcal{R}$. The general solution of the Clairaut differential equation can be immediately found as

$$f(A) = c_1 A + \frac{c_1^2}{2}\mu^2, \tag{20.42}$$

where c_1 is an arbitrary constant of integration, which must be determined from some appropriately chosen initial conditions. In the next order of approximation for $f(A)$ we obtain the Clairaut-type equation

$$-\frac{\mu^2}{2}\left[f'(A)\right]^2 + \frac{1}{24}\left(\mu^2 + 6\xi\right)\left[f'(A)\right]^4 = Af'(A) - f(A), \tag{20.43}$$

with the general solution

$$f(A) = c_1 A + \frac{c_1^2}{2}\mu^2 - \frac{1}{4}\left(\frac{\mu^2}{6} + \xi\right)c_1^4, \tag{20.44}$$

where c_1 is an arbitrary constant of integration.

Hence, by continuing this iterative process, we reach the conclusion that for the adopted Higgs-type functional form of the scalar field potential, we generally have

$$f(A) = c_1 A + c_2(c_1, \mu, \xi), \tag{20.45}$$

where the arbitrary integration constant c_1 must be obtained from some appropriately formulated physical requirements, or initial/boundary conditions. It is important to point out that this representation corresponds to a functional representation of f of the form $f(\mathcal{R}) = \mathcal{R} + \Lambda_{\text{eff}}$, in which an effective cosmological constant given by $\Lambda_{\text{eff}} = c_2(c_1, \mu, \xi)$ also appears.

It is important to point out that in the present approach the numerical values of the arbitrary constant c_2 are dependent on the parameters (μ, ξ) of the adopted Higgs-type potential. On the other hand the constant c_2 is essentially generated by the existence of the coupling. Hence, in order to analyze in detail the effects of the coupling on the stellar structure, in our numerical investigations we will modify the values of the potential parameters in a given range. We also present the standard general relativistic result for each investigated stellar structure in the framework of the hybrid metric-Palatini gravitational theory (corresponding to a particular choice of the equation of state of the high-density matter). The general relativistic limit is obtained by cutting off the coupling by imposing the limits $\mu \to 0$ and $\xi \to 0$, respectively. Hence our numerical investigations could give a relevant description of the effects on the structure and properties of the high-density compact astrophysical objects of the variation of the gravitational couplings in this class of modified theories of gravity.

20.3.2 Stiff Fluid Stars

As a first example of stellar models in hybrid metric-Palatini gravity we consider compact objects described in dimensionless variables by the equation of state $P = \theta$. This equation of state is called the stiff (Zeldovich) equation of state, and from a physical point of view it gives the upper limit for the equation of state of a hot and extremely dense nucleonic gas. Presently it is thought that nuclear matter can be described by the stiff equation of state at densities about ten times bigger than the nuclear density, that is, at matter densities greater than 10^{17} g/cm^3, and at temperatures $T = (\rho/\sigma)^{1/4} > 10^{13}$ K, where by σ we have denoted the radiation constant [462]. For the stiff equation of state of the dense matter the speed of sound is given by $c_s^2 = \partial P/\partial\theta = 1$. Hence an important property of this equation of state is that the speed of matter perturbations (the speed of sound) cannot surpass the speed of light, which is the maximum speed by which particles can travel.

The stiff matter equation of state plays an important role in theoretical astrophysics. With the use of the static spherically symmetric Einstein field equations, of the principle of causality, and of Le Chatelier's principle, in [435] it was shown, by using numerical methods, that the maximum mass of the equilibrium configuration of a high-density neutron star cannot surpass the upper limit of $3.2M_\odot$. To find this fundamental astrophysical result it was postulated that at high stellar densities the neutron matter can be described by the stiff fluid equation of state $p = \rho c^2$. The numerical value of the absolute maximum mass of a stable neutron star provides a fundamental criterion for observationally distinguishing neutron or other types of compact/exotic stars from black holes.

In Fig. 20.1 we have represented the mass–radius relations for stiff fluid stars in both standard GR and hybrid metric-Palatini gravity theory. The basic physical properties of this class of stars have been obtained by numerically integrating for the stiff fluid equation of state the star structure equations, given by Eqs. (20.11), (20.17), and (20.18), respectively. To obtain the results presented in the plots, we have chosen for the coefficients μ and ξ in the Higgs potential the values $\mu = 1.5 \times 10^{-5}$ cm^{-1}, and $\xi = 9.5 \times 10^{-10}$ cm^{-2}, respectively. Then, the central

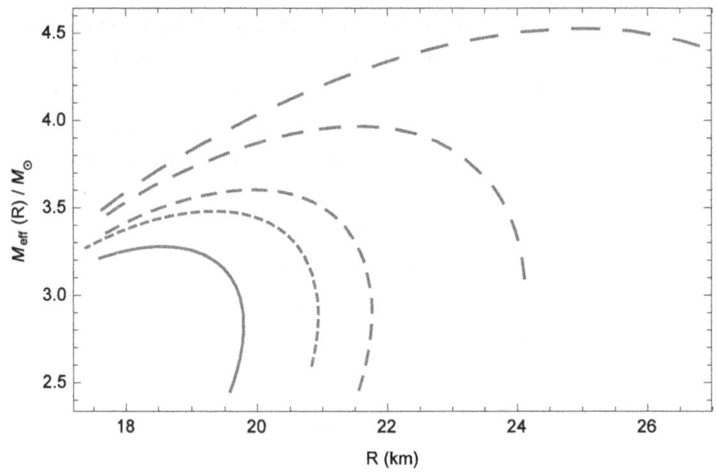

Figure 20.1 Mass–radius relation for stiff fluid stars in hybrid metric-Palatini gravity theory, for $\mu = 1.5\times10^{-5}$ cm^{-1}, $\xi = 9.5\times10^{-10}$ cm^{-2}, and for different values of $\Phi(0)$ and $\Phi'(0)$: $\Phi \equiv 0$, $\Phi'(0) \equiv 0$ (standard general relativistic limit) (solid curve), $\Phi(0) = 0.16$, $\Phi'(0) = -5.83 \times 10^{-16}$ cm^{-1} (dotted curve), $\Phi(0) = 0.24$, $\Phi'(0) = -3.18 \times 10^{-16}$ cm^{-1} (short dashed curve), $\Phi(0) = 0.30$, $\Phi'(0) = -1.83\times10^{-16}$ cm^{-1} (dashed curve), and $\Phi(0) = 0.32$, $\Phi'(0) = -0.66\times 10^{-16}$ cm^{-1} (long dashed curve).

values of the scalar field $\Phi(0)$ have been varied, and thus this procedure generates several sequences of stable stiff fluid stellar models. In the central density interval considered the derivative of the scalar field Φ' at the center of the star was computed to be in the range $\Phi'(0) = -5.83 \times 10^{-16}$ cm^{-1} and $\Phi'(0) = -0.66 \times 10^{-16}$ cm^{-1}. To compare the structure of the stellar-type objects in standard GR and in hybrid metric-Palatini gravity, the mass–radius relation for stiff fluid stars, obtained as the solution of the general relativistic mass continuity and TOV equations (20.34) and (20.35), respectively, is also presented.

For the hybrid metric-Palatini stellar models, with $\Phi \neq 0$, the central density was varied between the values 3.1×10^{14} g/cm^3, and 2.9×10^{15} g/cm^3, respectively. For the standard general relativistic model the central density is in the range 3.1×10^{14} g/cm^3 and 2.2×10^{15} g/cm^3. For these stellar sequences the obtained maximum masses are $M_{max} = 3.278M_\odot$ (which gives the standard general relativistic maximum mass value [435]), $M_{max} = 3.459M_\odot$, $M_{max} = 3.61M_\odot$, $M_{max} = 3.819M_\odot$, and $M_{max} = 3.971M_\odot$, respectively.

As one can see from Fig. 20.1, stars satisfying the stiff fluid equation of state are much more massive in hybrid metric-Palatini as compared with their general relativistic counterparts. However, an increase in the central density to a value of $\rho_c = 10^{16}$ g/cm^3 will significantly affect the mass of the star, decreasing it to $M = 4.27M_\odot$. On the other hand, a central density of the order of $\rho_c = 10^{17}$ g/cm^3 gives for the hybrid metric-Palatini gravity star a mass of around $M = 1.35M_\odot$ only.

20.3.3 Radiation Fluid Stars

The equation of state $P = \theta/3$ describes radiation fluid stars. The possibility that stars made of photons, and therefore obeying the radiation equation of state, could exist, has already been intensively investigated in the astrophysical literature. In [453] numerical solutions of Einstein's gravitational field equation, representing spherically symmetric, static stars made of a photon gas were obtained. On the other hand, as it was pointed out in [358], an interesting class of massive stellar objects, called *radiation pressure supported stars* (RPSS), can be obtained even within the framework of classical Newtonian gravity. The generalization of this class of radiation-dominated stellar objects to standard GR are called *relativistic radiation pressure supported stars* (RRPSS). In [203] it was proposed that the gravitational collapse of massive matter clouds, which may end in a very high-density phase, could lead to the formation of RRPSS. At sufficiently large cosmological redshifts $z \gg 1$, the trapped radiation flux always reaches the Eddington luminosity, and this property is independent of the details of the contraction process.

The properties of the radiation fluid stars in hybrid metric-Palatini gravity have been obtained for the equation of state $P = \rho c^2/3$ (dimensionless form $P = \theta/3$), by numerically integrating the star structure equations (20.11), (20.17) and (20.18), and (20.27)–(20.29), respectively. For the parameters of the Higgs potential we have used the same values as in the case of the stiff fluid star. To compare the structure of the stars in standard GR and hybrid metric-Palatini gravity, we have also constructed the solution of the general relativistic mass continuity and TOV equations (20.36) and (20.37), respectively.

In order to obtain the mass–radius relation for the radiation fluid stars we have used the same initial values for the scalar field and its derivative as in the stiff fluid case. Similarly to the previous analysis, we have again varied the central value $\Phi(0)$ of the scalar field. The derivative Φ' of the scalar field at the center of the star, $\Phi'(0)$, was worked out to be between $\Phi'(0) = -1.39 \times 10^{-17}$ cm^{-1} and $\Phi'(0) = -2.67 \times 10^{-16}$ cm^{-1} for the $\Phi(0)$ values considered. When the matter density reached the surface value $\rho = 2 \times 10^{14}$ g/cm^3, we stopped the numerical integration. The mass-radius relations for hybrid metric-Palatini stars and for standard general relativistic stars are depicted in Fig. 20.2. For all curves the central density varies between 3.1×10^{14} g/cm^3 and 2.95×10^{15} g/cm^3, respectively. The maximum obtained stellar masses corresponding to the adopted set of initial conditions and scalar field potential parameters are $M_{max} = 2.039 M_{\odot}$, $M_{max} = 2.098 M_{\odot}$, $M_{max} = 2.182 M_{\odot}$, $M_{max} = 2.297 M_{\odot}$, and $M = 2.459 M_{\odot}$, respectively.

In both hybrid metric-Palatini gravity and standard GR the global structure of the radiation fluid compact stars is comparable to the stiff fluid stars, respectively. The scalar field of the hybrid metric-Palatini gravity is a monotonically decreasing function of the dimensionless radial coordinate η, and on the star's

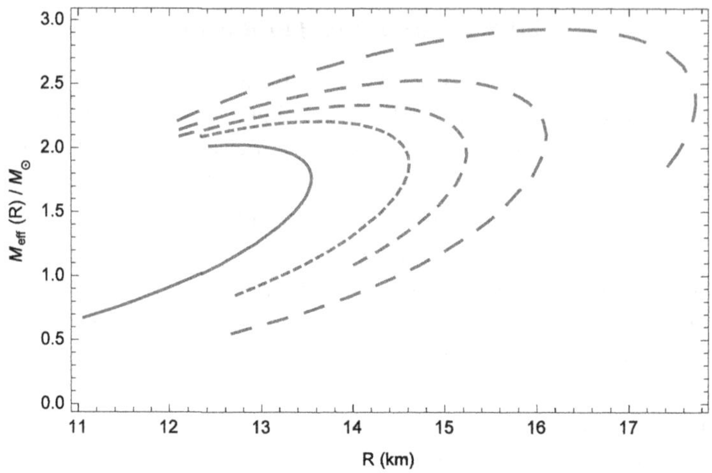

Figure 20.2 Mass–radius relation for radiation fluid stars in hybrid metric-Palatini gravity theory, for $\mu = 1.5 \times 10^{-5}$ cm^{-1}, $\xi = 9.5 \times 10^{-10}$ cm^{-2}, and for different values of $\Phi(0)$ and $\Phi'(0)$: $\Phi \equiv 0$, $\Phi' \equiv 0$ (standard general relativistic limit) (solid curve), $\Phi(0) = 0.12$, $\Phi'(0) = -1.39 \times 10^{-17}$ cm^{-1} (dotted curve), $\Phi(0) = 0.17$, $\Phi'(0) = -2.37 \times 10^{-16}$ cm^{-1} (short dashed curve), $\Phi(0) = 0.22$, $\Phi'(0) = -3.01 \times 10^{-16}$ cm^{-1} (dashed curve), and $\Phi(0) = 0.27$, $\Phi'(0) = -2.67 \times 10^{-16}$ cm^{-1} (long dashed curve).

surface it reaches the value zero. On the other hand inside the star the scalar field potential has negative values, and it vanishes on the vacuum surface. In both hybrid metric-Palatini gravity and standard GR the radiation fluid stars are less massive, as compared to the stiff fluid counterparts. Still, radiation fluid stars in hybrid metric-Palatini gravity are much more massive than the corresponding general relativistic massive compact objects.

20.3.4 Quark Stars

As initially suggested in [510], and according to a large number of theoretical particle physics arguments, the strange quark matter, formed of u, d, and s quarks, may be the most energetically favorable state of baryonic matter in nature. For the formation of the strange matter in an astrophysical environment two possibilities have been proposed: the first implies the possibility of a quark-hadron phase transition in the early Universe, while the second way could be the conversion at ultrahigh densities of neutron matter into strange matter inside neutron stars. The eventuality of the existence in nature of stars made of quark matter was initially proposed in [266, 70]. The equation of state of the quark matter can be derived, from a theoretical point of view, from the fundamental Lagrangian of quantum chromodynamics (QCD), by using field theoretical methods [503]. The weakening of the quark–quark interaction at short distances, due to the asymptotic freedom of the theory, is an important prediction of QCD.

At a temperature T and at a chemical potential μ_f, by using thermal theory, the energy density ρ and the pressure p of a quark-gluon plasma can be calculated under the assumption that the interactions of quarks and gluons are sufficiently small. After neglecting quark masses, the equation of state of quark matter is given in the first-order perturbation theory by [510, 132],

$$\rho = \sum_{i=u,d,s,c;e^-,\mu^-} \rho_i + B, p + B = \sum_{i=u,d,s,c;e^-,\mu^-} p_i, \tag{20.46}$$

where B, called the bag constant, is defined as the difference between the energy densities of the perturbative and non-perturbative QCD vacuum. Therefore, it follows that the equation of state for quark matter is given by the Massachusetts Institute of Technology (MIT) bag model equation of state, which can be written as [510, 132]

$$p = \frac{1}{3}\left(\rho c^2 - 4Bc^2\right). \tag{20.47}$$

The equation of state (20.47) defines the equation of state of a free gas of massless particles, with corrections due to the trace anomaly of QCD, as well as corrections appearing as a result of the inclusion of perturbative interactions. All these corrections are always negative, and at a given temperature, when the strong interaction coupling constant is of the order of $\alpha_s = 0.5$, they reduce the energy density of the quark-gluon plasma by a factor of about two [503]. The MIT bag model equation of state becomes in the dimensionless variables introduced in Eq. (20.25)

$$P = \frac{1}{3}\left(\theta - 4b\right), \tag{20.48}$$

where $b = B/\rho_c$.

To obtain the basic physical properties of quark stars in hybrid metric-Palatini gravity, by using the MIT bag model equation of state (20.47), we integrate numerically the star's structure equations given by Eqs. (20.11), (20.17), and (20.18), and (20.27)–(20.29), respectively. For the physical free parameters of the Higgs potential we have adopted the same values as in the case of the stiff and radiation fluid stars, respectively. In order to compare the structure of the quark stars in standard GR and in hybrid metric-Palatini gravity, we have also obtained the corresponding solution of the general relativistic mass continuity and TOV equations, given by Eqs. (20.36) and (20.37), respectively. In all considered cases the numerical integration stops when the condition $P = 0$ is reached. The mass–radius relations for the quark stars in both hybrid metric-Palatini gravity and standard GR are depicted in Fig. 20.3. For the bag constant B, describing quark matter properties, we have adopted the standard value $B = 10^{14}$ g/cm^3 [132]. The central density of the star has been varied between the numerical values 4.1×10^{14} g/cm^3 and 3.2×10^{16} g/cm^3, respectively.

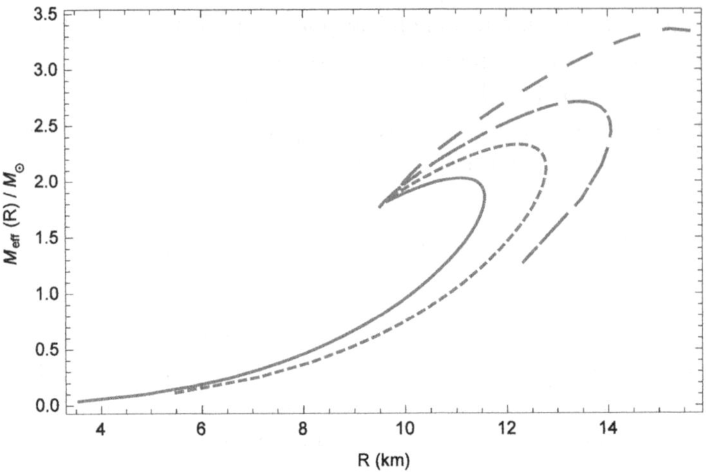

Figure 20.3 Mass–radius relation for quark stars in hybrid metric-Palatini gravity theory, for $\mu = 1.5 \times 10^{-5}$ cm^{-1}, $\xi = 9.5 \times 10^{-10}$ cm^{-2}, and for different values of $\Phi(0)$ and $\Phi'(0)$: $\Phi \equiv 0$, $\Phi' \equiv 0$ (standard general relativistic limit) (solid curve), $\Phi(0) = 0.14$, $\Phi'(0) = -1.69 \times 10^{-16}$ cm^{-1} (dotted curve), $\Phi(0) = 0.17$, $\Phi'(0) = -2.85 \times 10^{-16}$ cm^{-1} (short dashed curve), $\Phi(0) = 0.20$, $\Phi'(0) = -2.65 \times 10^{-16}$ cm^{-1} (dashed curve), and $\Phi(0) = 0.23$, $\Phi'(0) = -2.47 \times 10^{-16}$ cm^{-1} (long dashed curve).

To numerically integrate the field equations in their dimensionless form we have adopted the initial conditions $\theta(0) = 1$, $M_{\rm eff}(0) = 0$, and $\Phi(0) = 0.47$. In the dimensionless representation $(d\Phi/d\eta)|_{\eta=0}$ is a function depending on $\Phi(\eta)$. For concreteness we have fixed the numerical value of the bag constant as $b = 0.047$.

For the chosen values of the parameters of the potential and for the initial conditions for the scalar field the maximum masses of the quark stars are $M_{max} = 2.029 M_\odot$, $M_{max} = 2.2303 M_\odot$, $M_{max} = 2.401 M_\odot$, $M_{max} = 2.551 M_\odot$, and $M_{max} = 2.713 M_\odot$, respectively. Inside the star, the scalar field is a monotonically decreasing function of η. However, it is important to point out that for most of the adopted values of the parameters of the Higgs potential, in the case of the quark stars, the scalar field does not vanish on the star's surface. On the other hand, the scalar field potential has negative values inside the star, a situation similar to the previously considered stellar structures, but generally it has a nonzero value on the star's surface. The quark stars are less massive in both hybrid metric-Palatini gravity and standard GR, for the adopted range of the physical parameters, as compared to the stiff and radiation fluid stars. However, again quark hybrid metric-Palatini stars have much bigger masses as compared to their general relativistic counterparts. On the other hand the structure of the global parameters (mass and radius) of the quark stars in both hybrid metric-Palatini gravity and standard GR shows significant deviations when contrasted to the stiff and radiation fluid stars, respectively.

20.3.5 Bose–Einstein Condensate Stars

As a consequence of the superfluid properties of the neutron matter, a significant number of compact astrophysical objects (neutron stars) may have their interior matter content in a condensed form, represented by a Bose–Einstein condensate [131, 203]. A Newtonian, nonrelativistic Bose–Einstein condensate in a gravitational field can be characterized as an ideal gas, whose pressure and density are related by a barotropic equation of state of the form $p = p(\rho)$. The equation of state of the condensate depends generally on two important physical parameters: the mass of the condensate particle m_c and the scattering length a [72]. In the case of a Bose–Einstein condensate with quartic nonlinearity, the equation of state is a polytropic one, with the polytropic index n having the value $n = 1$ [72, 131] so that,

$$p\left(\rho\right) = K\rho^{2}, \tag{20.49}$$

with

$$K = \frac{2\pi\hbar^{2}a}{m_{c}^{3}} = 0.1856 \times 10^{5}\left(\frac{a}{1\ \mathrm{fm}}\right)\left(\frac{m_{c}}{2m_{n}}\right)^{-3}, \tag{20.50}$$

where $m_n = 1.6749 \times 10^{-24}$ g denotes the mass of the neutron. Compact very high-density stellar objects, having superfluid cores, in which particles form Cooper pairs, having masses of the order of two neutron masses, and scattering length of the order of 10–20 fm, respectively, can have maximum masses of the order of $2M_{\odot}$, maximum central densities of the order of $0.1 - 0.3 \times 10^{16}$ g/cm^3, and minimum radii in the range of 10–20 km [131]. The equation of state (20.49) takes, in the dimensionless variables introduced in Eq. (20.25), the dimensionless form

$$P(\theta) = k\theta^{2}, \tag{20.51}$$

where $k = K\rho_c/c^2$.

By numerically integrating the star's structure equations (20.11), (20.17), (20.18), and (20.27)–(20.29), respectively, for the index $n = 1$ polytropic equation of state, we have obtained the global properties of the Bose–Einstein condensate stars in hybrid metric-Palatini gravity. For the parameters μ and ξ of the Higgs potential we have adopted the same values as in the case of our previous studies of the stiff, radiation fluid and quark stars, respectively. We have adopted the value $m_c = m_n$ for the mass of the condensate particle. For each adopted values of the model parameters the numerical integration stops at $\rho = \rho_c/60$. For all cases the central density of the condensate star varies in the range 2.1×10^{13} g/cm^3 and 6.43×10^{15} g/cm^3. In order to compare the global structure of the Bose–Einstein condensate stars in both standard GR and hybrid metric-Palatini gravity we have also computed, and depicted, the corresponding numerical solutions of the standard general relativistic structure equations (20.36) and (20.37).

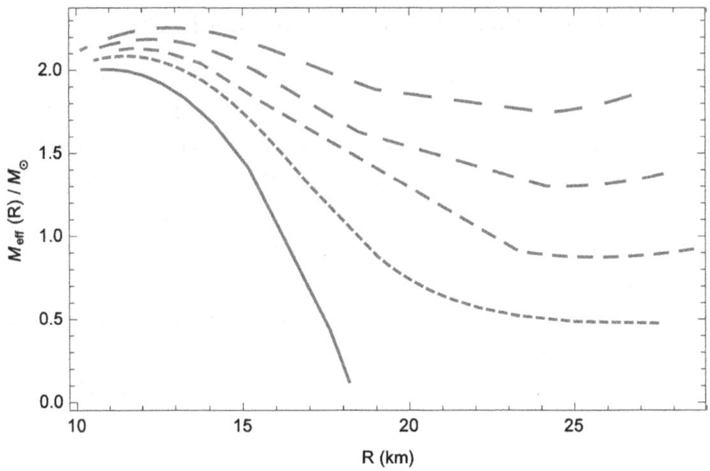

Figure 20.4 Mass–radius relation for Bose–Einstein condensate stars in hybrid metric-Palatini gravity theory, for $\mu = 1.5 \times 10^{-5}$ cm^{-1}, $\xi = 9.5 \times 10^{-10}$ cm^{-2}, $\Phi'(0) = -1.8 \times 10^{-16}$ to -5.7×10^{-16} cm^{-1} for the range of the considered central densities, and for different values of $\Phi(0)$ and $\Phi'(0)$: $\Phi \equiv 0$, $\Phi' \equiv 0$ (standard general relativistic limit) (solid curve), $\Phi(0) = 0.06$, $\Phi'(0) = -0.38 \times 10^{-16}$ cm^{-1} (dotted curve), $\Phi(0) = 0.09$, $\Phi'(0) = -0.83 \times 10^{-16}$ cm^{-1} (short dashed curve), $\Phi(0) = 0.12$, $\Phi'(0) = -1.38 \times 10^{-16}$ cm^{-1} (dashed curve), and $\Phi(0) = 0.15$, $\Phi'(0) = -1.98 \times 10^{-16}$ cm^{-1} (long dashed curve).

In Fig. 20.4 we present the comparison of the mass–radius relations for Bose-Einstein condensate stars in both hybrid metric-Palatini gravity and GR.

To numerically integrate the dimensionless set of the structure equations for the Bose–Einstein condensate stars we have adopted the initial conditions $\theta(0) = 1$, $M_{\text{eff}}(0) = 0$, $\Phi(0) = 0.27$, and where $(d\Phi/d\eta)|_{\eta=0}$ is a function depending on $\Phi(0)$, the potential parameters, and the central pressure. The value of the coefficient k in the polytropic equation of state has been fixed to the value $k = 0.10$,

For the considered range of parameters the maximum obtained stable stellar masses are $M_{max} = 2.009 M_{\odot}$, $M_{max} = 2.078 M_{\odot}$, $M_{max} = 2.119 M_{\odot}$, $M_{max} = 2.171 M_{\odot}$, and $M_{max} = 2.239 M_{\odot}$, respectively.

In the case of Bose–Einstein condensate stars the matter pressure P (or, equivalently, its energy density θ) vanishes on the star's surface, which gives the boundary condition $\theta(\eta_S) = 0$, for the computation of the dimensionless radius of the star η_S. Inside the star the scalar field has a complex behavior, which is strongly dependent on the physical parameters of the Higgs-type potential. Φ is an increasing function inside the star for the first set of numerical parameters, while for the next set of parameter values it is a decreasing function of η. Inside the Bose–Einstein condensate stars, and for the adopted values of the parameters of the Higgs potential, the scalar field does not vanish on the star's surface, a situation that is similar to the case of the quark stars. A complex mathematical pattern corresponds to the scalar field potential evolution, which takes negative values inside the star, and has an increasing/decreasing behavior, which is fully determined by the numerical values of the parameters μ and ξ. The scalar field potential does not vanish on the star's surface for the considered range of parameter values.

The Bose–Einstein condensate stars are, for the considered range of the physical parameters of the Higgs-type potential, less massive in both hybrid metric-Palatini gravity and standard GR, as compared to the stiff, radiation fluid and quark stars, respectively. However, Bose–Einstein condensate hybrid metric-Palatini stars are more massive than their general relativistic analogues, a situation that is similar to all previously considered cases. However, the difference between the stellar masses are significantly reduced. Therefore, in both hybrid metric-Palatini gravity and standard GR the global structure of the Bose–Einstein condensate stars demonstrate again the existence of important differences with respect to the properties of the stiff and radiation fluid stars, as well as with the quark stars.

20.4 Stellar Models with Fixed Functional Forms of the Scalar Field

In the following we investigate stellar models in hybrid metric-Palatini gravity in which the form of the scalar field is fixed initially. As a first example of such an astrophysical model, we consider the case in which the scalar field Φ is determined by the second-order differential equation

$$f\left(\Phi(\eta)\right) = \frac{d^2\Phi}{d\eta^2} + \frac{1}{4}\left(1 + \frac{3}{1-e^{\Phi}}\right)\left(\frac{d\Phi}{d\eta}\right)^2 = 0. \tag{20.52}$$

There are two solutions to Eq. (20.52), given by

$$\Phi = \Phi_0 = \text{constant}, \tag{20.53}$$

and

$$\Phi = \ln\left[1 + (\alpha\eta + \beta)^4\right], \tag{20.54}$$

respectively, where

$$\alpha = \frac{1}{4}\frac{e^{\Phi_0}}{\left(e^{\Phi_0} - 1\right)^{3/4}}\Phi_0', \qquad \beta = \left(e^{\Phi_0} - 1\right)^{1/4}, \tag{20.55}$$

and $\Phi_0 = \Phi\left(0\right)$, $\Phi_0' = \left(d\Phi/d\eta\right)|_{\eta=0}$. In the following discussion, we will adopt these two functional forms for Φ, and we will investigate the geometrical, physical and astrophysical properties of the corresponding stellar models.

20.4.1 Effective Quark Star Models – The Case Φ = Constant

The generalized dimensionless Klein–Gordon equation Eq. (20.29) gives, in the simple case Φ = constant, the following linear relation between the thermodynamic pressure and matter density

$$P = \frac{1}{3}\left(\theta - 4B_{\Phi}\right), \tag{20.56}$$

where

$$-4B_{\Phi} = e^{\Phi_0}\left[u\left(\Phi\right) - \frac{du\left(\Phi\right)}{d\Phi}\right]\Big|_{\Phi=\Phi_0}. \tag{20.57}$$

The equation of state (20.56) has the interesting property of having the same form as the MIT bag model equation of state, $p = \left(\rho c^2 - 4B\right)/3$, where B represents the bag constant, and which describes quark matter confined inside a bag [510]. Hence, in the hybrid metric-Palatini gravitational theory with constant scalar field, B_{Φ} plays the role, and can be interpreted, as an effective bag constant induced by the scalar coupling. Hence, for the case of the constant scalar field the mass continuity and the hydrostatic equilibrium equations describing the structure of massive astrophysical objects become

$$\frac{dM_{\rm eff}}{d\eta} = \frac{\eta^2}{2}e^{-\Phi_0}\left(\theta + B_0\right), \tag{20.58}$$

$$\frac{d\theta}{d\eta} = -\frac{2e^{-\Phi_0}\left(\theta - B\right)\left(\theta - \alpha\right)\eta^2 + 3M_{\rm eff}/\eta}{3\eta\left(1 - 2M_{\rm eff}/\eta\right)}, \tag{20.59}$$

where $B_0 = e^{\Phi_0}u\left(\Phi_0\right)/2$ and $\alpha = 4B_{\Phi} + B_0$. In the following we will adopt again for the scalar field potential the Higgs form. Moreover, we assume that for a stable stellar configuration, the scalar field is in the minimum of the potential, which implies that $dV(\phi)/d\phi|_{\phi=\phi_0} = 0$, giving $\phi_0 = \pm\mu/\sqrt{\xi}$ and $V\left(\phi_0\right) = -\mu^4/4\xi$.

Hence $e^{\Phi_0} = 1 + \phi_0 = 1 \pm \mu/\sqrt{\xi}$, and

$$\begin{aligned} U\left(\Phi\right)|_{\phi=\phi_0} &= e^{-\Phi}V\left(e^{\Phi} - 1\right)|_{\phi=\phi_0} = \frac{V}{\left(1+\phi_0\right)} \\ &= -\left(\frac{\mu^4}{4\xi}\right)\left(1 \pm \frac{\mu}{\sqrt{\xi}}\right), \end{aligned} \tag{20.60}$$

respectively, and thus we find

$$B_{\Phi} = a^2\left(\mu^4/4\xi\right) = c^2\mu^4/32\pi G_0\rho_c\xi, \qquad B_0 = -a^2\left(\mu^4/2\xi\right) = -2B_{\Phi}, \tag{20.61}$$

and $\theta_0 = 10B_{\Phi}$.

In Fig. 20.5 the mass–radius relations of the hybrid metric-Palatini stars with a constant scalar field are depicted, for different values of the effective bag constant B_{Φ}. In each analyzed case the numerical integration stops at a surface density $\rho = 4\times10^{14}$ g/cm^3, and thus the general relativistic quark star model, described by the standard MIT bag model, can be superimposed on the known curve.

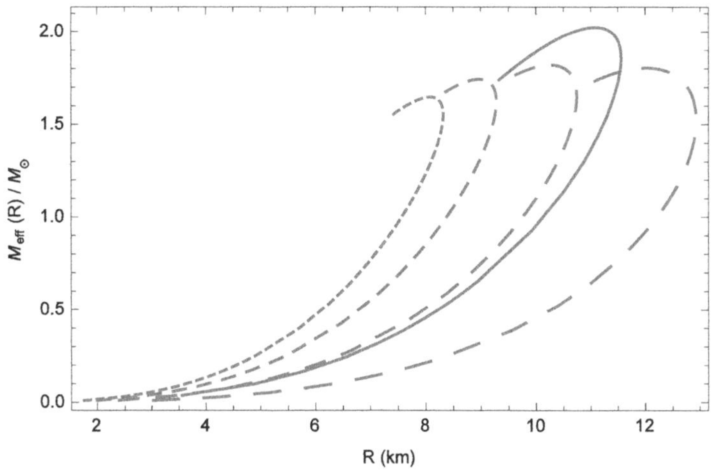

Figure 20.5 Mass–radius relation for hybrid metric-Palatini gravity stars for a constant scalar field, for $\mu = 1.5 \times 10^{-5}$ cm^{-1}, $\xi = 9.5 \times 10^{-10}$ cm^{-2}, and for different values of Φ: $\Phi \equiv 0$ (standard general relativistic quark star model) (solid curve), $\Phi = 0.39$ (dotted curve), $\Phi = 0.44$ (short dashed curve), $\Phi = 0.47$ (dashed curve), and $\Phi = 0.49$ (long dashed curve).

The central density of the compact object was modified in the range 4.1×10^{14} g/cm^3 and 8.5×10^{15} g/cm^3. The maximum masses of the hybrid metric-Palatini stars with constant scalar field obtained for the adopted set of parameters are $M_{max} = 2.025 M_{\odot}$, $M_{\max} = 1.608 M_{\odot}$, $M_{max} = 1.663 M_{\odot}$, $M_{max} = 1.770 M_{\odot}$, and $M_{max} = 1.632 M_{\odot}$, respectively.

We have also compared the effective hybrid metric-Palatini quark star model with the general relativistic standard quark star model, described by the MIT bag model equation of state. The maximum mass of the "analogue" quark star of the hybrid metric-Palatini gravity is much smaller than the mass of the ordinary general relativistic quark stars. It turns out that the constant scalar field in the minimum of the Higgs-type potential does not give a significant contribution to the gravitational properties of the compact object. This inference is, of course, dependent on the numerical values adopted for the model parameters. Thus, for example, by modifying the numerical values of the constant scalar field, "analogue" quark star models in hybrid metric-Palatini gravity with different global properties can be easily obtained.

20.4.2 The Case $\Phi = \ln\left[1 + (\alpha\eta + \beta)^4\right]$

As a second example of stellar models with a fixed functional form we consider the case in which for the scalar field Φ we adopt the functional form given by Eq. (20.54). Then it follows that the structure equations describing the interior of hybrid metric-Palatini stars with scalar field given by $\Phi = \ln\left[1 + (\alpha\eta + \beta)^4\right]$ take the form

$$\frac{dM_{\rm eff}}{d\eta}=-\frac{24\alpha M_{\rm eff}(\alpha\eta+\beta)^3-\eta\left[16\alpha(\alpha\eta+\beta)^3+2\eta\theta+\eta u\left((\alpha\eta+\beta)^4+1\right)\right]}{4\left[(3\alpha\eta+\beta)(\alpha\eta+\beta)^3+1\right]}, \tag{20.62}$$

$$\frac{d\theta}{d\eta}=-\frac{\theta}{3\eta\left(\eta-2M_{\rm eff}\right)\left[(3\alpha\eta+\beta)\times(\alpha\eta+\beta)^3+1\right]}\times$$
$$\times\Big\{12M_{\rm eff}\Big[(3\alpha\eta+\beta)(7\alpha\eta+\beta)(\alpha\eta+\beta)^2+1\Big]$$
$$+\eta^2\left[-24\alpha(5\alpha\eta+2\beta)(\alpha\eta+\beta)^2+2\eta\theta-3\eta u\left((\alpha\eta+\beta)^4+1\right)\right]\Big\}, \tag{20.63}$$

$$\frac{du}{d\eta}=\frac{4\left(\eta-2M_{\rm eff}\right)(\alpha\eta+\beta)}{\eta\left[(\alpha\eta+\beta)^4+1\right]^2}\Bigg\{-12\alpha^2+\frac{\eta u(\alpha\eta+\beta)^2\left[(\alpha\eta+\beta)^4+1\right]}{\eta-2M_{\rm eff}}$$
$$+\frac{3\alpha(\alpha\eta+\beta)}{\eta\theta\left(\eta-2M_{\rm eff}\right)\left[(3\alpha\eta+\beta)(\alpha\eta+\beta)^3+1\right]}\times$$
$$\times\Bigg\{2\eta^3\theta^2+\eta\left[\eta-2M_{\rm eff}\right]\left[(3\alpha\eta+\beta)(\alpha\eta+\beta)^3+1\right]\frac{d\theta}{d\eta}$$
$$+\theta\left[(\alpha\eta+\beta)^4+1\right]\left(-8\eta+12M_{\rm eff}+\eta^3u\right)\Bigg\}\Bigg\}. \tag{20.64}$$

In the above equations we have adopted the radiation fluid equation of state $P=\theta/3$ to describe the dense matter inside the stellar object. To integrate the system of nonlinear ordinary differential equations (20.62)–(20.64) we adopt the boundary conditions $M_{\rm eff}(0)=0$, $\theta(0)=1$, $u(0)=u_0$, and $\theta(\eta_S)=0$, respectively. In the ordinary dimensional physical coordinates the scalar field is represented in the form

$$\Phi(r)=\ln\left[1+(Ar+C)^4\right]. \tag{20.65}$$

In Fig. 20.6 we present the mass–radius relation for hybrid metric-Palatini stars with the interior scalar field defined by Eq. (20.65). During the numerical integration of the structure equations we have fixed the value $A=100$ cm^{-1}, while the constant C in the scalar field was obtained from the initial condition $\Phi(0)=\Phi_0$. The dimensionless central density was modified in the range 0.049 to 0.729. The numerical values of the maximum masses of the stellar objects in hybrid metric-Palatini gravity, with scalar field given by Eq. (20.65), obtained for this set of model parameters are $M_{max}=2.027M_\odot$ (standard general relativistic model), $M_{max}=2.013M_\odot$, $M_{max}=2.206M_\odot$, $M_{max}=2.419M_\odot$, $M_{max}=2.655M_\odot$, and $M_{max}=2.916M_\odot$, respectively.

As one can see from the plots in Fig. 20.6, the hybrid metric-Palatini stars with the scalar field represented as $\Phi=\ln\left[1+(\alpha\eta+\beta)^4\right]$ are more massive than their corresponding general relativistic stellar objects. The variation of the scalar field parameters has an important effect on the effective mass $M_{\rm eff}$ of the star. An important change in the mass of the star occurs as a result of a

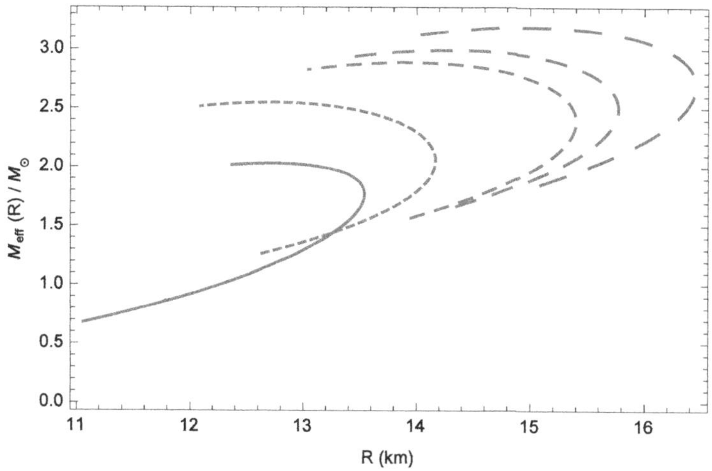

Figure 20.6 Mass–radius relation for hybrid metric-Palatini gravity stars for the scalar field $\Phi(r) = \ln\left[1 + (Ar + B)^4\right]$, for $A = 110$ cm $^{-1}$, $U(0) = 2\times10^{-3}$ cm^{-2}, and for different values of $\Phi(0)$: $\Phi(0) = 2.05$ (dotted curve), $\Phi(0) = 2.15$ (short dashed curve), $\Phi(0) = 2.29$ (dashed curve), and $\Phi = 2.33$ (long dashed curve). The solid curve represents the standard general relativistic radiation fluid star model.

small variation in the numerical values of α and β. Since the masses of stars in this model are of the order of two solar masses, it follows that these masses, as well as the corresponding radii, are also much bigger than those of the general relativistic fluid stars. The dynamics of the scalar field potential u inside the star also shows a strong dependence on the numerical values of the parameters α and β. When α takes small values, the potential is practically constant inside the star, having only positive numerical values. The scalar field potential becomes a monotonically decreasing function of the radial coordinate, once α begins to increase, and it also changes sign inside the star. The potential also shows a similar behavior when one investigates its functional dependence on Φ, and it becomes a decreasing function of the radial coordinate, taking negative values outside a small region around the stellar core.

20.5 Conclusions

The hybrid metric-Palatini gravitational theory, which attempts to explain gravitational phenomena on both local and large scales through a single unified formalism, combines in an interesting way the theoretical elements of the metric and Palatini $f(R)$ theories. In the present chapter, following the presentation in [154], we have performed a study of the global physical properties of dense compact objects in this approach to gravity. To study the interior structure of the high-density stellar-type objects we have used its scalar–tensor-type formulation, which represents important characteristics of hybrid metric-Palatini gravity. However, it is important to point out that due to the coupling between the scalar field and the geometry in the gravitational action, hybrid metric-Palatini gravity differs fundamentally from the Brans–Dicke-type theories. In static spherically symmetric geometries the coupling between matter and geometry generates a rather complicated set of interior field equations, represented by a set of highly nonlinear second-order differential equations, whose solutions can be obtained numerically only by the extensive use of numerical integration algorithms. The basic equations describing the structure of compact objects in hybrid metric-Palatini gravity have been derived, as a first step in our study. Thus, we have obtained and formulated in a concise mathematical form the mass continuity equation, the generalized hydrostatic equilibrium equation (the TOV equation), and the generalized Klein–Gordon equation, which in the present case describes the coupling of the geometric curvature and matter with the scalar field. The self-interaction potential V of the equivalent scalar field is an important physical parameter that essentially determines the basic physical properties of the stars. We have assumed, in the present study, that the potential is of the Higgs type, a choice that is suggested and supported by the important role such potentials play in the standard model of the elementary particles, and in the physical description of the strong interactions. One can of course also adopt other functional and mathematical forms of the scalar field potential (exponential, hyperbolic, power-law, etc.), and these alternative choices will generate compact object models having different global and astrophysical properties as compared to those outlined in the present chapter.

To close the system of nonlinear structure differential equations of the star, once the scalar field potential is specified, we must also specify the equation of state of the dense matter, or, alternatively, the functional form of the scalar field. In the framework of the first approach, after adopting the equation of state of the dense matter, we have investigated the global astrophysical properties of the star by numerically integrating its structure equations. We have considered four classes of stellar models, described by the stiff fluid, radiation fluid, strange quark matter, and Bose–Einstein condensate superfluid matter equations of state, respectively. We have effectively built up hybrid metric-Palatini stellar models in all of these four cases, and we have compared these modified-gravity models with their general relativistic counterparts. As a general result of our analysis we have found that the hybrid metric-Palatini stars are much more massive than their corresponding standard general relativistic stars, for all these four equations of state of the dense astrophysical matter. For example, in the case of the stiff fluid equation of state, hybrid metric-Palatini stars have a mass about five times bigger than masses of the general relativistic stars with the same central density. Strange quark stars described by the bag model equation of state have masses around two times bigger than the corresponding general relativistic stars, while superfluid Bose–Einstein condensate stars are about 1.4x

Table 20.1. *Comparison between the maximum masses of stars in GR (mass denoted as* $M_{max}^{GR}/M_{\odot}$*) and in hybrid metric-Palatini (mass denoted as* $M_{max}^{HMP}/M_{\odot}$*) obtained for the four equations of state considered in the present study. To numerically integrate the hybrid metric-Palatini structure equations we have adopted for the parameters of the Higgs-type potential the numerical values* $\mu = 1.5 \times 10^{-5}$ cm^{-1} *and* $\xi = 9.5 \times 10^{-10}$ cm^{-2}*, respectively, while the central value of the scalar field has been taken as* $\Phi(0) = 0.32$ cm^{-1}*. For each case we have computed the central value of* $(d\Phi/dr)|_{r=0}$*, which is a function of* $\Phi(0)$*, of the potential parameters, and of the central densities and pressures, by using Eq. (20.31). For both the MIT bag model describing quark stars, and for the BEC equations of state of superfluid stars, the maximum mass occurs at the point of minimum central density.*

Equation of State	$\Phi(0)$	$M_{max}^{GR}/M_{\odot}$	$M_{max}^{HMP}/M_{\odot}$
MIT Bag Model	0.32	2.031	4.363
Stiff fluid	0.32	3.283	3.974
Radiation fluid	0.32	2.261	3.670
BEC	0.32	2.237	4.980

more heavy. Of course, there is a strong dependence of the mass of the star on its central density, with high central density stars having significantly smaller gravitational masses. On the other hand, the wide mass range of the stable hybrid metric-Palatini stars suggests the possibility that stellar mass black holes, having gravitational masses in the range of $3.8M_{\odot}$ and $6M_{\odot}$, respectively, may be in fact massive hybrid metric-Palatini stars (for the case of the quarks stars in the color-flavor locked phase, such a possibility was investigated in the framework of standard GR in [296]). Table 20.1 presents a comparison of the numerical values of the maximum masses of stellar objects in the hybrid metric-Palatini gravity and in standard general relativistic.

In the second possible approach to the study of the stellar objects in the hybrid metric-Palatini gravity we have investigated two types of solutions of the gravitational field equations. In the first case, we have assumed that the scalar field is in the minimum of the Higgs potential, and takes a constant value, which depends on the parameters of the potential. Interestingly enough, via the Klein–Gordon equation, the constant form of the scalar field allows to fully determine the equation of state of the matter inside the star, which has a form similar to the MIT bag model equation of state, which was intensively applied for the description of the properties of the hypothetical strange quark stars. The presence of the bag constant B in the equation of state can be interpreted, from a simple physical point of view, as determining the quarks to remain confined into a small spherical region of space (the bag), having a radius $r = a$. In this case the confining potential is zero inside the bag, $V(r) = 0$ for $r < a$, while the vacuum pressure B on the walls of the bag balances the pressure of the quarks, thus making the hadrons stable. For the formation of the quark stars several possible astrophysical mechanisms have been proposed. One intensively discussed scenario is that they could come into existence after a supernova explosion, during the gravitational collapse of the massive core of the exploding star [203]. The supernova explosion may lead to the formation of deconfined quark matter by inducing a first- or second-order phase transition. The core of a proto-neutron or of a massive neutron star may also represent a favorable medium for the conversion of neutron matter to quark matter, as it has been already pointed out in [132]. In low-mass X-ray binaries, neutron stars can also become quark stars by accreting enough cosmic matter that could lead to a mass increase and trigger a phase transition [132]. Therefore, from a physical point of view, one cannot rule out the possibility that in hybrid metric-Palatini gravity a phase transition can take place under extreme astrophysical and gravitational conditions (supernova explosions, gamma ray bursts, accretion, etc.), with such a phase transition triggered by the presence of the scalar field with a Higgs-type self-interaction potential. After the possible occurrence of such a phase transition, the massive star ends in a minimum of the Higgs potential, becoming either a "true," or an "analogue" quark star.

Recently, high-precision astronomical observations have found many stellar mass black hole candidates, with at least seven of them having masses bigger than 5 solar masses, $M \geq 5M_{\odot}$. Up to now at least 20 stellar mass black holes have been observed, and their masses lie in the range from 3.8 and 6 solar masses. On the other hand, statistical astronomical and astrophysical estimations provide for the total number of stellar mass black holes (in both binary systems and isolated) a value of the order of hundreds of millions for our galaxy only (see [296] and references therein). Therefore we cannot rule out automatically the possibility that stellar mass black holes could be ordinary baryonic stars, dominated by modified-gravity effects. As we have already seen, hybrid metric-Palatini stars may have significantly larger masses than standard general relativistic neutron stars, and thus they could be possible theoretical candidates for stellar mass black holes. A possibility of observationally distinguishing ordinary hybrid metric-Palatini stars from standard general relativistic stellar mass black holes could be based on the study of the electromagnetic astrophysical signatures of the thin accretion disks formed around rapidly rotating hybrid metric-Palatini stars, and Kerr-type black holes, respectively. We expect that the radiation properties of the thin accretion disks around modified-gravity stars and general relativistic black holes could be very different [227]. Therefore the electromagnetic emission properties of the stars themselves, and of the accretion disk around them, may offer essential astrophysical information that may provide the key signatures necessary to differentiate ordinary black holes from modified-gravity stars.

The existence of neutron stars with masses of the order of $2M_{\odot}$ has also been confirmed by high precision astronomical observations of the neutron star mass distribution [19, 160, 255]. The Black Widow pulsar B1957+20, an eclipsing binary millisecond pulsar, having a mass estimated observationally to be in the range $1.6-2.4M_{\odot}$, is one important example of such a star [487]. It is important to point out that in the framework of standard GR a range of 2–2.4 solar masses is very difficult to explain by using the ordinary and presently accepted stellar models, including standard neutron star models, as well as exotic models that assume the existence of quark or kaon stars. On the other hand it is important to point out that we can easily explain these observed stellar mass values once we model massive gravitating objects as hybrid metric-Palatini stars. As we have seen in the present chapter, hybrid metric-Palatini stars indeed possesses a very complicated internal stellar dynamics, which is closely associated with an equally complex interior structure. The presence of the coupling between the scalar field, geometry and matter gives the main contribution, and explains these important deviations from standard GR. The geometry-matter coupling effects can lead to a number of distinctive astrophysical signatures for high-density cosmic objects. Even if such effects are present, their observational detection can prove to be an extremely challenging task.

21

Hybrid Gravity Traversable Wormholes

Traversable wormholes, essentially shortcuts is space-time, are solutions to the Einstein field equation that violate the classical energy conditions. They are primarily useful as "gedanken-experiments" that probe the foundations of General Relativity. Wormhole solutions are obtained by solving the Einstein field equation in the reverse direction, where an interesting and exotic space-time metric is considered, then the matter source responsible for the respective geometry is found. An interesting implication is that they allow "effective" superluminal travel, although the speed of light is not surpassed *locally*, and generate closed timelike curves, with the associated causality violations. In this chapter, we consider wormhole solutions in the hybrid metric-Palatini gravitational theory.

21.1 Field Equations and Energy Conditions

To analyze traversable wormholes, we assume for simplicity a static and spherically symmetric geometry given by the following line element, in curvature coordinates [362]

$$ds^2 = -e^{2\Phi(r)}dt^2 + \frac{dr^2}{1 - b(r)/r} + r^2(d\theta^2 + \sin^2\theta d\phi^2)\,. \tag{21.1}$$

The coordinate r is non-monotonic in that it decreases from $+\infty$ to a minimum value r_0, representing the location of the wormhole throat, and then increases from r_0 to $+\infty$. The proper circumference of a circle of fixed r is given by $2\pi r$. The two metric functions $\Phi(r)$ and $b(r)$ are functions of the radial coordinate only. The former is denoted the redshift function and must be finite everywhere in order to avoid the presence of event horizons. The function $b(r)$ is denoted the shape function, as it determines the shape of the wormhole geometry and must obey the flaring-out condition of the throat r_0, which is translated by $(b - b'r)/b^2 > 0$ [362]. At the throat, we have $b(r_0) = r_0$, and the condition $b'(r_0) < 1$ is imposed. We will see below that the flaring-out condition, through the Einstein field equation, entails the violation of the classical energy conditions. It is possible to construct asymptotically flat space-times, in which $b(r)/r \to 0$ and $\Phi \to 0$ as $r \to \infty$.

Using the Einstein field equation, we verify the following stress-energy scenario

$$\rho(r) = \frac{1}{8\pi}\frac{b'}{r^2}\,, \tag{21.2}$$

$$p_r(r) = \frac{1}{8\pi}\left[-\frac{b}{r^3} + 2\left(1 - \frac{b}{r}\right)\frac{\Phi'}{r}\right], \tag{21.3}$$

$$p_t(r) = \frac{1}{8\pi}\left(1 - \frac{b}{r}\right)\left[\Phi'' + (\Phi')^2 - \frac{b'r - b}{2r^2(1-b/r)}\Phi' - \frac{b'r - b}{2r^3(1-b/r)} + \frac{\Phi'}{r}\right]. \tag{21.4}$$

where $\rho(r)$ is the energy density, $p_r(r)$ is the radial pressure, and $p_t(r)$ is the pressure measured in the tangential directions, orthogonal to the radial direction. It is not essential that the wormhole be sustained by a negative energy density, as often mentioned in the literature, for if $b'(r) > 0$ this implies that $\rho(r) > 0$. However, note that evaluated at the throat the radial pressure is necessarily negative, i.e., $p_r(r_0) = -1/8\pi r_0^2$.

Adding Eqs. (21.2) and (21.3), one gets

$$\rho(r) + p_r(r) = \frac{1}{8\pi}\left[\frac{b'r - b}{r^3} + 2\left(1 - \frac{b}{r}\right)\frac{\Phi'}{r}\right], \tag{21.5}$$

where taking into account the finite character of $\Phi(r)$, and evaluated at the throat's vicinity, one immediately arrives at the following relation:

$$\rho(r_0) + p_r(r_0) = \frac{1}{8\pi}\frac{b'(r_0) - 1}{r_0^2}\,. \tag{21.6}$$

Note that taking into account the flaring-out condition, Eq. (21.6) implies that $\rho(r_0) + p_r(r_0) < 0$. This condition violates the null energy condition (NEC), which asserts that for any null vector k^μ, we have $T_{\mu\nu}k^\mu k^\nu \geq 0$ (recall that for a diagonal energy-momentum tensor, $T^\mu{}_\nu = \mathrm{diag}[-\rho(r), p_r(r), p_t(r), p_t(r)]$; this implies $\rho + p_r \geq 0$ and $\rho + p_t \geq 0$). In fact, wormhole solutions violate all of the pointwise classical energy conditions. Matter that violates the NEC is denoted as *exotic matter*.

Now, it is interesting to note that the NEC arises when one refers back to the Raychaudhuri equation, which is given by

$$\frac{d\theta}{d\tau} = -\frac{1}{2}\theta^2 - \sigma_{\mu\nu}\sigma^{\mu\nu} + \omega_{\mu\nu}\omega^{\mu\nu} - R_{\mu\nu}k^\mu k^\nu\,, \tag{21.7}$$

where θ, $\sigma^{\mu\nu}$, and $\omega_{\mu\nu}$ are, respectively, the expansion, shear, and rotation associated to the congruence defined by the null vector field k^μ. $R_{\mu\nu}$ is the Ricci tensor. In deducing the Raychaudhuri equation, no reference to any gravitational field equations was made, and as such it is a purely geometric statement. For any hypersurface orthogonal congruences, we have $\omega_{\mu\nu} \equiv 0$ and the shear is a "spatial" tensor with $\sigma^2 \equiv \sigma_{\mu\nu}\sigma^{\mu\nu} \geq 0$. Thus, the condition for attractive gravity reduces to $R_{\mu\nu}k^\mu k^\nu \geq 0$, which ensures that geodesic congruences focus within a finite value of the parameter labeling points on the geodesics. In Einstein's GR, contracting both sides of the Einstein equation with any null vector k^μ, one can write the above condition in terms of the energy-momentum tensor, given by $T_{\mu\nu}k^\mu k^\nu \geq 0$.

For completeness, we consider the weak energy condition (WEC) and the strong energy condition (SEC). More specifically, the WEC states that for any timelike vector U^μ, we have $T_{\mu\nu}U^\mu U^\nu \geq 0$. Note that the quantity $T_{\mu\nu}U^\mu U^\nu$ can be physically interpreted as the energy density measured by any timelike observer with four-velocity U^μ. Thus, the WEC requires that this quantity to be positive. In terms of a diagonal energy-momentum tensor, this provides $\rho \geq 0$ and $\rho + p_r \geq 0$ (and $\rho + p_t \geq 0$). By continuity, the WEC implies the NEC.

The SEC asserts that the following inequality $\left(T_{\mu\nu} - \frac{T}{2}\, g_{\mu\nu}\right) U^\mu U^\nu \geq 0$ holds, where T is the trace of the stress energy tensor. In terms of the diagonal stress energy tensor the SEC reads, $\rho + p_i \geq 0$ and $\rho + \sum_i p_i \geq 0$ ($\forall\, i$ where $i = r, t$). The SEC implies the NEC but not necessarily the WEC.

As mentioned above, in modified theories of gravity, the gravitational field equations can be rewritten in terms of an effective Einstein equation, given by $G_{\mu\nu} = \kappa^2 T^{\rm eff}_{\mu\nu}$, where $T^{\rm eff}_{\mu\nu}$ is an effective energy-momentum tensor containing the matter energy-momentum tensor, $T_{\mu\nu}$, and the geometric quantities, arising from the specific modified theory of gravity considered. In modified gravity, the positivity condition $R_{\mu\nu}k^\mu k^\nu \geq 0$ in the Raychaudhuri equation provides a generalization of the NEC given by $T^{\rm eff}_{\mu\nu}k^\mu k^\nu \geq 0$, through the modified gravitational field equation. Here the necessary condition to have a wormhole geometry is the violation of the generalized NEC, i.e., $T^{\rm eff}_{\mu\nu}k^\mu k^\nu < 0$. It is interesting to note that now one may, in principle, impose that the matter energy-momentum tensor satisfies the standard NEC, $T_{\mu\nu}k^\mu k^\nu \geq 0$, while the respective generalized NEC is necessarily violated, $T^{\rm eff}_{\mu\nu}k^\mu k^\nu < 0$, in order to ensure the flaring-out condition.

Thus, in modified theories gravity, one may impose that the matter threading the wormhole throat satisfies all of the energy conditions, and it is the higher-order curvature terms, which may be interpreted as a gravitational fluid, that support these nonstandard wormhole geometries.

21.2 Wormholes in Generalized Modified Gravity

It is useful to deduce the generic conditions that should be imposed in finding wormhole solutions in modified theories of gravity [239]. For this purpose, consider the following generalized gravitational field equations for a large class of modified theories of gravity

$$g_1(\Psi^i)(G_{\mu\nu} + H_{\mu\nu}) - g_2(\Psi^j)\, T_{\mu\nu} = \kappa^2\, T_{\mu\nu}\,, \tag{21.8}$$

where the functions $g_i(\Psi^j)$ $(i = 1, 2)$ are multiplicative factors that modify the geometrical sector of the field equations, in which Ψ^j denote generically curvature invariants or gravitational fields such as scalar fields; the term $g_2(\Psi^i)$ is due to the coupling of the curvature invariants or the scalar fields with the matter energy-momentum tensor, $T_{\mu\nu}$. In addition to this, the quantity $H_{\mu\nu}$ is an additional geometric term that includes the geometrical modifications of the modified gravitational theory under consideration.

As before, one may rewrite this Eq. (21.8) as an effective field equation, $G_{\mu\nu} = \kappa^2 T^{\text{eff}}_{\mu\nu}$, where $T^{\text{eff}}_{\mu\nu}$ is given by

$$T^{\text{eff}}_{\mu\nu} \equiv \frac{1 + \bar{g}_2(\Psi^j)}{g_1(\Psi^i)} T_{\mu\nu} - \bar{H}_{\mu\nu} \,, \tag{21.9}$$

in which $\bar{g}_2(\Psi^j) = g_2(\Psi^j)/\kappa^2$ and $\bar{H}_{\mu\nu} = H_{\mu\nu}/\kappa^2$ are defined for notational convenience. For the modified theories of gravity, the violation of the generalized NEC, $T^{\text{eff}}_{\mu\nu} k^\mu k^\nu < 0$, implies the following restriction

$$\frac{1 + \bar{g}_2(\Psi^j)}{g_1(\Psi^i)} T_{\mu\nu} k^\mu k^\nu < \bar{H}_{\mu\nu} k^\mu k^\nu \,. \tag{21.10}$$

Note that the standard violation of the NEC for matter threading the wormhole in GR is recovered, i.e., $T_{\mu\nu} k^\mu k^\nu < 0$, for $g_1(\Psi^j) = 1$, $g_2(\Psi^j) = 0$, and $H_{\mu\nu} = 0$.

Imposing the additional condition $[1 + \bar{g}_2(\Psi^j)]/g_1(\Psi^i) > 0$, then one obtains a general bound for the normal matter threading the wormhole, given by

$$0 \leq T_{\mu\nu} k^\mu k^\nu < \frac{g_1(\Psi^i)}{1 + \bar{g}_2(\Psi^j)} \bar{H}_{\mu\nu} k^\mu k^\nu \,. \tag{21.11}$$

In modified gravity, wormhole solutions also violate the generalized SEC, i.e., $T^{\text{eff}}_{\mu\nu} U^\mu U^\nu < -\frac{1}{2} T^{\text{eff}}$, which implies the following bound on the matter stress-energy tensor

$$\frac{1 + \bar{g}_2(\Psi^j)}{g_1(\Psi^i)} \left[T_{\mu\nu} U^\mu U^\nu - \frac{1}{2} T \right] < \bar{H}_{\mu\nu} U^\mu U^\nu - \frac{1}{2} \bar{H}. \tag{21.12}$$

One may impose that the latter condition is fulfilled even if the matter stress-energy tensor satisfies the usual SEC, $T_{\mu\nu} U^\mu U^\nu - \frac{1}{2} T \geq 0$, or the WEC, $T_{\mu\nu} U^\mu U^\nu \geq 0$.

In order for normal matter to satisfy the WEC, in particular, to have a positive energy density, one also needs to demand the following inequality

$$T_{\mu\nu} U^\mu U^\nu = \frac{g_1(\Psi^i)}{\kappa^2 + g_2(\Psi^j)} \left(G_{\mu\nu} + H_{\mu\nu} \right) U^\mu U^\nu \geq 0 \,. \tag{21.13}$$

By demanding that $T_{\mu\nu} U^\mu U^\nu \geq 0$ entails a restriction on the geometry arising from the modified gravity under consideration. As mentioned above, if the normal matter is given by a diagonal stress-energy tensor, one can physically interpret $T_{\mu\nu} U^\mu U^\nu$ as the energy density measured by any timelike observer with four-velocity U^μ. This definition is extremely useful, as, using local Lorentz transformations, it is possible to show that $T_{\mu\nu}u^\mu u^\nu \geq 0$ implies that the energy density is positive in all local frames of reference. Thus, the standard WEC imposes that $\rho \geq 0$ and $\rho + p_i \geq 0$ (where $i = r, t$).

21.3 Wormholes in Hybrid Metric-Palatini Gravity

Consider the specific case of hybrid metric-Palatini gravity, in the scalar–tensor representation, where the field equation (17.24) can also be written as an effective Einstein field equation, i.e., $G_{\mu\nu} = \kappa^2 T^{\rm eff}_{\mu\nu}$, where the effective energy-momentum tensor is given by Eq. (17.25), which we rewrite here for self-completeness

$$T^{\rm eff}_{\mu\nu} = \frac{1}{1+\phi}\left\{T_{\mu\nu} - \frac{1}{\kappa^2}\left[\frac{1}{2}g_{\mu\nu}\left(V + 2\Box\phi\right) + \nabla_\mu\nabla_\nu\phi \right.\right.$$
$$\left.\left. - \frac{3}{2\phi}\partial_\mu\phi\,\partial_\nu\phi + \frac{3}{4\phi}g_{\mu\nu}(\partial\phi)^2\right]\right\}. \tag{21.14}$$

As outlined in the previous section, when studying wormhole geometries in modified theories of gravity, it is the effective stress-energy tensor that violates the NEC at the throat, namely, $T^{\rm eff}_{\mu\nu}k^\mu k^\nu|_{r_0} < 0$. This inequality provides the following constraint in hybrid metric-Palatini gravity [107]

$$T^{\rm eff}_{\mu\nu}k^\mu k^\nu|_{r_0} = \frac{1}{1+\phi}\left\{T_{\mu\nu}k^\mu k^\nu - \frac{1}{\kappa^2}\left[k^\mu k^\nu\nabla_\mu\nabla_\nu\phi \right.\right.$$
$$\left.\left. - \frac{3}{2\phi}k^\mu k^\nu\,\partial_\mu\phi\,\partial_\nu\phi\right]\right\}\Big|_{r_0} < 0. \tag{21.15}$$

Furthermore, assuming that $1 + \phi > 0$ and that standard matter satisfies the energy conditions, in particular the NEC, one finds the generic constraint for hybrid metric-Palatini wormhole geometries

$$0 < T_{\mu\nu}k^\mu k^\nu|_{r_0} < \frac{1}{\kappa^2}\left[k^\mu k^\nu\nabla_\mu\nabla_\nu\phi - \frac{3}{2\phi}k^\mu k^\nu\,\partial_\mu\phi\,\partial_\nu\phi\right]\Big|_{r_0}. \tag{21.16}$$

Using the metric (21.1), the effective Einstein field equation provides the following gravitational field equations

$$\kappa^2\rho(r) = \frac{b'}{r^2}(1+\phi) - \left(1-\frac{b}{r}\right)\left[\phi'' - \frac{3(\phi')^2}{4\phi}\right] + \frac{\phi'}{2r}\left(b' + \frac{3b}{r} - 4\right) - \frac{V}{2}, \tag{21.17}$$

$$\kappa^2 p_r(r) = \left[-\frac{b}{r^3} + \frac{2\Phi'}{r}\left(1-\frac{b}{r}\right)\right](1+\phi) + \phi'\left(\Phi' + \frac{2}{r} + \frac{3\phi'}{4\phi}\right)\left(1-\frac{b}{r}\right) + \frac{V}{2}, \tag{21.18}$$

$$\kappa^2 p_t(r) = \left[\left(\Phi'' + (\Phi')^2 + \frac{\Phi'}{r}\right)\left(1 - \frac{b}{r}\right) + \frac{b - b'r}{2r^3}\left(1 + r\Phi'\right)\right](1+\phi)$$
$$+ \left[\phi'' + \phi'\Phi' + \frac{3(\phi')^2}{4\phi}\right]\left(1 - \frac{b}{r}\right) + \frac{\phi'}{r}\left(1 - \frac{b + rb'}{2r}\right) + \frac{V}{2}. \tag{21.19}$$

The effective Klein–Gordon equation (17.28) is given by

$$\left[\phi'' + \phi'\Phi' - \frac{(\phi')^2}{2\phi} + \frac{3\phi'}{2r}\right]\left(1 - \frac{b}{r}\right) + \frac{\phi'}{2r}(1 + b')\frac{\phi}{3}\left[2V - (1+\phi)V_\phi\right] = \frac{\phi\kappa^2}{3}T. \tag{21.20}$$

Equations (21.17)–(21.20) provide four independent equations, for seven unknown quantities, i.e. $\rho(r)$, $p_r(r)$, $p_t(r)$, $\Phi(r)$, $b(r)$, $\phi(r)$, and $V(r)$. Therefore, the system of equations can be closed by reducing the number of unknown functions in assuming suitable conditions. We will consider specific solutions in the next section.

21.4 Specific Examples

21.4.1 Solution I

As a first case and for simplicity, consider a zero redshift function, $\Phi(r) = 0$, and a specific choice for a power-law scalar field, given by

$$\phi(r) = \phi_0 \left(\frac{r_0}{r}\right)^\alpha . \tag{21.21}$$

In order to simplify the modified Klein–Gordon equation, consider $2V - (1 + \phi)V_\phi = 0$, which yields the potential

$$V(\phi) = V_0(1+\phi)^2 . \tag{21.22}$$

Substituting these choices into the expressions for the energy-momentum tensor components, one obtains the following expression of the energy-momentum tensor trace

$$T = \frac{1}{2\kappa^2 r^2}\left\{3\alpha\,\phi_0\left(\frac{r_0}{r}\right)^\alpha\left[1 - b' + \left(1 - \frac{b}{r}\right)(1-\alpha)\right]\right.$$
$$\left. + \left[V_0 r^2\left(1 + \phi_0\left(\frac{r_0}{r}\right)^\alpha\right) - b'\right]\left[1 + \phi_0\left(\frac{r_0}{r}\right)^\alpha\right]\right\}. \tag{21.23}$$

Finally, substituting these expressions into the modified Klein–Gordon equation, i.e., Eq. (21.20), which simplifies to the following ordinary differential equation

$$(4 - 3\alpha)b' + 3\alpha\left[1 + (1-\alpha)\left(1 - \frac{b}{r}\right)\right] - 4V_0 r^2\left[1 + \phi_0\left(\frac{r_0}{r}\right)^\alpha\right] = 0, \tag{21.24}$$

which provides the following solution for the shape function

$$b(r) = \left[\frac{2V_0\phi_0 r^3}{3\alpha^2 - 8\alpha + 6} \left(\frac{r_0}{r}\right)^\alpha + \frac{3\alpha(\alpha-2)r}{3\alpha^2 - 6\alpha + 4} \right.$$
$$\left. + \frac{4V_0 r^3}{3(\alpha^2 - 4\alpha + 4)} + r^{\frac{3\alpha(\alpha-1)}{3\alpha-4}} C \right], \tag{21.25}$$

where C is a constant of integration, which is fixed by the boundary condition $b(r_0) = r_0$, and is given by

$$C = r_0 r_0^{\frac{-3\alpha(\alpha-1)}{3\alpha-4}} \left\{ 1 - \left[\frac{2V_0\phi_0 r_0^2}{3\alpha^2 - 8\alpha + 6} + \frac{3\alpha(\alpha-2)}{3\alpha^2 - 6\alpha + 4} + \frac{4V_0 r_0^2}{3(\alpha^2 - 4\alpha + 4)} \right] \right\}.$$

For the case of $\alpha = 1$, the shape function is given by

$$b(r) = \frac{4V_0}{3}\left(r^3 - r_0^3\right) + 2r_0 V_0\phi_0 \left(r^2 - r_0^2\right) + 4r_0 - 3r. \tag{21.26}$$

For this case the energy-momentum tensor profile is given by

$$\rho(r) = \frac{1}{6\kappa^2 r^4}\left\{21V_0 r^4 + 28\phi_0 V_0 r^3 + 6r^2(V_0 r_0^2\phi_0^2 - 3) + \phi_0 r_0^2\left[V_0 r_0^2(2 + 3\phi_0) - 6\right]\right\}, \tag{21.27}$$

$$p_r(r) = -\frac{1}{6\kappa^2 r^4}\Big\{5V_0 r^4 + 4\phi_0 V_0 r^3 - 6r^2(V_0 r_0^2\phi_0^2 + 3)$$
$$-4r_0 r(2V_0 r_0^2 + 3\phi_0 V_0 r_0^2 - 6 - 3\phi_0)$$
$$+\phi_0 r_0^2\left[V_0 r_0^2(2 + 3\phi_0) - 6\right]\Big\}, \tag{21.28}$$

$$p_t(r) = -\frac{1}{6\kappa^2 r^4}\Big\{5V_0 r^4 + 2\phi_0 V_0 r^3$$
$$-2r_0(2V_0 r_0^2 + 3\phi_0 V_0 r_0^2 - 6 - 3\phi_0)$$
$$-\phi_0 r_0^2\left[V_0 r_0^2(2 + 3\phi_0) - 6\right]\Big\}. \tag{21.29}$$

This solution is not asymptotically flat, so that, in principle, we need to match the interior wormhole solution to an exterior vacuum space-time, at a junction interface. To avoid the presence of exotic matter on the thin shell, one may also impose, in principle, that the surface stresses, lying on the junction interface, satisfy the energy conditions.

21.4.2 Solution II: Asymptotically Flat Space-Time

As a second solution, consider now the choices for the following metric functions and the scalar field, given by

$$\Phi(r) = \Phi_0\left(\frac{r_0}{r}\right)^\alpha, \qquad b(r) = r_0\left(\frac{r_0}{r}\right)^\beta, \qquad \phi(r) = \phi_0\left(\frac{r_0}{r}\right)^\gamma, \tag{21.30}$$

respectively, where $\alpha > 0$, $\beta > -1$ and $\gamma > 0$ are constant parameters. General solutions for these specific cases can be found, however, they are extremely lengthy, so that without a significant loss of generality, we consider the specific choices for the following parameters: $\alpha = 0$, $\beta = 0$, and $\gamma = 3$.

As before, let us insert these functions into the field equations, deduce the trace of stress-energy tensor, given by

$$T = \frac{1}{2\kappa^2 r^2}\left\{4V(r) + 9\phi_0\left(\frac{r_0}{r}\right)^3\left[1 - 2\frac{r_0}{r}\right]\right\}, \tag{21.31}$$

and substitute the latter into the modified Klein–Gordon equation, which simplifies to

$$2r_0^3 V'(r) + 27\phi_0\left(\frac{r}{r_0}\right)^6\left[1 - 2\left(\frac{r_0}{r}\right)\right] = 0\,. \tag{21.32}$$

This ordinary differential equation yields the following solution for the potential in parametric form, $V(r)$

$$V(r) = \frac{9\phi_0}{10r_0^2}\left(\frac{r_0}{r}\right)^5\left[5\left(\frac{r_0}{r}\right) - 3\right]. \tag{21.33}$$

From the definition of the scalar field, Eq. (21.30), and with $\gamma = 3$, one obtains that $r_0/r = (\phi/\phi_0)^{1/3}$, so that the potential $V(\phi)$ is finally given by

$$V(\phi) = \frac{9\phi_0}{2r_0^2}\left(\frac{\phi}{\phi_0}\right)^{5/3}\left[\left(\frac{\phi}{\phi_0}\right)^{1/3} - \frac{3}{5}\right]. \tag{21.34}$$

The stress-energy tensor components have the following profile

$$\rho(r) = \frac{21\phi_0}{10\kappa^2 r_0^2}\left(\frac{r_0}{r}\right)^5\left[1 - \frac{5}{7}\left(\frac{r_0}{r}\right)\right],$$

$$p_r(r) = -\frac{1}{\kappa^2 r_0^2}\left(\frac{r_0}{r}\right)^3\left\{1 + \frac{3\phi_0}{5}\left(\frac{r_0}{r}\right)^2\left[1 - \frac{5}{6}\left(\frac{r_0}{r}\right)\right]\right\},$$

$$p_t(r) = \frac{1}{2\kappa^2 r_0^2}\left(\frac{r_0}{r}\right)^3\left\{1 + \frac{9\phi_0}{5}\left(\frac{r_0}{r}\right)^2\left[1 - \frac{10}{9}\left(\frac{r_0}{r}\right)\right]\right\},$$

which tend to zero as $r \to \infty$. Note that assuming $\phi_0 > 0$, the energy density is positive throughout the space-time, and at the throat takes the value $\rho(r_0) = 3\phi_0/(5\kappa^2 r_0^2)$. The NEC at the throat takes the form $\rho(r_0) + p_r(r_0) = (\phi_0 - 2)/(2\kappa^2 r_0^2)$, which is positive, by imposing $\phi_0 > 2$; it is an easy matter to verify that this specific example satisfies Eq. (21.16).

21.5 Concluding Remarks

As mentioned above the fundamental ingredient in wormhole physics is the flaring-out condition at the throat which, in classical GR, entails the violation of the null energy condition. As the violation of the energy conditions is often considered a problematic issue, it is useful to minimize these violations [325]. In fact, it is possible to carry this out though several means. First, one could apply the cut-and-paste procedure by constructing thin-shell wormholes [309, 322, 323, 327, 490, 491, 492], where the exotic matter is now concentrated at the junction interface [307, 308, 326, 417]. We refer the reader to [366] for a recent, novel approach in considering the construction of generic spherically symmetric thin-shell traversable wormhole space-times, where it was shown in full generality that the stability of the wormhole is equivalent to choosing suitable properties for the exotic material residing on the wormhole throat.

Another approach to avoiding the use of exotic matter, i.e., normal matter violating the NEC, is in the context of modified theories of gravity. Here, as mentioned above, it is the higher-order curvature terms, which may be interpreted as a gravitational fluid, that support these nonstandard wormhole geometries [239]. Thus, we explicitly show that wormhole geometries can be theoretically constructed without the presence of exotic matter, but are sustained in the context of modified gravity. This approach has been carried out in the literature in a plethora of solutions, namely, in the curvature-matter coupling [364, 365], $f(R)$ gravity [329], in modified teleparralel gravity [77], higher-dimensional evolving wormholes [518], in Einstein–Gauss–Bonnet gravity [350], in third-order Lovelock gravity [517], amongst others.

22
Conclusions

In Part III of this book we have presented a hybrid metric-Palatini framework for theories of gravity, and have tested the new theories it entails using a number of theoretical consistency checks and observational constraints. From the field theory perspective, we found that the hybrid metric-Palatini or $f(X)$ class of theories, where $X = R + \kappa^2 T$, enjoys a similar special status amongst the more general hybrid metric-Palatini gravitational theories as the $f(R)$ theories within the narrower framework of purely metric gravity [511]. This is so because when one excludes theories inhabited by ghost-like, superluminally propagating and otherwise pathological degrees of freedom, there is evidence, as shown in Section 17.5, that the $f(X)$ family is singled out as the only viable form of an action one can construct using the metric (and thus the metric Levi–Civita connection) and an independent "Palatini connection." The underlying reason is that in the special case of $f(X)$ actions the higher derivatives in the gravity sector can be separated into a scalar mode, thus avoiding an Ostrogradskian instability. Indeed, $f(X)$ gravity represents a generic case within the one-parameter family of the Algebraic Scalar–Tensor theories (recall Eq. (17.21)), at one end of which lies the pure Palatini $f(\mathcal{R})$ (wherein the scalar field is a function of the stress-energy trace T) and at the other end the pure metric $f(R)$ (where the field is a function of the metric curvature R). Furthermore, the propagating degrees of freedom have proven to be healthy also on curved backgrounds as confirmed also by our cosmological perturbation analysis in Section 18.2. Concerning the Cauchy problem, it was shown that in this class of theories the initial value problem can always be well formulated and well posed depending on the adopted matter sources.

Having established the theoretical consistency and interest on the hybrid metric-Palatini $f(X)$ family of theories, we considered applications in which these theories provide gravitational alternatives to dark energy. As shown by our post-Newtonian analysis in Section 17.3, the hybrid theories are promising in this respect as they can avoid the local gravity constraints but modify the cosmological dynamics at large scales. This is simply because, as a scalar–tensor theory, the hybrid $f(X)$ gravity is characterized by an evolving Brans–Dicke coupling, which allows to introduce potentially large deviations from GR in the past (and future), as long as the coupling at the present epoch is strong enough to hide the field from the local gravity experiments. In contrast, in the metric $f(R)$ models the Brans–Dicke coupling is a finite constant and one needs to invoke some of the various "screening mechanisms" (workings of which remain to be studied in the hybrid theories, but for a recent paper see [488]) in order to reconcile the solar system experiments with cosmology.

Cosmological perturbations have been also analyzed in these models up to the linear order [104, 224, 316], and the results imply that the formation of large-scale structure in the aforementioned accelerating cosmologies is viable, though it exhibits subtle features that might be detectable in future experiments. In Section 18.2, we derived the full perturbations equations and, extracting their Newtonian limit, describing the observable scales of the matter power spectrum, the growth of matter overdensities was shown to be modified by a time-dependent effective fifth force that is expected to modify the redshift evolution of the growth rate of perturbations. We also note that numerical studies of the perturbations imply that the difference of the gravitational potentials can exhibit oscillations at higher redshifts even when the background expansion and the full lensing potential are indistinguishable from the standard ΛCDM predictions [316]. Such features could potentially be observed in cross-correlations of the matter and lensing power spectra, but detailed calculations of the cosmic microwave background anisotropies and other perturbation observables remain to be carried out. This is especially worthwhile in view of the potential of the forthcoming Euclid mission to experimentally test different accelerating cosmological models by their impact to the large-scale structure [17].

At an effective level, the $f(X)$ modifications involve both (the trace of) the matter stress energy and (the Ricci scalar of) the metric curvature, and from this point of view it appears appealing to speculate on the possible relevance of these theories to both the problems of dark energy and dark matter, in a unified theoretical framework and without distinguishing a priori, matter and geometric sources. Various aspects of dark matter phenomenology from astronomical to galactic and extragalatic scales were also discussed. The generalized virial theoreom can acquire, in addition to the contribution from the baryonic masses, effective contributions of geometrical origin to the total gravitational potential energy, which may account for the well-known virial theorem mass discrepancy in clusters of galaxies. In the context of galactic rotation curves, the scalar-field modified relations between the various physical quantities such as tangential velocities of test particles around galaxies, Doppler frequency shifts and stellar dispersion velocities were derived. More recently, observational data of stellar motion near the galactic center was compared with simulations of the hybrid gravity theory, which turned out particularly suitable to model star dynamics.

Yet, to promote the $f(X)$ theory into a convincing alternative to particle dark matter, one should produce also the cosmological successes of the ΛCDM model without the CDM component.

Though wormhole solutions have already been found in these theories [107], the nature of possible black hole solutions remains an outstanding open question. Though no-hair theorems for scalar–tensor theories exclude the simplest nontrivial solutions, their assumptions are restrictive. Another interesting though as-yet unstudied issue is the strong field regime of hybrid gravity and the constraints that one can put on the theories from astrophysical data such as measurements of binary pulsars. Finally, the possible relevance of the hybrid gravity framework to the early universe cosmology has not been explored in any detail. The Einstein frame formulation of the scalar–tensor description Eq. (17.30) could provide a useful starting point to investigate how the inflaton potential changes due to finite Ω_A, in order to understand how the "hybrid" nature of space-time modifies the standard predictions of, for instance, the $\mathcal{R}^2$ model of inflation.

To conclude, while the physics of the metric and the Palatini versions of $f(R)$ gravity have been uncovered in exquisite detail in a great variety of different contexts [34, 99, 103, 120, 140, 157, 324, 382, 383, 396], those studies largely wait to be extended for the hybrid $f(X)$ version of the theory. We believe the results this far, as reported in this work, provide compelling motivation for the further exploration of these particular theories.

References

[1] Aad, G., et al. [ATLAS and CMS Collaborations]. 2015. Combined measurement of the Higgs Boson mass in *pp* collisions at $\sqrt{s} = 7$ and 8 TeV with the ATLAS and CMS experiments. *Phys. Rev. Lett.*, **114**, 191803.

[2] Abbott, B. P., et al. [LIGO Scientific and Virgo Collaborations]. 2016. Observation of gravitational waves from a binary black hole merger. *Phys. Rev. Lett.*, **116**, 061102.

[3] Abbott, B. P., et al. [LIGO Scientific and Virgo Collaborations]. 2016. GW151226: Observation of gravitational waves from a 22-solar-mass binary black hole coalescence. *Phys. Rev. Lett.*, **116**, 241103. arXiv:1606.04855 [gr-qc].

[4] Abbott, B. P., et al. [LIGO Scientific and Virgo Collaborations]. 2017. GW170104: Observation of a 50-solar-mass binary black hole coalescence at redshift 0.2. *Phys. Rev. Lett.*, **118**, 221101.

[5] Abbott, B. P., et al. [LIGO Scientific and Virgo Collaborations]. 2017. GW170814: A three-detector observation of gravitational waves from a binary black hole coalescence. *Phys. Rev. Lett.*, **119**, 141101.

[6] Abbott, B. P., et al. [LIGO Scientific and Virgo Collaborations]. 2017. GW170817: Observation of gravitational waves from a binary neutron star inspiral. *Phys. Rev. Lett.*, **119**, 161101.

[7] Abbott, B. P., et al. [LIGO Scientific and Virgo Collaborations]. 2017. GW170608: Observation of a 19-Solar-Mass binary black hole coalescence. *Astrophys. J.*, **851**, L35.

[8] Abebe, A., Abdelwahab, M., de la Cruz-Dombriz, A., and Dunsby, P. K. S. 2012. Covariant gauge-invariant perturbations in multifluid *f(R)* gravity. *Class. Quant. Grav.*, **29**, 135011.

[9] Abramo, L. R., Batista, R. C., Liberato, L., and Rosenfeld, R. 2007. Structure formation in the presence of dark energy perturbations. *J. Cosmo. Astropart. Phys.*, **0711**, 012.

[10] Ade, P. A. R., et al. 2014. Planck 2013 results. XXII. Constraints on inflation. *Astron. Astrophys.*, **571**, A22. arXiv:1303.5082 [astro-ph.CO].

[11] Ade, P. A. R., et al. 2015. Planck Collaboration, Planck 2015 results. XIII. Cosmological parameters. *Astron. Astrophys.*, **594**, A13.

[12] Aldrovandi, R., and Pereira, J. G. 2013. *Teleparallel Gravity: An Introduction*, (Springer: Dordrecht).

[13] Allemandi, G., Borowiec, A., Francaviglia, M., and Odintsov, S. D. 2005. Dark energy dominance and cosmic acceleration in first order formalism. *Phys. Rev. D*, **72**, 063505.

[14] Alvarenga, F. G., de la Cruz-Dombriz, A., Houndjo, M. J. S., Rodrigues, M. E., and Sáez-Gómez, D. 2013. Dynamics of scalar perturbations in *f(R,T)* gravity. *Phys. Rev. D*, **87**, 103526.

[15] Alvarenga, F. G., Houndjo, M. J. S., Monwanou, A. V., and Orou, J. B. C. 2013. Testing some *f(R,T)* gravity models from energy conditions. *J. Mod. Phys.*, **4**, 130.

[16] Amendola, L. 2000. Coupled quintessence. *Phys. Rev. D*, **62**, 043511.

[17] Amendola, L., Appleby, S., Bacon, D., Baker, T. M., Baldi, M., and Bartolo, N. 2013. Cosmology and fundamental physics with the Euclid satellite. *Living Rev. Rel.*, **16**, 6.

[18] Amendola, L., Enqvist, K., and Koivisto, T. 2011. Unifying Einstein and Palatini gravities. *Phys. Rev. D*, **83**, 044016.

[19] Antoniadis, J., et al. 2013. A massive pulsar in a compact relativistic binary. *Science*, **340**, 448–449.

[20] Aoki, K., Maeda, K.-i., and Tanabe, M. 2016. Relativistic stars in bigravity theory. *Phys. Rev. D*, **93**, 064054.

[21] Aramaki, T., et al. 2016. Review of the theoretical and experimental status of dark matter identification with cosmic-ray antideuterons. *Phys. Rep.*, **618**, 1–37.

[22] Arbuzova, E. V., Dolgov, A. D., and Reverberi, L. 2012. Cosmological evolution in R2 gravity. *J. Cosmo. Astropart. Phys.*, **2012**, 049.

[23] Armendáriz-Picón, C. 2002. Predictions and observations in theories with varying couplings. *Phys. Rev. D*, **66**, 064008.

[24] Arnaud, M. 2005. X-ray observations of clusters of galaxies. In F. Melchiorri and Y. Rephaeli, eds., *Background Microwave Radiation and Intracluster Cosmology*, Vol. 159, Proceedings of the International School of Physics "Enrico Fermi", (Amsterdam: IOS Press), 77.

[25] Arun, K., Gudennavar, S. B., and Sivaram, C. 2017. Dark matter, dark energy, and alternate models: A review. *Adv. Space Res.*, **60**, 166–186.

[26] Aviles, A., and Cervantes-Cota, J. L. 2011. Dark degeneracy and interacting cosmic components. *Phys. Rev. D*, **83**, 023510.

[27] Ayuso, I., Jimenez, J. B., and de la Cruz-Dombriz, A. 2015. Consistency of universally nonminimally coupled $f(R, T, R_{\mu\nu}T^{\mu\nu})$ theories. *Phys. Rev. D*, **91**, 104003.

[28] Babichev, E., Koyama, K., Langlois, D., Saito, R., and Sakstein, J. 2016. Relativistic Stars in Beyond Horndeski Theories. *Class. Quant. Grav.*, **33**, 235014.

[29] Baer, H., Choi, K.-Y., Kim, J. E., and Roszkowski, L. 2015. Dark matter production in the early Universe: Beyond the thermal WIMP paradigm. *Phys. Rep.*, **555**, 1–60.

[30] Baez, J., and Muniain, J. P. 2008. *Gauge Fields, Knots and Gravity*, (New Jersey, London, Hong Kong, Singapore: World Scientific, 2008).

[31] Balaguera-Antolinez, A., Böhmer, C. G., and Nowakowski, M. 2006. Scales set by the cosmological constant. *Class. Quant. Grav.*, **23**, 485–496.

[32] Balaguera-Antolinez, A., Mota, D. F., and Nowakowski, M. 2007. Astrophysical configurations with background cosmology: Probing dark energy at astrophysical scales. *Mon. Not. R. Astron. Soc.*, **382**, 621–640.

[33] Baldi, M., Pettorino, V., Robbers, G., and Springel, V. 2010. Hydrodynamical N-body simulations of coupled dark energy cosmologies. *Mon. Not. Roy. Astron. Soc.*, **403**, 1684–1702.

[34] Bamba, K., Capozziello, S., Nojiri, S., and Odintsov, S. D. 2012. Dark energy cosmology: The equivalent description via different theoretical models and cosmography tests. *Astrophys. Space Sci.*, **342**, 155–228.

[35] Bamba, K., Nojiri, S. 'i., and Odintsov, S. D. 2008. Inflationary cosmology and the late-time accelerated expansion of the universe in non-minimal Yang-Mills-*F(R)* gravity and non-minimal vector-*F(R)* gravity. *Phys. Rev. D*, **77**, 123532.

[36] Bamba, B., and Odintsov, S. D. 2008. Inflation and late-time cosmic acceleration in non-minimal Maxwell-$F(R)$ gravity and the generation of large-scale magnetic fields. *J. Cosmo. Astropart. Phys.*, **0804**, 024.

[37] Bamba, K., and Odintsov, S. D. 2015. Inflationary cosmology in modified gravity theories. *Symmetry*, **7**, 220.

[38] Bambi, C., and Barausse, E. 2011. Constraining the quadrupole moment of stellar-mass black hole candidates with the continuum fitting method. *Astrophys. J.*, **731**, 121.

[39] Barrow, J. D., and Ottewill, A. C. 1983. The stability of general relativistic cosmological theory. *J. Phys. A: Math. Gen.*, **16**, 2757.

[40] Baykal, A., and Delice, O. 2011. A unified approach to variational derivatives of modified gravitational actions. *Class. Quant. Grav.*, **28**, 015014.

[41] Baykal, A., and Delice, O. 2013. Multi-scalar-tensor equivalents for modified gravitational actions. *Phys. Rev. D*, **88**, 084041.

[42] Bekenstein, J. D. 2004. Relativistic gravitation theory for the modified Newtonian dynamics paradigm. *Phys. Rev. D*, **70**, 083509.

[43] Bennett, C. L., et al. 2003. First year Wilkinson Microwave Anisotropy Probe (WMAP) observations: Preliminary maps and basic results. *Astrophys. J. Suppl.*, **148**, 1.

[44] Bertolami, O., Boehmer, C. G., Harko, T., and Lobo, F. S. N. 2007. Extra force in *f(R)* modified theories of gravity. *Phys. Rev. D*, **75**, 104016.

[45] Bertolami, O., and Ferreira, R. Z. 2012. Traversable Wormholes and Time Machines in non-minimally coupled curvature-matter $f(R)$ theories. *Phys. Rev. D*, **85**, 104050.

[46] Bertolami, O., Frazão, P., and Páramos, J. 2010. Accelerated expansion from a non-minimal gravitational coupling to matter. *Phys. Rev. D*, **81**, 104046.

[47] Bertolami, O., Frazão, P., and Páramos, J. 2011. Reheating via a generalized non-minimal coupling of curvature to matter. *Phys. Rev. D*, **83**, 044010.

[48] Bertolami, O., Frazão, P., and Páramos, J. 2012. Mimicking dark matter in galaxy clusters through a non-minimal gravitational coupling with matter. *Phys. Rev. D*, **86**, 044034.

[49] Bertolami, O., Frazão, P., and Páramos, J. 2013. Cosmological perturbations in theories with non-minimal coupling between curvature and matter. *J. Cosmo. Astropart. Phys.*, **1305**, 029.

[50] Bertolami, O., Pedro, F. G., and Le Delliou, M. 2007. Dark energy-dark matter interaction and putative violation of the Equivalence Principle from the Abell Cluster A586. *Phys. Lett. B*, **654**, 165.

[51] Bertolami, O., Gomes, C., and Lobo, F. S. N. 2017. Gravitational waves in theories with a non-minimal curvature-matter coupling. arXiv:1706.06826 [gr-qc].

[52] Bertolami, O., Harko, T., Lobo, F. S. N., and Páramos, J. 2008. Non-minimal curvature-matter couplings in modified gravity. arXiv:0811.2876 [gr-qc].

[53] Bertolami, O., Lobo, F. S. N., and Páramos, J. 2008. Nonminimal coupling of perfect fluids to curvature. *Phys. Rev. D*, **78**, 064036.

[54] Bertolami, O., March, R., and Páramos, J. 2013. Solar system constraints to nonminimally coupled gravity. *Phys. Rev. D*, **88**, 064019.

[55] Bertolami, O., and Martins, A. 2012. On the dynamics of perfect fluids in non-minimally coupled gravity. *Phys. Rev. D*, **85**, 024012.

[56] Bertolami, O., and Páramos, J. 2008. Do *f(R)* theories matter? *Phys. Rev. D*, **77**, 084018.

[57] Bertolami, O., and Páramos, J. 2008. On the non-trivial gravitational coupling to matter. *Class. Quant. Grav.*, **25**, 245017.

[58] Bertolami, O., and Páramos, J. 2010. Mimicking dark matter through a non-minimal gravitational coupling with matter. *J. Cosmo. Astropart. Phys.*, **1003**, 009.

[59] Bertolami, O., and Páramos, J. 2014. Minimal extension of General Relativity: Alternative gravity model with non-minimal coupling between matter and curvature. *Int. J. Geom. Meth. Mod. Phys.*, **11**, 1460003.

[60] Bertolami, O., and Páramos, J. 2014. Modified Friedmann equation from nonminimally coupled theories of gravity. *Phys. Rev. D*, **89**, 044012.

[61] Bertolami, O., and Páramos, J. 2015. Homogeneous spherically symmetric bodies with a nonminimal coupling between curvature and matter: The choice of the Lagrangian density for matter. *Gen. Rel. Grav.*, **47**, 1835.

[62] Bertolami, O., Páramos, J., and Turyshev., S. 2008. General theory of relativity: Will it survive the next decade? *Astrophys. Space Sci. Libr.*, **349**, 27.

[63] Bertolami, O., and Sequeira, M. C. 2009. Energy conditions and stability in *f(R)* theories of gravity with non-minimal coupling to matter. *Phys. Rev. D*, **79**, 104010.

[64] Bildhauer, S. 1989. Transport equations for freely propagating photons in curved space-times: A derivation by Wigner transformation. *Class. Quant. Grav.*, **6**, 1171–1187.

[65] Binney, J., and Tremaine, S. 2008. *Galactic Dynamics.* (Princeton: Princeton University Press).

[66] Bisabr, Y. 2012. Modified gravity with a non-minimal gravitational coupling to matter. *Phys. Rev. D*, **86**, 044025.

[67] Bisabr, Y. 2013. Non-minimal gravitational coupling of phantom and big rip singularity. *Gen. Rel. Grav.*, **45**, 1559–1566.

[68] Biswas, T., Gerwick, E., Koivisto, T., and Mazumdar, A. 2012. Towards singularity and ghost free theories of gravity. *Phys. Rev. Lett.*, **108**, 031101.

[69] Biswas, T., Koivisto, T., and Mazumdar, A. 2013. Nonlocal theories of gravity: The flat space propagator. arXiv:1302.0532 [gr-qc].

[70] Bodmer, A. R. 1971. Collapsed nuclei. *Phys. Rev. D*, **4**, 1601–1606.

[71] Boehmer, C. G., De Risi, G., Harko, T., and Lobo, F. S. N. 2010. Classical tests of general relativity in brane world models. *Class. Quant. Grav.*, **27**, 185013.

[72] Boehmer, C. G., and Harko, T. 2007. Can dark matter be a Bose-Einstein condensate? *J. Cosmo. Astropart. Phys.*, **0706**, 025.

[73] Boehmer, C. G., and Harko, T. 2008. Physics of dark energy particles. *Found. Phys.*, **38** (2008), 216–227.

[74] Boehmer, C. G., Harko, T., and Lobo, F. S. N. 2008. Dark matter as a geometric effect in *f(R)* gravity. *Astropart. Phys.*, **29**, 386.

[75] Boehmer, C. G., Harko, T., and Lobo, F. S. N. 2008. Generalized virial theorem in *f(R)* gravity. *J. Cosmo. Astropart. Phys.*, **0803**, 024.

[76] Boehmer, C. G., Harko, T., and Lobo, F. S. N. 2008. Solar system tests of brane world models. *Class. Quant. Grav.*, **25**, 045015.

[77] Boehmer, C. G., Harko, T., and Lobo, F. S. N. 2012. Wormhole geometries in modified teleparralel gravity and the energy conditions. *Phys. Rev. D*, **85**, 044033.

[78] Boehmer, C. G., Lobo, F. S. N., and Tamanini, N. 2013. Einstein static Universe in hybrid metric-Palatini gravity. *Phys. Rev. D*, **88**, 104019.

[79] Boisseau, B., Esposito-Farese, G., Polarski, D., and Starobinsky, A. A. 2000. Reconstruction of a scalar tensor theory of gravity in an accelerating universe. *Phys. Rev. Lett.*, **85**, 2236.

[80] Bojowald, M. 2011. *Quantum Cosmology: A Fundamental Description of the Universe*, (New York, Dordrecht, Heidelberg, London: Springer).

[81] Bojowald, M. 2015. Quantum cosmology: A review. *Reports on Progress in Physics*, **78**, 023901.

[82] Bondi, H. 1952. On spherically symmetrical accretion. *Mon. Not. R. Acad. Sci.*, **112**, 195–204.

[83] Borka, B., Capozziello, S., Jovanovic, P., and Jovanovic, V. B. 2016. Probing hybrid modified gravity by stellar motion around Galactic Center. *Astropart. Phys.* **79**, 41.

[84] Borowiec, A., Capozziello, S., de Laurentis, M., et al. 2015. Invariant solutions and Noether symmetries in hybrid gravity. *Phys. Rev. D*, **91**, 023517.

[85] Borriello, A., and Salucci, P. 2001. The dark matter distribution in disc galaxies. *Mon. Not. R. Astron. Soc.*, **323**, 285.

[86] Boylan-Kolchin, M., Bullock, J. S., and Kaplinghat, M. 2011. Too big to fail? The puzzling darkness of massive Milky Way subhaloes. *Mon. Not. R. Astron. Soc.*, **415**, L40–L44.

[87] Boylan-Kolchin, M., Bullock, J. S., and Kaplinghat, M. 2012. The Milky Way's bright satellites as an apparent failure of ΛCDM. *Mon. Not. R. Astron. Soc.*, **422**, 1203–1218.

[88] Brans, C., and Dicke, R. H. 1961. Mach's principle and a relativistic theory of gravitation. *Phys. Rev.*, **124**, 925–935.

[89] Brax, P., van de Bruck, C., Mota, D. F., Nunes, N. J., and Winther, H. A. 2010. Chameleons with field dependent couplings. *Phys. Rev. D*, **82**, 083503.

[90] Brax, P. 2012. Screened modified gravity. *Acta Phys. Polon. B*, **43**, 2307.

[91] Brax, P., Davis, A.-C., Li, B., and Winther, H. A. 2012. A unified description of screened modified gravity. *Phys. Rev. D*, **86**, 044015.

[92] Brihaye, Y., Cisterna, A., and Erices, C. 2016. Boson stars in biscalar extensions of Horndeski gravity. *Phys. Rev. D*, **93**, 124057.

[93] Brown, J. D. 1993. Action functionals for relativistic perfect fluids. *Class. Quant. Grav.*, **10**, 1579–1606.

[94] Buchdahl, H. A. 1959. General relativistic fluid spheres. *Phys. Rev.*, **116**, 1027–1034.

[95] Buchdahl, H. A. 1970. Non-linear Lagrangians and cosmological theory. *Mon. Not. Roy. Astron. Soc.*, **150**, 1.

[96] Cai, Y. F., Capozziello, S., De Laurentis, M., and Saridakis, E. N. 2016. $f(T)$ teleparallel gravity and cosmology. *Rept. Prog. Phys.*, **79**, 106901.

[97] Calvao, M. O., Lima, J. A. S., and Waga, I. 1992. On the thermodynamics of matter creation in cosmology. *Phys. Lett. A*, **162**, 223.

[98] Carloni, S., Koivisto, T., and Lobo, F. S. N. 2015. A dynamical system analysis of hybrid metric-Palatini cosmologies. *Phys. Rev. D*, **92**, 064035.

[99] Capozziello, S. 2002. Curvature quintessence. *Int. J. Mod. Phys. D*, **11**, 483–492.

[100] Capozziello, C., Cardone, V. F., and Troisi, A. 2006. Dark energy and dark matter as curvature effects. *J. Cosmo. Astropart. Phys.*, **0608**, 001.

[101] Capozziello, C., Cardone, V. F., and Troisi, A. 2007. Low surface brightness galaxies rotation curves in the low energy limit of R^n gravity: no need for dark matter? *Month. Not. R. Astron. Soc.*, **375**, 1423–1440.

[102] Capozziello, S., Carloni, S., and Troisi, A. 2003. Quintessence without scalar fields. *Recent Res. Dev. Astron. Astrophys.*, **1**, 625.

[103] Capozziello, S., and De Laurentis, M. 2011. Extended theories of gravity. *Phys. Rept.*, **509**, 167.

[104] Capozziello, S., Harko, T., Koivisto, T. S., Lobo, F. S. N., and Olmo, G. J. 2013. Cosmology of hybrid metric-Palatini *f(X)*-gravity. *J. Cosmo. Astropart. Phys.*, **1304**, 011.

[105] Capozziello, S., Harko, T., Koivisto, T. S., Lobo, F. S. N., and Olmo, G. J. 2013. Galactic rotation curves in hybrid metric-Palatini gravity. *Astropart. Phys.*, **50–52**, 65–75.

[106] Capozziello, S., Harko, T., Koivisto, T. S., Lobo, F. S. N., and Olmo, G. J. 2013. The virial theorem and the dark matter problem in hybrid metric-Palatini gravity. *J. Cosmo. Astropart. Phys.*, **07**, 024.

[107] Capozziello, S., Harko, T., Koivisto, T. S., Lobo, F. S. N., and Olmo, G. J. 2013. Wormholes supported by hybrid metric-Palatini gravity. *Phys. Rev. D*, **86**, 127504.

[108] Capozziello, S., Harko, T., Koivisto, T. S., Lobo, F. S. N., and Olmo, G. J. 2015. Hybrid metric-Palatini gravity. *Universe*, **1**, 199.

[109] Capozziello, S., Harko, T., Lobo, F. S. N., and Olmo, G. J. 2013. Hybrid modified gravity unifying local tests, galactic dynamics and late-time cosmic acceleration. *Int. J. Mod. Phys. D*, **22**, 1342006.

[110] Capozziello, S., Harko, T., Lobo, F. S. N., and Olmo, G. J., and Vignolo, S. 2014. The Cauchy problem in hybrid metric-Palatini *f(X)*-gravity. *Int. J. Geom. Meth. Mod. Phys.*, **11**, 1450042.

[111] Capozziello, S., and Vignolo, S. 2009. The Cauchy problem for metric-affine *f(R)*-gravity in presence of perfect-fluid matter. *Class. Quant. Grav.*, **26**, 175013.

[112] Capozziello, S., and Vignolo, S. 2009. On the well formulation of the initial value problem of metric–affine *f(R)*-gravity. *Int. J. Geom. Meth. Mod. Phys.*, **6**, 985–1001.

[113] Capozziello, S., and Vignolo, S. 2011. The Cauchy problem for metric-affine *f(R)*-gravity in presence of a Klein-Gordon scalar field. *Int. J. Geom. Meth. Mod. Phys.*, **8**, 167–176.

[114] Capozziello, S., and Vignolo, S. 2012. The Cauchy problem for *f(R)*-gravity: An overview. *Int. J. Geom. Meth. Mod. Phys.*, **9**, 1250006.

[115] Cardenas, V. H. 2012. Dark energy, matter creation and curvature. *The European Physical Journal C*, **72**, 2149.

[116] Carlberg, R. G., Yee, H. K. C., and Ellingson, E. 1997. The average mass and light profiles of galaxy clusters. *Astrophys. J.*, **478**, 462–475.

[117] Carlip, S. 2008. Is quantum gravity necessary? *Class. Quant. Grav.*, **25**, 154010.

[118] Carlson, E. D., Machacek, M. E., and Hall, J. L. 1992. Self-interacting dark matter. *Astrophys. J.*, **398**, 43–52.

[119] Carroll, S. M. 2014. *Spacetime And Geometry: An Introduction To General Relativity*, (Harlow, UK: Pearson Education Limited, 2014)

[120] Carroll, S. M., Duvvuri, V., Trodden, M., and Turner, M. S. 2004. Is cosmic speed-up due to new gravitational physics? *Phys. Rev. D*, **70**, 043528.

[121] Cartan, E. 1922. Sur une généralisation de la notion de courbure de Riemann et les espaces *à* torsion. *C. R. Acad. Sci. (Paris)*, **174**, 593601.

[122] Cartan, E. 1925. Sur les variétés à connexion affine et la théorie de la relativité généralisée (deuxième partie). *Ann. Ec. Norm. Sup.*, **42**, 17–28.

[123] Casas, J. A., García-Bellido, J., and Quiros, M. 1992. Scalar-tensor theories of gravity with Phi-dependent masses. *Class. Quant. Grav.*, **9**, 1371.

[124] Castañeda Valle, D., and Mielke, E. W. 2016. Relativistic soliton collisions of axion type dark matter. *Phys. Lett. B*, **758**, 93–97.

[125] Castel-Branco, N., Paramos, J., and March, R. 2014. Perturbation of the metric around a spherical body from a nonminimal coupling between matter and curvature. *Phys. Lett. B*, **735**, 25.

[126] Celani, D. C. F., Pinto-Neto, N., and Vitenti, S. D. P. 2017. Particle creation in bouncing cosmologies. *Phys. Rev. D*, **95**, 023523.

[127] Cembranos, J. A. R. 2009. Dark matter from R^2 gravity. *Phys. Rev. Lett.*, **102**, 141301.

[128] Chakraborty, S. 2013. An alternative $f(R,T)$ gravity theory and the dark energy problem. *Gen. Rel. Grav.*, **45**, 2039.

[129] Chakraborty, S. 2014. Is emergent universe a consequence of particle creation process? *Phys. Lett. B*, **732**, 81–84.

[130] Charmousis, C., Copeland, E. J., Padilla, A., and Saffin, P. M. 2012. Self-tuning and the derivation of a class of scalar-tensor theories. *Phys. Rev. D*, **85**, 104040.

[131] Chavanis, P. H., and Harko, T. 2012. Bose-Einstein condensate general relativistic stars. *Phys. Rev. D*, **86**, 064011.

[132] Cheng, K. S., Dai, Z. G., and Lu, T. 1998. Strange stars and related astrophysical phenomena. *Int. J. Mod. Phys. D*, **7**, 139–176.

[133] Chen, C.-M., Harko, T., and Mak, M. K. 2000. Bianchi type I cosmologies in arbitrary dimensional dilaton gravities. *Phys. Rev. D*, **62**, 124016.

[134] Choquet-Bruhat, Y. 1962. Cauchy problem, in L. Witten, ed., *Gravitation: An Introduction to Current Research*, (New York: John Wiley & Sons).

[135] Choquet-Bruhat, Y. 2009. *General Relativity and the Einstein Equations*, (New York: Oxford University Press).

[136] Chudaykin, A., Gorbunov, D., and Tkachev, I. 2016. Dark matter component decaying after recombination: Lensing constraints with Planck data. *Phys. Rev. D*, **94**, 023528.

[137] Ciarlet, P., and Lions, J., eds. 1993. *Handbook of Numerical Analysis*, Vol. 3, (Amsterdam: Elsevier).

[138] Cisterna, A., Delsate, T., Ducobu, L., and Rinaldi, M. 2016. Slowly rotating neutron stars in the nonminimal derivative coupling sector of Horndeski gravity. *Phys. Rev. D*, **93**, 084046.

[139] Cisterna, A., Delsate, T., and Rinaldi, M. 2015. Neutron stars in general second order scalar-tensor theory: The case of nonminimal derivative coupling. *Phys. Rev. D*, **92**, 044050.

[140] Copeland, E. J., Sami, M., and Tsujikawa, S. 2006. Dynamics of dark energy. *Int. J. Mod. Phys. D*, **15**, 1753–1936.

[141] Courteau, S., et al. 2014. Galaxy masses. *Rev. Mod. Phys.*, **86**, 47–119.

[142] Cui, Y. 2015. A review of WIMP baryogenesis mechanisms. *Mod. Phys. Lett. A*, **A 30**, 1530028.

[143] Damour, T. 1996. Testing the equivalence principle: Why and how? *Class. Quant. Grav.*, **13**, A33.

[144] Damour, T. 2001. Questioning the equivalence principle. *Compt. Rend. Acad. Sci. Ser. IV Phys. Astrophys.*, **2**, 1249.

[145] Damour, T. 2012. Theoretical aspects of the equivalence principle. *Class. Quant. Grav.*, **29**, 184001.

[146] Damour, T., and Donoghue, J. F. 2010. Equivalence principle violations and couplings of a light dilaton. *Phys. Rev. D*, **82**, 084033.

[147] Damour, T., and Esposito-Farese, G. 1992. Tensor multiscalar theories of gravitation. *Class. Quant. Grav.*, **9**, 2093.

[148] Damour, T., and Esposito-Farese, G. 1996. Tensor-scalar gravity and binary-pulsar experiments. *Phys. Rev. D*, **54**, 1474–1491.

[149] Damour, T., Gibbons, G. W., and Gundlach, C. C. 1990. Dark matter, time-varying G, and a dilaton field. *Phys. Rev. Lett.*, **64**, 123.

[150] Damour, T., Piazza, F., and Veneziano, G. 2002. Violations of the equivalence principle in a dilaton-runaway scenario. *Phys. Rev. D*, **66**, 046007.

[151] Damour, T., and Polyakov, A. M. 1994. The string dilation and a least coupling principle. *Nuc. Phys. B*, **423**, 532.

[152] Damour, T., and Polyakov, A. M. 1994. String theory and gravity. *Gen. Rel. & Grav.*, **26**, 1171.

[153] Damour, T., and A. Vilenkin, A. 1996. String theory and inflation. *Phys. Rev. D*, **53**, 2981.

[154] Danila, B., Harko, T., Lobo, F. S. N., and Mak, M. K. 2017. Hybrid metric-Palatini stars. *Phys. Rev. D*, **95**, 044031.

[155] Das, D., and Banerjee, N. 2008. Brans–Dicke scalar field as a chameleon. *Phys. Rev. D*, **78**, 043512.

[156] Delgaty, M. S. R., and Lake, K. 1998. Physical acceptability of isolated, static, spherically symmetric, perfect fluid solutions of Einstein's equations. *Comput. Phys. Commun.*, **115**, 395–415.

[157] De Felice, A., and Tsujikawa, S. 2010. $f(R)$ theories. *Living Rev. Rel.*, **13**, 3.

[158] De la Cruz-Dombriz, A., Dobado, A., and Maroto, A. L. 2008. On the evolution of density perturbations in *f(R)* theories of gravity. *Phys. Rev. D*, **2008**, *77*, 123515.

[159] de Laix, A. A., Scherrer, R. J., and Schaefer, R. K. 1995. Constraints on self-interacting dark matter. *Astrophys. J.*, **452**, 495–500.

[160] Demorest, P., Pennucci, T., Ransom, S., Roberts, M., and Hessels, J. 2010. A two-solar-mass neutron star measured using Shapiro delay. *Nature*, **467**, 1081–1083.

[161] de Rham, C. 2014. Massive gravity. *Living Rev. Rel.*, **17**, 7.

[162] de Rham, C., Hofmann, S., Khoury, J., and Tolley, A. J. 2008. Cascading gravity and degravitation. *J. Cosmol. Astropart. Phys.*, **0802**, 011.

[163] De Risi, G., Harko, T., and Lobo, F. S. N. 2012. Solar system constraints on local dark matter density. *J. Cosmo. Astropart. Phys.*, **07**, 047.

[164] Deser, S., and Gibbons, G. W. 1998. Born-Infeld-Einstein actions? *Class. Quan. Grav.*, **15**, L35.

[165] de Sitter, W. 1917. On the relativity of inertia: Remarks concerning Einstein's latest hypothesis. *Koninklijke Nederlandsche Akademie van Wetenschappen Proceedings*, **19**, 1217–1225.

[166] de Sitter, W. 1917. Einstein's theory of gravitation and its astronomical consequences. Third paper. *Mon. Not. Roy. Astron. Soc.*, **78**, 3–28.

[167] Dewitt, B. S. 1967. Quantum theory of gravity. I. The canonical theory. *Phys. Rev.*, **160**, 1113–1148.

[168] Dewitt, B. S. 1967. Quantum theory of gravity. II. The manifestly covariant theory. *Phys. Rev.*, **162**, 1195–1239.

[169] Dewitt, B. S. 1967. Quantum Theory of Gravity. III. Applications of the covariant theory. *Phys. Rev.*, **162**, 1239–1256.

[170] Dirac, P. A. M. 1996. *General Theory of Relativity*, (Princeton: Princeton University Press).

[171] Dodelson, S. 2003. *Modern Cosmology*, (Amsterdam: Elsevier)

[172] Dolgov, A. D., and Kawasaki, M. 2003. Can modified gravity explain accelerated cosmic expansion? *Phys. Lett. B*, **573**, 1.

[173] Drummond, I. T., and Hathrell, S. J. 1980. QED Vacuum polarization in a background gravitational field and its effect on the velocity of photons. *Phys. Rev. D*, **22**, 343.

[174] Dubrovin, B. A., Fomenko, A. T., and Novikov, S. P. 1985. *Modern Geometry–Methods and Applications: Part II: The Geometry and Topology of Manifolds*, 2nd edition, (New York: Springer Verlag).

[175] Dubrovin, B. A., Fomenko, A. T., and Novikov, S. P. 1990. *Modern Geometry–Methods and Applications: Part III: Introduction to Homology Theory*, 2nd edition, (New York: Springer Verlag).

[176] Dubrovin, B. A., Fomenko, A. T., and Novikov, S. P. 1992. *Modern Geometry–Methods and Applications: Part I: The Geometry of Surfaces, Transformation Groups, and Fields*, 2nd edition, (New York: Springer Verlag).

[177] Duruisseau, J. P., Kerner, R., and Eysseric, P. 1983. Non-Einsteinian gravitational Lagrangians assuring cosmological solutions without collapse. *Gen. Rel. Grav.*, **15**, 797–807.

[178] Dvali, G. R., Gabadadze, G., and Porrati, M. 2000. 4-D gravity on a brane in 5-D Minkowski space. *Phys. Lett. B*, **485**, 208.

[179] Dzhunushaliev, V. 2015. Nonperturbative quantization: Ideas, perspectives, and applications. arXiv:1505.02747.

[180] Dzhunushaliev, V., Folomeev, V., Kleihaus, B., and Kunz, J. 2014. Modified gravity from the quantum part of the metric. *Eur. Phys. J. C*, **74**, 2743.

[181] Dzhunushaliev, V., Folomeev, V., Kleihaus, B., and Kunz, J. 2015. Modified gravity from the nonperturbative quantization of a metric. *Eur. Phys. J. C*, **75**, 157.

[182] Eddington, A. S. 2010. *The Mathematical Theory of Relativity*, (Cambridge: Cambridge University Press).

[183] Einstein, A. 1917. Kosmologische Betrachtungen zur allgemeinen Relativitätstheorie. *Sitzungsberichte der Königlich Preussischen Akademie der Wissenschaften (Berlin)*, **1917**, 142–152.

[184] Einstein, A. 1928. *Sitz. Preuss. Akad. Wiss. p.* 217; ibid, 224; Unzicker, A., and Case, T. 2005. Translation of Einstein's attempt of a unified field theory with teleparallelism (transcription).

[185] Einstein, A. 1988. *The Meaning of Relativity: Including the Relativistic Theory of the Non-Symmetric Field*, 5th edition, (Princeton: Princeton University Press).

[186] Faraoni, V. 2004. *Cosmology in Scalar-Tensor Gravity*, (Dordrecht: Kluwer Academic Publishers).

[187] Faraoni, V. 2007. De Sitter space and the equivalence between $f(R)$ and scalar-tensor gravity. *Phys. Rev. D*, **75**, 067302.

[188] Faraoni, V. 2007. A viability criterion for modified gravity with an extra force. *Phys. Rev. D*, **76**, 127501.

[189] Faraoni, V. 2009. The Lagrangian description of perfect fluids and modified gravity with an extra force. *Phys. Rev. D*, **80**, 124040.

[190] Faraoni, V., and Nadeau, S. 2005. The stability of modified gravity models. *Phys. Rev. D*, **72**, 124005.

[191] Ferraro, R., and Fiorini, F. 2007. Modified teleparallel gravity: Inflation without inflaton. *Phys. Rev. D*, **75**, 084031.

[192] Flanagan, E. E. 2003. Higher order gravity theories and scalar tensor theories. *Class. Quant. Grav.*, **21**, 417–426.

[193] Fock, V. 1959. *The Theory of Space, Time and Gravitation*, (Oxford: Pergamon Press).

[194] Fomenko, A. T., and Mishchenko, A. S. 2009. *A Short Course in Differential Geometry and Topology*, (Cambridge: Cambridge Scientific Publishers)

[195] Fourés-Bruhat, Y. 1958. *Bulletin de la Societe Mathematique de France,* **86**, 155.

[196] Friedman, A. 1922. Über die krümmung des raumes. *Zeitschrift für Physik*, **10**, 377–386.

[197] Frieman, J. A., Hill, C. T., Stebbins, A., and Waga, I. 1995. Cosmology with ultra-light pseudo-Nambu-Goldstone bosons. *Phys. Rev. Lett.*, **75**, 20772080.

[198] Fujii, Y., and K. I. Maeda. 2003. *The Scalar-Tensor Theory of Gravitation*, (Cambridge: Cambridge University Press).

[199] Fulling, S. A. 1989. *Aspects of Quantum Field Theory in Curved Space-Time*, (Cambridge: Cambridge University Press).

[200] Gariel J., and le Denmat, G. 1995. Matter creation and bulk viscosity in early cosmology. *Physics Letters A*, **200**, 11–16.

[201] Gasperini, M., Piazza, F., Veneziano, G. 2002. Quintessence as a run-away dilaton. *Phys. Rev. D*, **65**, 023508.

[202] Georgiou, A. 2003. A virial theorem for rotating charged perfect fluids in general relativity. *Class. Quantum Grav.*, **20**, 359–367.

[203] N. K. Glendenning, *Compact Stars, Nuclear Physics, Particle Physics and General Relativity*, (New York: Springer, 2000).

[204] Goenner, H. F. M. 1984. Theories of gravitation with nonminimal coupling of matter and the gravitational field. *Found. Phys.*, **14**, 865881.

[205] Golovnev, A., Karciauskas, M., and Nyrhinen, H. J. 2015. ADM analysis of gravity models within the framework of bimetric variational formalism. *J. Cosmo. Astropart. Phys.*, **2015**, 021.

[206] Gorini, V., Kamenshchik, A. Y., Moschella U., and Pasquier, V. 2004. Tachyons, scalar fields, and cosmology. *Phys. Rev. D*, **69**, 123512.

[207] Gottlober, S., Schmidt, H. J., and Starobinsky, A. A. 1990. Sixth-order gravity and conformal transformations. *Class. Quant. Grav.*, **7**, 893.

[208] Gou, L., et al. 2014. Confirmation via the continuum-fitting method that the spin of the black hole in Cygnus X-1 is extreme. *Astrophys. J.*, **790**, 29.

[209] Gourgoulhon E., and Bonazzola, S. 1994. A virial identity applied to relativistic stellar models. *Class. Quantum Grav.*, **11**, 1775–1784.

[210] Grøn, O., and Hervik, S. 2010. *Einstein's General Theory of Relativity: With Modern Applications in Cosmology*, (New York: Springer Verlag).

[211] Haghani, Z., Harko, H., Lobo, F. S. N., Sepangi, H. R., and Shahidi, S. 2013. Further matters in space-time geometry: $f(R, T, R_{\mu\nu}T^{\mu\nu})$ gravity. *Phys. Rev. D*, **88**, 044023.

[212] Haghani, Z., Harko, H., Sepangi, H. R., and Shahidi, S. 2014. Matter may matter. *Int. J. Mod. Phys. D*, **23**, 1442016.

[213] Haghi, H., Bazkiaei, A. E., Hasani Zonoozi, A., and Kroupa, P. 2016. Declining rotation curves of galaxies as a test of gravitational theory. *Mon. Not. Roy. Astron. Soc.*, **458**, 4172–4187.

[214] Harko, T. 2003. Gravitational collapse of a Hagedorn fluid in Vaidya geometry. *Phys. Rev. D*, **68**, 064005.

[215] Harko, T. 2008. Modified gravity with arbitrary coupling between matter and geometry. *Phys. Lett. B*, **669**, 376.

[216] Harko, T. 2009. Matter accretion by brane-world black holes. *Journal of the Korean Physical Society*, **54**, 2583–2594.

[217] Harko, T. 2010. Galactic rotation curves in modified gravity with non-minimal coupling between matter and geometry. *Phys. Rev. D*, **81**, 084050.

[218] Harko, T. 2010. The matter Lagrangian and the energy-momentum tensor in modified gravity with non-minimal coupling between matter and geometry. *Phys. Rev. D*, **81**, 044021.

[219] Harko, T. 2011. Evolution of cosmological perturbations in Bose-Einstein condensate dark matter. *Mon. Not. Roy. Astron. Soc.*, **413**, 3095–3104.

[220] Harko, T. 2014. Thermodynamic interpretation of the generalized gravity models with geometry-matter coupling. *Phys. Rev. D*, **90**, 044067.

[221] Harko, T., and Cheng, K. S. 2000. Collapsing strange quark matter in Vaidya geometry. *Phys. Lett. A*, **266**, 249–253.

[222] Harko, T., and Cheng, K. S. 2007. The virial theorem and the dynamics of clusters of galaxies in the brane world models. *Phys. Rev. D*, **76**, 044013.

[223] Harko, T., Koivisto, T. S., and Lobo, F. S. N. 2011. Palatini formulation of modified gravity with a nonminimal curvature-matter coupling. *Mod. Phys. Lett. A*, **26**, 1467–1480.

[224] Harko, T., Koivisto, T. S., Lobo, F. S. N., Olmo, G. J. 2012. Metric-Palatini gravity unifying local constraints and late-time cosmic acceleration. *Phys. Rev. D*, **85**, 084016.

[225] Harko, T., Kovács, Z., and Lobo, F. S. N. 2008. Electromagnetic signatures of thin accretion disks in wormhole geometries. *Phys. Rev. D*, **78**, 084005.

[226] Harko, T., Kovács, Z., and Lobo, F. S. N. 2009. Can accretion disk properties distinguish gravastars from black holes? *Class. Quant. Grav.*, **26**, 215006.

[227] Harko, T., Kovács, Z., and Lobo, F. S. N. 2009. Testing Horava-Lifshitz gravity using thin accretion disk properties. *Phys. Rev. D*, **80**, 044021.

[228] Harko, T., Kovács, Z., and Lobo, F. S. N. 2009. Thin accretion disks in stationary axisymmetric wormhole spacetimes. *Phys. Rev. D*, **79**, 064001.

[229] Harko, T., Kovács, Z., and Lobo, F. S. N. 2010. Thin accretion disk signatures in dynamical Chern-Simons modified gravity. *Class. Quant. Grav.*, **27**, 105010.

[230] Harko, T., Kovács, Z., and Lobo, F. S. N. 2011. Solar system tests of Horava–Lifshitz gravity. *Proc. Roy. Soc. Lond. A Math. Phys. Eng. Sci.*, **467**, 1390–1407.

[231] Harko, T., Kovács, Z., and Lobo, F. S. N. 2011. Thin accretion disk signatures of slowly rotating black holes in Horava gravity. *Class. Quant. Grav.*, **28**, 165001.

[232] Harko, T. and Lake, M. J. 2014. Null fluid collapse in brane world models. *Phys. Rev. D*, **89**, 064038.

[233] Harko, T., and Lobo, F. S. N. 2010. $f(R, L_m)$ gravity. *Eur. Phys. J. C*, **70**, 373.

[234] Harko, T., and Lobo, F. S. N. 2012. Generalized dark gravity. *Int. J. Mod. Phys. D*, **21**, 1242019.

[235] Harko, T., and Lobo, F. S. N. 2012. Geodesic deviation, Raychaudhuri equation, and tidal forces in modified gravity with an arbitrary curvature-matter coupling. *Phys. Rev. D*, **86**, 124034.

[236] Harko, T., and Lobo, F. S. N. 2013. Irreversible thermodynamic description of interacting dark energy-dark matter cosmological models. *Phys. Rev. D*, **87**, 044018.

[237] Harko, T., and Lobo, F. S. N. 2014. Generalized curvature-matter couplings in modified gravity. *Galaxies*, **2**, 410.

[238] Harko, T., Lobo, F. S. N., and Mak, M. K. 2014. Arbitrary scalar field and quintessence cosmological models. *Eur. Phys. J. C*, **74**, 2784.

[239] Harko, T., Lobo, F. S. N., Mak, M. K., and Sushkov, S. V. 2013. Modified-gravity wormholes without exotic matter. *Phys. Rev. D*, **87**, 067504.

[240] Harko, T., Lobo, F. S. N., Mak, M. K., and Sushkov, S. V. 2013. Structure of neutron, quark, and exotic stars in Eddington-inspired Born-Infeld gravity. *Phys. Rev. D* **88**, 044032.

[241] Harko, T., Lobo, F. S. N., Mimoso, J. P., and Pavón, D. 2015. Gravitational induced particle production through a nonminimal curvature-matter coupling. *Eur. Phys. J. C*, **75**, 386.

[242] Harko, T., Lobo, F. S. N., and Minazzoli, O. 2013. Extended $f(R, L_m)$ gravity with generalized scalar field and kinetic term dependences. *Phys. Rev. D*, **87**, 047501.

[243] Harko, T., and Lobo, F. S. N., Nojiri, S. 'i., and Odintsov, S. D. 2011. $f(R,T)$ gravity. *Phys. Rev. D*, **84**, 024020.

[244] Harko, T., Lobo, F. S. N., Otalora, G., and Saridakis, E. N. 2014. $f(T, \mathcal{T})$ gravity and cosmology. *J. Cosmol. Astropart. Phys.*, **1412**, 021.

[245] Harko, T., Lobo, F. S. N., Otalora, G., and Saridakis, E. N. 2014. Nonminimal torsion-matter coupling extension of $f(T)$ gravity. *Phys. Rev. D*, **89**, 124036.

[246] Harko T., and Mak, M. K. 1999. Particle creation in varying speed of light cosmological models. *Class. Quantum Grav.*, **16**, 2741–2752.

[247] Harko T., and Mak, M. K. 2016. Exact power series solutions of the structure equations of the general relativistic isotropic fluid stars with linear barotropic and polytropic equations of state. *Astrophysics and Space Science*, **361**, 1–19.

[248] Harrison, E. R. 1965. Cosmology without general relativity. *Ann. Phys.*, **35**, 437–446.

[249] Harvey, D., Massey, R., Kitching, T., Taylor, A., and Tittley, E. 2015. The nongravitational interactions of dark matter in colliding galaxy clusters. *Science*, **347**, 1462–1465.

[250] Hawking, S. W., and Ellis, G. F. R. 1973. *The large scale structure of space-time*, (Cambridge: Cambridge University Press).

[251] Hawking, S. W., and Page, D. N. 1986. Operator ordering and the flatness of the universe. *Nucl. Phys. B*, **264**, 185–196.

[252] Hayashi, K., and Shirafuji, T. 1979. New general relativity. *Phys. Rev. D*, **19**, 3524. (Addendum: ibid. D **24**, 3312 [1982]).

[253] Hinshaw, G., et al. (WMAP Collaboration). 2013. First-year Wilkinson Microwave Anisotropy Probe (WMAP) observations: The A angular power spectrum. *Astrophys. J. Suppl.*, **148**, 135.

[254] Horndeski, G. W. 1974. Second-order scalar-tensor field equations in a four-dimensional space. *Int. J. Theor. Phys.*, **10**, 363.

[255] Horvath, J. E., and Valentim, R. 2017. The masses of neutron stars. In A. W. Alsabti and P. Murdin, eds., *Handbook of Supernovae*, (New York: Springer) arXiv:1607.06981.

[256] Houndjo, M. J. S. 2012. Reconstruction of $f(R,T)$ gravity describing matter dominated and accelerated phases. *Int. J. Mod. Phys. D*, **21**, 1250003.

[257] Houndjo, M. J. S., Batista, C. E. M., Campos, J. P., and Piattella, O. F. 2013. Finite-time singularities in $f(R,T)$ gravity and the effect of conformal anomaly. *Can. J. Phys.*, **91**, 548. arXiv:1203.6084 [gr-qc].

[258] Houndjo, M. J. S., and Piattella, O. F. 2012. Reconstructing $f(R,T)$ gravity from holographic dark energy. *Int. J. Mod. Phys. D*, **21**, 1250024.

[259] Hu, B. L. 1982. Vacuum viscosity description of quantum processes in the early universe. *Phys. Lett. A*, **90**, 375–380.

[260] Hu, W., and Sawicki, I. 2007. Models of $f(R)$ Cosmic acceleration that evade solar-system tests. *Phys. Rev. D*, **76**, 064004.

[261] Huang, R. -N. 2013. The Wheeler–DeWitt equation of $f(R, L_m)$ gravity in minisuperspace. arXiv:1304.5309 [gr-qc].

[262] Hulse, R. A., and Taylor, J. H. 1975. Discovery of a pulsar in a binary system. *Astrophys. J. Lett.*, **195**, L51–L53.

[263] Ibragimov, N. H. 2015. *Tensors and Riemannian Geometry*, (Berlin/Boston: Higher Education Press and Walter de Gruyter).

[264] Iorio, L. 2015. Editorial for the Special Issue: 100 Years of Chronogeometrodynamics: The Status of the Einstein's Theory of Gravitation in Its Centennial Year. *Universe*, **1**, 38–81.

[265] Iorio, L. 2015. Gravitational anomalies in the solar system? *Int. J. Mod. Phys. D*, **24**, 1530015.

[266] Itoh, N. 1970. Hydrostatic Equilibrium of Hypothetical Quark Stars. *Progress of Theoretical Physics*, **44**, 291–292.

[267] Jackson, J. C. 1970. The dynamics of clusters of galaxies in universes with non-zero cosmological constant, and the virial theorem mass discrepancy. *Month. Not. R. Astr. Soc.*, **148**, 249–260.

[268] Jamil, M., Momeni, D., Raza, M., and Myrzakulov, R. 2012. Reconstruction of some cosmological models in $f(R, T)$ gravity. *Eur. Phys. J. C*, **72**, 1999.

[269] Jauzac, M., et al. 2016. The extraordinary amount of substructure in the *Hubble Frontier Fields* cluster Abell 2744. *Month. Not. R. Astr. Soc.*, **463**, 3876–3893.

[270] Jesus, J. F., Oliveira, F. A., Basilakos, S., and Lima, J. A. S. 2011. Newtonian perturbations on models with matter creation. *Phys. Rev. D*, **84**, 063511.

[271] Jimenez, J. B., Golovnev, A., Karciauskas, M., and Koivisto, T. S. 2012. The bimetric variational principle for General Relativity. *Phys. Rev. D*, **86**, 084024.

[272] Jimenez, J. B., and Koivisto, T. S. 2014. Extended Gauss-Bonnet gravities in Weyl geometry. *Class. Quant. Grav.*, **31**, 135002.

[273] Joshi, P. S. 1993. *Global Aspects in Gravitation and Cosmology*, (Oxford: Clarendon Press).

[274] Jou, D., Casas-Vázquez, J., and Lebon, G. 2010. *Extended Irreversible Thermodynamics*, (Berlin: Springer Verlag).

[275] Joyce, A., Jain, B., Khoury, J., and Trodden, M. 2015. Beyond the Cosmological Standard Model. *Phys. Rept.* **568**, 1–98.

[276] Karukes, E. V., and Salucci, P. 2017. The universal rotation curve of dwarf disk galaxies. *Mon. Not. Roy. Astron. Soc.*, **465**, 4703–4722.

[277] Katsuragawa, T., Nojiri, S., Odintsov, S. D., and Yamazaki, M. 2016. Relativistic stars in dRGT massive gravity. *Phys. Rev. D*, **93**, 124013.

[278] Kerner, R. 1982. Cosmology without singularity and nonlinear gravitational Lagrangians. *Gen. Rel. Grav.*, **14**, 453–469.

[279] Khoury, J., and Weltman, A. 2004. Chameleon fields: Awaiting surprises for tests of gravity in space. *Phys. Rev. Lett.*, **93**, 171104.

[280] Khoury, J., and Weltman, A. 2004. Chameleon cosmology. *Phys. Rev. D*, **69**, 044026.

[281] Kiani, F., and Nozari, K. 2014. Energy conditions in $F(T, \Theta)$ gravity and compatibility with a stable de Sitter solution. *Phys. Lett. B*, **728**, 554.

[282] Kibble, T. W. B., and Randjbar-Daemi, S. 1980. Non-linear coupling of quantum theory and classical gravity. *J. Phys. A: Math. Gen.*, **13**, 141–148.

[283] Kleihaus, B., Kunz, J., Mojica, S., and Zagermann, M. 2016. Rapidly rotating neutron stars in dilatonic Einstein-Gauss-Bonnet theory. *Phys. Rev. D*, **93**, 064077.

[284] Koivisto, T. 2006. Covariant conservation of energy momentum in modified gravities. *Class. Quant. Grav.*, **23**, 4289–4296.

[285] Koivisto, T. 2006. The matter power spectrum in $f(R)$ gravity. *Phys. Rev. D*, **73**, 083517.

[286] Koivisto, T. S. 2009. Cosmology of modified (but second order) gravity. *AIP Conf. Proc.*, **1206**, 79–96.

[287] Koivisto, T. S. 2011. On new variational principles as alternatives to the Palatini method. *Phys. Rev. D*, **83**, 101501.

[288] Koivisto, T. S. 2011. The post-Newtonian limit in C-theories of gravitation. *Phys. Rev. D*, **84**, 121502.

[289] Koivisto, T., and Kurki-Suonio, H. 2006. Cosmological perturbations in the Palatini formulation of modified gravity. *Class. Quant. Grav.*, **23**, 2355.

[290] Koivisto, T. S., Mota, D. F., and Sandstad, M. 2013. Novel aspects of C-theories in cosmology. Unpublished. arXiv:1305.4754.

[291] Koivisto, T. S., Mota, D. F., and Zumalacarregui, M. 2012. Screening modifications of gravity through disformally coupled fields. *Phys. Rev. Lett.*, **109**, 241102.

[292] Koivisto, T. S., Saridakis, E. N., and Tamanini, N. 2015. Scalar fluid theories: Cosmological perturbations and large-scale structure. *J. Cosmo. Astropart. Phys.*, **1509**, 047.

[293] Koivisto, T. S., and Tamanini, N. 2013. Ghosts in pure and hybrid formalisms of gravity theories: A unified analysis. *Phys. Rev. D*, **87**, 104030.

[294] Kolb, E. W., and Turner, M. S. 1989. *The Early Universe*, (Redwood City, CA: Addison-Wesley Company)

[295] Kong, L., Li, Z., and Bambi, C. 2014. Constraints on the spacetime geometry around 10 stellar-mass black hole candidates from the disk's thermal spectrum. *Astrophys. J.*, **797**, 78.

[296] Kovács, Z., Cheng, K. S., and Harko, T. 2009. Can stellar mass black holes be quark stars? *Mon. Not. Roy. Astron. Soc.*, **400**, 1632–1642.

[297] Kovács, Z., Cheng, K. S., and Harko, T. 2009. Thin accretion disks around neutron and quark stars. *Astron. Astrophys.*, **500**, 621–631.

[298] Kovács, Z., and Harko, T. 2010. Can accretion disk properties observationally distinguish black holes from naked singularities? *Phys. Rev. D*, **82**, 124047.

[299] Kramer, D., Stephani, H., MacCallum, M., and Herlt, E. 1980. *Exact Solutions of Einstein's Field Equations*, (Cambridge: Cambridge University Press).

[300] Kubis, S., and Kutschera, M. 1997. Nuclear matter in relativistic mean field theory with isovector scalar meson. *Phys. Lett. B*, **399**, 191–195.

[301] Lake, K. 2004. Galactic potentials. *Phys. Rev. Lett.*, **92**, 051101.

[302] Landau, L. D., and Lifshitz, E. M. 1994. *The Classical Theory of Fields*, Fourth Revised English Edition, (Amsterdam: Butterworth-Heinemann).

[303] Landau, L. D., and Lifshitz, E. M. 1998. *The Classical Theory of Fields*, (Oxford: Butterworth-Heinemann).

[304] Landau, L. D., and Lifshitz, E. M. 2000. *Non-relativistic Quantum Mechanics*, (Beijing: World Publishing Company).

[305] Leanizbarrutia, I., Lobo, F. S. N., and Saez-Gomez, D. 2017. Crossing SNe Ia and BAO observational constraints with local ones in hybrid metric-Palatini gravity. *Phys. Rev. D*, **95**, 084046.

[306] Lemaître, G. 1931. Expansion of the universe: A homogeneous universe of constant mass and increasing radius accounting for the radial velocity of extra-galactic nebulae. *Mon. Not. Roy. Astron. Soc.*, **91**, 483–490.

[307] Lemos, J. P. S., and Lobo, F. S. N. 2004. Plane symmetric traversable wormholes in an anti-de Sitter background. *Phys. Rev. D*, **69**, 104007.

[308] Lemos, J. P. S., and Lobo, F. S. N. 2008. Plane symmetric thin-shell wormholes: Solutions and stability. *Phys. Rev. D*, **78**, 044030.

[309] Lemos, J. P. S., Lobo, F. S. N., and Quinet de Oliveira, S. 2003. Morris-Thorne wormholes with a cosmological constant. *Phys. Rev. D*, **68**, 064004.

[310] Leray, J. 1953. *Hyperbolic Differential Equations*, (Princeton: Institute for Advanced Study).

[311] Levkov, D., Rebbi, C., and Rubakov, V. A. 2002. Tunneling in quantum cosmology: Numerical study of particle creation. *Phys. Rev. D*, **66**, 083516.

[312] Li, B., and Barrow, J. D. 2011. N-body simulations for coupled scalar field cosmology. *Phys. Rev. D*, **83**, 024007.

[313] Li, M., Li, X.-D., Wang, S., and Wang, Y. 2013. Dark energy: A brief review. *Frontiers of Physics*, **8**, 828–846.

[314] Lima, J. A. S., and Germano, A. S. M. 1992. On the equivalence of bulk viscosity and matter creation. *Physics Letters A*, **170**, 373–378.

[315] Lima, J. A. S., Zanchin, V., and Brandenberger, R. 1997. On the Newtonian cosmology equations with pressure. *Mon. Not. Roy. Astron. Soc.*, **291**, L1–L4.

[316] Lima, N. A. 2014. Dynamics of linear perturbations in the hybrid metric-Palatini gravity. *Phys. Rev. D*, **89**, 083527.

[317] Lima, N. A., and Smer-Barreto, V. 2016. Constraints on hybrid metric-Palatini models from background evolution. *Astrophys. J.*, **818**, 186.

[318] Linder, E. V. 2010. Einstein's other gravity and the acceleration of the universe. *Phys. Rev. D*, **81**, 127301.

[319] Lindquist, R. W. 1966. Relativistic transport theory. *Ann. Phys., NY*, **37**, 487–518.

[320] Liu, X., Harko, T., and Liang, S.-D. 2016. Cosmological implications of modified gravity induced by quantum metric fluctuations. *Eur. Phys. J. C*, **76**, 420.

[321] Llinares, C., and Mota, D. 2013. Releasing scalar fields: cosmological simulations of scalar-tensor theories for gravity beyond the static approximation. *Phys. Rev. Lett.*, **110**, 161101.

[322] Lobo, F. S. N. 2004. Surface stresses on a thin shell surrounding a traversable wormhole. *Class. Quant. Grav.*, **21**, 4811.

[323] Lobo, F. S. N. 2005. Energy conditions, traversable wormholes and dust shells. *Gen. Rel. Grav.*, **37**, 2023.

[324] Lobo, F. S. N. 2009. The dark side of gravity: Modified theories of gravity. In *Dark Energy: Current Advances and Ideas*, (Kerala: Research Signpost), 173–204.

[325] Lobo, F. S. N. 2008. Exotic solutions in General Relativity: Traversable wormholes and "warp drive" spacetimes. *Class. Quant. Grav.*, 1-78, arXiv:0710.4474 [gr-qc].

[326] Lobo, F. S. N., and Crawford, P. 2004. Linearized stability analysis of thin shell wormholes with a cosmological constant. *Class. Quant. Grav.*, **21**, 391.

[327] Lobo, F. S. N., and Crawford, P. 2005. Stability analysis of dynamic thin shells. *Class. Quant. Grav.*, **22**, 4869.

[328] Lobo, F. S. N., and Harko, T. 2012. Extended $f(R, L_m)$ theories of gravity. Proceedings of the Thirteenth Marcel Grossmann Meeting, Stockholm University, Sweden. arXiv:1211.0426 [gr-qc].

[329] Lobo, F. S. N., and Oliveira, M. A. 2009. Wormhole geometries in $f(R)$ modified theories of gravity. *Phys. Rev. D*, **80**, 104012.

[330] Logunov, A. A. 2001. *The Theory of Gravity*, (Moscow: Nauka).

[331] Ludwick, K. J. 2017. The viability of phantom dark energy: A review. *Mod. Phys. Lett. A*, **32**, 1730025.

[332] Ma, C.-P., and Bertschinger, E. 1995. Cosmological perturbation theory in the synchronous and conformal Newtonian gauges. *Astrophys. J.*, **455**, 7–25.

[333] Maartens, R. 1996. Causal thermodynamics in relativity. Lectures given at the Hanno Rund Workshop on Relativity and Thermodynamics, South Africa. arXiv:astro-ph/9609119.

[334] Maartens, R. 2004. Brane world gravity. *Living Rev. Rel.*, **7**, 7.

[335] Maartens, R., and Maharaj, S. D. 1990. Collision-free gases in spatially homogeneous space-times. *J. Math. Phys.*, **22**, 595–607

[336] Maartens, R, and Mendez, V. 1997. Nonlinear bulk viscosity and inflation. *Phys. Rev. D*, **55**, 1937–1942.

[337] Mak, M. K., Dobson Jr., P. N., and Harko, T. 2000. Maximum mass-radius ratio for compact general relativistic objects in Schwarzschild–de Sitter geometry. *Mod. Phys. Lett. A*, **15**, 2153–2158.

[338] Mak, M. K., and Harko, T. 1999. Cosmological particle production in five-dimensional Kaluza-Klein theory. *Class. Quantum Grav.*, **16**, 4085–4099.

[339] Mak, M. K., and Harko, T. 2004. Can the galactic rotation curves be explained in brane world models? *Phys. Rev. D*, **70**, 024010.

[340] Maluf, J. W. 2013. The teleparallel equivalent of general relativity. *Annalen Phys.*, **525**, 339.

[341] Mamon, G. A., and Lokas, E. L. 2005. Dark matter in elliptical galaxies – II. Estimating the mass within the virial radius. *Mon. Not. R. Astron. Soc.*, **363**, 705.

[342] Martín-Moruno, P., Nunes, N. J., and Lobo, F. S. N. 2015. Horndeski theories self-tuning to a de Sitter vacuum. *Phys. Rev. D*, **91**, 8, 084029.

[343] Mashhoon, B. 1973. Tidal gravitational radiation. *Astrophys. J.*, **185**, 83.

[344] Mashhoon, B. 1975. On tidal phenomena in a strong gravitational field. *Astrophys. J.*, **197**, 705.

[345] Mashhoon, B., and Theiss, D. S. 1982. Relativistic tidal forces and the possibility of measuring them. *Phys. Rev. Lett.*, **49**, 1542.

[346] Massey, R., et al. 2007. Dark matter maps reveal cosmic scaffolding. *Nature*, **445**, 286–290.

[347] Matsumoto, S., Mukhopadhyay, S., and Sming Tsai, Y.-L. 2016. Effective theory of WIMP dark matter supplemented by simplified models: Singlet-like Majorana fermion case. *Phys. Rev. D*, **94**, 065034.

[348] McCrea, W. H. 1951. Relativity theory and the creation of matter. *Proc. Roy. Soc. Lond. A Math. Phys. Eng. Sci.*, **206**, 562–575.

[349] McCrea, W. H., and Milne, E. A. 1934. Newtonian universes and the curvature of space. *Quart. J. Math. Oxford*, **5**, 73–80.

[350] Mehdizadeh, M. R., Zangeneh, M. K., and Lobo, F. S. N. 2015. Einstein-Gauss-Bonnet traversable wormholes satisfying the weak energy condition. *Phys. Rev. D*, **91**, 084004.

[351] Michel, F. C. 1972. Accretion of matter by condensed objects. *Astrophys. Space Science*, **15**, 153–160.

[352] Milgrom, M. 1983. A modification of the Newtonian dynamics as a possible alternative to the hidden mass hypothesis. *Astrophys. J.*, **270**, 365.

[353] Milne, E. A. 1934. A Newtonian expanding universe. *Quart. J. Math. Oxford*, **5**, 64–72.

[354] Minamitsuji, M., and Silva, H. O. 2016. Relativistic stars in scalar-tensor theories with disformal coupling. *Phys. Rev. D*, **93**, 124041.

[355] Minazzoli, O. 2012. The γ parameter in Brans-Dicke-like (light-)Scalar-Tensor theory with a universal scalar/matter coupling and a new decoupling scenario. eprint, arXiv:1208.2372.

[356] Minazzoli, O. 2013 Conservation laws in theories with universal gravity/matter coupling. *Phys. Rev. D*, **88**, 027506.

[357] Minazzoli, O., and Harko, T. 2012. New derivation of the Lagrangian of a perfect fluid with a barotropic equation of state. *Phys. Rev. D*, **86**, 087502.

[358] Mitra, A. and Glendenning, N. K. 2010. Likely formation of general relativistic radiation pressure supported stars or "eternally collapsing objects". *Mon. Not. Roy. Astron. Soc. Lett.*, **404**, L50–L54.

[359] Modak, S. K., and Singleton, D. 2012. Inflation with a graceful exit and entrance driven by Hawking radiation. *Phys. Rev. D*, **86**, 123515.

[360] Mohseni, M. 2009. Non-geodesic motion in $f(\mathcal{G})$ gravity with non-minimal coupling. *Phys. Lett. B*, **682**, 89.

[361] Möller, C. 1961. Conservation laws and absolute parallelism in General Relativity. *Mat. Fys. Skr. Dan. Vid. Selsk.*, **1**, 10.

[362] Morris, M. S., and Thorne, K. S. 1988. Wormholes in spacetime and their use for interstellar travel: A tool for teaching General Relativity. *Am. J. Phys.*, **56**, 395.

[363] Pellegrini, C., and Plebanski, J. 1963. Tetrad fields and gravitational fields. *Mat. Fys. Skr. Dan. Vid. Selsk.*, **2**, 4.

[364] Montelongo Garcia, N., and Lobo, F. S. N. 2010. Wormhole geometries supported by a nonminimal curvature-matter coupling. *Phys. Rev. D*, **82**, 104018.

[365] Montelongo Garcia, N., and Lobo, F. S. N. 2011. Nonminimal curvature-matter coupled wormholes with matter satisfying the null energy condition. *Class. Quant. Grav.*, **28**, 085018.

[366] Montelongo Garcia, N., Lobo, F. S. N., and Visser, M. 2012. Generic spherically symmetric dynamic thin-shell traversable wormholes in standard general relativity. *Phys. Rev. D*, **86**, 044026.

[367] Moraes, P. H. R. S., Arbañil, J. D. V., and Malheiro, M. 2016. Stellar equilibrium configurations of compact stars in $f(R,T)$ theory of gravity. *J. Cosmo. Astropart. Phys.*, **06**, 005.

[368] Mortonson, M. J., Weinberg, D. H., and White, M. 2014. Dark energy: A short review. In K. A. Olive et al. (Particle Data Group), eds., *2014 Review of Particle Physics*. arXiv: 1401.0046.

[369] Mukhanov, V. F., and Winitzki, S. 2010. *Introduction to Quantum Effects in Gravity*, (Cambridge: Cambridge University Press).

[370] Muller, V., Schmidt, H. J., and Starobinsky, A. A. 1990. Power-law inflation as an attractor solution for inhomogeneous cosmological models. *Class. Quant. Grav.*, **7**, 1163–1168.

[371] Multamaki, T., and Vilja, I. 2006. Spherically symmetric solutions of modified field equations in $f(R)$ theories of gravity. *Phys. Rev. D*, **74**, 064022.

[372] Muñoz, J. B., Kovetz, E. D., Dai, L, and Kamionkowski, M. 2016. Lensing of fast radio bursts as a probe of compact dark matter. *Phys. Rev. Lett.*, **117**, 091301.

[373] Murphy, G. L. 1973. Big-bang model without singularities. *Phys. Rev. D*, **8**, 4231–4233.

[374] Naidu, R. L., Reddy, D. R. K., Ramprasad, T., and Ramana, K. V. 2013. Bianchi type-V bulk viscous string cosmological model in $f(R,T)$ gravity. *Astrophys. Space Sci.*, **348**, 247.

[375] Navarro, J. F., Frenk, C. S., and White, S. D. M. 1996. The structure of cold dark matter halos. *Astrophys. J.*, **462**, 563–575.

[376] Nesseris, S. 2009. Matter density perturbations in modified gravity models with arbitrary coupling between matter and geometry. *Phys. Rev. D*, **79**, 044015.

[377] Nojiri, S., and Odintsov, S. D. 2003. Modified gravity with negative and positive powers of the curvature: Unification of the inflation and of the cosmic acceleration. *Phys. Rev. D*, **68**, 123512.

[378] Nojiri, S., and Odintsov, S. D. 2003. Where new gravitational physics comes from: M theory? *Phys. Lett. B*, **576**, 5.

[379] Nojiri, S., and Odintsov, S. D. 2004. Dark energy and cosmic speed-up from consistent modified gravity. *PoS WC*, **2004**, 024.

[380] Nojiri, S., and Odintsov, S. D. 2004. Gravity assisted dark energy dominance and cosmic acceleration. *Phys. Lett. B*, **599**, 137.

[381] Nojiri, S., and Odintsov, S. D. 2004. Modified gravity with ln R terms and cosmic acceleration. *Gen. Rel. Grav.*, **36**, 1765.

[382] Nojiri, S., and Odintsov, S. D. 2007. Introduction to modified gravity and gravitational alternative for dark energy. *Int. J. Geom. Meth. Mod. Phys.*, **4**, 115–146.

[383] Nojiri, S., and Odintsov, S. D. 2011. Unified cosmic history in modified gravity: from $F(R)$ theory to Lorentz non-invariant models. *Phys. Rept.*, **505**, 59.

[384] Nojiri, S., and Odintsov, S. D. 2007. Unifying inflation with LambdaCDM epoch in modified $f(R)$ gravity consistent with solar system tests. *Phys. Lett. B*, **657**, 238. arXiv:0707.1941 [hep-th].

[385] Nojiri, S., Odintsov, S. D., and Oikonomou, V. K. 2017. Modified gravity theories on a nutshell: Inflation, bounce and late-time evolution. *Phys. Rept.*, **692**, 1–104.

[386] Nojiri, S., Odintsov, S. D., and Sasaki, M. 2005. Gauss-Bonnet dark energy. *Phys. Rev. D*, **71**, 123509.

[387] Nucamendi, U., Salgado, M., and Sudarsky, D. 2001. Alternative approach to the galactic dark matter problem. *Phys. Rev. D*, **63**, 125016.

[388] Nunes, R. C., and Pan, S. 2016. Cosmological consequences of an adiabatic matter creation process. *Mon. Not. Roy. Astron. Soc.*, **459**, 673–682.

[389] Obukhov, Y. N., and Puetzfeld, D. 2013. Conservation laws in gravitational theories with general nonminimal coupling. *Phys. Rev. D*, **87**, 081502.

[390] Odintsov, S. D., and Sáez-Gómez, D. 2013. $f(R,T,R_{\mu\nu}T^{\mu\nu})$ gravity phenomenology and ΛCDM universe. *Phys. Lett. B*, **725**, 437.

[391] Ohanian, H. C. 1976. *Gravitation and Spacetime*, (New York: Norton).

[392] Olmo, G. J. 2005. The gravity Lagrangian according to solar system experiments. *Phys. Rev. Lett.*, **95**, 261102.

[393] Olmo, G. J. 2005. Post-Newtonian constraints on $f(R)$ cosmologies in metric and Palatini formalism. *Phys. Rev. D*, **72**, 083505.

[394] Olmo, G. J. 2007. Violation of the Equivalence Principle in modified theories of gravity. *Phys. Rev. Lett.*, **98**, 061101.

[395] Olmo, G. J. 2008. Hydrogen atom in Palatini theories of gravity. *Phys. Rev. D*, **77**, 084021.

[396] Olmo, G. J. 2011. Palatini approach to modified gravity: $f(R)$ theories and beyond. *Int. J. Mod. Phys. D*, **20**, 413–462.

[397] Olmo, G. J., and Sanchis-Alepuz, H. 2011. Hamiltonian formulation of Palatini $f(R)$ theories a la Brans-Dicke. *Phys. Rev. D*, **83**, 104036.

[398] Olmo, G. J., and Rubiera-Garcia, D. 2015. Brane-world and loop cosmology from a gravity-matter coupling perspective. *Phys. Lett. B*, **740**, 73.

[399] Overduin, J. M., and Wesson, P. S. 1997. Kaluza-Klein gravity. *Phys. Repts.*, **283**, 303.

[400] Overduin, J. M., and Wesson, P. S. 2004. Dark matter and background light. *Phys. Repts.*, **402**, 267–406.

[401] Ozel, F., and Freire, P. 2016. Masses, radii, and equation of state of neutron stars. *Astron. Astrophys.*, **54**, 401–440.

[402] Pace, F., Waizmann, J.-C., and Bartelmann, M. 2010. Spherical collapse model in dark energy cosmologies. *Mon. Not. Roy. Astron. Soc.*, **406**, 1865–1874.

[403] Padmanabhan, T. 2003. Cosmological constant – the weight of the vacuum. *Phys. Rept.*, **380**, 235–320.

[404] Page, D. N., and Thorne, K. S. 1974. Disk-accretion onto a black hole: Time-averaged structure of accretion disk. *Astrophys. J.*, **191**, 499–506.

[405] Parker, L. 1968. Particle creation in expanding universes. *Phys. Rev. Lett.*, **21**, 562–564.

[406] Parker, L. 2009. Quantized fields and particle creation in expanding universes. II. *Phys. Rev. D*, **3**, 346–356.

[407] Parker, L. E., and Toms, D. J. 2009. *Quantum Field Theory in Curved Spacetime-Quantized Fields and Gravity*, (Cambridge: Cambridge University Press).

[408] Pauli, W. 1958. *Theory of Relativity*, (London: Pergamon Press).

[409] Peebles, P. J. E. 1993. *Principles of Physical Cosmology*, (Princeton: Princeton University Press).

[410] Peebles, P. J. E., and Ratra, B. 2003. The cosmological constant and dark energy. *Rev. Mod. Phys.*, **75**, 559–606.

[411] Penzias, A. A., and Wilson, R. W. 1965. A measurement of excess antenna temperature at 4080 Mc/s. *Astrophys. J.*, **142**, 419.

[412] Perlmutter, S., Turner, M. S., White, M. J. 1999. Constraining dark energy with SNe Ia and large scale structure. *Phys. Rev. Lett.*, **83**, 670–673.

[413] Perlmutter, S., et al. 1999. Measurements of Omega and Lambda from 42 high redshift supernovae. *Astrophys. J.*, **517**, 565–586.

[414] Persic, M., Salucci, P., and Stel, F. 1996. The universal rotation curve of spiral galaxies – I. The dark matter connection. *Mon. Not. R. Astron. Soc.*, **281**, 27–47.

[415] Pigozzo, C., Carneiro, S., Alcaniz, J. S., Borges, H. A., Fabris, J. C. 2016. Evidence for cosmological particle creation? *J. Cosmo. Astropart. Phys.*, **1605**, 022.

[416] Pogosian, L., and Silvestri, A. 2008. Pattern of growth in viable $f(R)$ cosmologies. *Phys. Rev. D*, **77**, 023503.

[417] Poisson, E., and Visser, M. 1995. Thin-shell wormholes: Linearization stability. *Phys. Rev. D*, **52**, 7318.

[418] Poplawski, N. J. 2006. A Lagrangian description of interacting dark energy. arXiv:gr-qc/0608031.

[419] Prigogine, I., and Géhéniau, J. 1986. Entropy, matter, and cosmology. *Proc. Natl. Acad. Sci. USA*, **83**, 6245–6249.

[420] Prigogine, I., and Géhéniau, J., Gunzig, E., and Nardone, P. 1988. Thermodynamics of cosmological matter creation. *Proc. Natl. Acad. Sci. USA*, **85**, 7428–7432.

[421] Pritchard, J. R., and Loeb, A. 2012. 21-cm cosmology. *Rept. Prog. Phys.*, **75**, 086901.

[422] Puetzfeld, D., and Obukhov, Y. N. 2008. On the motion of test bodies in theories with non-minimal coupling. *Phys. Rev. D*, **78**, 121501.

[423] Puetzfeld, D., and Obukhov, Y. N. 2013. Covariant equations of motion for test bodies in gravitational theories with general nonminimal coupling. *Phys. Rev. D*, **87**, 044045.

[424] Puetzfeld, D., and Obukhov, Y. N. 2013. Equations of motion in gravity theories with nonminimal coupling: a loophole to detect torsion macroscopically? *Phys. Rev. D*, **88**, 064025.

[425] Pun, C. S. J., Kovács, Z., and Harko, T. 2008. Thin accretion disks in $f(R)$ modified gravity models. *Phys. Rev. D*, **78**, 024043.

[426] Pun, C. S. J., Kovács, Z., and Harko, T. 2008. Thin accretion disks onto brane world black holes. *Phys. Rev. D*, **78**, 084015.

[427] Raigorodski, L. D., Stavrions, P. C., and Balan, V. 1999. *Introduction to the Physical Principles of Differential Geometry*, (Athens: University of Athens).

[428] Ram, S., and Priyanka. 2013. Some Kaluza-Klein cosmological models in $f(R,T)$ gravity theory. *Astrophys. Space Sci.*, **347**, 389.

[429] Ramos, R. O., dos Santos, M. V., and Waga, I. 2014. Matter creation and cosmic acceleration. *Phys. Rev. D*, **89**, 083524.

[430] Ratra, B., and Peebles, P. J. E. 1988. Cosmological consequences of a rolling homogeneous scalar field. *Phys. Rev. D*, **37**, 3406–3427.

[431] Read, J. I., Iorio, G., Agertz, O., and Fraternali, F. 2016. Understanding the shape and diversity of dwarf galaxy rotation curves in ΛCDM. *Mon. Not. Roy. Astron. Soc.*, **462**, 3628–3645.

[432] Reiprich, T. H., and Boehringer, H. 2002. The Mass function of an X-ray flux-limited sample of galaxy clusters. *Astrophys. J.*, **567**, 716.

[433] Reis, R. R. 2003. Domain of validity of the evolution of perturbations in Newtonian cosmology with pressure. *Phys. Rev. D*, **67**, 087301.

[434] Resco, M.A., de la Cruz-Dombriz, A., Llanes-Estrada, F. J., and Zapatero Castrillo, V. 2016. On neutron stars in $f(R)$ theories: Small radii, large masses and large energy emitted in a merger. *Phys. Dark Univ.*, **13**, 147–161.

[435] Rhoades, C. E., and Ruffini, R. 1974. Maximum mass of a neutron star. *Phys. Rev. Lett.*, **32**, 324–327.

[436] Riess, A. G., et al. 1998. Observational evidence from supernovae for an accelerating universe and a cosmological constant. *Astron. J.*, **116**, 1009–1038.

[437] Riess, A. G., et al. 2001. The farthest known supernova: Support for an accelerating universe and a glimpse of the epoch of deceleration. *Astrophys. J.*, **560**, 49–71.

[438] Riess, A. G., et al. (Supernova Search Team Collaboration). 2004. Type Ia supernova discoveries at $z > 1$ from the Hubble Space Telescope: Evidence for past deceleration and constraints on dark energy evolution. *Astrophys. J.*, **607**, 665–687.

[439] Robertson, H. P. 1935. Kinematics and world-structure. *Astrophys. J.*, **82**, 284.

[440] Rosa, J. L., Carloni, S., Lemos, J. P. S., and Lobo, F. S. N. 2017. Cosmological solutions in generalized hybrid metric-Palatini gravity. *Phys. Rev. D*, **95**, 124035.

[441] Rubin, V. C., Thonnard, N., and Ford Jr., W. K. 1980. Rotational properties of 21 SC galaxies with a large range of luminosities and radii, from NGC 4605 /R = 4kpc/ to UGC 2885 /R = 122 kpc/. *Astrophys. J.*, **238**, 471–487.

[442] Salgado, M. 2006. The Cauchy problem of scalar tensor theories of gravity. *Class. Quant. Grav.*, **23**, 4719–4742.

[443] Salopek, D. S., and Bond, J. R. 1990. Nonlinear evolution of long-wavelength metric fluctuations in inflationary models. *Phys. Rev. D*, **42**, 3936–3962.

[444] Salucci, P., et al. 2007. The universal rotation curve of spiral galaxies. II The dark matter distribution out to the virial radius. *Mon. Not. Roy. Astron. Soc.*, **378**, 41–47.

[445] Salucci, P., and Persic, M. 1997. Dark halos around galaxies. In M. Persic and P. Salucci, eds., *Dark and Visible Matter in Galaxies*, ASP Conference Series, **117**, 1.

[446] Sandstad, M., Koivisto, T. S., and Mota, D. F. 2013. Non-locality of the C- and D-theories. *Class. Quant. Grav.*, **30**, 155005.

[447] Santos, A. F. 2013. Gödel solution in $f(R,T)$ gravity. *Mod. Phys. Lett. A*, **28**, 1350141.

[448] Saxton, C. J., and Ferreras, I. 2010. Polytropic dark haloes of elliptical galaxies. *Mon. Not. Roy. Astron. Soc.*, **405**, 77–90.

[449] Saxton, C. J. 2013. Galaxy stability within a self-interacting dark matter halo. *Mon. Not. Roy. Astron. Soc.*, **430**, 1578–1598.

[450] Sawicki, I., and Bellini, E. 2015. Limits of quasistatic approximation in modified-gravity cosmologies. *Phys. Rev. D*, **92**, 084061.

[451] Schmidt, H.-J. 1990. New exact solutions for power-law inflation Friedmann models. *Astron. Nachr.*, **311**, 165–168.

[452] Seliger, R. L., and Whitham, G. B. 1968. Variational principles in continuum mechanics. *Proc. R. Soc. (London)*, **A 305**, 1–25.

[453] Schmidt, H.-J., and Homann, F. 2000. Photon stars. *Gen. Rel. Grav.*, **32**, 919–931.

[454] Schrödinger, E. 1939. The proper vibrations of the expanding universe. *Physica*, **6**, 899–912.

[455] Schrödinger, E. 1950. *Space-Time Structure*, (Cambridge: Cambridge University Press).

[456] Schuecker, P., Boehringer, H., Arzner, K., and Reiprich, T. H. 2001. Cosmic mass functions from Gaussian stochastic diffusion processes. *Astron. Astrophys.*, **370**, 715–728.

[457] Schutz, B. 1970. Perfect fluids in General Relativity: Velocity potentials and a variational principle. *Phys. Rev. D*, **2**, 2762–2773.

[458] Schwabe, B., Niemeyer, J. C., and Engels, J. F. 2016. Simulations of solitonic core mergers in ultralight axion dark matter cosmologies. *Phys. Rev. D*, **94**, 043513.

[459] Sham, Y.-H., Lin, L.-M., and Leung, P. T. 2014. Testing universal relations of neutron stars with a nonlinear matter-gravity coupling theory. *Astrophys. J.*, **781**, 66.

[460] Shapiro, I. I., Smith, W. B., Ash, M. E., and Herrick, S. 1971 General Relativity and the orbit of Icarus. *Astron. J.*, **76**, 588.

[461] Shapiro, I. I., Counselman, C. C., and King, R. W. 1976. Verification of the Principle of Equivalence for massive bodies. *Phys. Rev. Lett.*, **36**, 555–558.

[462] Shapiro, S. L., and Teukolsky, S. A. 1983. *Black Holes, White Dwarfs and Neutron Stars: The Physics of Compact Objects*, (New York: John Wiley & Sons).

[463] Sharif, M., Rani, S., and Myrzakulov, R. 2013. Analysis of $F(R,T)$ gravity models through energy conditions. *Eur. Phys. J. Plus.*, **128**, 123.

[464] Sharif, M., and Zubair, M. 2012. Thermodynamics in $f(R,T)$ Theory of Gravity. *J. Cosmo. Astropart. Phys.*, **1203**, 028. (Erratum-ibid. **05**, E01 [2013].)

[465] Sharif, M., and Zubair, M. 2013. Energy conditions constraints and stability of power law solutions in $f(R,T)$ gravity. *J. Phys. Soc. Jap.*, **82**, 014002.

[466] Sharif, M., and Zubair, M. 2014. Reconstruction and stability of $f(R,T)$ gravity with Ricci and modified Ricci dark energy. *Astrophys. Space Sci.*, **349**, 529.
[467] Sharif, M., and Zubair, M. 2014. Study of Bianchi I anisotropic model in $f(R,T)$ gravity. *Astrophys. Space Sci.*, **349**, 457.
[468] Sotiriou, T. P. 2007. Modified actions for gravity: Theory and phenomenology. PhD thesis. arXiv:0710.4438 [gr-qc].
[469] Sotiriou, T. P. 2008. The viability of theories with matter coupled to the Ricci scalar. *Phys. Lett. B*, **664**, 225.
[470] Sotiriou, T. P., and Faraoni, V. 2008. Modified gravity with R-matter couplings and (non-)geodesic motion. *Class. Quant. Grav.*, **25**, 205002.
[471] Sotiriou, T. P., and Faraoni, V. 2010. $f(R)$ theories of gravity. *Rev. Mod. Phys.*, **82**, 451.
[472] Sotiriou, T. P., and Liberati, S. 2007. Metric-affine $f(R)$ theories of gravity. *Annals Phys.*, **322**, 935.
[473] Spergel, D. N., et al. 2007. Wilkinson Microwave Anisotropy Probe (WMAP) three year results: Implications for cosmology. *Astrophys. J. Suppl.*, **170**, 377–408.
[474] Starobinsky, A. A. 1980. A new type of isotropic cosmological models without singularity. *Phys. Lett. B*, **91**, 99–102.
[475] Steigl, R., and Hinterleitner, F. 2006. Factor ordering in standard quantum cosmology. *Class. Quant. Grav.*, **23**, 3879–3893.
[476] Strigari, L. E. 2013. Galactic searches for dark matter. *Phys. Rept.*, **531**, 1–88.
[477] Tamanini, N., and Boehmer, C. G. 2013. Generalized hybrid metric-Palatini gravity. *Phys. Rev. D*, **87**, 084031.
[478] Tamanini, N., and Koivisto, T. S. 2013. Consistency of nonminimally coupled $f(R)$ gravity. *Phys. Rev. D*, **88**, 064052.
[479] Teyssandier, P., and Tourrenc, Ph. 1983. The Cauchy problem for the $R + R^2$ theories of gravity without torsion. *J. Math. Phys.*, **24**, 2793.
[480] Thakur, S., Sen, A. A. 2013. Can structure formation distinguish ΛCDM from nonminimal $f(R)$ gravity?," *Phys. Rev. D*, **88**, 044043.
[481] Thakur, S., Sen, A. A., and Seshadri, T. R. 2011. Non-minimally coupled $f(R)$ cosmology. *Phys. Lett. B*, **696**, 309.
[482] Thorne, K. S. 1974. Disk-accretion onto a black hole. II. Evolution of the hole. *Astrophys. J.*, **191**, 507–520.
[483] Tian, D. W., and Booth, I. 2014. Lessons from $f(R, R_c^2, R_m^2, L_m)$ gravity: smooth Gauss-Bonnet limit, energy-momentum conservation and nonminimal coupling. *Phys. Rev. D*, **90**, 024059.
[484] Torres, D. 2002. Accretion disc onto a static non-baryonic compact object. *Nucl. Phys. B*, **626**, 377–394.
[485] Vakili, B. 2010. Classical and quantum dynamics of a perfect fluid scalar-metric cosmology. *Physics Letters B*, **688**, 129–136.
[486] Vakili, B. 2010. Quadratic quantum cosmology with Schutz' perfect fluid. *Class. Quant. Grav.*, **27**, 025008.
[487] Van Kerkwijk, M. H., Breton, R. P., and Kulkarni, S. R. 2011. Evidence for a massive neutron star from a radial-velocity study of the companion to the black-widow pulsar PSR B1957+20. *Astrophys. J.*, **728**, 95.
[488] Vargas dos Santos, M., Alcaniz, J. S., Mota, D. F., and S. Capozziello, S. 2017. Screening mechanisms in hybrid metric-Palatini gravity. *Phys. Rev. D*, **97**, 104010. arXiv:1712.03831 [gr-qc].
[489] Velten, H., and Wamba, E. 2012. Power spectrum for the Bose-Einstein condensate dark matter. *Phys. Lett. B*, **709**, 1–5.

[490] Visser, M. 1989. Traversable wormholes: Some simple examples. *Phys. Rev. D*, **39**, 3182.

[491] Visser, M. 1989. Traversable wormholes from surgically modified Schwarzschild space-times. *Nucl. Phys. B*, **328**, 203.

[492] Visser, M. 1995. *Lorentzian Wormholes: From Einstein to Hawking*, (New York: Springer).

[493] Wald, R. M. 1984. *General Relativity*, (Chicago: Chicago University Press).

[494] Walker, A. G. 1937. On Milne's theory of world-structure. *Proc. Lon. Math. Soc.*, **42**, 90127.

[495] Wands, D. 1994. Extended gravity theories and the Einstein–Hilbert action. *Class. Quant. Grav.*, **11**, 269.

[496] Wang, J., and Liao, K. 2012. Energy conditions in $f(R, L(m))$ gravity. *Class. Quant. Grav.*, **29**, 215016.

[497] Wang, J., Wu, Y.-B., Guo, Y.-X., Yang, W.-Q., and Wang, L. 2010. Energy conditions and stability in generalized $f(R)$ gravity with arbitrary coupling between matter and geometry. *Phys. Lett. B*, **689**, 133.

[498] Wang, J., Wu, Y.-B., Guo, Y.-X., Qi, F., Zhao, Y.-Y., and Sun, X.-Y. 2010. Conditions and instability in $f(R)$ gravity with non-minimal coupling between matter and geometry. *Eur. Phys. J. C*, **69**, 541.

[499] Watson, G. N. 1922. *A Treatise on the Theory of Bessel Functions*, (Cambridge: Cambridge University Press).

[500] Wegg, C., Gerhard, O., and Portail, M. 2016. MOA-II galactic microlensing constraints: The inner Milky Way has a low dark matter fraction and a near maximal disk. *Mon. Not. Roy. Astron. Soc.*, **463**, 557–570.

[501] Weinberg, S. 1972. *Gravitation and Cosmology: Principles and Applications of the General Theory of Relativity*, (New York: Wiley).

[502] Weinberg, S. 1989. The cosmological constant problem. *Rev. Mod. Phy.*, **61**, 1–23.

[503] Weinberg, S. 1996. *The Quantum Theory of Fields*, Vol. II, (Cambridge: Cambridge University Press).

[504] Weinberg, D. H., et al. 2013. Observational probes of cosmic acceleration. *Phys. Rep.*, **530**, 87–255.

[505] Wetterich, C. 1988. Cosmology and the fate of dilatation symmetry. *Nucl. Phys. B*, **302**, 668–696.

[506] Wetterich, C. 1995. An asymptotically vanishing time-dependent cosmological "constant". *Astron. Astrophys.*, **301**, 321.

[507] Weyl, H. 1952. *Space, Time, Matter*, (New York: Dover Publications).

[508] Whitt, B. 1984. Fourth-order gravity as general relativity plus matter. *Phys. Lett. B*, **145**, 176.

[509] Will, C. M. 2014. The Confrontation between General Relativity and experiment. *Living Rev. Rel.*, **17**, 4.

[510] Witten, E. 1984. Cosmic separation of phases. *Phys. Rev D*, **30**, 272–285.

[511] Woodard, R. P. 2007. Avoiding dark energy with 1/r modifications of gravity. *Lect. Notes Phys.*, **720**, 403–433.

[512] Wu, Y.-B., Zhao, Y.-Y., Lu, J.-W., Zhang, X., Zhang, C.-Y., and Qiao, J.-W. 2014. Five-dimensional generalized $f(R)$ gravity with curvature-matter coupling. *Eur. Phys. J. C*, **74**, 2791.

[513] Xu, M.-X., Harko, T., and Liang, S.-D. 2016. Quantum cosmology of $f(R, T)$ gravity. *Eur. Phys. J. C*, **76**, 449.

[514] Yang, R.-J. 2016. Effects of quantum fluctuations of metric on the universe. *Phys. Dark Univ.*, **13**, 87–91.

[515] Yuan, Y. F., Narayan, R., and Rees, M. J. 2004. Constraining alternate models of black holes: Type I X-ray bursts on accreting fermion-fermion and boson-fermion stars. *Astrophys. J.*, **606**, 1112–1124.
[516] Zakharov, V. I. 1970. Linearized gravitation theory and the graviton mass. *JETP Lett.*, **12**, 312.
[517] Zangeneh, M. K., Lobo, F. S. N., and Dehghani, M. H. 2015. Traversable wormholes satisfying the weak energy condition in third-order Lovelock gravity. *Phys. Rev. D*, **92**, 124049.
[518] Zangeneh, M. K., Lobo, F. S. N., and Riazi, N. 2014. Higher-dimensional evolving wormholes satisfying the null energy condition. *Phys. Rev. D*, **90**, 024072.
[519] Zeldovich, Ya. B. 1970. Particle production in cosmology. *Sov. Phys. JETP Lett.*, **12**, 307–312; *JETP Letters*, **12** (1970), 307–310.
[520] Zhang, S. N., Cui, W., and Chen, W. 1997. Black hole spin in X-ray binaries: Observational consequences. *Astrophys. J.*, **482**, L155–L158.
[521] Zlatev, I., Wang, L., and Steinhardt, P. J. 1999. Quintessence, cosmic coincidence, and the cosmological constant. *Phys. Rev. Lett.*, **82**, 896–899.
[522] Zurek, K. M. 2014. Asymmetric dark matter: Theories, signatures, and constraints. *Phys. Rep.*, **537**, 91–121.
[523] Zwillinger, D. 1989. *Handbook of Differential Equations*, 3rd edition, (San Diego: Academic Press).

Index